FUNDAMENTALS OF NETWORK ANALYSIS AND SYNTHESIS

PRENTICE-HALL NETWORKS SERIES

Robert W. Newcomb, *editor*

ANDERSON AND VONGPANITLERD
Network Analysis and Synthesis: A Modern Systems Theory Approach

ANDERSON AND MOORE
Linear Optimal Control

ANNER
Elementary Nonlinear Electronic Circuits

CARLIN AND GIORDANO
Network Theory: An Introduction to Reciprocal and Nonreciprocal Circuits

CHIRLIAN
Integrated and Active Synthesis and Analysis

DEUTSCH
Systems Analysis Techniques

HERRERO AND WILLONER
Synthesis of Filters

HOLTZMAN
Nonlinear System Theory: A Functional Analysis Approach

HUMPHREYS
The Analysis, Design, and Synthesis of Electrical Filters

KIRK
Optimal Control Theory: An Introduction

KOHONEN
Digital Circuits and Devices

MANASSE, ECKERT, AND GRAY
Modern Transistor Electronics Analysis and Design

NEWCOMB
Active Integrated Circuit Synthesis

PEIKARI
Fundamentals of Network Analysis and Synthesis

SAGE
Optimum Systems Control

SU
Time-Domain Synthesis of Linear Networks

VAN DER ZIEL
Introductory Electronics

VAN VALKENBURG
Network Analysis

FUNDAMENTALS OF NETWORK ANALYSIS AND SYNTHESIS

BEHROUZ PEIKARI
Southern Methodist University

Electrical Engineering Department

PRENTICE-HALL, INC.
Englewood Cliffs, New Jersey

Library of Congress Cataloging in Publication Data

PEIKARI, BEHROUZ.
Fundamentals of network analysis and synthesis.

(Prentice-Hall networks series)
1. Electric networks. 2. Electronic data processing—Electric networks. I. Title.
TK454.2.P43 621.381'1 73-15563
ISBN 0-13-341321-7

10 9 8 7 6 5 4 3 2 1

Printed in the United States of America

PRENTICE-HALL INTERNATIONAL, INC., *London*
PRENTICE-HALL OF AUSTRALIA, PTY. LTD., *Sydney*
PRENTICE-HALL OF JAPAN, INC., *Tokyo*
PRENTICE-HALL OF CANADA, LTD., *Toronto*
PRENTICE-HALL OF INDIA PRIVATE LTD., *New Delhi*

To my sons *Behzad*, *Cyrus*, and *Darius*

Contents

Preface, *x*

Network Elements 1

1.1 Introduction, *1*
1.2 Two-Terminal Resistors, *5*
1.3 Two-Terminal Capacitors, *15*
1.4 Two-Terminal Inductors, *23*
1.5 Multiterminal Network Elements, *29*
Problems, *43*

Network Characterization 2

2.1 Introduction, *48*
2.2 Introductory Matrix Algebra, *50*
2.3 Norm of Vectors and Matrices, *59*
2.4 Linear Versus Nonlinear n Ports, *63*
2.5 Time-Varying Versus Time-Invariant n Ports, *72*
2.6 Passive Versus Active n Ports, *77*
2.7 Causal and Nonanticipative Networks, *81*
2.8 Stable and Unstable Networks, *84*
Problems, *87*

Network Graph Theory 3

3.1 Notations and Definitions, *91*
3.2 Incidence Matrix and Kirchhoff's Current Law, *96*
3.3 Loop Matrix and Kirchhoff's Voltage Law, *104*
3.4 Interrelationship Between Matrices of a Graph, *112*
3.5 Tellegen's Theorem and Its Application, *116*
Problems, *122*

Analysis of Linear Time-Invariant Networks 4

4.1 Introduction, *126*
4.2 Direct Analysis Methods, *127*
4.3 Nodal Analysis, *139*
4.4 Loop Analysis, *142*
4.5 Analysis of Networks Containing Dependent Sources, *148*
4.6 Sinusoidal Steady-State Analysis, *156*
4.7 Network Functions, *161*
Problems, *180*

State-Variable Representation of Networks 5

5.1 Introduction, *186*
5.2 Preliminary Considerations, *190*
5.3 State-Variable Formulation of Proper Networks, *199*
5.4 Concept of State and Order of Complexity of a Network, *216*
5.5 State-Variable Formulation of General Networks, *221*
Problems, *234*

Time-Domain Solution of State Equations 6

6.1 Introduction, *242*
6.2 Solution of Linear Time-Invariant State Equations, *243*
6.3 Solution of Linear Time-Varying State Equations, *259*
6.4 Solution of Nonlinear State Equations, *267*
Problems, *276*

Computer-Aided Network Analysis 7

7.1 Introduction, *282*
7.2 Computer-Aided Analysis of Linear Resistive Networks, *284*
7.3 Computer Solution of Sinusoidal Steady-State Response of Linear Time-Invariant Networks, *300*
7.4 Computer Solution of Nonlinear Resistive Networks, *303*
7.5 Computer Solution of State Equations, *314*
Problems, *334*

Passive and Active Network Synthesis 8

8.1 Introduction, *336*
8.2 Positive Real Functions and Matrices, *337*
8.3 Driving-Point Synthesis of Reactance Functions, *342*
8.4 Synthesis of *RC* and *RL* Networks, *348*
8.5 Synthesis of General *RLC* Networks (Brune Synthesis), *352*
8.6 Bott-Duffin Synthesis (Transformerless Synthesis), *361*
8.7 Active Network Synthesis, *365*
8.8 Sensitivity Considerations in Network Design, *378*
Problems, *383*

Computer-Aided Network Design 9

9.1 Introduction, *387*
9.2 Iterative Methods in Computer-Aided Design, *388*
9.3 Mean Squared Error Optimization, *392*
9.4 General Optimization Methods, *402*
9.5 Application of Adjoint Network in Computing the Gradient, *406*
9.6 Computer-Aided Design of Linear Time-Varying Networks, *410*
Problems, *415*

Small-Signal Analysis and Design of Nonlinear Networks 10

10.1 Introduction, *418*
10.2 Small-Signal Analysis of Autonomous Networks, *419*
10.3 Small-Signal Analysis of Nonautonomous Networks, *427*
10.4 Frozen-Operating-Point Method of Small-Signal Analysis, *431*
10.5 Design of Nonlinear Networks, *439*
Problems, *446*

Stability Analysis of Linear and Nonlinear Networks 11

11.1 Introduction, *450*
11.2 Zero-Input Stability, *451*
11.3 Equilibrium Stability of Linear Autonomous Networks, *457*
11.4 Equilibrium Stability of Nonlinear Autonomous Networks, *462*
11.5 Lyaponov Stability of Nonautonomous Networks, *468*
11.6 Bounded Input—Bounded Output Stability, *473*
Problems, *479*

Appendix: Laplace Transform

A.1 Introduction, *482*
A.2 Basic Definition and Convergence Properties of Laplace Transform, *482*
A.3 Inverse Laplace Transform, *489*

Preface

This book provides the advanced undergraduate and beginning graduate student in electrical engineering with a comprehensive treatment of the fundamental topics in network theory. Notwithstanding the inclusion of recent advances in circuit analysis and design, the materials are developed in such a manner that the only prerequisite for a course based on this text is an elementary knowledge of circuit analysis. The book covers, in a unified fashion, the analysis and synthesis of passive linear time-invariant networks, as well as active, nonlinear, and time-varying networks.

Because the role of digital computers has become increasingly vital in the analysis and design of networks, the presentation is such that it is readily adaptable to computer implementation. In addition, two chapters are devoted entirely to computer-aided analysis and design. The matrix formulation of the analysis problem and the implementation of computers for its solution enable the student to analyze the large-scale networks that are encountered in communication nets, complex power systems, and the like. Throughout the book, basic concepts are carefully defined and numerous examples and exercises are provided.

The content of the book can be divided into three principal parts. The first part, which consists of the first seven chapters, focuses on the fundamental concepts and techniques of network analysis. Chapter 1 offers a general description and classification of various circuit elements, including some multiterminal elements such as gyrators, transistors, operational amplifiers, and convertors. Small-signal and large-signal properties of nonlinear elements and their application are also discussed in this chapter. Chapter 2 considers the characterization and classification of n-port networks. In this chapter, after an introductory discussion on matrix algebra, the concepts of linearity, time-invariance, passivity, causality, and stability are defined. Several examples and counter-examples are worked out to stimulate the understanding of these important concepts. Chapter 3 is an introduction to graph theory in which matrices of a graph, their interrelationship, and their application in large-scale network analysis is presented. A section on Tellegen's theorem and its implications is also included. Chapter 4 is devoted exclusively to the analysis of linear time-invariant networks. Matrix formulation of nodal analysis, loop analysis, and sinusoidal steady-state analysis of n-port passive and active networks are discussed. Open-circuit impedance, short-circuit admittance, transfer function, and scattering matrices are also covered. Chapter 5 covers the state variable representation of linear, nonlinear, time-invariant, time-varying, and active networks with an emphasis on obtaining the state equations by inspection. In addition, the concept of state and order of complexity are defined. Topological formulas are used to develop a general procedure for obtaining the state equations of large-scale networks, which are useful in computer-aided analysis programs. Chapters 6 and 7 are concerned with the solution of state equations. Chapter 6 includes various analytical methods of the solution of state equations, a section on the function of a matrix, impulse response and weighting function of linear time-varying networks, and a discussion on the existence and uniqueness of the solution of nonlinear state equations. Chapter 7 presents a summary of various numerical methods of network analysis such as Gauss-Seidel, Newton-Raphson, Runge-Kutta, predictor-corrector, Gaussian elimination, and LU transformation as well as sample computer programs to illustrate the application of these methods.

The second part of this book, which consists of Chapters 8 and 9, considers classical network synthesis and contemporary computer-aided design. Chapter 8 covers positive real functions and matrices and Foster, Cauer, Brune, and Bott-Duffin synthesis techniques. Fundamentals of active network design are presented and several illustrative examples of active filter design are worked out. Chapter 9 covers various iterative techniques including steepest descent and Fletcher-Powell and Director-Rohrer methods in conjunction with the optimal design of passive and active networks.

The last part of the book deals with two special topics in network theory: small-signal analysis of nonlinear networks and the stability of networks and systems. Chapter 10 presents concepts of the equilibrium point of autonomous networks and the operating point of nonautonomous networks, together with the perturbation analysis of these networks. Included is a brief discussion on the design of nonlinear networks operated under the small-signal mode. Chapter 11 considers basic definitions of stability such as bounded-input, bounded-output, Lyaponov, and asymptotic stability of networks and systems. Several simple rules for testing the stability of networks are given.

Problems are included at the end of each chapter. These problems range from simple application of the derived results to more subtle questions on the fundamental concepts. It is assumed that the reader is already familiar with the Laplace transform technique. For convenience, however, a summary of the main results of this technique and a table of Laplace transform pairs are given in the Appendix.

The development of this book was made possible by the help and constant encouragement of Professor A. P. Sage, Head of the Electrical Engineering Department at Southern Methodist University, to whom I would like to express my appreciation. I am indebted to Professors R. W. Newcomb and C. A. Desoer, who made a number of constructive comments on the format of the book, and to Professors M. D. Srinath and S. C. Gupta, who read portions of the manuscript and made helpful suggestions. Some sections of Chapter 10 are based on the author's research under the sponsorship of the Air Force Office of Scientific Research, the National Science Foundation, and the National Aeronautic and Space Administration. I am grateful to my graduate assistants, P. S. Chang, C. C. Lin, and A. K. Agrawal, who helped to draw rough drafts of some of the figures, develop computer programs, and prepare solutions to the problems. Thanks are also due to Carolyn Hughes and Mary Lou Caruthers who typed the manuscript. I wish to thank most of all my wife, Patricia, for proofreading the entire manuscript and for her patience throughout the course of this work.

BEHROUZ PEIKARI

1 Network Elements

1.1 INTRODUCTION

In electrical network theory we concern ourselves with four basic manifestations of electricity. These are in the form of the following variables: (1) voltage, $v(t)$, measured in volts; (2) current, $i(t)$, measured in amperes; (3) charge, $q(t)$, measured in coulombs; and (4) flux, $\phi(t)$, measured in webers, where the independent variable t denotes the time of observation. The study of the interrelationships among these variables and the physical laws governing their behavior is the subject of network theory. These laws are divided into two different categories. First are those laws that are independent of the nature of the network elements with which these quantities are associated—these are called *universal relationships*; second are the laws that depend on the network elements. The objective of this chapter is to classify these network elements and state the relationship among the variables. In addition to the variables mentioned previously, there are two more variables that are of extreme importance in the study of network theory. These are (5) energy, $E(t)$, measured in joules, and (6) power, $p(t)$, measured in watts.

The universal relationships among these variables, which do not depend on the network elements, are

$$i(t) = \frac{dq(t)}{dt} \tag{1.1-1}$$

$$v(t) = \frac{d\phi(t)}{dt} \tag{1.1-2}$$

$$p(t) = v(t)i(t) \tag{1.1-3}$$

$$E(t) = \int_{-\infty}^{t} v(\tau)i(\tau)\, d\tau \tag{1.1-4}$$

where τ represents the integration variable. These relations hold for *any* network element, whether it is linear, nonlinear, time-invariant, time-varying, lumped, distributed, passive, or active. The preceding equations can also be written in the form

$$q(t) = \int_{-\infty}^{t} i(\tau)\, d\tau \tag{1.1-5}$$

$$\phi(t) = \int_{-\infty}^{t} v(\tau)\, d\tau \tag{1.1-6}$$

$$p(t) = \frac{dE(t)}{dt} \tag{1.1-7}$$

$$E(t) = \int_{-\infty}^{t} p(\tau)\, d\tau \tag{1.1-8}$$

In these equations we have made the reasonable assumption that the value of the corresponding variables is zero at $-\infty$. To study the second category of laws governing the variables mentioned, we must consider specific network elements. In this book we concern ourselves primarily with *lumped* network elements. For some network elements the variables i, v, q, and ϕ are functions of two independent variables, the observation time t and the observation position x. Such elements are called *distributed parameter* network elements. As an example, consider the transmitting antenna shown in Fig. 1.1-1. The

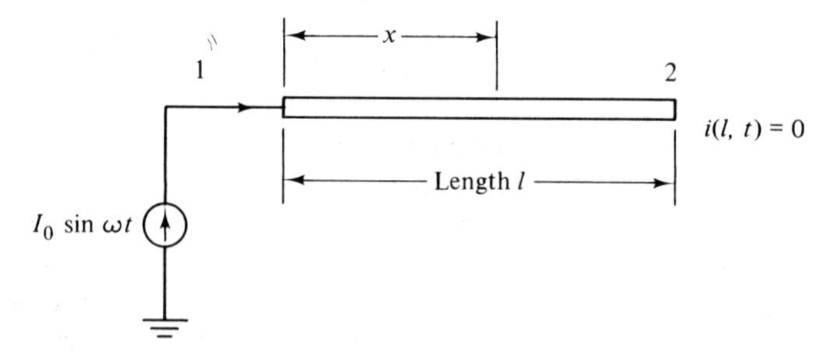

Figure 1.1-1 Example of a distributed network element.

transmitted signal is designated by the sinusoidal current source in series with the antenna. The current entering terminal 1 is $I_0 \sin \omega t$ and the current leaving terminal 2 is zero; the current is propagated in space in the form of electromagnetic waves. A typical current distribution along the antenna is given by

$$i(x, t) = I_0 \sin \beta(l - x) \sin \omega t \tag{1.1-9}$$

where I_0, β, l, and ω are constants.

The voltage $v(x, t)$ is generally related to $i(x, t)$ through a *partial differential equation*. Networks containing one or more distributed parameter elements are called distributed parameter networks.

If, on the other hand, the variables i, v, q, and ϕ are not functions of the observation variable x, the corresponding element is said to be lumped. The symbolic representation of a two-terminal network element is shown in Fig. 1.1-2. Let the terminals n_1 and n_2 represent the nodes of a network and let

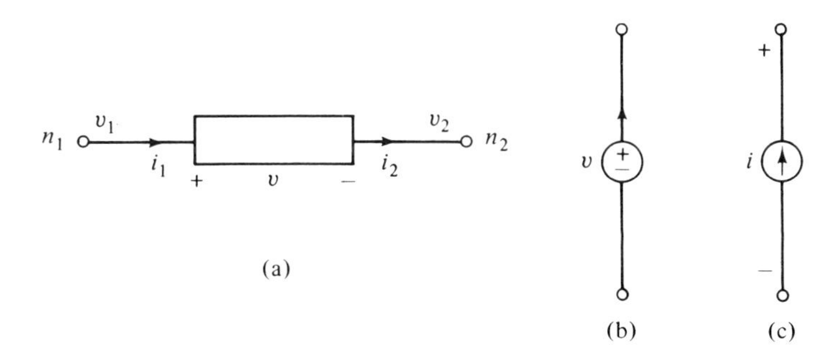

Figure 1.1-2 Symbolic representation of (a) a two-terminal lumped network element; (b) an independent voltage source; (c) an independent current source.

the currents entering and leaving the element be i_1 and i_2; then from the preceding discussion a more explicit definition of lumped elements is

Definition 1.1-1 (Lumped Element). A network element is said to be *lumped* if the instantaneous current entering one terminal is equal to the instantaneous current leaving the other terminal.

For the network element shown in Fig. 1.1-2(a) this implies that

$$i_1(t) = i_2(t) = i(t) \quad \text{for all } t \tag{1.1-10}$$

where $i(t)$ is called the element current.

Similarly, let voltages v_1 and v_2 denote the potentials of the nodes n_1 and n_2 of the given network with respect to an arbitrary reference. Then to the lumped element connected to these nodes there corresponds a unique potential difference $v_1 - v_2$ called the terminal potential difference, which is denoted by v; that is,

$$v(t) = v_1(t) - v_2(t) \tag{1.1-11}$$

Before proceeding with the discussion of the polarity convention, let us define the independent voltage and current sources. A two-terminal network element is called an *independent voltage source* if the voltage across it is independent from the current through it or, for that matter, from any other circuit variable.

For example, a generator that supplies the electricity of a large city can, for many practical purposes, be considered as an independent voltage source, since the voltage across it remains $v_{max} \sin(120\pi t)$ volts regardless of the amount of current drawn from it. The symbolic representation of an independent voltage source is shown in Fig. 1.1-2(b); the waveform and the amplitude of the independent voltage source remain the same regardless of the network to which it has been connected.

A two-terminal network element is called an *independent current source* if the current through it is independent from the voltage across it or any other variable in the circuit.

The symbolic representation of an independent current source is shown in Fig. 1.1-2(c); the waveform and the amplitude of the current source $i(t)$ are independent from the network it is connected to.

Throughout this text the reference direction for the voltage is indicated by the plus and minus signs located at the terminals of the element and the reference direction for the current is shown by an arrow through it. Furthermore, the following *associated reference directions* will be employed:

1. For a nonsource element (such as resistors, capacitors, inductors, etc.) the positive current enters the element at the node with plus (+) polarity and leaves the element at the node with the minus (—) polarity.
2. For a source element (voltage source or current source) the positive current enters the negative polarity (—) and leaves the positive polarity (see Fig. 1.1-2).

It should be mentioned that the reference directions are chosen arbitrarily, and they need not represent the actual direction of the current flowing in the element. If the actual current flowing in the element is in the same direction as the chosen direction, the current is said to be positive; otherwise, it is said to be negative.

Before discussing various circuit elements, let us introduce the important concept of passivity of a network element. Generally speaking, a two-terminal element is passive if it absorbs energy; more precisely,

Definition 1.1-2 (Passive Element). A two-terminal element is said to be *passive* if and only if

$$E(t) \triangleq \int_{-\infty}^{t} v(\tau)i(\tau)\, d\tau \geq 0 \qquad (1.1\text{-}12)$$

for all $t \geq -\infty$ and for all possible combinations of $i(t)$ and $v(t)$.

Furthermore, we assume that the element is initially relaxed in the sense that the energy stored on it at $-\infty$ is zero; $E(-\infty) = 0$. A two-terminal element is said to be *active* if it is not passive; that is, $E(t) < 0$ for *some* admissible $i(t)$, $v(t)$ or $t \geq -\infty$. An active element generally supplies electric energy to the network to which it is connected. Next we shall discuss three of the most important two-terminal network elements: resistors, capacitors and inductors.

1.2 TWO-TERMINAL RESISTORS

A two-terminal element is called a resistor if the current through it, $i(t)$, and the voltage across it, $v(t)$, are related through an algebraic relation $f(v(t), i(t), t) = 0$. This implies that the characteristic of a resistor can be represented as a curve in the i-v plane. Note that this definition classifies diodes, switches, and most other two-terminal elements without memory as resistors. The symbolic representation of a general resistor is shown in Fig. 1.2-1(a). Depending on the shape of their i-v characteristic, resistors can be classified in several categories; some of these categories are discussed next.

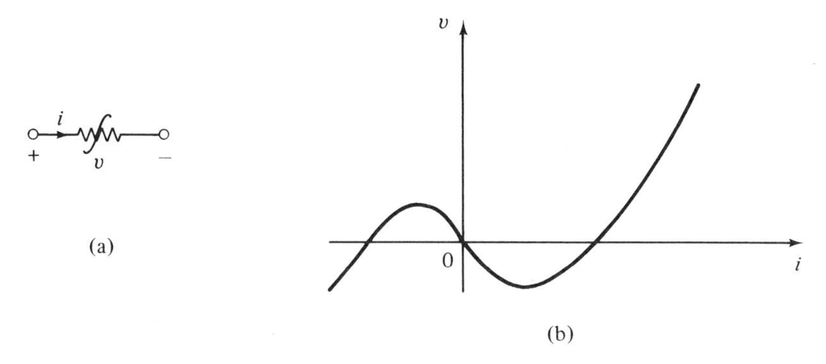

Figure 1.2-1 (a) Symbolic representation of a two-terminal nonlinear resistor; (b) typical i-v characteristic of a current-controlled resistor.

Current-Controlled Resistors. A two-terminal resistor is called *current controlled* if the voltage across it can be written as a single-valued function of the current through it; that is,

$$v(t) = f(i(t), t) \tag{1.2-1}$$

This implies that for each i in the domain of $f(\cdot, t)$ there exists at most one v in its range for each $t \geq 0$. A typical i-v curve of a current-controlled resistor is shown in Fig. 1.2-1(b). The independent variable t appearing as the second argument of $f(\cdot, \cdot)$ implies that the resistor whose *i-v* relation is given by (1.2-1) is *time varying*.

Voltage-Controlled Resistors. A two-terminal element is called a *voltage-controlled* resistor if the current through it can be expressed as a single-valued function of the voltage across it; that is,

$$i(t) = g(v(t), t) \tag{1.2-2}$$

An example of such a resistor is the tunnel diode, which is used in a variety of electronic circuits. Figure 1.2-2(a) shows the characteristics of such a diode. Notice that the tunnel diode shown in Fig. 1.2-2(a) is not a current-

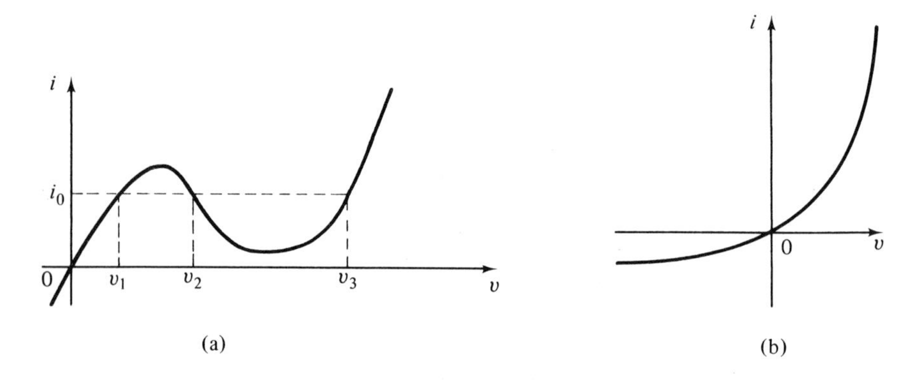

Figure 1.2-2 (a) Characteristic of a tunnel diode; (b) characteristic of a *pn*-junction transistor diode.

controlled resistor, since to some values of the current, say i_0, there correspond three different values of the voltage v; that is, the voltage cannot be expressed as a single-valued function of current.

Monotonic Resistors. A two-terminal resistor is said to be *monotonic* if it is both voltage controlled *and* current controlled. The *i-v* characteristics of such resistors are monotonically increasing or monotonically decreasing curves. An example of this class of resistors is the widely used *pn*-junction diode, whose *i-v* characteristic is shown in Fig. 1.2-2(b) and whose relation is given by

$$i(t) = \alpha(e^{\beta v(t)} - 1) \tag{1.2-3}$$

where α and β are nonzero constants. It is clear that (1.2-3) can be written as

$$v(t) = \frac{1}{\beta} \log_e\left[\frac{1}{\alpha} i(t) + 1\right] \tag{1.2-4}$$

That is, the voltage can be expressed as a unique function of the current.

In general, the equation of such a resistor is written as

$$v = f(i, t) \tag{1.2-5}$$

where $f(\cdot, t)$ is an invertible function for each $t \geq 0$. This means that the inverse of $f(\cdot, t)$ exists and is unique. Denoting this inverse by $g(\cdot, t)$, equation (1.2-5) can be written as

$$i = g(v, t) \tag{1.2-6}$$

If the function f does not depend on t, the resistor is called *time invariant*. The v-i relation of a time-invariant resistor can then be written as

$$v(t) = f(i(t)) \tag{1.2-7}$$

A special case of great importance and practical use of monotonic resistors is when $f(\cdot, t)$ is a linear function. More specifically, a two-terminal resistor is called *linear* if $f(\cdot, t)$, defined in (1.2-5), is a linear function for each $t \geq 0$. The equation of a linear resistor is normally written in the form

$$v(t) = R(t)i(t) \tag{1.2-8}$$

where $R(t)$, the *resistance* of the resistor at time t, denotes the slope of the v-i curve, which in this case is a straight line going through the origin. Figure 1.2-3 shows the characteristic of a linear time-varying resistor for several values of t.

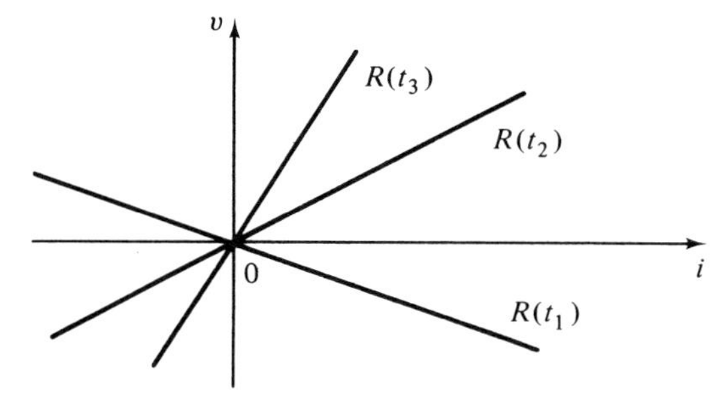

Figure 1.2-3 Characteristic of linear time-varying resistor.

An important subclass of the linear resistors defined in (1.2-8) is when $R(t)$ is constant; more precisely, a wo-terminal resistor is *linear time invariant* if its v-i equation can be written as

$$v(t) = Ri(t)$$

where R is a constant and is measured in *ohms*. It should be mentioned at this stage that in reality no resistor is linear or time invariant. However, depending on their applications, many resistors can be approximated as linear and/or time-invariant resistors.

The energy absorbed by a resistor can be found using (1.1-4) and its v-i relation; for a current-controlled resistor this becomes

$$E(t) = \int_{-\infty}^{t} i(\tau) f(i(\tau), \tau)\, d\tau, \qquad t > -\infty \tag{1.2-9}$$

and for a voltage-controlled resistor (1.1-4) can be written as

$$E(t) = \int_{-\infty}^{t} v(\tau)g(v(\tau), \tau)\, d\tau, \qquad t > -\infty \tag{1.2-10}$$

According to Definition 1.1-2, a resistor is passive if $E(t) \geq 0$ for all $t > -\infty$ and for all possible i and v. However, if the initial time of the application of the input is t_0, the energy absorbed or delivered by the resistor from time t_0 to t is given by

$$E(t, t_0) = \int_{t_0}^{t} i(\tau)f(i(\tau)\tau)\, d\tau \tag{1.2-11}$$

Thus, a resistor is passive if $E(t, t_0) \geq 0$ for *all* $t \geq t_0$, for *all* initial times t_0, and for *all* possible $v(t)$, $i(t)$ pairs. Accordingly, a resistor is active if $E(t, t_0) < 0$ for *some* $t \geq t_0$ or for *some* $i(t)$ or $v(t)$.

Exercise 1.2-1. Show that a nonlinear time-varying resistor is passive if and only if its v-i characteristic remains in the first and third quadrant for all t. [*Hint:* For the "if" part show that the integrand in (1.2-9) or (1.2-10) is always positive. For the "only if" part assume that a portion of the v-i characteristic lies in the second or fourth quadrant; then choose appropriate voltage and currents so that the corresponding integrands become negative.]

We say that a passive resistor absorbs energy and an active resistor may deliver energy to other network elements. A good example of an active resistor is a dc source or a battery. Consider the battery shown in Fig. 1.2-4;

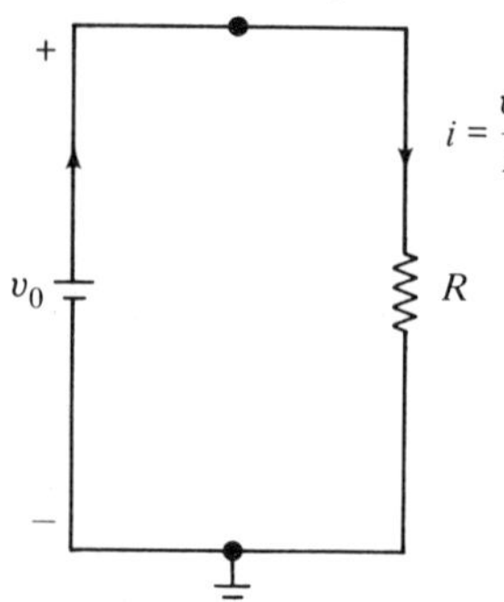

Figure 1.2-4 Direct current source with voltage v_0 can be considered as an active resistor supplying energy to the passive resistor with resistance $R > 0$.

assume that the resistor in series has a resistance R. Then the energy delivered to the resistor by the battery from t_0 to t is $[(t - t_0)/R]v_0^2$.

Small-Signal Behavior of Nonlinear Resistors. Consider a nonlinear time-invariant resistor whose v-i relation is given by

$$v(t) = f(i(t)) \tag{1.2-12}$$

where $f(\cdot)$ is a function whose derivative with respect to its argument, i, is continuous. Let the input $i(t)$ consist of a time-varying bias source $I(t)$ and a *small signal*, $\delta i(t)$ (see Fig. 1.2-5); then

$$i(t) = I(t) + \delta i(t) \tag{1.2-13}$$

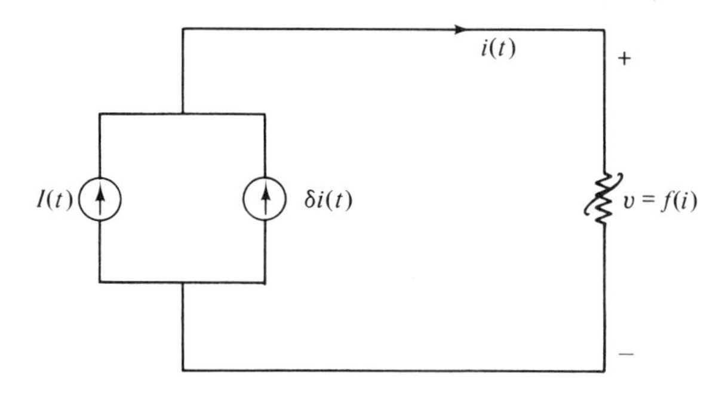

Figure 1.2-5 Nonlinear resistor that behaves as a linear resistor as far as the small signal $\delta i(t)$ is concerned.

By the small-signal $\delta i(t)$ we mean that the amplitude of the signal $\delta i(t)$ is much smaller than the amplitude of the bias source $I(t)$; that is,

$$|\delta i(t)| \ll |I(t)| \quad \text{for all } t \tag{1.2-14}$$

For example, in a transistor receiver, $\delta i(t)$ can be considered as the current generated by the signal received through the antenna and $I(t)$ to be the current supplied by the batteries used as bias sources. In this case, $\delta i(t)$ may be on the order of 10^{-6} A whereas $I(t)$ might be on the order of 10^{-3} A.

Let us denote the voltage corresponding to the bias source $I(t)$ by $V(t)$; then the voltage $v(t)$ can be written as

$$v(t) = V(t) + \delta v(t) \tag{1.2-15}$$

where $\delta v(t)$ is a *small* voltage corresponding to the small signal $\delta i(t)$. Now replacing v and i from (1.2-15) and (1.2-13) into (1.2-12), we obtain

$$V(t) + \delta v(t) = f(I(t) + \delta i(t)) \tag{1.2-16}$$

If $\delta i(t)$ is zero, its corresponding small-signal voltage $\delta v(t)$ will also be zero; then (1.2-16) becomes

$$V(t) = f(I(t)) \tag{1.2-17}$$

Using a Taylor expansion around $I(t)$, we can write (1.2-16) as

$$V(t) + \delta v(t) = f(I(t)) + f'(I(t))\cdot\delta i(t) + \tfrac{1}{2}f''(I(t))\cdot\delta i^2(t) + \cdots \tag{1.2-18}$$

From (1.2-17) and (1.2-18) we get

$$\delta v(t) = f'(I(t))\cdot\delta i(t) + \tfrac{1}{2}f''(I(t))\delta i^2(t) + \cdots \tag{1.2-19}$$

Since $\delta i(t)$ is small and $f(\cdot)$ is smooth, we can neglect the second- and higher-order terms in (1.2-19) and write

$$\delta v(t) = f'(I(t)) \cdot \delta i(t) \tag{1.2-20}$$

Notice that $f'(I(t))$ is a time-varying scalar function that does not depend on $\delta i(t)$; then (1.2-20) represents a linear *time-varying* resistor whose resistance is $f'(I(t))$, the slope of $f(\cdot)$ at $I(t)$. This is an interesting and useful result; as far as the small signals $\delta i(t)$ and $\delta v(t)$ are concerned, the nonlinear resistor of Fig. 1.2-5 behaves as a linear time-varying resistor with resistance $f'(I(t))$. This also suggests a method of constructing linear time-varying resistors from nonlinear time-invariant resistors and a time-varying bias source.

Example 1.2-1. Consider a nonlinear resistor whose v-i relation is given by

$$v = f(i) \triangleq \tfrac{1}{3}i^3 + 2i$$

Let the bias be a current source $I(t) = \sin \omega t$; then the small-signal equivalent element is a linear time-varying resistor whose resistance, $R(t)$, is given by

$$R(t) = f'(I(t)) = 2 + \sin^2 \omega t$$

The v-i relation of the small-signal equivalent resistor is therefore

$$\delta v(t) = (2 + \sin^2 \omega t)\, \delta i(t)$$

Remark 1.2-1. If the bias of the nonlinear resistor is a dc source or a battery, the small-signal equivalent resistor is a linear time-invariant resistor whose resistance is equal to the slope of the nonlinear characteristic at the operating point (V, I). As an example, consider the v-i curve shown in Fig. 1.2-6. If the nonlinearity is given by

$$v = f(i) = i^2, \qquad i \geq 0$$

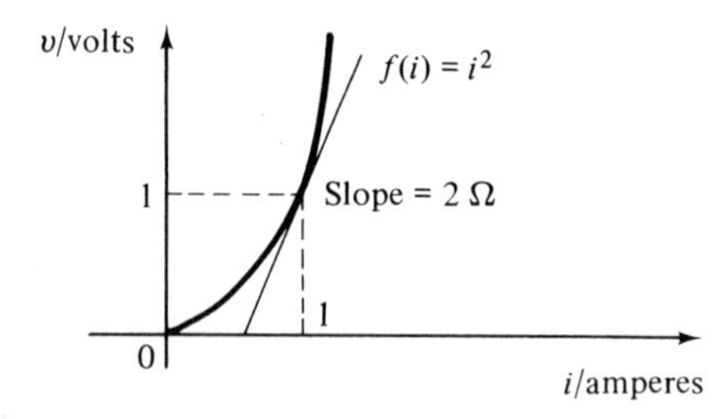

Figure 1.2-6 The slope of the v-i characteristic represents the small-signal equivalent resistance of the nonlinear resistor.

and the dc bias is 1 A, then the small-signal equivalent resistor is a time-invariant resistor whose resistance is given by

$$R = f'(1) = 2\,\Omega$$

Our small-signal analysis suggests a method of designing linear adjustable resistors using nonlinear resistors and an adjustable bias source. This method of adjusting the resistance of linear resistors eliminates the need for mechanically adjustable resistors, which inherently introduce some random noise to the circuit.

Applications of Nonlinear and Time-Varying Resistors. It was mentioned earlier that resistors can be used to convert electric energy to heat. This is not, however, the most interesting property of resistors. Linear time-varying and nonlinear resistors are used extensively in a variety of circuits, such as rectifiers, frequency multipliers, modulators, limiters, automatic sorting devices, and so on. In the following we shall briefly mention some of the most common applications of such resistors.

Pulse Generator. Consider the tunnel diode shown in the simple circuit of Fig. 1.2-7(a) with its *i-v* characteristic shown in Fig. 1.2-7(b). If the current of the current generator is given by $i(t) = I[1 + 0.5 \sin (2\pi/T)t]$, then the voltage across the diode is as shown in Fig. 1.2-8.

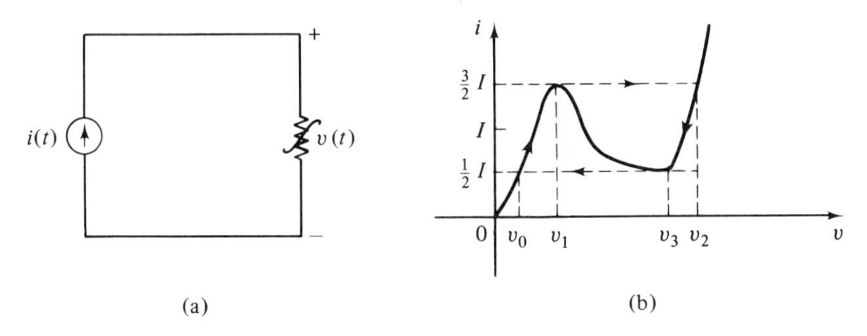

Figure 1.2-7 (a) Pulse generator circuit; (b) characteristic of the tunnel diode.

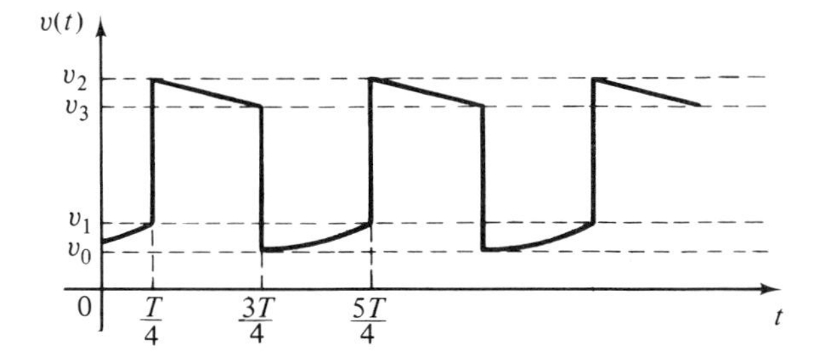

Figure 1.2-8 Voltage across the tunnel diode due to the current $I\left[1 + 0.5 \sin \left(\frac{2\pi}{T}\right)t\right]$.

Rectification. Diodes have been used for many years for rectification; rectifiers come in many different forms and sizes. Here we mention only the simplest diode circuit. Consider the circuit shown in Fig. 1.2-9(a); the voltage source $v_1(t)$ is given by

$$v_1(t) = v_0 \sin \frac{2\pi}{T} t$$

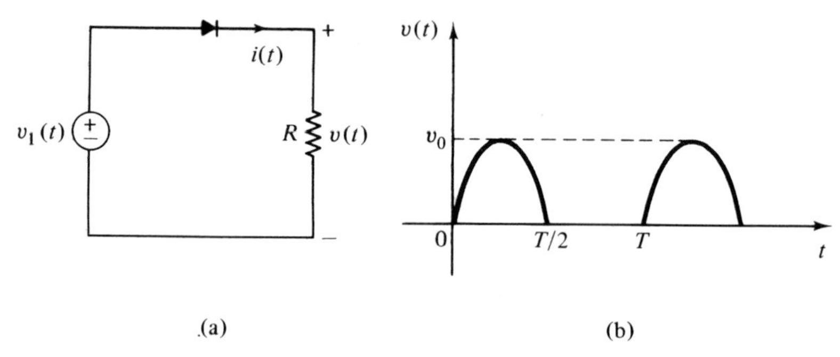

Figure 1.2-9 (a) Simple rectifier circuit; (b) voltage across the resistor R due to a sinusoidal voltage.

The rectifier consists of a linear resistor in series with an *ideal diode* whose *i-v* characteristic and symbolic representation are shown in Figs. 1.2-10(a) and (c), respectively. The *i-v* characteristic of the combination of the ideal diode and the linear resistor is given in Fig. 1.2-10(b). Notice that the current

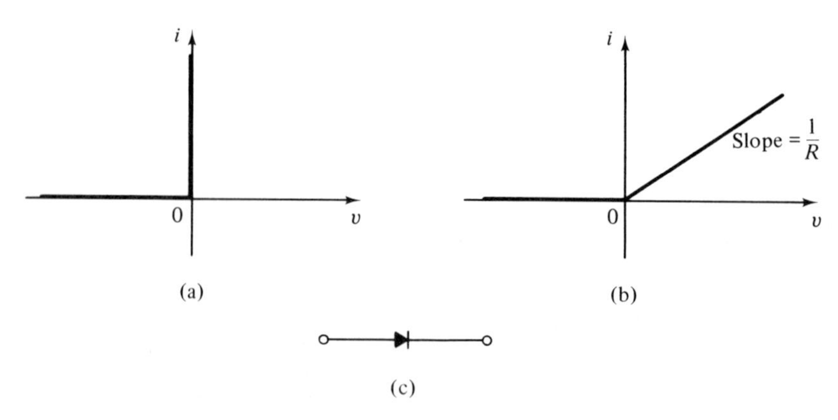

Figure 1.2-10 (a) *i-v* characteristic of an ideal diode; (b) *i-v* characteristic of the rectifier of Fig. 1.2-9(a); (c) symbolic representation of an ideal diode.

through the circuit is given by

$$i(t) = \frac{v_0}{R} \sin \frac{2\pi}{T} t \qquad \text{for } v_1(t) \geq 0$$

and

$$i(t) = 0 \qquad \text{for } v_1(t) < 0$$

Consequently, the voltage across the linear resistor is as shown in Fig. 1.2-9(b).

Frequency Multipliers. Linear time-varying and nonlinear resistors can be used to convert a low-frequency input signal to an output signal whose frequency is a multiple of the input frequency. This is shown by an example.

Example 1.2-2. Consider the current source $i(t)$ in series with the nonlinear resistor shown in Fig. 1.2-11. Let the v-i relation of the nonlinear resistor

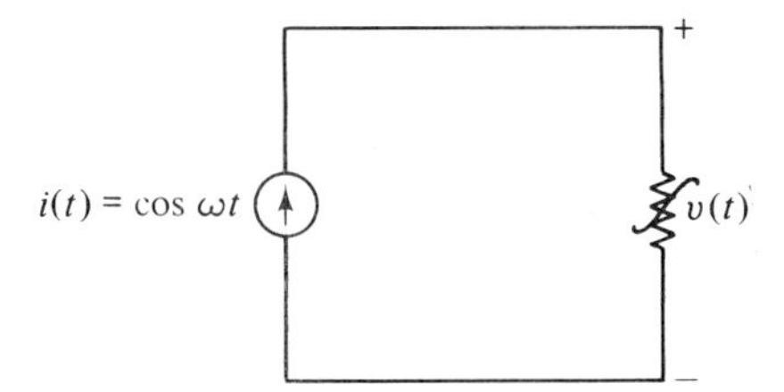

Figure 1.2-11 Frequency-multiplier resistor.

be given by

$$v(t) = f(i(t)) \triangleq 2i^2(t) - 1$$

The voltage across the resistor is therefore

$$v(t) = 2\cos^2 \omega t - 1 = \cos 2\omega t$$

Hence, the frequency of the output is twice the frequency of the input. In a similar fashion we can show that by properly choosing the nonlinear resistor any multiple of the input frequency can be obtained.

Exercise 1.2-2. Find the v-i relation of a nonlinear resistor so that the voltage across it has a frequency four times greater than the frequency of the current through it. Assume that the current through the resistor is $\cos \omega t$. [Ans: $v = 8i^4 - 8i^2 + 1$.]

Signal Generation. A generalization of the basic idea used in obtaining frequency multipliers can be stated by considering that the voltage across a current-controlled resistor is a function of the current through it. For a given signal $i(t)$, under certain conditions a nonlinear characteristic $f(\cdot)$ can be found so that any desired signal $v(t)$ is realized. This idea is made clear in the following example.

Example 1.2-3. Consider the nonlinear resistor shown in Fig. 1.2-12(a). Let the input be the current source, which is given by $i(t) = e^{-t}$ for $t \geq 0$

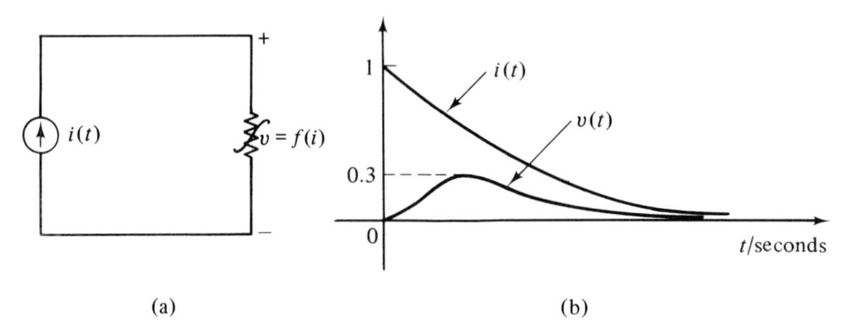

Figure 1.2-12 (a) Nonlinear resistor used in generating an arbitrary voltage waveform $v(t)$; (b) $i(t)$ is the current through and $v(t)$ is the voltage across the nonlinear resistor.

and $i(t) = 0$ for $t < 0$. The problem is to find the nonlinear function $f(\cdot)$ so that the output $v(t)$ is as given in Fig. 1.2-12(b).

Solution

The output $v(t)$ can be approximated by the function

$$v(t) = 0.8(e^{-t} - e^{-3t})$$

Since $e^{-t} = i$, the nonlinear function $f(\cdot)$ is given by

$$f(i) = 0.8(i - i^3) \tag{1.2-21}$$

Remark 1.2-2. It should be mentioned that it is not always possible to generate a desired signal $v(t)$ from a given signal $i(t)$ by a nonlinear time-invariant resistor. The following counterexample illustrates this point. Take the given current source $i(t)$ as

$$i(t) = \sin t$$

and take the desired output voltage to be

$$v(t) = e^{-t}, \qquad t \geq 0$$

Then no nonlinear *current-controlled* resistor can be found so that the following equation holds:

$$v(t) = f(i(t)), \qquad t \geq 0$$

The pair (v, i) given here is sometimes said to be a *nonadmissible* signal pair of a current-controlled resistor, whereas the (v, i) pair given in Example 1.2-3 is said to be an *admissible* signal pair of the nonlinear resistor given by equation (1.2-21).

Exercise 1.2-3. Find a necessary condition under which a pair $(v(t), i(t))$ is an admissible signal pair of a time-invariant current-controlled resistor. [Hint: For the given functions $v(t)$ and $i(t)$, consider a graphical method of obtaining $f(\cdot)$ so that $v = f(i)$.]

Before concluding this section, it should be mentioned that an important application of nonlinear resistors is in the design of linear time-varying networks. As was mentioned earlier, a nonlinear time-invariant resistor will behave as a linear time-varying resistor if it is operated in the small-signal mode with a time-varying bias source.

1.3 TWO-TERMINAL CAPACITORS

A two-terminal network element is called a capacitor if the voltage, $v(t)$, across it and the electric charge, $q(t)$, on it are related together through an algebraic relation. Hence the characteristic of a capacitor can be represented as a curve in the q-v or v-q plane. In contrast with resistors, capacitors store energy, and for this reason they are classified as *energy-storing* network elements. The relation between the current through a capacitor, $i(t)$, and the voltage across it, $v(t)$, can be obtained using equations (1.1-1) or (1.1-5). We shall study this relation in detail for different capacitors. As in the case of resistors, capacitors can be classified in three different categories: charge controlled, voltage controlled, and monotone.

A two-terminal network element is called a *charge-controlled* capacitor if the voltage across it can be expressed as a single-valued function of its electric charge; that is,

$$v(t) = h(q, t) \tag{1.3-1}$$

Figure 1.3-1(a) shows a symbolic representation of a nonlinear capacitor and Fig. 1.3-1(b) shows a typical v-q curve of a charge-controlled capacitor for two instances, t_1 and t_2.

A two-terminal network element is called a *voltage-controlled* capacitor if the charge on it can be expressed as a single-valued function of the voltage across it; that is,

$$q(t) = f(v, t) \tag{1.3-2}$$

Finally, a capacitor is called *monotone* if it is both charge controlled and voltage controlled. The v-q characteristic of such a capacitor is a monotonically increasing or decreasing curve in the v-q plane. In this case, functions $h(\cdot, t)$ and $f(\cdot, t)$ have inverses, and $h(x, t) = f^{-1}(x, t)$, for all x and for all t.

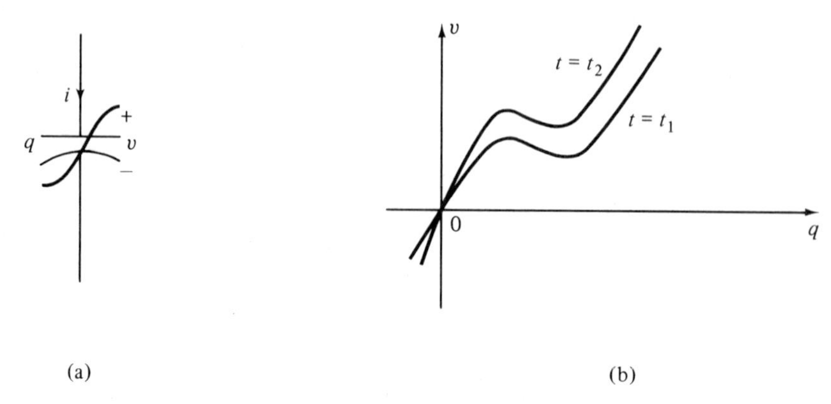

Figure 1.3-1 (a) Symbolic representation of a nonlinear capacitor; (b) v-q characteristic of a nonlinear time-varying capacitor.

Exercise 1.3-1. The v-q relation of a nonlinear time-invariant capacitor is given by

$$v = \tanh q \tag{1.3-3}$$

Show that this capacitor is both charge controlled and voltage controlled. [*Hint:* Show that the derivative of the given nonlinear characteristic is always positive.]

An important class of monotonic capacitors is the class of linear capacitors. The v-q characteristic of a linear capacitor is a straight line going through the origin.

The q-v relation of linear capacitors can then be written as

$$q(t) = C(t)v(t) \tag{1.3-4}$$

where $C(t)$ is called the capacitance of the linear capacitor at time t and is measured in farads. If $C(t)$ is constant, the corresponding capacitor will be linear time invariant.

v-i **Relationship of Two-Terminal Capacitors.** Consider a two-terminal voltage-controlled capacitor whose q-v relation is given by

$$q(t) = f(v(t), t) \tag{1.3-5}$$

Using the basic relations (1.1-1), (1.3-5), and the chain rule, the current through the capacitor can be found to be

$$i(t) = \frac{dq}{dt} = \frac{\partial f(v, t)}{\partial v}\cdot\frac{dv}{dt} + \frac{\partial f(v, t)}{\partial t} \tag{1.3-6}$$

where $[\partial f(v, t)]/\partial v$ and $[\partial f(v, t)]/\partial t$ denote partial derivatives of $f(\cdot, \cdot)$ with respect to v and t, respectively.

Since the current and voltage of a capacitor are related through a differential equation, the capacitors are sometimes classified as *dynamical elements*.

Example 1.3-1. The q-v relation of a nonlinear time-varying capacitor is given by

$$q = (1 + 0.5 \sin t)v^3 \tag{1.3-7}$$

Let the voltage across this capacitor be

$$v(t) = \sin \omega t \tag{1.3-8}$$

Find the current through this capacitor (see Fig. 1.3-2).

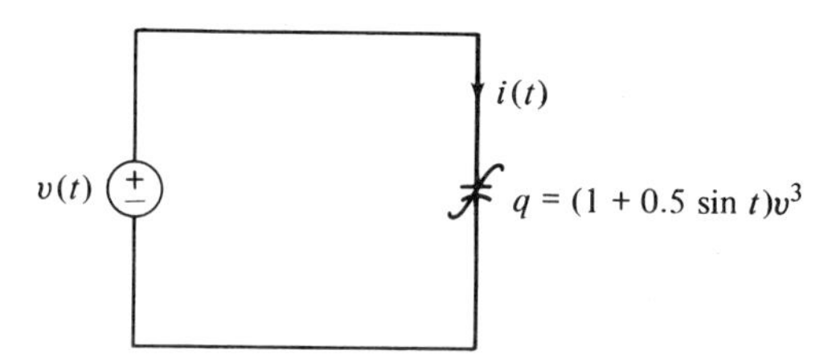

Figure 1.3-2 Nonlinear capacitor in series with a voltage source.

Solution

From (1.3-6) and (1.3-7) we obtain

$$i(t) = 3(1 + 0.5 \sin t)v^2 \cdot \frac{dv}{dt} + (0.5 \cos t)v^3 \tag{1.3-9}$$

Replacing $v(t)$ in (1.3-9) from (1.3-8) results in

$$i(t) = 3\omega(1 + 0.5 \sin t) \sin^2 \omega t \cos \omega t + 0.5 \sin^3 \omega t \cos t$$

For special forms of $f(\cdot, t)$, equation (1.3-6) can be put in a more simple form. Some of these cases are as follows:

1. Nonlinear time-invariant voltage-controlled capacitor:

$$q = f(v): \qquad i(t) = \frac{df}{dv} \cdot \frac{dv}{dt} \tag{1.3-10}$$

2. Linear time-varying capacitor:

$$q = C(t)v: \qquad i(t) = C(t)\frac{dv}{dt} + v(t)\frac{dC}{dt} \tag{1.3-11}$$

3. Linear time-invariant capacitor:

$$q = Cv: \qquad i(t) = C\frac{dv}{dt} \tag{1.3-12}$$

Exercise 1.3-2. The q-v relation of a nonlinear time-invariant capacitor is given by

$$q = \frac{v}{1 + |v|}$$

Show that this capacitor is both charge controlled and voltage controlled and find the current through it in terms of the voltage $v(t)$.
[Ans: $i(t) = (1/(1 + |v|)^2)\cdot(dv(t)/dt)$.]

Energy Stored in a Capacitor. It was mentioned earlier that capacitors are energy-storing elements, and the energy stored in a capacitor at time t is given by

$$E(t) = \int_{-\infty}^{t} v(\tau)i(\tau)\,d\tau \tag{1.3-13}$$

This equation can be written in the form

$$E(t) = \int_{t_0}^{t} v(\tau)i(\tau)\,d\tau + \int_{-\infty}^{t_0} v(\tau)i(\tau)\,d\tau \tag{1.3-14}$$

If we denote

$$E(t, t_0) \triangleq \int_{t_0}^{t} v(\tau)i(\tau)\,d\tau \tag{1.3-15}$$

then $E(t, t_0)$ is the net energy delivered to the capacitor from t_0 to t, and

$$E(t_0) \triangleq \int_{-\infty}^{t_0} v(\tau)i(\tau)\,d\tau \tag{1.3-16}$$

is the energy stored on the capacitor prior to t_0.

Since the current $i(t)$ is the derivative of the charge, the net energy delivered to a capacitor, $E(t, t_0)$, can be written as

$$E(t, t_0) = \int_{t_0}^{t} v(\tau)\frac{dq(\tau)}{d\tau}\,d\tau \tag{1.3-17}$$

For a charge-controlled capacitor (1.3-17) has an interesting interpretation in the v-q plane; to see this, let us assume that the v-q relation of the capacitor under study is given by

$$v(t) = f(q(t)) \tag{1.3-18}$$

Then, with a simple change of variable, (1.3-17) can be written as

$$E(t, t_0) = \int_{q(t_0)}^{q(t)} f(q)\,dq \tag{1.3-19}$$

Hence, $E(t, t_0)$ is numerically equal to the shaded area in Fig. 1.3-3. This observation suggests a simple graphical method of computing the energy stored on nonlinear capacitors.

Passive and Active Capacitors. As in the case of resistors, a capacitor is passive if it is incapable of delivering energy at any time.

If the initial stored energy $E(t_0)$ on a capacitor is zero, then from (1.1-4) and (1.3-14) it is clear that the capacitor is passive if and only if

$$E(t, t_0) = \int_{t_0}^{t} v(\tau)i(\tau)\,d\tau \geq 0 \tag{1.3-20}$$

for all $t \geq t_0$, for all t_0, and for all admissible signal pairs (v, i).

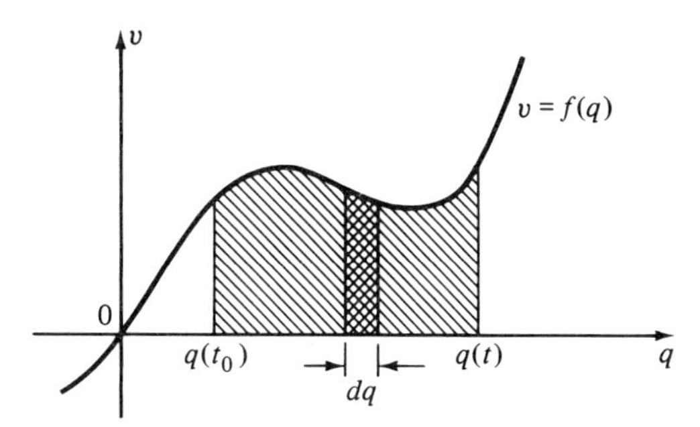

Figure 1.3-3 Energy stored in the capacitor is designated by the shaded region.

If $E(t, t_0)$ is negative for some $t \geq t_0$ or for some admissible signal pair, the capacitor is said to be active.

For a time-invariant voltage-controlled capacitor with q-v relation

$$q = f(v)$$

we have

$$i = f'(v) \cdot \frac{dv}{dt} \tag{1.3-21}$$

If we assume that $v(-\infty) = 0$, after a simple change of variable, (1.3-13) can be written as

$$E(t) = \int_0^{v(t)} vf'(v)\, dv \tag{1.3-22}$$

Hence, a sufficient condition for such a capacitor to be passive is that

$$f'(v) \geq 0 \quad \text{for all } v \tag{1.3-23}$$

This is because in the first and fourth quadrants of the v-q plane both v and dv are positive; hence, the integrand in (1.3-22) will be positive. In the second and third quadrants, since both v and dv are negative, the integrand is still positive.

Exercise 1.3-3. Consider a time-invariant charge-controlled capacitor with v-q relation

$$v = f(q)$$

Show that this capacitor is passive if $f(q)$ remains in the first and third quadrants. Assume that $q(-\infty) = 0$.

For a linear time-varying capacitor with q-v relation

$$q(t) = C(t)v(t)$$

since

$$i(t) = \dot{C}(t)v(t) + C(t)\dot{v}(t) \tag{1.3-24}$$

where $\dot{C}(t)$ and $\dot{v}(t)$ denote the derivative of $C(t)$ and $v(t)$, respectively, equation (1.3-13) can be written as

$$E(t) = \int_{-\infty}^{t} [\dot{C}(\tau)v^2(\tau) + C(\tau)v(\tau)\dot{v}(\tau)]\, d\tau$$

Exercise 1.3-4. Show that a linear time-varying capacitor is passive if *and only if*

$$\dot{C}(t) \geq 0 \quad \text{and} \quad C(t) \geq 0 \quad \text{for all } t \geq -\infty$$

Assume that at $t = -\infty$ the capacitor voltage and current are zero.

Exercise 1.3-5. Show that a linear time-invariant capacitor is passive if and only if its capacitance C is positive.

From Exercise 1.3-4 it is clear that to obtain an active capacitor it is sufficient to make either $C(t)$ or $\dot{C}(t)$ negative and hence extract electric energy from the capacitor. This idea has been applied successfully to design *parametric amplifiers.* Owing to high reliability, light weight, and low random noise, these amplifiers are widely used in space communication systems. Figure 1.3-4 shows a simple parametric amplifier. The time-varying capacitor

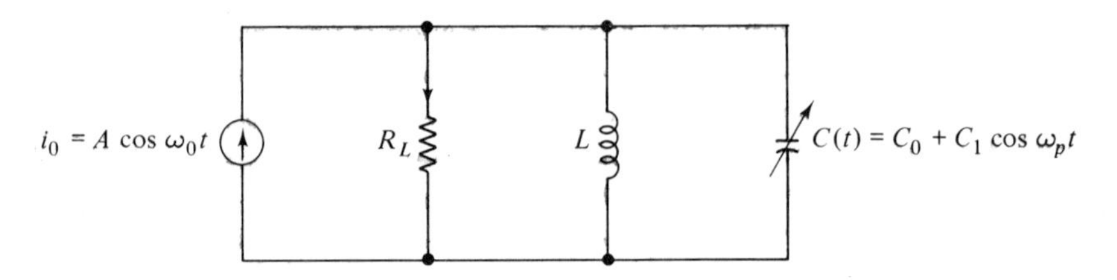

Figure 1.3-4 Simplified parametric amplifier.

has a capacitance $C(t) = C_0 + C_1 \cos \omega_p t$, where ω_p is called the *pump* frequency and C_0 and C_1 are constants. Note that $\dot{C}(t) = -C_1\omega_p \sin \omega_p t$, which is negative for $n\pi/\omega_p \leq t \leq [\pi(n+1)]/\omega_p$. The signal to be amplified is represented as the current source i_0, and the output signal is the current through the load resistor R_L. By appropriately choosing ω_0, ω_p, C_0, and C_1, it is possible to pump an arbitrary large current through R_L.

The linear time-varying capacitor $C(t)$ used in parametric amplifiers, like most other time-varying capacitors used in electronic devices, is constructed by operating a nonlinear capacitor with a time-varying bias source. The basic idea underlying this technique is discussed next.

Small-Signal Behavior of Nonlinear Capacitors. Consider a nonlinear time-invariant voltage-controlled capacitor whose q-v relation is given by

$$q = f(v) \tag{1.3-25}$$

As in the case of nonlinear resistors, assume that the voltage v consists of a time-varying bias $V(t)$ and a small signal, $\delta v(t)$. The bias $V(t)$ will induce an *operating point*, $[V(t), Q(t)]$, which satisfies

$$Q(t) = f(V(t)) \tag{1.3-26}$$

and the small signal $\delta q(t)$ will cause a small perturbation, $\delta q(t)$, in $Q(t)$ so that

$$q(t) = Q(t) + \delta q(t) \tag{1.3-27}$$

and

$$v(t) = V(t) + \delta v(t) \tag{1.3-28}$$

In a similar manner as in the case of resistors, by replacing $q(t)$ and $v(t)$ from equations (1.3-27) and (1.3-28) in (1.3-25) and expanding $f(\cdot)$ around $V(t)$, we can show that $\delta v(t)$ and $\delta q(t)$ satisfy the equation

$$\delta q(t) \simeq f'(V(t))\, \delta v(t) \tag{1.3-29}$$

where $f'(V(t))$ denotes the derivatives of $f(\cdot)$ evaluated at $V(t)$. It is clear that $f'(V(t))$ is a time-varying scalar function, which we denote by

$$C(t) \triangleq f'(V(t)) \tag{1.3-30}$$

Then (1.3-29) can be written as

$$\delta q(t) = C(t)\, \delta v(t)$$

That is, the nonlinear capacitor behaves as a linear time-varying capacitor with capacitance $C(t) = f'(V(t))$. Figure 1.3-5 shows the two equivalent capacitors.

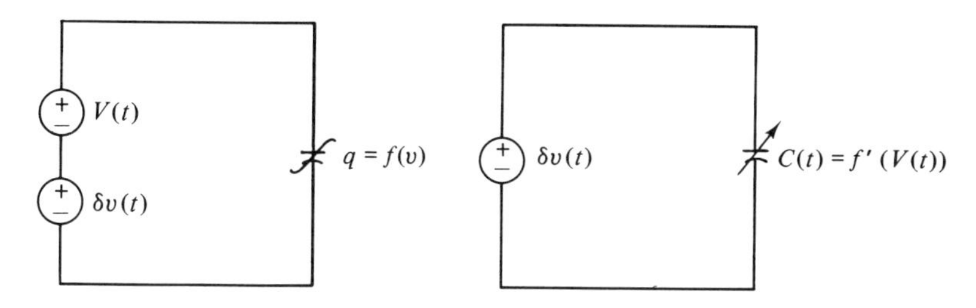

Figure 1.3-5 Nonlinear capacitor and its small-signal equivalent.

If the bias is time invariant, then $f'(V)$ is constant and the small-signal equivalent capacitor is a capacitor with capacitance $C = f'(V)$, and the q-v relation is

$$\delta q = C\, \delta v \tag{1.3-31}$$

The capacitance C corresponds to the slope of the q-v curve at $v = V$ and sometimes is called the *incremental capacitance* of the nonlinear capacitor. Commercial nonlinear capacitors whose small-signal equivalent is a linear time-varying capacitor are known as *varactor diodes*. A comprehensive treatment of the application of such nonlinear capacitors is given in [3].

Example 1.3-2. Consider a nonlinear capacitor with q-v relation

$$q = f(v) \triangleq v + \tanh v \tag{1.3-32}$$

Let the bias $V(t)$ be

$$V(t) = \sin t \tag{1.3-33}$$

Then the capacitance of the small-signal equivalent capacitor is

$$f'(V(t)) = 2 - \tanh^2 V(t)$$

or, equivalently,

$$C(t) = 2 - \tanh^2 \sin t$$

Example 1.3-3. For the nonlinear capacitor whose q-v relation is

$$q = v^3$$

let the bias be a 1-V battery; then the incremental capacitor is

$$C = 3 \cdot (1)^2 = 3 \text{ F}$$

(See Fig. 1.3-6.)

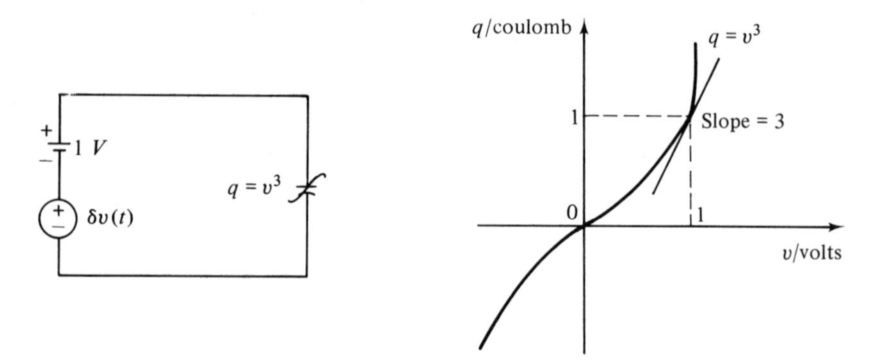

Figure 1.3-6 Nonlinear capacitor and its incremental capacitance.

Example 1.3-4. Electronic Tuning Devices. Tuning circuits are used in most communication devices when it is desired to discriminate among signals on the basis of their frequency content. A conventional tuning circuit consists of a resistor, an inductor, and an adjustable capacitor, which are connected in series. The capacitor is generally adjusted mechanically by changing the area of the parallel plates comprising the capacitor. This method of adjusting the capacitor is not only bulky and expensive, but also introduces random noise to the device. In modern tuning circuits, the mechanically adjustable capacitor is replaced by a nonlinear capacitor in series with an adjustable voltage source [see Fig. 1.3-7(a)].

Let the q-v relation of the capacitor be $q = \frac{1}{2}kv^2$ [see Fig. 1.3-7(b)], where k is a positive constant. Then the incremental capacitor has a capacitance

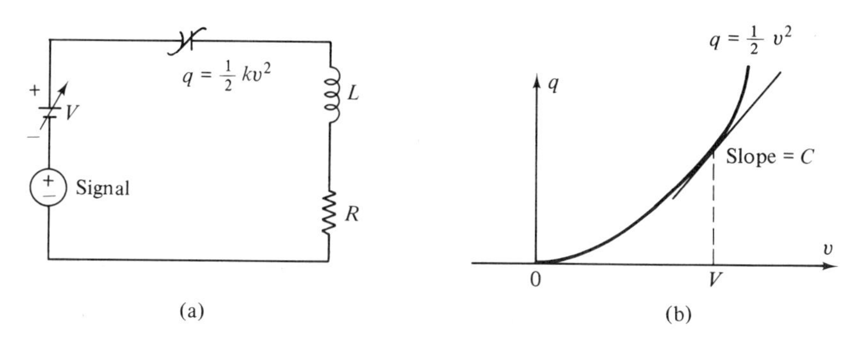

Figure 1.3-7 (a) Electronic tuning circuit; (b) q-v characteristic of the capacitor used in (a).

$C = kV$. Consequently, by adjusting the dc source V, the circuit can be tuned in the signal with the desired frequency.

The small-signal behavior of nonlinear capacitors is one of the most useful properties of such network elements. In many communication networks this property is used to obtain AM and FM signals and electronically adjustable filters whose delay or bandwidth is a function of time. [1]

Exercise 1.3-6. Consider a time-invariant voltage-controlled capacitor whose q-v relation is

$$q = f(v)$$

Let the bias voltage be

$$V(t) = \sin \omega t$$

Find the nonlinear function $f(\cdot)$ so that the small-signal equivalent capacitor $C(t)$ is

$$C(t) = 1 + 0.5 \sin^2 \omega t$$

[Ans: $f(v) = v + \frac{1}{6}v^3$.]

1.4 TWO-TERMINAL INDUCTORS

A two-terminal network element is said to be an inductor if the current, $i(t)$, and the magnetic flux, $\phi(t)$, through it are related by an algebraic relation. This implies that the flux and the current can be represented as a curve in the ϕ-i plane. The symbolic representation of a nonlinear inductor is shown in Fig. 1.4-1(a). Inductors can also be classified in several categories. In the following we define some of the most important ones.

A two-terminal network element is said to be a *flux-controlled* inductor if the current through it can be expressed as a function of its magnetic flux,

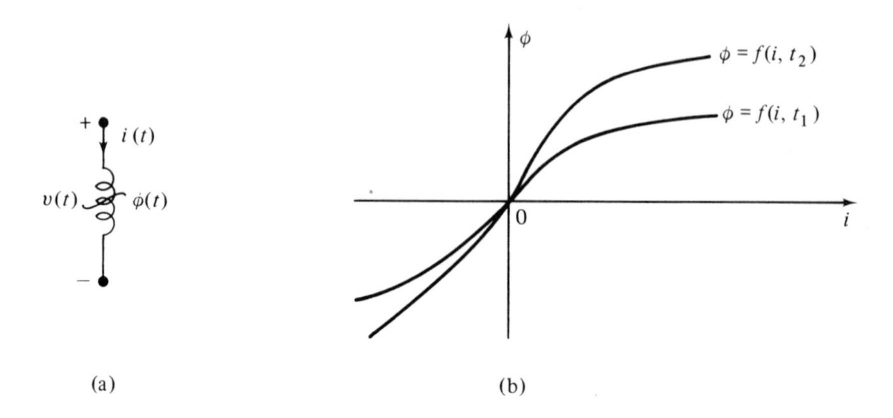

Figure 1.4-1 (a) Symbolic representation of a nonlinear inductor; (b) characteristic of a nonlinear monotonic inductor.

that is, if

$$i(t) = h(\phi(t), t) \tag{1.4-1}$$

where $h(\cdot, t)$ is a single-valued function.

If the flux on an inductor can be expressed as a function of the current through it, that is, if

$$\phi(t) = f(i(t), t) \tag{1.4-2}$$

then the inductor is called *current controlled.*

Finally, a two-terminal inductor is both flux controlled and current controlled if the functions $f(\cdot, t)$ and $h(\cdot, t)$ defined by (1.4-2) and (1.4-1), respectively, are invertible and

$$h^{-1}(\cdot, t) = f(\cdot, t) \quad \text{for all } t \tag{1.4-3}$$

This means that the ϕ-i curve of the inductor under consideration is either monotonically increasing or monotonically decreasing [see Fig. 1.4-1(b)].

Exercise 1.4-1. The flux through a nonlinear inductor is

$$\phi = \frac{2i}{1 + |i|}$$

Show that this inductor is both flux controlled and current controlled.

In contrast to nonlinear resistors and capacitors, most practical inductors are neither flux controlled nor current controlled. This is due to the hysteresis phenomenon caused by inserting an iron core in the magnetic coil. A typical ϕ-i characteristic of such an inductor is shown in Fig. 1.4-2. It should be mentioned that this hysteresis loop corresponds to a sinusoidal excitation only; for other forms of excitation, the loop will be different.

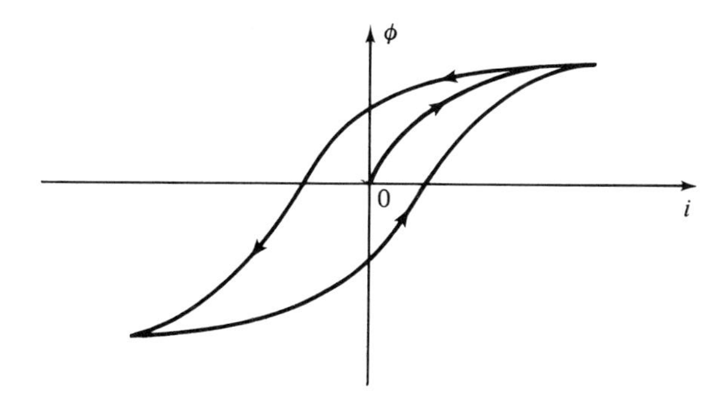

Figure 1.4-2 Typical hysteresis loop.

A two-terminal inductor is said to be *time invariant* if its ϕ-i relation is not an implicit function of time. Hence, for time-invariant inductors (1.4-1) and (1.4-2) can be written as

$$i = h(\phi) \tag{1.4-4}$$

and

$$\phi = f(i) \tag{1.4-5}$$

If the flux and the current through an inductor are related through a linear function, the corresponding inductor is called *linear*.

In this case ϕ and i are related by

$$\phi = L(t)i \tag{1.4-6}$$

where $L(t)$ is called the inductance of the linear inductor at time t and is measured in henries. If $L(t)$ is constant, the linear inductor under consideration is also *time invariant*.

v-i Relation of Two-Terminal Inductors. To obtain the v-i relations for various inductors from the ϕ-i characteristic, we use the basic equation (1.1-2); that is,

$$v(t) = \frac{d\phi(t)}{dt}$$

1. For a current-controlled inductor whose ϕ-i relation is given by (1.4-2), this becomes

$$v(t) = \frac{\partial f(i, t)}{\partial i} \cdot \frac{di}{dt} + \frac{\partial f(i, t)}{\partial t} \tag{1.4-7}$$

For example, if the current through an inductor is $i(t) = i_0 \sin \omega t$ and if its ϕ-i relation is $\phi = \sqrt[3]{i}$, then the voltage across this inductor is:

$$v(t) = (\omega/3)i_0^{1/3}(\sin \omega t)^{-2/3} \cdot \cos \omega t$$

Exercise 1.4-2. The ϕ-i relation of a particular nonlinear inductor is

$$\phi = e^{-t}i^2 + \sin \omega t$$

Let the current through it be

$$i(t) = \cos \omega t$$

Find the voltage across this inductor.

2. For a flux-controlled inductor the v-i relation can be obtained by differentiating (1.4-1) with respect to t; this gives

$$\frac{di}{dt} = \frac{\partial h(\phi, t)}{\partial \phi} \cdot \frac{d\phi}{dt} + \frac{\partial h(\phi, t)}{\partial t} \tag{1.4-8}$$

This equation, however, is in general a nonlinear equation in ϕ and $\dot{\phi}$ for which usually a unique solution does not exist.

3. For a linear time-varying inductor with inductance $L(t)$, we get

$$v(t) = i\frac{dL(t)}{dt} + L(t)\frac{di}{dt} \tag{1.4-9}$$

4. The voltage across a linear time-invariant inductor with inductance L is the familiar relation

$$v(t) = L\frac{di}{dt} \tag{1.4-10}$$

Exercise 1.4-3. Consider a nonlinear time-invariant inductor whose i-ϕ relation is

$$i = \phi^3$$

Let the voltage across this inductor be

$$\begin{aligned} v(t) &= e^{-t} \quad \text{for } t \geq 0 \\ &= 0 \quad \text{for } t < 0 \end{aligned}$$

Find the current through this inductor.
[Ans: $i(t) = e^{-3t}$ for $t \geq 0$ and $i(t) = -1$ for $t < 0$.]

Energy Stored in an Inductor. An inductor, like a capacitor, is an energy-storing element; the energy stored at time t in an inductor is

$$E(t) = \int_{t_0}^{t} v(\tau)i(\tau)\, d\tau + E(t_0) \tag{1.4-11}$$

where $E(t_0)$ represents the energy stored up to time t_0. The energy delivered to an inductor from t_0 to t is, therefore,

$$E(t, t_0) = \int_{t_0}^{t} v(\tau)i(\tau)\, d\tau \tag{1.4-12}$$

For a time-invariant flux-controlled inductor whose i-ϕ relation is $i = h(\phi)$, $E(t, t_0)$ can be written as

$$E(t, t_0) = \int_{\phi(t_0)}^{\phi(t)} h(\phi)\, d\phi \tag{1.4-13}$$

This integral is numerically equal to the shaded area in the i-ϕ plane of Fig. 1.4-3.

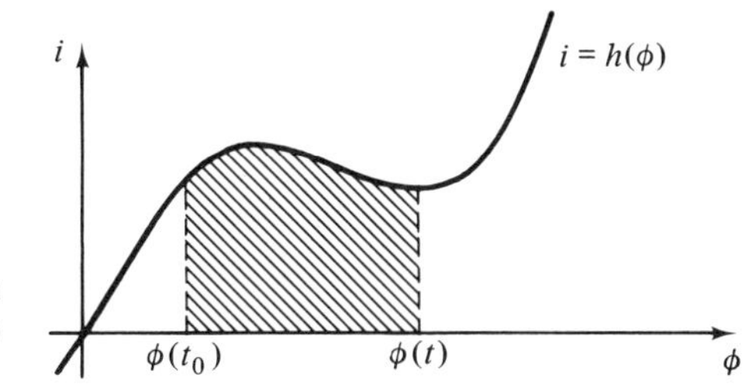

Figure 1.4-3 The energy stored in the inductor from t_0 to t is equal to the area of the shaded region.

As in the case of resistors and capacitors, inductors can be divided into two categories, those that absorb and store energy (passive) and those that deliver energy to the rest of the circuit to which they are connected (active). More precisely, a two-terminal inductor is said to be *passive* if and only if

$$E(t) = \int_{-\infty}^{t} v(\tau)i(\tau)\,d\tau \geq 0 \tag{1.4-14}$$

for *all* t and *all* admissible signal pairs (v, i). Also, an inductor is active if for *some* t or for *some* admissible signal pair $E(t)$ is negative.

Exercise 1.4-4. Consider a nonlinear time-invariant flux-controlled inductor whose i-ϕ relation is given by (1.4-4). Show that a sufficient (but not necessary) condition for this inductor to be passive is that its i-ϕ characteristic remain in the first and third quadrants. Assume that $\phi(-\infty) = 0$.

In the particular case of linear time-varying inductors, since

$$v(t) = L(t)\frac{di}{dt} + i(t)\frac{dL}{dt} \tag{1.4-15}$$

we have

$$E(t) = \int_{t_0}^{t} \left[L(\tau)i(\tau)\frac{di}{d\tau} + i^2(\tau)\frac{dL}{d\tau} \right] d\tau + E(t_0) \tag{1.4-16}$$

Then from the conclusion of Exercise 1.3-4 we state that: A linear time-varying inductor is passive if and only if

$$L(t) \geq 0 \quad \text{and} \quad \dot{L}(t) \geq 0 \quad \text{for all } t$$

If the inductor is linear time invariant with inductance L, (1.4-16) will reduce to

$$E(t) = \tfrac{1}{2}L[i^2(t) - i^2(t_0)] + \tfrac{1}{2}Li^2(t_0) = \tfrac{1}{2}Li^2(t) \tag{1.4-17}$$

Consequently, a linear time invariant inductor is passive if (and only if) its inductance L is positive. Note that the energy stored on a linear time-invariant inductor from t_0 to t_1 is

$$E(t_1, t_0) = \frac{L}{2}[i^2(t_1) - i^2(t_0)]$$

That is, $E(t_1, t_0)$ can be determined from the knowledge of the current through the inductor at t_0 and t_1 only.

Small-Signal Analysis of Nonlinear Inductors. In a similar manner as in the case of nonlinear capacitors, it can be shown that a nonlinear inductor driven by a time-varying bias will behave as a linear time-varying inductor. Consider a current-controlled inductor whose ϕ-i relation is

$$\phi = f(i) \tag{1.4-18}$$

Let the current source consist of a bias $I(t)$ and a small-signal $\delta i(t)$; that is,

$$i(t) = I(t) + \delta i(t) \tag{1.4-19}$$

Then the magnetic flux $\phi(t)$ will consist of $\Phi(t)$ [which corresponds to $I(t)$] and $\delta\phi(t)$ [which corresponds to $\delta i(t)$] so that

$$\Phi(t) = f(I(t)) \tag{1.4-20}$$

and

$$\phi(t) = \Phi(t) + \delta\phi(t) \tag{1.4-21}$$

If the function $f(\cdot)$ is sufficiently smooth, with a similar procedure as in the case of resistors we can show that the small-signal equivalent inductor has a ϕ-i relation given by

$$\delta\phi(t) \cong f'(I(t))\, \delta i(t) \tag{1.4-22}$$

where $f'(I(t))$ represents a scalar time function that is equal to the instantaneous inductance of the linear inductor at time t. Typically, $I(t)$, the bias, is a slowly varying current source, and the signal $\delta i(t)$ is a high-frequency signal.

Inductors in general have a wide variety of applications in electrical networks. Inductors with saturation-type nonlinearity are used in computers as memory-storage devices. Pulse compressors and networks with time-varying delay or bandwidth use linear time-varying inductors. Nonlinear inductors are also used for frequency conversion and the design of linear time-varying networks [1]. The following exercise suggests a method of designing frequency multipliers using nonlinear inductors.

Exercise 1.4-5. Consider a nonlinear time-invariant inductor whose ϕ-i relation is

$$\phi = f(i)$$

Let the input to this inductor be a current source

$$i(t) = \cos t$$

Find the nonlinear function $f(\cdot)$ so that the voltage across the inductor is $\sin 3t$.
[Ans: $f(i) = i - \frac{4}{3}i^3$.]

1.5 MULTITERMINAL NETWORK ELEMENTS

In the preceding sections we discussed the basic two-terminal elements. In this section we shall briefly mention some of the most important three- and four-terminal network elements and point out some of their applications. A complete description of all the multiterminal network elements and their properties is beyond the scope of this book; the interested reader can refer to existing texts [2–4].

In the discussion of lumped two-terminal elements we concerned ourselves with two variables, the voltage across the element and the current through it. In the case of three-terminal elements, however, there are six variables that we should take into account (see Fig. 1.5-1); these are i_1, i_2, i_3, v_{12}, v_{13}, and v_{32}, where v_{jk} denotes the voltage drop from node n_j to n_k.

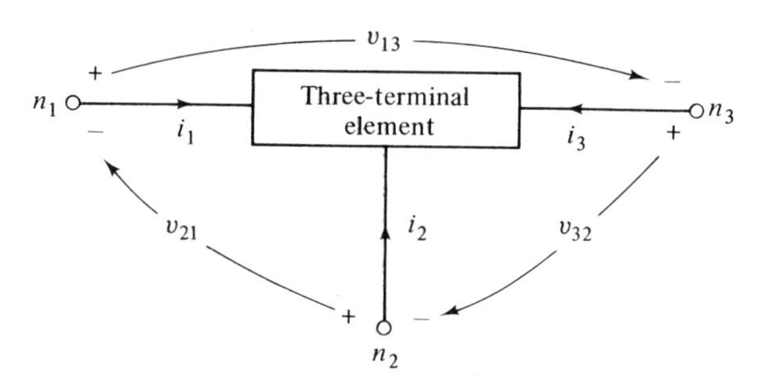

Figure 1.5-1 Symbolic representation of a three-terminal network element.

Using the generalized Kirchhoff's current and voltage laws (KCL and KVL), it can be shown that only four of the six variables given are linearly independent. More precisely, by KCL and KVL we have

$$i_1 + i_2 + i_3 = 0 \tag{1.5-1}$$

and

$$v_{21} + v_{13} + v_{32} = 0 \tag{1.5-2}$$

Hence, for a three-terminal element there are only two independent currents and two independent voltages. In general, for an n-terminal network element we can choose at most $n - 1$ independent currents and $n - 1$ independent voltages.

As in the case of two-terminal elements, multiterminal elements can be classified in three important categories: multiterminal resistors, multiterminal capacitors, and multiterminal inductors. A network element is called an *n-terminal resistor* if its terminal voltages and terminal currents are related algebraically.

Examples of multiterminal resistors are vacuum triodes and transistors. A network element is called an *n-terminal capacitor* if its terminal voltages and charges are related algebraically.

Notice that in this case we can specify $n - 1$ independent voltages and $n - 1$ independent charges; the charges and currents are, of course, related through equation (1.1-1). A network element is called an *n-terminal inductor* if the currents and the magnetic fluxes through its terminals are related by an algebraic function. In this case the magnetic fluxes and voltages are related by (1.1-2). Next we give a brief discussion of some of the most common three- and four-terminal resistors; the emphasis will be on their representation and application.

Vacuum-Tube Triode. The symbolic representation of a triode is given in Fig. 1.5-2; the four independent variables are chosen as the grid current,

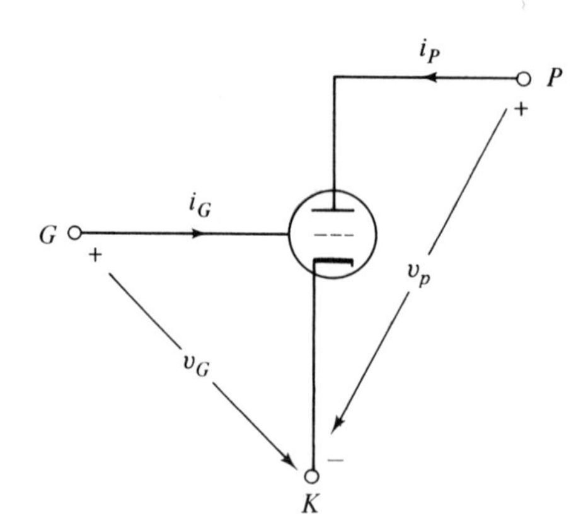

Figure 1.5-2 Symbolic representation of a triode.

i_G, the grid voltage, v_G, the plate current, i_P, and the plate voltage, v_P. It is customary to present the *v-i* characteristics of a triode in two groups. First, i_G versus v_G with v_P kept constant, and second, i_P versus v_P with v_G kept constant. The first group of curves is called *input characteristics* and the second group of curves is called *output characteristics* of the triode. These sets of curves are shown in Fig. 1.5-3.

Vacuum-tube triodes have been used in countless varieties of electronic

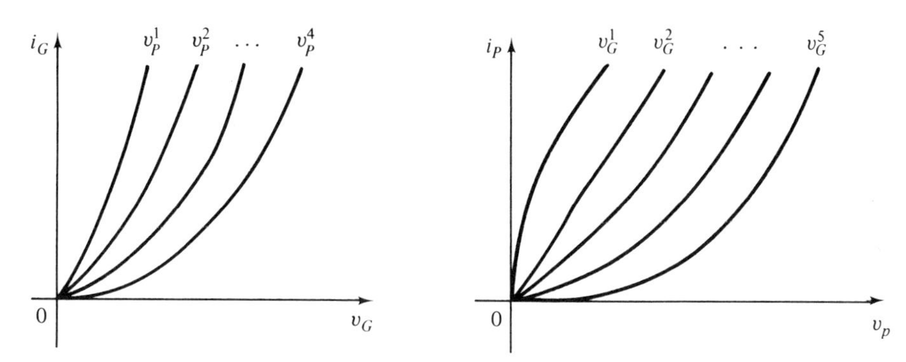

Figure 1.5-3 Characteristics of a vacuum-tube triode.

components for many decades. In recent years, however, they have been replaced by transistors.

Transistor. Another common three-terminal resistor is a transistor. As in the case of triodes, the characteristics of a transistor are usually given in two groups, input and output characteristics. A wide variety of transistors are commercially available. Here, as an illustration, we consider a field-effect transistor (FET). The symbolic representation of an FET is shown in Fig. 1.5-4.

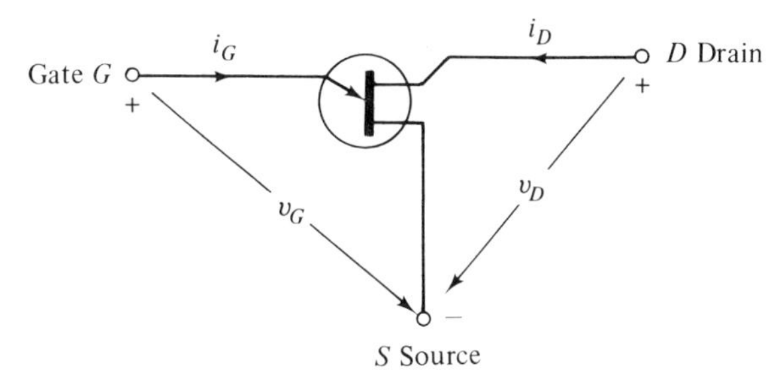

Figure 1.5-4 Symbolic representation of an FET.

Among six possible variables, the four independent variables are chosen as gate current i_G, gate voltage v_G, drain current i_D, and the drain voltage v_D. Input and output characteristics of a typical field-effect transistor are shown in Fig. 1.5-5.

Next we consider an important four-terminal resistor, the gyrator, and briefly mention some of its properties.

Gyrator. A four-terminal resistor that is very useful in both practical circuit design and theoretical network synthesis is the gyrator, which was first intro-

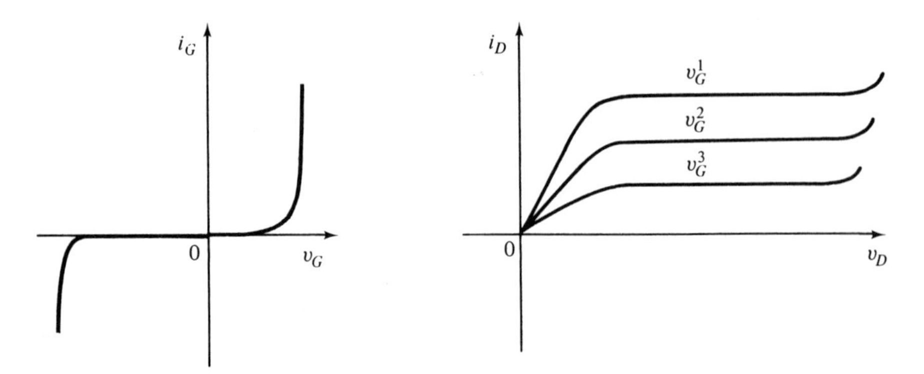

Figure 1.5-5 Characteristics of an FET.

duced by Tellegen in 1948. The symbolic representation of an ideal gyrator is shown in Fig. 1.5-6. The ideal gyrator has the property that it "gyrates"

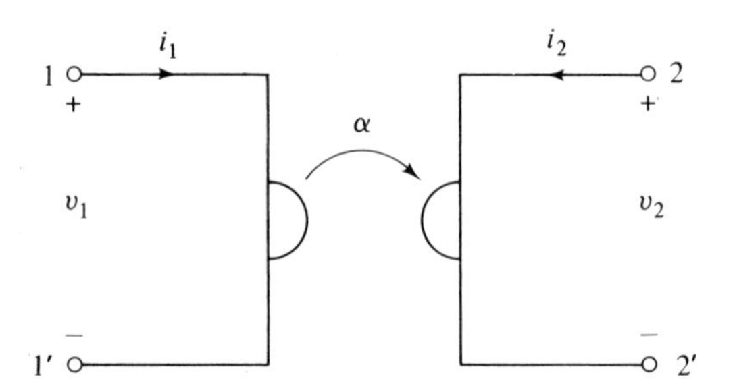

Figure 1.5-6 Symbolic representation of a gyrator.

the *current* of one port into the *voltage* of the other port and vice versa; that is,

$$v_1 = -\alpha i_2 \tag{1.5-3}$$

and

$$v_2 = \alpha i_1 \tag{1.5-4}$$

The coefficient α, which has the dimension of a resistor, is called the gyration constant. Gyrators can be constructed using vacuum tubes, transistors, or operational amplifieıs. A simplified transistor gyrator is shown in Fig. 1.5-7. More precise schematics of gyrators are given in other texts [2, 4, and 5].

Example 1.5-1 (Simulation of Inductors). An important application of a gyrator is in simulation of inductors. To clarify this point consider the gyrator shown in Fig. 1.5-8, which is put in tandem with a capacitor with capacitance C. Note that, according to our polarity convention, v_2 and i_1 are related by

$$i_2 = -C\frac{dv_2}{dt}$$

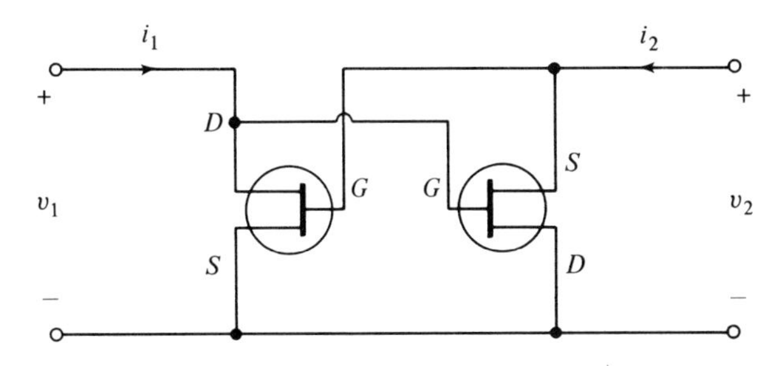

Figure 1.5-7 Transistor realization of a gyrator.

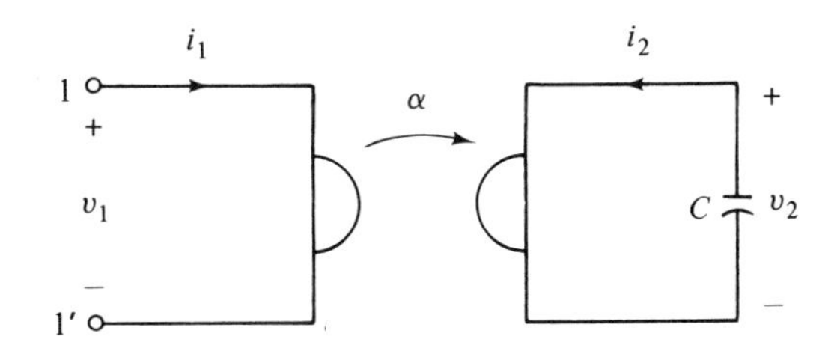

Figure 1.5-8 Gyrator circuit for simulating an inductor.

Using this equation together with (1.5-3), we get

$$v_1 = \alpha C \frac{dv_2}{dt}$$

But from (1.5-4), $v_2 = \alpha i_1$; thus,

$$v_1 = \alpha^2 C \frac{di_1}{dt} \tag{1.5-5}$$

Consequently, as far as the first port is concerned, the circuit shown behaves as a linear time-invariant inductor with inductance $L = \alpha^2 C$.

Since inductors are inherently bulky and difficult to adjust, the above property of gyrators can be used to simulate inductors required in transistor and integrated circuits.

Exercise 1.5-1. Use an inductor and a gyrator to simulate a capacitor.

Some other properties of ideal gyrators will be discussed in later chapters. Let us now discuss another important four-terminal network element.

Coupled Inductor. The symbolic representation of a coupled inductor is shown in Fig. 1.5-9. Such inductors are typically made of two coils, called *primary* and *secondary coils*. For current-controlled coupled inductors, the flux ϕ in the first coil is a function of the currents in the first *and* second coils, i_1 and i_2. Similarly, the flux in the second coil is a function of the currents in

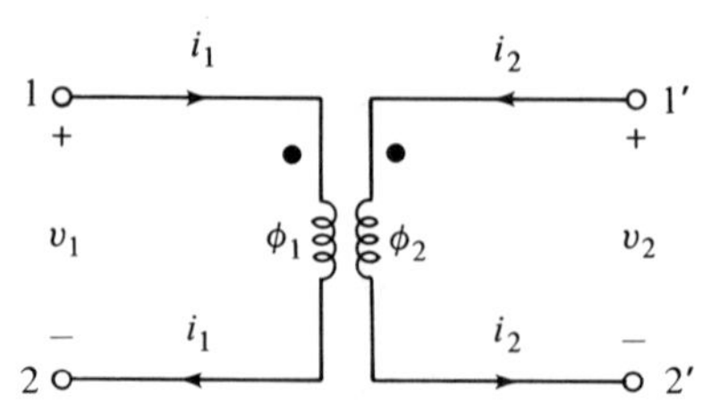

Figure 1.5-9 Symbolic representation of a coupled inductor.

the first and second coils. Most practical inductors, under large input currents, show some nonlinear behavior. Then, if the coupled inductor under consideration is nonlinear, its ϕ-i relations are

$$\phi_1 = f_1(i_1, i_2) \tag{1.5-6}$$

$$\phi_2 = f_2(i_1, i_2) \tag{1.5-7}$$

The voltages v_1 and v_2 shown in Fig. 1.5-9 can then be obtained in terms of i_1 and i_2 using (1.1-2):

$$v_1 = \frac{d\phi_1}{dt} = \frac{\partial f_1}{\partial i_1}\cdot\frac{di_1}{dt} + \frac{\partial f_1}{\partial i_2}\cdot\frac{di_2}{dt} \tag{1.5-8}$$

$$v_2 = \frac{d\phi_2}{dt} = \frac{\partial f_2}{\partial i_1}\cdot\frac{di_1}{dt} + \frac{\partial f_2}{\partial i_2}\cdot\frac{di_2}{dt} \tag{1.5-9}$$

Hence, given the characteristics of a current-controlled coupled inductor, the corresponding input and output voltages can be computed using the previous equations. If the coupled inductor under study is linear, the fluxes ϕ_1 and ϕ_2 can be written as

$$\phi_1 = L_{11}i_1 + M_{12}i_2 \tag{1.5-10}$$

$$\phi_2 = L_{22}i_2 + M_{21}i_1 \tag{1.5-11}$$

where L_{11} and L_{22} are called the *self-inductances* of the first and second inductors respectively, and M_{12} and M_{21} are called the *mutual inductance* between the first and second coil. If $M_{12} = M_{21} \triangleq \pm M$, the coupled inductor is called *reciprocal.* The mutual inductance M is assumed to be positive. If the windings of the coils are in the same direction, M appears with positive sign; otherwise, M appears with negative sign. For simplicity, a pair of dots is used to indicate this convention. If both i_1 and i_2 enter *or* leave the dots, M appears with a plus sign. If one current leaves a dot and the other enters a dot, M appears with negative sign. An important special case of linear coupled inductors is the ideal transformer, which we discuss next.

Ideal Transformer. The schematic representation of an ideal transformer is shown in Fig. 1.5-10. In this figure, i_1 and v_1 are the current through and voltage across the first coil, respectively. Similarly, i_2 and v_2 are the current and voltage of the second coil. Let the ratio of the number of the windings in

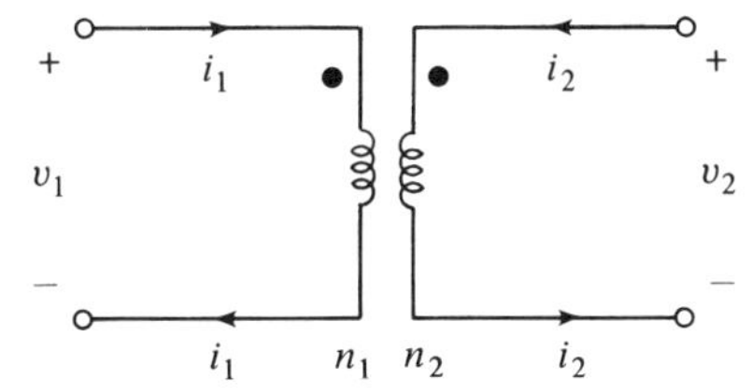

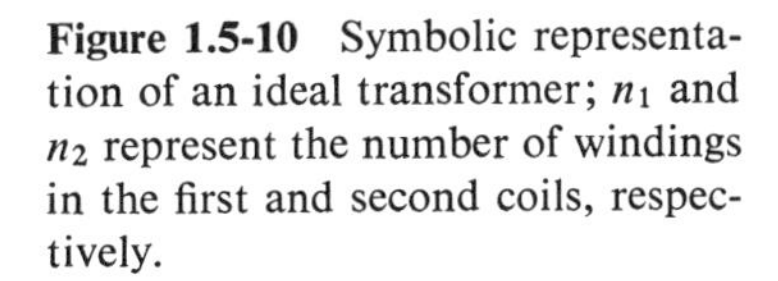
Figure 1.5-10 Symbolic representation of an ideal transformer; n_1 and n_2 represent the number of windings in the first and second coils, respectively.

the first coil to that of the second coil be n_1/n_2; then the *v-i* relation of the ideal transformer is given as follows:

1. If both i_1 and i_2 enter the dots,

$$\frac{v_1}{v_2} = \frac{n_1}{n_2} \tag{1.5-12}$$

$$\frac{i_1}{i_2} = -\frac{n_2}{n_1} \tag{1.5-13}$$

2. If one of the currents leaves a dot and the other enters,

$$\frac{v_1}{v_2} = -\frac{n_1}{n_2} \tag{1.5-14}$$

$$\frac{i_1}{i_2} = \frac{n_2}{n_1} \tag{1.5-15}$$

Note that in ideal transformers self- and mutual inductances will not enter the *v-i* relations. An important application of ideal transformers is in *impedance scaling*; if the second terminal of the transformer shown in Fig. 1.5-10 is connected to a resistor R, the input resistance of the first terminal is $(n_1/n_2)^2R$. (Show this.)

Dependent Sources. In modeling transistor circuits it is sometimes necessary to introduce a voltage or current source whose waveform depends on some other voltage or current in the circuit. For example, the field-effect transistor shown in Fig. 1.5-4 can be modeled as the network shown in Fig. 1.5-11.

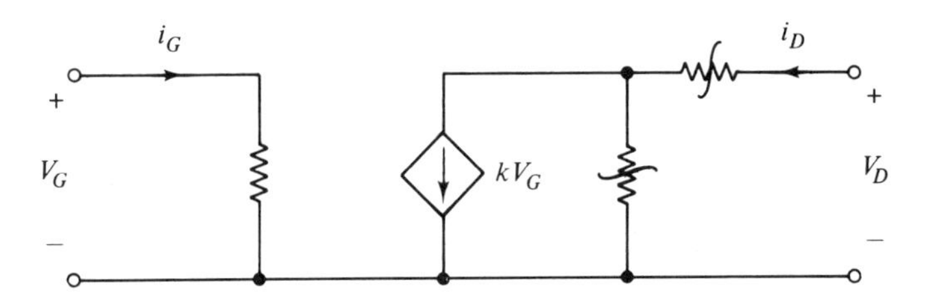

Figure 1.5-11 Model of an FET.

Notice that the current through the current source depends on the gate voltage V_G.

Dependent sources can be categorized in four groups:

1. *Voltage-controlled voltage source* is a voltage source whose waveform and amplitude depend on the voltage across some other terminals in the network. The symbolic representation of this four-terminal element is shown in Fig. 1.5-12(a).

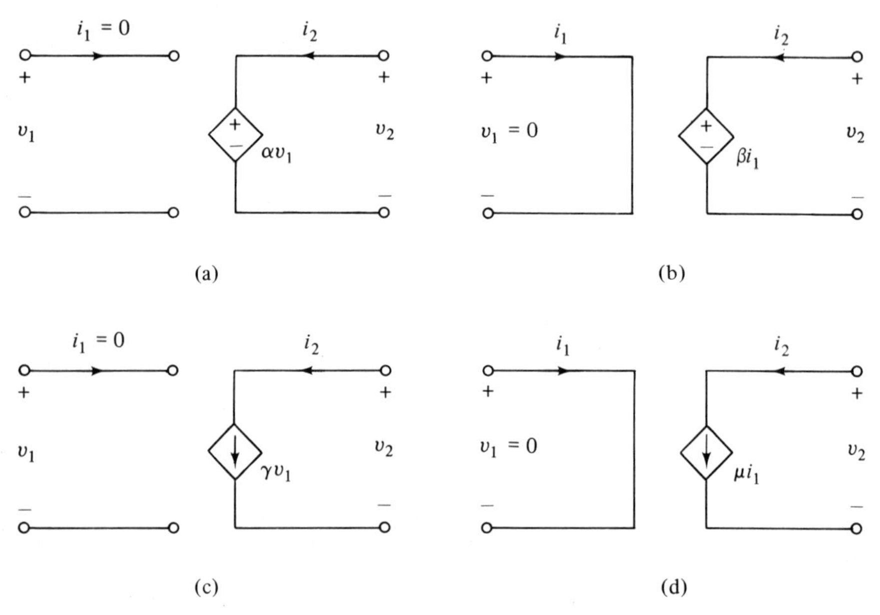

Figure 1.5-12 (a) Voltage-controlled voltage source; (b) current-controlled voltage source; (c) voltage-controlled current source; (d) current-controlled current source.

2. *Current-controlled voltage source.* The voltage across this voltage source depends on the current through some other terminal; see Fig. 1.5-12(b).
3. *Voltage-controlled current source.* The current through this current source depends on the voltage across some other terminals of the network; see Fig. 1.5-12(c).
4. *Current-controlled current source.* This is a current source whose current depends on some other current in the network; see Fig. 1.5-12(d).

Note that α, β, γ, and δ given in Fig. 1.5-12 are scalar constants.

Operational Amplifiers. One of the most commonly used four-terminal elements is the *operational amplifier*. This element has applications in analog computers, logic circuits, feedback control systems, and many other electronic devices. In what follows, the basic operational amplifier is defined and some of its applications are discussed.

The symbolic representation of an operational amplifier is shown in Fig. 1.5-13(a). Its corresponding idealized v_2-v_1 characteristic is shown in

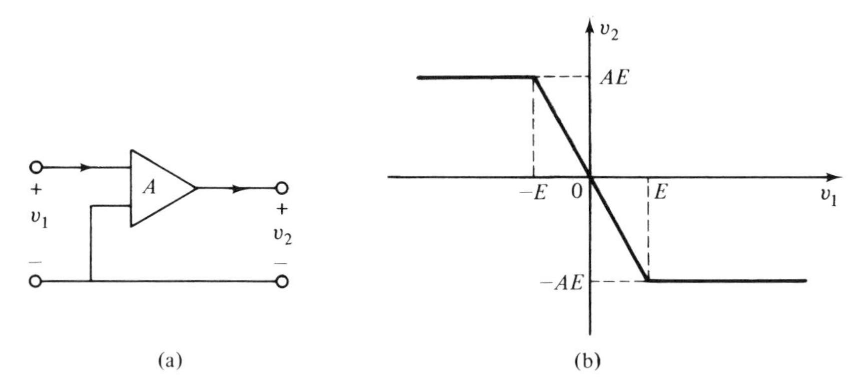

Figure 1.5-13 (a) Symbolic representation of an operational amplifier; (b) its idealized v_2-v_1 characteristic.

Fig. 1.5-13(b). Note that for $|v_1| \leq E$, the output voltage v_2 is a linear function of v_1:

$$v_2 = -Av_1 \tag{1.5-16}$$

where A is called the open-loop gain of the operational amplifier. For $|v_1| \geq E$, saturation occurs and the output remains constant. For actual operational amplifiers, the gain A ranges from 10,000 to 20,000. Thus, small perturbations on the input voltage cause large changes in the output. For this reason, the use of operational amplifiers in open-loop form is quite limited. The circuit realization of an open-loop operational amplifier is shown in Fig. 1.5-14(a). Ideally, $R_1 \longrightarrow \infty$ and $R_2 \longrightarrow 0$; however, in practical operational amplifiers, typical values are $R_1 = 50\ \text{k}\Omega$ and $R_2 = 100\ \Omega$.

To remedy the shortcomings of the open-loop model, the output voltage is usually fed back through a resistor R_f [see Fig. 1.5-14(b)]. To find the rela-

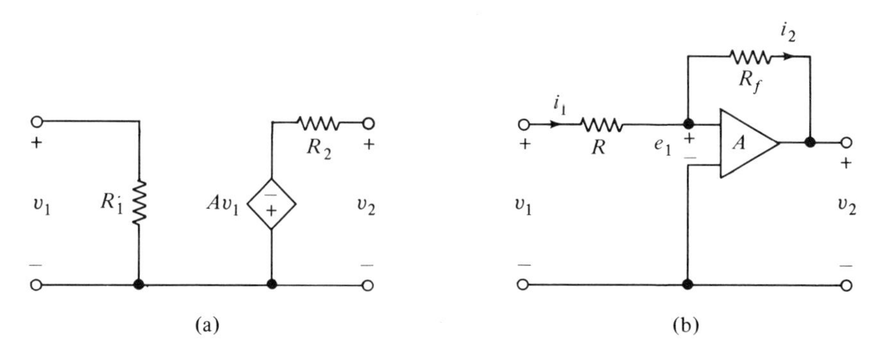

Figure 1.5-14 (a) Circuit realization of an open-loop operational amplifier; (b) typical feedback operational amplifier.

tionship between v_1 and v_2 for the feedback model, note that since ideally the input resistance of the open-loop operational amplifier is ∞, from Fig. 1.5-14(b) we can write

$$i_1 \simeq i_2 \tag{1.5-17}$$

Hence, from the figure it is clear that

$$\frac{v_1 - e_1}{R} = \frac{e_1 - v_2}{R_f} \tag{1.5-18}$$

But $v_2 = -Ae_1$ and, consequently,

$$e_1 = -\frac{v_2}{A}$$

Typical values for v_2 and A are 10 V and 20,000 respectively; that is, e_1 is much smaller than v_1 and v_2 and hence can be neglected from (1.5-18) without introducing any significant error. Thus,

$$v_2 = -\frac{R_f}{R} v_1 \tag{1.5-19}$$

Exercise 1.5-2. (a) The actual model of a feedback operational amplifier is shown in Fig. 1.5-15. Use the Kirchhoff current law at nodes n_1 and n_2 to

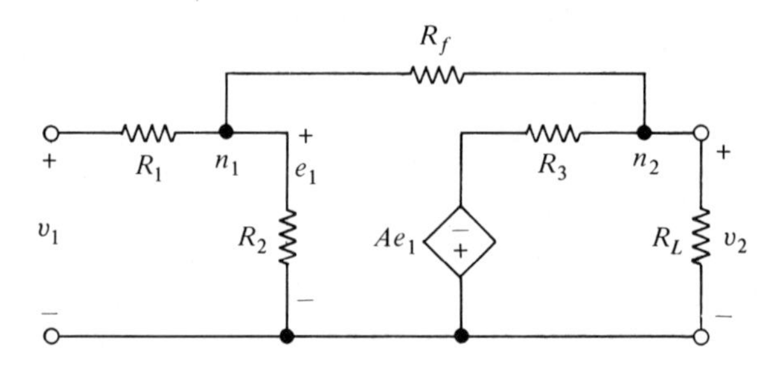

Figure 1.5-15 Model of a feedback operational amplifier.

show that

$$\frac{v_2}{v_1} = -\frac{R_f}{R_1} \frac{1}{1 + \epsilon} \tag{1.5-20}$$

where

$$\epsilon = \frac{\left(1 + \frac{R_3}{R_L} + \frac{R_3}{R_f}\right)\left(1 + \frac{R_f}{R_1} + \frac{R_f}{R_2}\right)}{A - \frac{R_3}{R_f}} \tag{1.5-21}$$

(b) For a typical operational amplifier $A = 20{,}000$, $R_1 = 10\ \text{k}\Omega$, $R_2 = 50\ \text{k}\Omega$, $R_3 = 100\ \Omega$, $R_f = 100\ \text{k}\Omega$, and $R_L = 10\ \text{k}\Omega$. For these values show that $\epsilon = 0.00065$.

This is a rather interesting conclusion; the gain of the feedback operational amplifier does not depend on the open loop gain A. It only depends on the ratio R_f/R. This implies that the feedback gain is "insensitive" to the fluctuation of the open-loop gain. Some of the important applications of the feedback operational amplifier are given in the following examples.

Example 1.5-2 (Integrator). Consider the feedback operational amplifier shown in Fig. 1.5-16(a). The feedback consists of a capacitor with capacitance

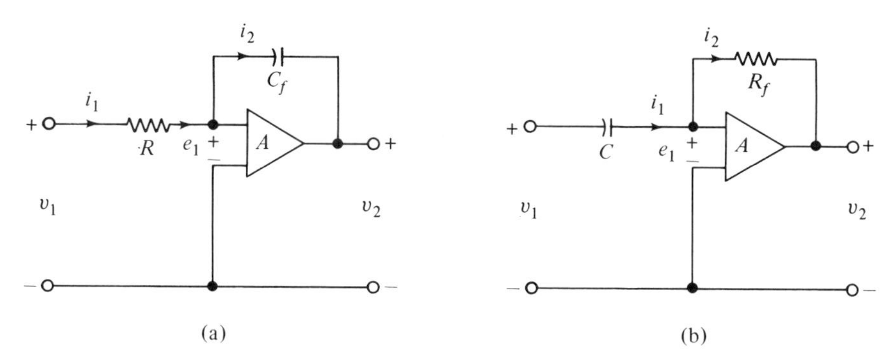

Figure 1.5-16 (a) Operational amplifier used as an integrator; (b) operational amplifier used as a differentiator.

C_f. Since $i_1 \simeq i_2$, we can write

$$\frac{v_1 - e_1}{R} = C_f \frac{d}{dt}(e_1 - v_2)$$

Since e_1 is negligible in comparison to v_1 and v_2, this equation yields

$$v_2(t) = -\frac{1}{C_f R}\int_0^t v_1(\tau)\,d\tau + v_2(0)$$

Assuming that $v_2(0) = 0$, this equation implies that the output voltage is proportional to the integral of the input voltage. Such integrators are extensively used in analog computers.

Example 1.5-3 (Differentiator). Figure 1.5-16(b) shows an operational amplifier used as a differentiator. As in the previous example, since $i_1 \simeq i_2$ and e_1 is negligible in comparison to v_1 and v_2, we have

$$v_2(t) = -CR_f \frac{d}{dt} v_1(t)$$

That is, the output voltage is proportional to the derivative of the input voltage. Such differentiators, however, are not commonly used in practice due to their susceptibility to random noise.

Example 1.5-4 (Convertors). Feedback operational amplifiers can be used to convert a voltage source to a current source, and vice versa. The schematic representations of voltage-to-current and current-to-voltage convertors are shown in Figs. 1.5-17(a) and (b), respectively.

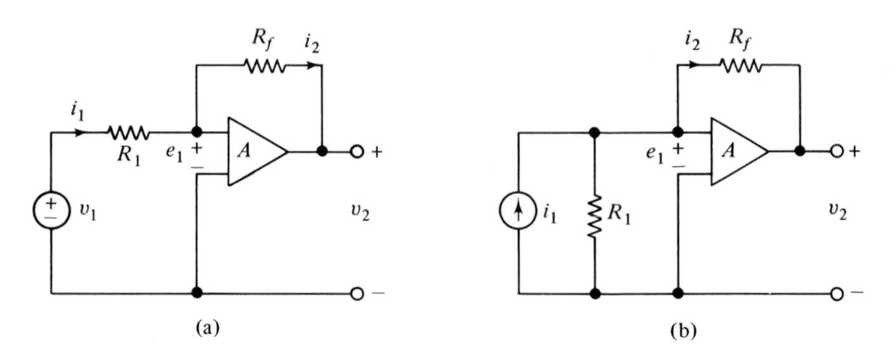

Figure 1.5-17 Operational amplifier used as (a) voltage-to-current convertor, and (b) as current-to-voltage convertor.

In Fig. 1.5-17(a), since $i_1 \simeq i_2$ and e_1 is much smaller than v_1, the current i_2 through the feedback can be written as

$$i_2 = \frac{1}{R_1} v_1$$

That is, i_2 is independent of v_2, and therefore the feedback path can be considered as an ideal current source whose current can be adjusted by the voltage source v_1 and resistor R_1. In Fig. 1.5-17(b) the resistance of R_1 is chosen to be much larger than R_f, so that $i_2 \simeq i_1$. Thus,

$$i_1 \simeq i_2 = \frac{e_1 - v_2}{R_f} \simeq -\frac{1}{R_f} v_2$$

That is,

$$v_2 = -R_f i_1$$

Hence, the output of the amplifier can be considered as an ideal voltage source whose voltage can be adjusted by R_f and the current source i_1. Of course, these sources are ideal only on a limited range.

Example 1.5-5 (Adders). Operational amplifiers can be used for adding several voltages $v_1, v_2, \ldots, v_n$ together. This application of the operational amplifier is shown in Fig. 1.5-18(a). Note that $i \cong i_1 + i_2 + \cdots + i_n$ or, equivalently,

$$\frac{e_1 - v_0}{R_f} \simeq \frac{v_1 - e_1}{R_1} + \frac{v_2 - e_1}{R_2} + \cdots + \frac{v_n - e_1}{R_n}$$

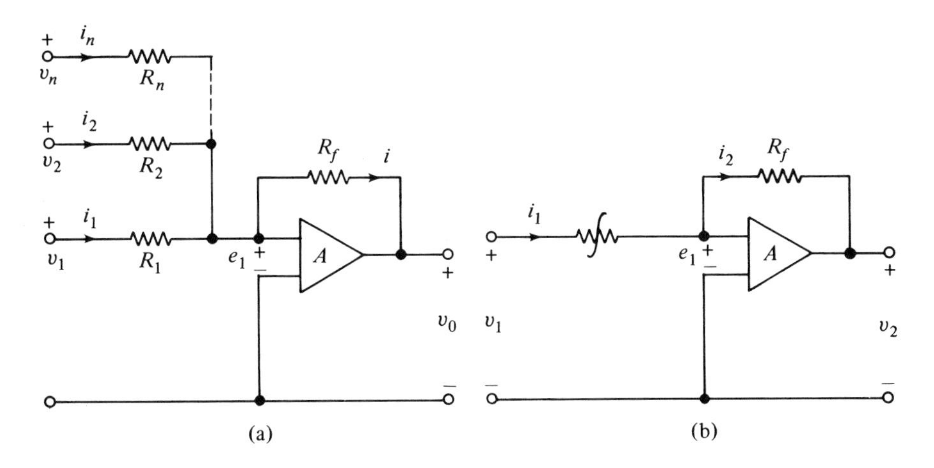

Figure 1.5-18 Operational amplifier used as (a) voltage adder, and (b) nonlinear amplifier.

Since e_1 is negligible, we get

$$v_0 = -R_f\left(\frac{v_1}{R_1} + \frac{v_2}{R_2} + \cdots + \frac{v_n}{R_n}\right)$$

In particular, if $R_1 = R_2 = \cdots = R_n = R_f$, then

$$v_0 = -(v_1 + v_2 + \cdots + v_n)$$

Example 1.5-6 (Nonlinear Amplifiers). If a nonlinear voltage-controlled resistor is placed in the forward path of an operational amplifier, the output voltage will be a nonlinear function of the input voltage. Consider the amplifier shown in Fig. 1.5-18(b). Since $i_1 \simeq i_2$, we can write

$$i_1 = \frac{e_1 - v_2}{R_f} = f(v_1 - e_1)$$

Neglecting e_1 and letting $R_f = 1$ we get

$$v_2 = -f(v_1)$$

That is, the output is a nonlinear function of the input. As a numerical example, let $i_1 = -k(v_1 - e_1)^3$; then the $v_2 - v_1$ characteristic of the resulting operational amplifier becomes

$$v_2 = kv_1^3$$

Nonlinear amplifiers are usually used to compensate for the nonlinearities inherent in some existing elements. For instance, in a thermocouple sensing device the voltage across the device may be given in terms of temperature as

$$v_1 = T^{1/3}$$

Now if the output of this device is connected in tandem to the nonlinear operational amplifier discussed, the output voltage v_2 will be a *linear* function of the temperature T; that is,

$$v_2 = kv_1^3 = kT$$

Exercise 1.5-3. Insert a current-controlled nonlinear resistor in the feedback of an operational amplifier and obtain the v_2-v_1 relations.
[Ans: $v_2 = f(v_1)$.]

Although in our discussions we used idealized operational amplifiers, it should be mentioned that due to recent advances in the construction of operational amplifiers, the actual operational amplifier results are good approximations of the ideal ones. An important application of operational amplifiers is in the design of *active filters*. This topic will be briefly mentioned in later chapters. Other applications of the operational amplifiers can be found in [4].

Convertors. Voltage-inversion negative-impedance convertors (VNIC) and current-inversion negative-impedance convertors (INIC) are among a class of useful four-terminal network elements classified as rotators [2].

The v-i relationship of a VNIC is given by

$$\begin{aligned} i_1 &= i_2 \\ v_1 &= -v_2 \end{aligned} \tag{1.5-22}$$

The symbolic representation of a VNIC is given in Fig. 1.5-19(a). While the

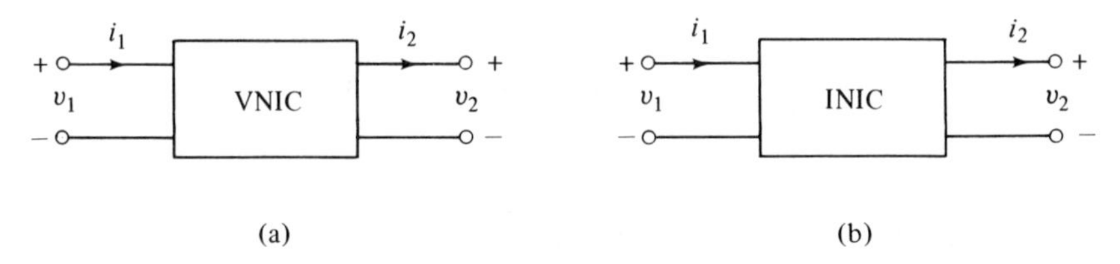

Figure 1.5-19 (a) Voltage-inversion negative-impedance convertor; (b) current-inversion negative-impedance convertor.

current entering and leaving the VNIC have the same direction, the input and output voltages will have opposite polarities. That is, the basic property of a VNIC is to invert the polarity of the input voltage.

The v-i relationship of an INIC, on the other hand, is given by

$$\begin{aligned} v_1 &= v_2 \\ i_1 &= -i_2 \end{aligned} \tag{1.5-23}$$

and its symbolic representation is given in Fig. 1.5-19(b). The basic property of the INIC is, therefore, to invert the direction of the current and keep the polarity of the voltage unchanged.

A large class of useful multiterminal network elements does not fall in the previously mentioned categories. Typical examples of these elements are logic circuits such as AND, OR, and the like. A comprehensive treatment of such elements can be found in [2] and [4].

To conclude this section, it should be mentioned that the passivity or activity of multiterminal elements can be discussed in a manner similar to those of two-terminal network elements. A general definition of passivity will be given in the next chapter that can be used to determine the passivity or activity of multiterminal elements when they are considered as a "black box."

REFERENCES

[1] PEIKARI, B., "On the Synthesis of Nonlinear Networks," *IEEE Trans. Circuit Theory*, **CT-17**, No. 4 (1970), 657–659.

[2] CHUA, L., *Introduction to Nonlinear Network Theory*, McGraw-Hill Book Company, New York, 1969.

[3] PENFIELD, P., and R. RAFUSE, *Varactor Applications*, The MIT Press, Cambridge, Mass., 1962.

[4] GRAME, J., G. TOBEY, and L. HUELSMAN, *Operational Amplifiers, Design and Application*, McGraw-Hill Book Company, New York, 1971.

[5] NEWCOMB, R., *Active Integrated Circuit Synthesis*, Prentice-Hall, Inc., Englewood Cliffs, N.J., 1968.

PROBLEMS

1.1 The current $i(t)$ through a two-terminal network element is

$$i(t) = 2e^{-|t|} \quad \text{for all } t$$

Find the charge accumulated on this element from $t = -\infty$ to $t = 1$ second.

1.2 The magnetic flux $\phi(t)$ through a nonlinear network element is

$$\phi(t) = te^{-t} \sin t \quad \text{for all } t$$

Determine the voltage across this element.

1.3 Figure P1.3 shows a crude model of a section of length dz of a transmission line. Let the flux ϕ and the charge q be

$$\phi = Li \quad \text{and} \quad q = Cv$$

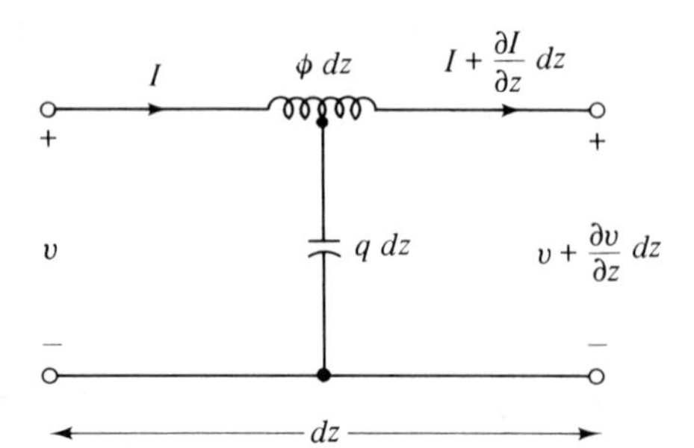

Figure P1.3

Show that the v-i relation of this distributed parameter element can be obtained by solving the following partial differential equations:

$$\frac{\partial v}{\partial z} = -L\frac{\partial I}{\partial t}$$

$$\frac{\partial I}{\partial z} = -C\frac{\partial v}{\partial t}$$

1.4 Assume that the element under consideration in Problem 1.1 is a linear capacitor with capacitance $C = 2\ F$. Find the energy stored on this capacitor from 0 to t.

1.5 Applying the voltage $v(t) = \sin \omega t$ across a two-terminal element, the current was found to be $i(t) = 1 + \sin \omega t$. Based on this information alone, can you determine whether this element is passive or not? If not, why?

1.6 The v-i relation of a current-controlled resistor is

$$v = 2i + \sin i$$

Determine whether this resistor is voltage controlled or not.

1.7 Show that a resistor whose v-i relation is

$$v = i + 1$$

is *not* linear.

1.8 Show that the resistor whose v-i relation is as given in Problem 1.6 is passive.

1.9 Prove that the nonlinear time-invariant resistor whose v-i relation is given in Fig. P1.9 is not passive.

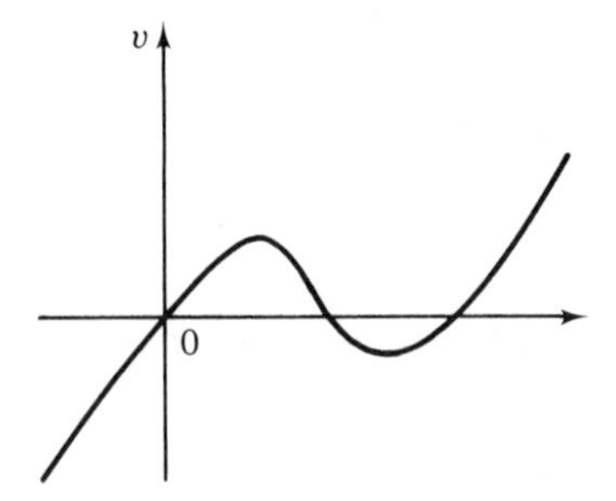

Figure P1.9

1.10 Consider the nonlinear time-invariant network shown in Fig. P1.10. Obtain the small-signal equivalent of this resistor.

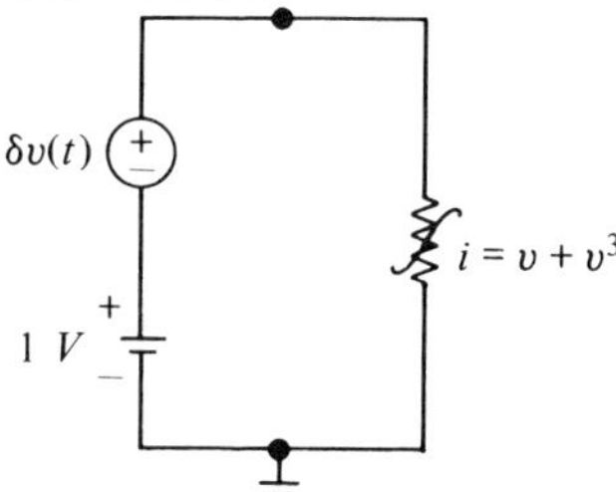

Figure P1.10

1.11 The small-signal equivalent conductance of a voltage-controlled resistor is

$$G(t) = 1 + \sin^3 t$$

Determine the i-v relation of this resistor if the bias voltage is $V(t) = \sin t$.

1.12 Obtain the v-i characteristics of a nonlinear resistor such that, for $v = \sin \omega t$, the current through the resistor is $i = |\sin \omega t|$.

1.13 Find the v-i relation of a nonlinear resistor so that the voltage across it has a frequency three times greater than the frequency of the current through it. Assume that the current through the resistor is $\sin \omega t$.

1.14 Use a graphical technique to obtain the v-i characteristics of the nonlinear resistor whose voltage is shown in Fig. P1.14(a) and whose corresponding current is shown in Fig. P1.14(b).

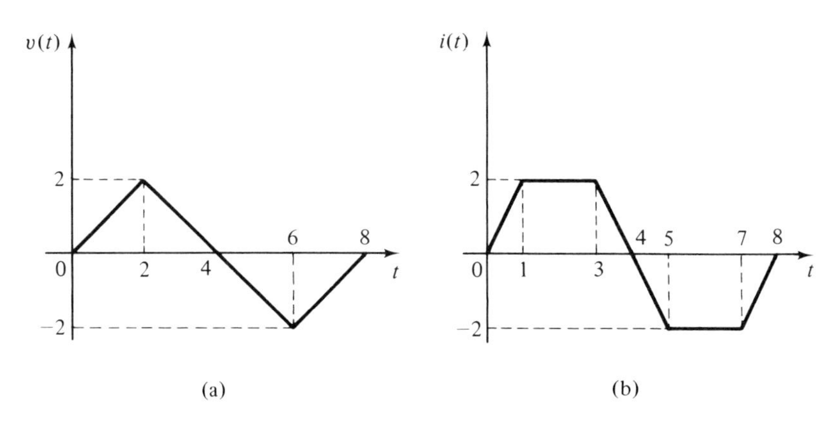

Figure P1.14

1.15 The q-v relation of a voltage-controlled capacitor is

$$q = \log_e v$$

Show that this capacitor is charge controlled as well.

1.16 The v-q relation of a capacitor is

$$v = 1 + q + q^2$$

Find the amount of energy required to charge this capacitor from $q(t_0) = 0$ to $q(t) = 1$ C.
[Ans: 11/6 joules]

1.17 The v-q relation of a capacitor is

$$v = q - q^3$$

Show that this capacitor is not passive.

1.18 The capacitance of a linear time-varying capacitor is

$$C(t) = t + 2\cos t$$

Determine whether this capacitor is passive or active.

1.19 The q-v characteristic of a nonlinear capacitor is

$$q = \frac{1}{1 - |v|}$$

Let the bias V be given by

$$V(t) = 0.5 \sin \omega t$$

Find the small-signal equivalent of this capacitor.

1.20 The ϕ-i relation of a nonlinear time-varying inductor is

$$i = \frac{1}{t} \tanh \phi$$

Let the current through this inductor be

$$i = \sin t$$

Find the voltage across this inductor.
[Ans: $v(t) = (t \cos t + \sin t)/(1 - t^2 \sin t)$]

1.21 Prove that a linear time-varying inductor with inductance $L(t)$ is passive if and only if

$$L(t) \geq 0 \quad \text{and} \quad \dot{L}(t) \geq 0 \quad \text{for all } t$$

1.22 Consider a current-controlled inductor with ϕ-i relation

$$\phi = f(i)$$

Let

$$i(t) = I(t) + \delta i(t)$$

where $I(t)$ is the bias source and $\delta i(t)$ is a small input signal. Denote the flux corresponding to $I(t)$ by $\Phi(t)$ and the flux corresponding to $\delta i(t)$ by $\delta\phi$. Use a Taylor expansion to show that the small-signal equivalent of this inductor has a ϕ-i relation as

$$\delta\phi = f'(I(t))\,\delta i$$

1.23 Use the v-i relation of an ideal transformer and that of two coupled inductors to show that the self- and mutual inductances of an ideal transformer are infinite.

1.24 Show that the circuits in Figs. P1.24(a) and (b) behave as gyrators. Find the gyration constant in each case.

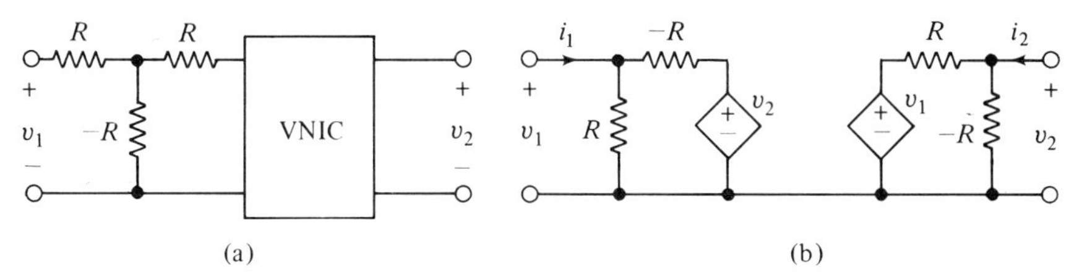

Figure P1.24

1.25 Show that the circuit in Fig. P1.25 is equivalent to a single capacitor. Find its capacitance in terms of L and α.

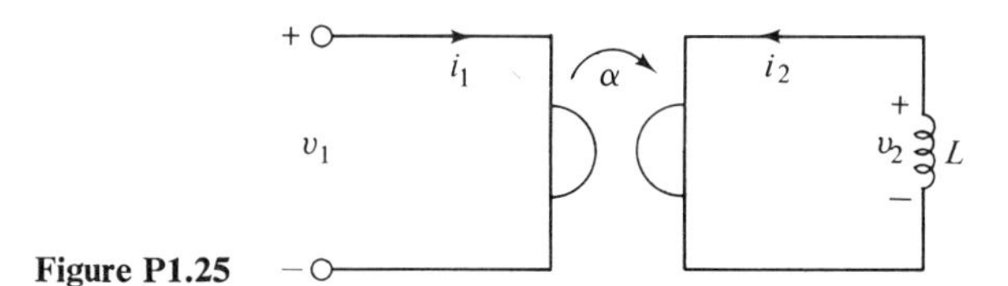

Figure P1.25

1.26 Consider the limiter circuit in Fig. P1.26 in which the diode is assumed to be an ideal diode. Plot the input–output characteristic (v_{out} versus v_{in}) of this circuit.

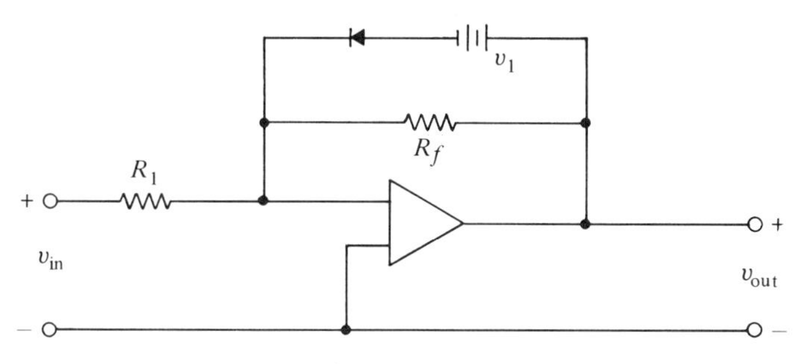

Figure P1.26

1.27 Consider the operational amplifier shown in Fig. 1.5-15. Use the typical values given in Exercise 1.5-2(b) to find the voltage e_1 if $v_2 = 2.1$ V. Is it reasonable to assume e_1 is negligible in comparison to v_1 and v_2?

1.28 The voltage-inversion negative-impedance convertor (VNIC) shown in Fig. P1.28 is terminated by an R-Ω resistor. Find the relation between v_1 and i_1.

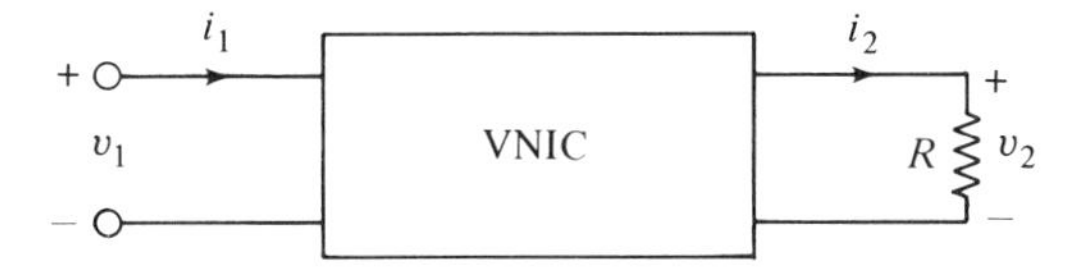

Figure P1.28

Network Characterization 2

2.1 INTRODUCTION

In Chapter 1 we introduced various network elements and discussed their basic properties. A *network* is an interconnection of such elements, which are put together to generate a desired output (response) from a given input. In general, a network may have several inputs and outputs; each input is connected to the network through a pair of terminals, and each output may be obtained through the same or a different pair of terminals. In this chapter we assume that the network under consideration is placed in a *black box* in the sense that our only access to it is through its terminals. To each terminal of this network we assign a voltage v and a current i; then from the constraints imposed upon the v-i relation we can characterize the network as linear or nonlinear, time varying or time invariant, passive or active, causal or noncausal, stable or unstable. In what follows we assume that the network under consideration is made up of lumped elements only, and the port voltage $v_k(t)$ and current $i_k(t)$ are real functions of time.

Consider the $2n$-terminal network shown in Fig. 2.1-1; denote the terminals by 1-1′, 2-2′, . . . , n-n'. A precise definition of a port is then given by

Definition 2.1-1 (Port). A pair of network terminals n_k and n'_k is said to form a *port* if the current through n_k is equal and in the opposite direction to the current through n'_k.

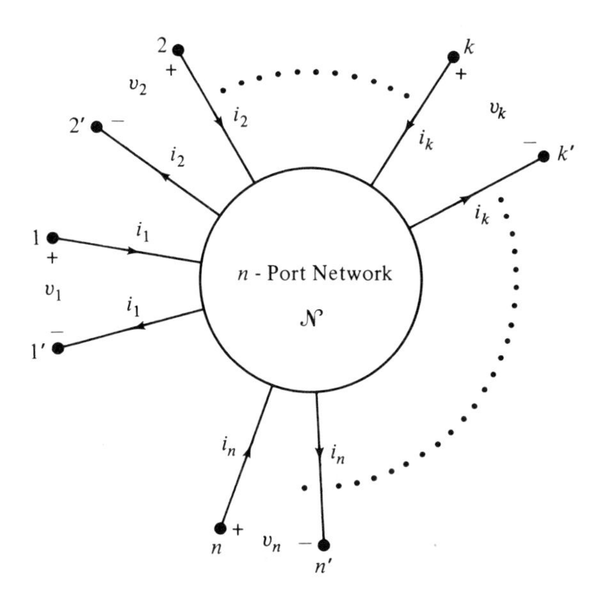

Figure 2.1-1 Symbolic representation of an n-port network.

Hence, a $2n$-terminal network is an n-port network if we can pair all its terminals to form n ports. Denote the current and the voltage corresponding to the port k by i_k and v_k respectively; then we say that the pair $[v_k, i_k]$ is an *admissible signal pair* if v_k and i_k are related through the constraints imposed by port k [1]. For example, consider the two-port network shown in Fig. 2.1-2; assume that the resistor R and the inductors are linear and time invariant. Then the pair $[v_1, i_1]$ is admissible to port 1 if

$$v_1 = Ri_1 + L_1\frac{di_1}{dt} + M\frac{di_2}{dt}$$

That is, the constraint of port 1 is that the voltage v_1 and the current i_1 must

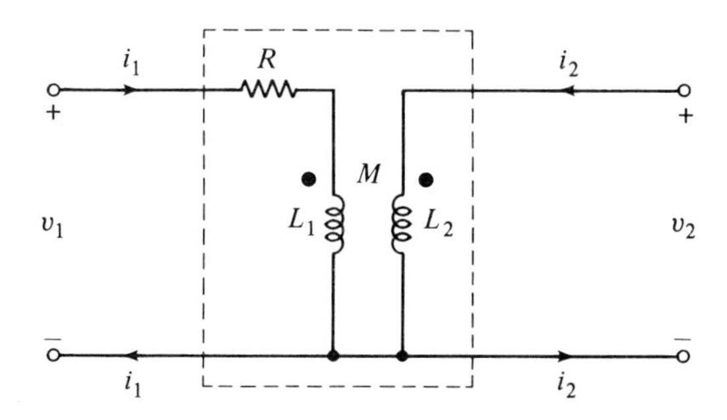

Figure 2.1-2 Simple two-port network.

satisfy this equation. Similarly, the pair $[v_2, i_2]$ is an admissible signal pair to port 2 if

$$v_2 = L_2\frac{di_2}{dt} + M\frac{di_1}{dt}$$

Throughout this chapter we concern ourselves mainly with the external behavior (or port behavior) of networks. To characterize n-port networks in a precise manner, we need to resort to matrix notation. For this reason we devote the next section to an introductory discussion on matrices. The reader who is familiar with matrix algebra may go directly to Section 2.3.

2.2 INTRODUCTORY MATRIX ALGEBRA

In the analysis of large-scale networks and systems it is generally necessary to solve a set of simultaneous algebraic and/or differential equations. The manipulation of such equations can be best handled if matrix notation is used. A *matrix* is a rectangular array of real or complex scalar functions arranged in *rows* and *columns*. A matrix $\mathbf{A}$ with m rows and n columns is written as

$$\mathbf{A} = \begin{bmatrix} a_{11} & a_{12} & a_{13} & \cdots & a_{1n} \\ a_{21} & a_{22} & a_{23} & \cdots & a_{2n} \\ \cdot & & & & \\ \cdot & & & & \\ \cdot & & & & \\ a_{m1} & a_{m2} & a_{m3} & \cdots & a_{mn} \end{bmatrix} \tag{2.2-1}$$

where a_{ij} denotes the element of $\mathbf{A}$ that is located at the ith row and jth column. This notation sometimes is written in the following more compact form:

$$\mathbf{A}_{m\times n} = (a_{ij}) \tag{2.2-2}$$

If $m = n$, the matrix is called a *square* matrix. If the matrix consists of only one row ($m = 1$), it is called a *row vector*; if it consists of only one column ($n = 1$), it is called a *column vector*. Throughout this text we use boldface capital letters to denote rectangular matrices and lowercase boldface letters to denote either a row or a column vector. The following are some examples of matrices.

$$\mathbf{A}(t) = \begin{bmatrix} 2 & 3t & -1 \\ 0 & 5 & e^{-t} \\ 1 & 2 & 0 \end{bmatrix}, \qquad \mathbf{r} = [1 + j \quad 2j \quad -j], \qquad \mathbf{x}(t) = \begin{bmatrix} x_1(t) \\ x_2(t) \\ x_3(t) \end{bmatrix}$$

Note that $\mathbf{A}(t)$ is a 3×3 square matrix some of whose elements are functions of time, $\mathbf{r}$ is a row vector whose elements are complex, and $\mathbf{x}(t)$ is a column vector whose elements are the unknown functions $x_1(t)$, $x_2(t)$, and $x_3(t)$. Before discussing the basic matrix operations, let us introduce some useful definitions.

A matrix $\mathbf{A} = (a_{ij})$ is called the *zero matrix* if $a_{ij} = 0$ for all i and j. A square matrix $\mathbf{D} = (d_{ij})$ is called a *diagonal matrix* if $d_{ij} = 0$ for $i \neq j$. If in addition all the diagonal elements are equal to 1 (i.e., if $d_{ii} = 1$), the matrix is called the *identity* (or *unit*) *matrix*; throughout we denote the identity matrix by **1**. A matrix $\mathbf{B} = (b_{ij})$ is said to be the *transpose* of $\mathbf{A} = (a_{ij})$ if $b_{ji} = a_{ij}$ for all i and j. The transpose of **A** is usually denoted by $\mathbf{A}^T$; for example, the transpose of the preceding matrices are

$$\mathbf{A}^T(t) = \begin{bmatrix} 2 & 0 & 1 \\ 3t & 5 & 2 \\ -1 & e^{-t} & 0 \end{bmatrix}, \qquad \mathbf{r}^T = \begin{bmatrix} 1+j \\ 2j \\ -j \end{bmatrix}, \qquad \mathbf{x}^T(t) = [x_1(t) \quad x_2(t) \quad x_3(t)]$$

Two matrices $\mathbf{A} = (a_{ij})$ and $\mathbf{B} = (b_{ij})$ which have the same number of rows and columns are said to be equal if their corresponding elements are equal, that is, if $a_{ij} = b_{ij}$ for all i and j.

Basic Operations on Matrices: Addition. The sum of two matrices $\mathbf{A} = (a_{ij})$ and $\mathbf{B} = (b_{ij})$ which have the same number of rows and columns is a matrix $\mathbf{C} = (c_{ij})$, where c_{ij} is the sum of a_{ij} and b_{ij}; that is,

$$\mathbf{C} = \mathbf{A} + \mathbf{B} \quad \text{if and only if} \quad c_{ij} = a_{ij} + b_{ij}$$

Notice that two matrices can be added together only if they are of the same size (i.e., they have the same number of rows and columns). Addition of matrices is both commutative and associative; that is, for any **A**, **B**, and **C** we have

$$\mathbf{A} + \mathbf{B} = \mathbf{B} + \mathbf{A} \tag{2.2-3}$$

and

$$(\mathbf{A} + \mathbf{B}) + \mathbf{C} = \mathbf{A} + (\mathbf{B} + \mathbf{C}) \tag{2.2-4}$$

Multiplication by Scalars. Let α be any scalar (real or complex) and $\mathbf{A} = (a_{ij})$ be an $m \times n$ matrix; then the product $\alpha\mathbf{A}$ is an $m \times n$ matrix whose ijth element is αa_{ij}.

Multiplication of Matrices. Let $\mathbf{A}_{m \times n} = (a_{ij})$ and $\mathbf{B}_{n \times p} = (b_{ij})$ be two matrices; the product of **A** and **B** is then an $m \times p$ matrix **C** defined by

$$\mathbf{C} = \mathbf{AB} = (c_{ij}) \tag{2.2-5}$$

whose elements c_{ij} are obtained from

$$c_{ij} = \sum_{k=1}^{n} a_{ik} b_{kj} \tag{2.2-6}$$

Note that the product of two matrices is only defined if the number of the columns in the left-hand matrix is equal to the number of the rows of the matrix in the right-hand side. For this reason, if the product **AB** is defined, it is not necessarily true that **BA** is defined. In the product **AB** the matrix **A** is said to be *postmultiplied* by **B** and the matrix **B** to be *premultiplied* by **A**.

Example 2.2-1. Let

$$\mathbf{A} = \begin{bmatrix} t & 2 & -1 \\ 0 & 5 & 6 \\ -1 & 2 & 5 \end{bmatrix} \quad \text{and} \quad \mathbf{B} = \begin{bmatrix} b_{11} & b_{12} \\ b_{21} & b_{22} \\ b_{31} & b_{32} \end{bmatrix}$$

Then

$$\mathbf{AB} = \begin{bmatrix} tb_{11} + 2b_{21} - b_{31} & tb_{12} + 2b_{22} - b_{32} \\ 5b_{21} + 6b_{31} & 5b_{22} + 6b_{32} \\ -b_{11} + 2b_{21} + 5b_{31} & -b_{12} + 2b_{22} + 5b_{33} \end{bmatrix}$$

whereas the product **BA** is not defined.

For two matrices **A** and **B**, even if both **AB** and **BA** are defined in general, the resulting matrices **AB** and **BA** are not necessarily equal; that is, in general, $\mathbf{AB} \neq \mathbf{BA}$. Thus, in multiplication, matrices are not commutative. They are, however, both distributive and associative; that is, for any **A**, **B**, and **C**, we have

$$\mathbf{A(B + C) = AB + AC} \tag{2.2-7}$$

and

$$\mathbf{(AB)C = A(BC) = ABC} \tag{2.2-8}$$

Note that the product of a matrix **A** by the zero matrix and the identity matrix (if these products are defined) are **0** and **A**, respectively. That is,

$$\mathbf{A \cdot 0 = 0} \quad \text{and} \quad \mathbf{A \cdot 1 = A}$$

The transpose of the product of several matrices is the product of the transpose of each matrix in reverse order; that is,

$$(\mathbf{ABCD})^T = \mathbf{D}^T\mathbf{C}^T\mathbf{B}^T\mathbf{A}^T \tag{2.2-9}$$

Matrix Inversion. Consider an $n \times n$ matrix **A**; then the $n \times n$ matrix $\mathbf{A}^{-1}$ (if it exists) is called the inverse of **A** if

$$\mathbf{A}^{-1}\mathbf{A} = \mathbf{A}\mathbf{A}^{-1} = \mathbf{1} \tag{2.2-10}$$

If $\mathbf{A}^{-1}$ exists, **A** is said to be *nonsingular*.

Exercise 2.2-1. Show that if **A**, **B**, and **C** are nonsingular matrices, then

$$(\mathbf{ABC})^{-1} = \mathbf{C}^{-1}\mathbf{B}^{-1}\mathbf{A}^{-1} \tag{2.2-11}$$

To compute the inverse of a square matrix, we need to obtain its determinant. Let us therefore define the determinant of a matrix and discuss some of its properties.

Consider an $n \times n$ matrix $\mathbf{A} = (a_{ij})$; the determinant of $\mathbf{A}$ is a scalar denoted by $\det \mathbf{A}$ and given by

$$\det \mathbf{A} = \sum_{i=1}^{n} (-1)^{i+j} a_{ij} M_{ij} \quad \text{for } j = 1, 2, \ldots, n \qquad (2.2\text{-}12)$$

where M_{ij}, in turn, is the determinant of a submatrix obtained from $\mathbf{A}$ by deleting the ith row and the jth column. M_{ij} is called the *minor* of a_{ij}. Furthermore, the determinant of a 2×2 matrix $\mathbf{A}$ is defined as

$$\det \begin{bmatrix} a_{11} & a_{12} \\ a_{21} & a_{22} \end{bmatrix} = a_{11}a_{22} - a_{21}a_{12} \qquad (2.2\text{-}13)$$

Thus, knowing the determinant of 2×2 matrices, the determinant of any square matrix can be computed.

Example 2.2-2. Consider the 3×3 matrix

$$\mathbf{A} = \begin{bmatrix} a_{11} & a_{12} & a_{13} \\ a_{21} & a_{22} & a_{23} \\ a_{31} & a_{32} & a_{33} \end{bmatrix}$$

In (2.2-12) let $j = 1$; then

$$M_{11} = a_{22}a_{33} - a_{23}a_{32}, \qquad M_{21} = a_{12}a_{33} - a_{13}a_{32}$$
$$M_{31} = a_{12}a_{23} - a_{13}a_{22}$$

Using (2.2-12), we obtain

$$\begin{aligned}\det \mathbf{A} = {} & a_{11}(a_{22}a_{33} - a_{23}a_{32}) - a_{21}(a_{12}a_{33} - a_{13}a_{32}) \\ & + a_{31}(a_{12}a_{23} - a_{13}a_{22})\end{aligned}$$

Note that depending on the nature of the elements of $\mathbf{A}$, the determinant of $\mathbf{A}$ may be a function of an independent variable t, a complex number, a real number, and so on. Sometimes, for compactness of notation, $\det \mathbf{A}$ is denoted by the Greek letter Δ; that is, $\Delta = \det \mathbf{A}$.

Remark 2.2-1. The determinants of the product of several matrices can be shown to be the product of the determinants of individual matrices; that is

$$\det(\mathbf{ABC}) = \det \mathbf{A} \cdot \det \mathbf{B} \cdot \det \mathbf{C} \qquad (2.2\text{-}14)$$

The proof of the above result is rather lengthy; the interested reader may refer to [2].

With this background in determinants we can now introduce a formula for computing the inverse of a square matrix. The inverse of an $n \times n$ matrix

$\mathbf{A} = (a_{ij})$ is

$$\mathbf{A}^{-1} = \frac{1}{\det \mathbf{A}} \begin{bmatrix} A_{11} & A_{12} & A_{13} & \cdots & A_{1n} \\ A_{21} & A_{22} & A_{23} & \cdots & A_{2n} \\ \cdot & & & & \\ \cdot & & & & \\ \cdot & & & & \\ A_{n1} & A_{n2} & A_{n3} & \cdots & A_{nn} \end{bmatrix}^T \tag{2.2-15}$$

where A_{ij} is called the *cofactor* of a_{ij} and is given as

$$A_{ij} = (-1)^{i+j} M_{ij} \tag{2.2-16}$$

in which M_{ij} is the minor of a_{ij} (i.e., the determinant of the submatrix obtained from $\mathbf{A}$ by deleting the ith row and the jth column).

From (2.2-15) it is clear that the inverse of $\mathbf{A}$ exists if and only if $\det \mathbf{A} \neq 0$. Thus, a necessary and sufficient condition that $\mathbf{A}$ be nonsingular is that $\det \mathbf{A} \neq 0$.

Example 2.2-3. Consider the matrix $\mathbf{A}$ given as

$$\mathbf{A} = \begin{bmatrix} -1 & 4 & 3 \\ 3 & 2 & -5 \\ 1 & 7 & 2 \end{bmatrix}$$

Its determinant is

$$\det \mathbf{A} = -1(4 + 35) - 3(8 - 21) + 1(-20 - 6) = -26$$

Its inverse is therefore

$$\mathbf{A}^{-1} = \frac{1}{-26} \begin{bmatrix} 39 & -11 & 19 \\ 13 & -5 & 11 \\ -26 & 4 & -14 \end{bmatrix}^T = \frac{-1}{26} \begin{bmatrix} 39 & 13 & -26 \\ -11 & -5 & 4 \\ 19 & 11 & -14 \end{bmatrix}$$

Some other important matrix operations are as follows:

Matrix Partitioning. For simplicity of manipulations, matrices are sometimes partitioned into submatrices. For example, a 4×4 matrix $\mathbf{A}$ can be written as

$$\mathbf{A} = \left[\begin{array}{cc|cc} a_{11} & a_{12} & a_{13} & a_{14} \\ a_{21} & a_{22} & a_{23} & a_{24} \\ \hline a_{31} & a_{32} & a_{33} & a_{34} \\ a_{41} & a_{42} & a_{43} & a_{44} \end{array}\right] = \left[\begin{array}{c|c} \mathbf{A}_1 & \mathbf{A}_2 \\ \hline \mathbf{A}_3 & \mathbf{A}_4 \end{array}\right]$$

where $\mathbf{A}_1$, $\mathbf{A}_2$, $\mathbf{A}_3$, and $\mathbf{A}_4$ are submatrices given by

$$\mathbf{A}_1 = \begin{bmatrix} a_{11} & a_{12} \\ a_{21} & a_{22} \end{bmatrix}, \qquad \mathbf{A}_2 = \begin{bmatrix} a_{13} & a_{14} \\ a_{23} & a_{24} \end{bmatrix},$$

$$\mathbf{A}_3 = \begin{bmatrix} a_{31} & a_{32} \\ a_{41} & a_{42} \end{bmatrix}, \qquad \mathbf{A}_4 = \begin{bmatrix} a_{33} & a_{34} \\ a_{43} & a_{44} \end{bmatrix}$$

In multiplying two matrices it is sometimes convenient to partition each matrix into appropriate submatrices and then multiply them together. To illustrate this procedure, consider the following example:

$$\left[\begin{array}{cc|c} a_{11} & a_{12} & a_{13} \\ a_{21} & a_{22} & a_{23} \\ \hline a_{31} & a_{32} & a_{33} \end{array}\right] \left[\begin{array}{c|c} b_{11} & b_{12} \\ b_{21} & b_{22} \\ \hline b_{31} & b_{32} \end{array}\right] = \left[\begin{array}{c|c} A_1 & A_2 \\ \hline A_3 & A_4 \end{array}\right] \left[\begin{array}{c|c} B_1 & B_2 \\ \hline B_3 & B_4 \end{array}\right]$$

$$= \left[\begin{array}{c|c} A_1B_1 + A_2B_3 & A_1B_2 + A_2B_4 \\ \hline A_3B_1 + A_4B_3 & A_3B_2 + A_4B_4 \end{array}\right]$$

$$= \left[\begin{array}{c|c} a_{11}b_{11} + a_{12}b_{21} + a_{13}b_{31} & a_{13}b_{32} \\ a_{21}b_{11} + a_{22}b_{21} + a_{23}b_{31} & a_{23}b_{32} \\ \hline a_{31}b_{11} + a_{32}b_{21} + a_{33}b_{31} & a_{33}b_{32} \end{array}\right]$$

Differentiation. If the elements of a matrix $\mathbf{A}$ are functions of an independent variable, say t, then the matrix can be differentiated with respect to t; that is,

$$\frac{d}{dt}\mathbf{A}(t) = \left(\frac{d}{dt}a_{ij}(t)\right)$$

If the matrix under consideration is a composite function of t, the chain rule applies:

$$\frac{d}{dt}\mathbf{A}(x(t)) = \frac{d\mathbf{A}}{dx}\cdot\frac{dx}{dt} \tag{2.2-17}$$

Furthermore, the derivative of the product of two matrices can be written as

$$\frac{d}{dt}[\mathbf{A}(t)\mathbf{B}(t)] = \frac{d\mathbf{A}(t)}{dt}\mathbf{B}(t) + \mathbf{A}(t)\frac{d\mathbf{B}(t)}{dt} \tag{2.2-18}$$

Note that in the right-hand side of equation (2.2-18) the order of the matrix products cannot, in general, be altered.

Integration. The integral of a matrix $\mathbf{A}(t)$ over an interval (t_0, t_1) is a matrix whose elements are the integral of the elements of the original matrix; that is,

$$\int_{t_0}^{t_1} \mathbf{A}(t)\,dt = \left(\int_{t_0}^{t_1} a_{ij}(t)\,dt\right) \tag{2.2-19}$$

One of the most important applications of matrices is in representation of simultaneous equations. In what follows we utilize matrix notations to study the solution of such equations.

Linear Algebraic Equations. Consider a set of n linear algebraic equations in n unknowns $x_1, x_2, \ldots, x_n$:

$$\begin{aligned} a_{11}x_1 + a_{12}x_2 + a_{13}x_3 + \ldots + a_{1n}x_n &= b_1 \\ a_{21}x_1 + a_{22}x_2 + a_{23}x_3 + \ldots + a_{2n}x_n &= b_2 \\ &\vdots \\ a_{n1}x_1 + a_{n2}x_2 + a_{n3}x_3 + \ldots + a_{nn}x_n &= b_n \end{aligned} \tag{2.2-20}$$

where a_{ij} and b_i are known scalars. These equations can be written in matrix form:

$$\begin{bmatrix} a_{11} & a_{12} & a_{13} & \ldots & a_{1n} \\ a_{21} & a_{22} & a_{23} & \ldots & a_{2n} \\ \cdot & & & & \\ \cdot & & & & \\ \cdot & & & & \\ a_{n1} & a_{n2} & a_{n3} & \ldots & a_{nn} \end{bmatrix} \begin{bmatrix} x_1 \\ x_2 \\ \cdot \\ \cdot \\ \cdot \\ x_n \end{bmatrix} = \begin{bmatrix} b_1 \\ b_2 \\ \cdot \\ \cdot \\ \cdot \\ b_n \end{bmatrix} \tag{2.2-21}$$

or in the compact form

$$\mathbf{Ax} = \mathbf{b} \tag{2.2-22}$$

where $\mathbf{A}$ is an $n \times n$ matrix whose elements are a_{ij}, and $\mathbf{x}$ and $\mathbf{b}$ are column vectors whose ith elements are x_i and b_i, respectively. To verify the validity of this representation, it is sufficient to carry out the matrix multiplication in the left-hand side of (2.2-21); we clearly obtain the set of equations given in (2.2-20). To obtain the solution of (2.2-22), all we have to do is to premultiply the equation by $\mathbf{A}^{-1}$ (if it exists). Thus, if $\mathbf{A}^{-1}$ exists, we have

$$\mathbf{x} = \mathbf{A}^{-1}\mathbf{b} \tag{2.2-23}$$

Example 2.2-4. Let us solve the following set of equations:

$$\begin{aligned} -x_1 + 4x_2 + 3x_3 &= 0 \\ 3x_1 + 2x_2 - 5x_3 &= 0 \\ x_1 + 7x_2 + 2x_3 &= 1 \end{aligned}$$

Putting the equations in matrix form, we get

$$\begin{bmatrix} -1 & 4 & 3 \\ 3 & 2 & -5 \\ 1 & 7 & 2 \end{bmatrix} \begin{bmatrix} x_1 \\ x_2 \\ x_3 \end{bmatrix} = \begin{bmatrix} 0 \\ 0 \\ 1 \end{bmatrix}$$

Matrix **A** in this equation is the same as in Example 2.2-3, for which we have already computed the inverse; then

$$\begin{bmatrix} x_1 \\ x_2 \\ x_3 \end{bmatrix} = \frac{-1}{26} \begin{bmatrix} 39 & 13 & -26 \\ -11 & -5 & 4 \\ 19 & 11 & -14 \end{bmatrix} \begin{bmatrix} 0 \\ 0 \\ 1 \end{bmatrix}$$

Carrying out the matrix multiplication in the right-hand side of the equation, we get

$$x_1 = 1, \qquad x_2 = -\tfrac{2}{13}, \qquad x_3 = \tfrac{7}{13}$$

From (2.2-23) it is clear that a unique solution to the equation $\mathbf{Ax} = \mathbf{b}$ exists if and only if **A** is nonsingular. The conditions under which a set of equations have a unique solution are discussed next.

Rank of a Matrix. Consider an $m \times n$ matrix $\mathbf{A} = (a_{ij})$; let the largest square submatrix of **A** with nonzero determinant be an $r \times r$ matrix; then the number r is called the *rank* of **A**. For example, the matrix given in Example 2.2-3 has rank 3, whereas the matrices

$$\mathbf{A} = \begin{bmatrix} 1 & 2 & -1 \\ 5 & 6 & 10 \\ -2 & -4 & 2 \end{bmatrix}, \qquad \mathbf{B} = \begin{bmatrix} 12 & 6 & 3 \\ 4 & 2 & 1 \\ -8 & -4 & -2 \end{bmatrix} \tag{2.2-24}$$

have ranks 2 and 1, since the largest submatrices of **A** and **B** with nonzero determinants are 2×2 and 1×1, respectively. From our definition it is then clear that an $n \times n$ matrix **A** is nonsingular if and only if it has rank n.

Linear Dependence of Vectors. Consider a set of row or column vectors $\mathbf{x}_1$, $\mathbf{x}_2, \ldots, \mathbf{x}_n$. These vectors are said to be *linearly dependent* if there exist scalar constants $\alpha_1, \alpha_2, \ldots, \alpha_n$ (not all zero) so that

$$\alpha_1\mathbf{x}_1 + \alpha_2\mathbf{x}_2 + \ldots + \alpha_n\mathbf{x}_n = \mathbf{0} \tag{2.2-25}$$

This implies that a set of vectors is dependent if at least one of them can be expressed in terms of the others. A set of vectors is said to be *linearly independent* if none of them can be expressed in terms of the others. There is a close relationship between the rank of a matrix and the number of its linearly independent rows or columns. This relationship is expressed in the following theorem.

Theorem 2.2-1. If the rank of a matrix **A** is r, then **A** has exactly r linearly independent rows and r linearly independent columns.

Consequently, a square matrix is nonsingular if and only if all its columns and all its rows are linearly independent.

Example 2.2-5. Consider the matrices **A** and **B** given in (2.2-24). In **A**, rows 1 and 2 are linearly independent, whereas rows 1 and 3 are linearly dependent, since row 3 can be obtained by multiplying row 1 by -2. Thus, matrix **A** has only two linearly independent rows, and therefore its rank is 2. Similarly, in matrix **B** rows 2 and 3 can be obtained from row 1; hence, the rank of **B** is 1.

Linear Independence of Simultaneous Equations. Consider a set of m linear equations in n unknowns $x_1, x_2, \ldots, x_n$:

$$\begin{aligned} a_{11}x_1 + a_{12}x_2 + \ldots + a_{1n}x_n &= b_1 \\ a_{21}x_1 + a_{22}x_2 + \ldots + a_{2n}x_n &= b_2 \\ &\vdots \\ a_{m1}x_1 + a_{m2}x_2 + \ldots + a_{mn}x_n &= b_m \end{aligned} \tag{2.2-26}$$

where a_{ij} and b_i are known coefficients. Let us rewrite these equations in the form

$$\begin{aligned} y_1{:}\quad & a_{11}x_1 + a_{12}x_2 + \ldots + a_{1n}x_n - b_1 = 0 \\ y_2{:}\quad & a_{21}x_1 + a_{22}x_2 + \ldots + a_{2n}x_n - b_2 = 0 \\ & \vdots \\ y_m{:}\quad & a_{m1}x_1 + a_{m2}x_2 + \ldots + a_{mn}x_n - b_m = 0 \end{aligned} \tag{2.2-27}$$

in which y_i denotes the ith equation. These equations are said to be *linearly dependent* if there exists a set of constants $\alpha_1, \alpha_2, \ldots, \alpha_m$ (not all zero) so that

$$\alpha_1 y_1 + \alpha_2 y_2 + \ldots + \alpha_m y_m = 0$$

Example 2.2-6. Consider the linear equations

$$\begin{aligned} y_1{:}\quad & x_1 + 2x_2 + 3x_3 = 1 \\ y_2{:}\quad & -x_1 + x_2 - 2x_3 = 2 \\ y_3{:}\quad & 2x_1 + 7x_2 + 7x_3 = 5 \end{aligned}$$

These equations are linearly dependent since, clearly, the following is true:

$$3y_1 + y_2 - y_3 = 0$$

In the preceding example we observe that the third equation gives no information concerning x_1, x_2, and x_3 which is not already contained in the first and the second equations. The third equation, therefore, plays no role in the solution of the set and, hence, can be deleted. In general, to obtain a unique solution for a set of m simultaneous equations given in (2.2-26), it is necessary that n of these equations be linearly independent. It can be shown that the number of linear independent equations in (2.2-26) is equal to the rank of the coefficient matrix representing these equations.

A useful method of checking the existence of a unique solution of a set of simultaneous equations is by examining the rank of its coefficient matrix. More specifically, let the set of equations given in (2.2-26) be represented in the form

$$\mathbf{A}\mathbf{x} = \mathbf{b}$$

where $\mathbf{A}$ is an $m \times n$ matrix whose ijth element is a_{ij}, and $\mathbf{x}$ and $\mathbf{b}$ are column vectors whose elements are x_i and b_i, respectively: then

Theorem 2.2-2. The set of simultaneous equations given in (2.2-26) has a *unique* solution if and only if the rank of its coefficient matrix $\mathbf{A}$ is n.

An equivalent statement of this theorem is the following:

Theorem 2.2-3. The set of simultaneous equations given in (2.2-26) has a *unique* solution if and only if the coefficient matrix $\mathbf{A}$ has n linearly independent rows.

The proofs of these theorems are not of immediate concern. These proofs together with other results concerning matrices and linear dependence can be found in [2]. Let us now present a brief discussion of the norm of vectors and matrices. The concept of norm will be extremely helpful in understanding the stability of networks, which we shall discuss in this and future chapters.

2.3 NORM OF VECTORS AND MATRICES

Loosely speaking, the concept of norm is a generalization of the notion of distance or length. In this section we shall define the norms of vectors and matrices, which will be useful in discussing the stability of networks. Let us begin by defining the norm of a vector

$$\mathbf{x} - (x_1, x_2, \ldots, x_n)^T \tag{2.3-1}$$

whose elements are scalar constants.

Definition 2.3-1 (Norm of Vectors). A number $|\mathbf{x}|$ is said to be the *norm* of a vector $\mathbf{x}$ if and only if

(a) $|\mathbf{x}| > 0$ for all $\mathbf{x} \neq \mathbf{0}$.
(b) $|\mathbf{x}| = 0$ if and only if $\mathbf{x} = \mathbf{0}$.

(c) $|\alpha\mathbf{x}| = |\alpha|\cdot|\mathbf{x}|$ for all scalars α ($|\alpha|$ denotes the absolute value of α).
(d) $|\mathbf{x}+\mathbf{y}| \leq |\mathbf{x}| + |\mathbf{y}|$ for any two vectors $\mathbf{x}$ and $\mathbf{y}$.

According to this general definition of norm, any of the following specific definitions qualify as a norm of $\mathbf{x}$ ($\triangleq$ means equal by definition):

$$|\mathbf{x}|_1 \triangleq \sum_{i=1}^{n} |x_i|, \quad \text{called } l_1 \text{ norm of } \mathbf{x} \tag{2.3-2}$$

$$|\mathbf{x}|_2 \triangleq \left(\sum_{i=1}^{n} |x_i|^2\right)^{1/2}, \quad \text{called } l_2 \text{ norm of } \mathbf{x} \tag{2.3-3}$$

$$|\mathbf{x}|_p \triangleq \left(\sum_{i=1}^{n} |x_i|^p\right)^{1/p}, \quad \text{called } l_p \text{ norm of } \mathbf{x}, \quad p \geq 1 \tag{2.3-4}$$

$$|\mathbf{x}|_\infty \triangleq \max_i |x_i|, \quad \text{called } l_\infty \text{ norm of } \mathbf{x} \tag{2.3-5}$$

Let us now show that the l_1 norm of $\mathbf{x}$ given by (2.3-2) indeed satisfies conditions (a) through (d) of Definition 2.3-1.

(a) For each scalar x_i, the absolute value $|x_i|$ is nonnegative and the sum of such nonnegative terms is nonnegative.
(b) If $\mathbf{x} = \mathbf{0}$, then $x_i = 0$ for $i = 1, 2, \ldots, n$; hence, $|\mathbf{x}| = \mathbf{0}$. Conversely, if $|\mathbf{x}| = \mathbf{0}$, since all $|x_i|$ are positive we must have $\mathbf{x} = \mathbf{0}$.
(c) By (2.3-2) we can write

$$|\alpha\mathbf{x}|_1 = \sum_{i=1}^{n} |\alpha x_i| = |\alpha| \sum_{i=1}^{n} |x_i| = |\alpha|\cdot|\mathbf{x}|_1$$

since $|\alpha x_i| = |\alpha|\cdot|x_i|$ for all scalars α and x_i.
(d) By (2.3-2) we have

$$|\mathbf{x}+\mathbf{y}|_1 = \sum_{i=1}^{n} |x_i + y_i|$$

and by triangle inequality [5] we have

$$|x_i + y_i| \leq |x_i| + |y_i| \quad \text{for each } i$$

Then $$|\mathbf{x}+\mathbf{y}|_1 \leq \sum_{i-1}^{n} |x_i| + \sum_{i=1}^{n} |y_i| \triangleq |\mathbf{x}|_1 + |\mathbf{y}|_1$$

Example 2.3-1. Let us take the respective norms of the vector

$$\mathbf{x} = [1 \quad -2 \quad 0]^T$$

By (2.3-2), (2.3-3), and (2.3-4), we have

$$|\mathbf{x}|_1 = 1 + 2 + 0 = 3$$

$$|\mathbf{x}|_2 = (1 + 4 + 0)^{1/2} = \sqrt{5}$$

and $$|\mathbf{x}|_\infty = \max\{1, 2, 0\} = 2$$

Exercise 2.3-1. Prove that the l_2 norm of $\mathbf{x}$ defined in (2.3-3) satisfies conditions (a) through (d) of Definition 2.3-1.

This definition of norm can be generalized to an $n \times n$ matrix $\mathbf{A}$ with elements a_{ij}. In this case, however, we must add an extra condition to Definition 2.3-1; more specifically,

Definition 2.3-2. (Norm of a Matrix). A number $|\mathbf{A}|$ is called the norm of the $n \times n$ matrix $\mathbf{A}$ if

(a) $|\mathbf{A}| > 0$ for all $\mathbf{A} \neq \mathbf{0}$.
(b) $|\mathbf{A}| = 0$ if and only if $\mathbf{A} = \mathbf{0}$.
(c) $|\alpha\mathbf{A}| = |\alpha| \cdot |\mathbf{A}|$ for all scalar α.
(d) $|\mathbf{A} + \mathbf{B}| \leq |\mathbf{A}| + |\mathbf{B}|$ for any two matrices $\mathbf{A}$ and $\mathbf{B}$.
(e) $|\mathbf{AB}| \leq |\mathbf{A}| \cdot |\mathbf{B}|$ for any two matrices $\mathbf{A}$ and $\mathbf{B}$.

According to this definition, the following qualify as norms for the $n \times n$ matrix $\mathbf{A} = (a_{ij})$:

$$|\mathbf{A}|_1 \triangleq \max_j \sum_{i=1}^{n} |a_{ij}|, \quad \text{called } l_1 \text{ norm of } \mathbf{A} \tag{2.3-6}$$

$$|\mathbf{A}|_\infty \triangleq \max_i \sum_{j=1}^{n} |a_{ij}|, \quad \text{called } l_\infty \text{ norm of } \mathbf{A} \tag{2.3-7}$$

Notice that to get the l_1 norm of $\mathbf{A}$ we must obtain the maximum of the sum of the absolute value of each column. Also, to get the l_∞ norm of $\mathbf{A}$, we must take the maximum of the sum of the absolute value of each row. For example, consider the matrix

$$\mathbf{A} = \begin{bmatrix} 1 & -1 & 2 \\ -1 & 0 & -5 \\ 0 & -1 & 3 \end{bmatrix}$$

For this matrix we have

$$|\mathbf{A}|_1 = \max\{2, 2, 10\} = 10$$

and

$$|\mathbf{A}|_\infty = \max\{4, 6, 4\} = 6$$

If the vector $\mathbf{x}$ is a function of the independent variable t, we must revise the definition of the norm to the following form:

Definition 2.3-3. Let $\mathbf{x}(t)$ be an n-column or -row vector whose elements are functions of the independent variable t; then various norms of $\mathbf{x}(t)$, denoted by $\|\mathbf{x}\|$, are given by

$$\|\mathbf{x}\|_1 \triangleq \int_{-\infty}^{\infty} |\mathbf{x}(t)|_1 \, dt, \quad \text{called } L_1 \text{ norm of } \mathbf{x} \tag{2.3-8}$$

$$||\mathbf{x}||_2 \triangleq \left[\int_{-\infty}^{\infty} |\mathbf{x}(t)|_2^2 \, dt\right]^{1/2}, \quad \text{called } L_2 \text{ norm of } \mathbf{x} \tag{2.3-9}$$

$$||\mathbf{x}||_p \triangleq \left[\int_{-\infty}^{\infty} |\mathbf{x}(t)|_p^p \, dt\right]^{1/p}, \quad \text{called } L_p \text{ norm of } \mathbf{x}, \quad p \geq 1 \tag{2.3-10}$$

$$||\mathbf{x}||_\infty \triangleq \sup_{-\infty < t < \infty} |\mathbf{x}(t)|_\infty, \quad \text{called sup norm of } \mathbf{x} \tag{2.3-11}$$

where $|\mathbf{x}(t)|_1$, $|\mathbf{x}(t)|_2$, $|\mathbf{x}(t)|_p$, and $|\mathbf{x}(t)|_\infty$ are the norms of $\mathbf{x}(t)$ for each t as given in Definition 2.3-1. The symbol "sup" denotes the *least upper bound* or the *supremum* of a function over its domain $-\infty < t < \infty$.

Notice that to take the norm of a vector function $\mathbf{x}(t)$ we must first take the norm of $\mathbf{x}(t)$ using Definition 2.3-1, which is a positive scalar function of t, and then use Definition 2.3-3 to obtain $||\mathbf{x}||$, which is a scalar constant.

Example 2.3-2. Let us take the norm of the vector function

$$\begin{cases} \mathbf{x}(t) = [e^{-t} \quad e^{-2t}]^T & \text{for } t \geq 0 \\ \mathbf{x}(t) = \mathbf{0} & \text{for } t < 0 \end{cases}$$

Then the l_1, l_2, and l_∞ norms are

$$|\mathbf{x}(t)|_1 = |e^{-t}| + |e^{-2t}| = e^{-t} + e^{-2t}, \qquad t \geq 0$$

$$|\mathbf{x}(t)|_2 = [|e^{-t}|^2 + |e^{-2t}|^2]^{1/2} = (e^{-2t} + e^{-4t})^{1/2}, \qquad t \geq 0$$

$$|\mathbf{x}(t)|_\infty = \max(e^{-t}, e^{-2t}) = e^{-t}, \qquad t \geq 0$$

Accordingly, the function norms are

$$||\mathbf{x}||_1 = \int_0^\infty (e^{-t} + e^{-2t}) \, dt = \tfrac{3}{2}$$

$$||\mathbf{x}||_2 = \left[\int_0^\infty (e^{-2t} + e^{-4t}) \, dt\right]^{1/2} = \sqrt{\tfrac{3}{4}}$$

and

$$||\mathbf{x}||_\infty = \sup_{0 \leq t < \infty} |e^{-t}| = \max_{0 \leq t < \infty} e^{-t} = 1$$

Hence, different norms of a given vector $\mathbf{x}(t)$ are different numbers.

For a given $n \times n$ matrix $\mathbf{A}(t)$ whose elements are functions of the independent variable t, the norm can be defined as

$$||\mathbf{A}||_p \triangleq \max_{|\mathbf{x}|_p = 1} ||\mathbf{A}\mathbf{x}||_p$$

Notice that in this equation $|\mathbf{x}|_p$ is the l_p norm of $\mathbf{x}$ and $||\mathbf{A}\mathbf{x}||_p$ is the L_p norm of $\mathbf{A}(t)\mathbf{x}$. The max must be taken over all the values of $\mathbf{x}$ for which $|\mathbf{x}|_p = 1$. These norms can be obtained using any of equations (2.3-8) through (2.3-11).

The matrix norm as defined is sometimes called the *induced* norm, since it is induced by the vector norm $|\mathbf{x}|$. The different norms of a matrix $\mathbf{A}(t)$

are, therefore,

$$||\mathbf{A}||_1 \triangleq \max_{|\mathbf{x}|_1=1} ||\mathbf{Ax}||_1$$

$$||\mathbf{A}||_2 \triangleq \max_{|\mathbf{x}|_2=1} ||\mathbf{Ax}||_2$$

$$||\mathbf{A}||_\infty \triangleq \max_{|\mathbf{x}|_\infty=1} ||\mathbf{Ax}||_\infty$$

The properties of a matrix norm will be extensively used in future chapters in the proof of the theorems and verification of the stability of a network for which its input–output relation is given. It should be mentioned at this time that there are many other norms that can be defined. For our purposes, however, the norms defined here will be adequate.

2.4 LINEAR VERSUS NONLINEAR *n* PORTS

In the analysis and synthesis of electrical networks there are three ways that a network can be characterized as linear:

First, a network might be called linear if it is comprised of linear network elements (with arbitrary initial conditions) and independent voltage or current sources. This definition of linearity, although intuitively plausible, is not necessarily useful as far as the input–output behavior of the network is concerned. This statement will be made clear later in this section by a counterexample; we shall show that the input–output relation of a network made of linear elements *and* independent sources is not usually linear in the sense that it fails to satisfy the superposition principle. Conversely, we show that the input–output relation of a network containing some nonlinear elements may satisfy the superposition property and can therefore be classified as linear.

Second, a network might be called linear if its input–output equations can be written in the form

$$\frac{d}{dt}\mathbf{x} = \mathbf{Ax} + \mathbf{Bu}, \qquad \mathbf{y} = \mathbf{Cx} + \mathbf{Du}$$

where $\mathbf{x}$ is a vector whose elements are capacitor charges and inductor fluxes, $\mathbf{u}$ is the input vector, $\mathbf{y}$ is the output vector, and $\mathbf{A}$, $\mathbf{B}$, $\mathbf{C}$, and $\mathbf{D}$ are matrices with appropriate dimensions. To verify the linearity of a network by this definition, we must have complete knowledge of the internal structure of the network. For this reason, this definition may not be useful as far as the port behavior of the network is concerned. We shall discuss this definition of linearity in detail in Chapter 5.

The third definition of linearity is given in terms of the input and output relation of the network. In what follows we give a precise description of this type of linearity.

Consider the network shown in Fig. 2.4-1. Let us label the inputs to this

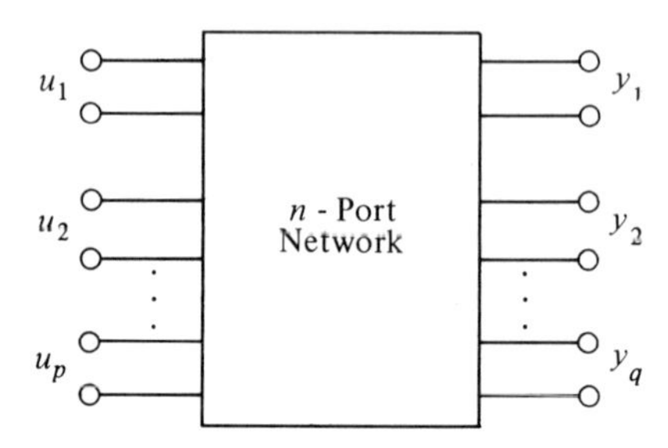

Figure 2.4-1 Symbolic representation of an n-port network with input vector **u** and output vector **y**.

network as $u_1, u_2, \ldots, u_p$ and the outputs as $y_1, y_2, \ldots, y_q$ Using the vector notation discussed in Section 2.2, we can consider the inputs as the elements of a p vector **u** and the outputs as the elements of a q vector **y**; that is,

$$\mathbf{u} = [u_1 \quad u_2 \quad \ldots \quad u_p]^T$$

and

$$\mathbf{y} = [y_1 \quad y_2 \quad \ldots \quad y_q]^T$$

where the transpose sign in the right-hand side of the equations indicates that **u** and **y** are column vectors. The elements of **u** and **y** may be voltages, currents, or a combination of both. We assume that the internal structure of the network is unknown to us and our only access to the network is through its ports. We further assume that the network under consideration is comprised of lumped circuit elements, such as capacitors, resistors, inductors, transistors, gyrators, operational amplifiers, and so on. Under these assumptions, the output vector **y** is related to the input vector **u** through an *integro-differential operator* $\mathfrak{N}$; that is,

$$\mathfrak{N}(\mathbf{u}, \mathbf{y}) = 0 \tag{2.4-1}$$

The operator $\mathfrak{N}$, in essence, is a set of simultaneous *integro-differential equations* that is obtained by applying Kirchhoff's laws to the network under study. It may contain various derivatives and/or integrals of both **u** and **y**. As an illustration, consider the single-input single-output network shown in Fig. 2.4-2. Let the v-i relation of the nonlinear resistor be given by $v_R = i_R^2$. Furthermore, assume that the initial conditions on the capacitor and the inductor are zero. Using Kirchhoff's voltage law and the voltage–current relationship of the elements, and since the current through the 1 Ω resistor is the same as the voltage across it (i.e., $i = y$), we get

$$\mathfrak{N}(u, y) = L\frac{d}{dt}y(t) + \frac{1}{C}\int_0^t y(\tau)\,d\tau + y^2(t) + y(t) - u(t) = 0 \tag{2.4-2}$$

This integro-differential equation is the constraint imposed on $u(t)$ and $y(t)$

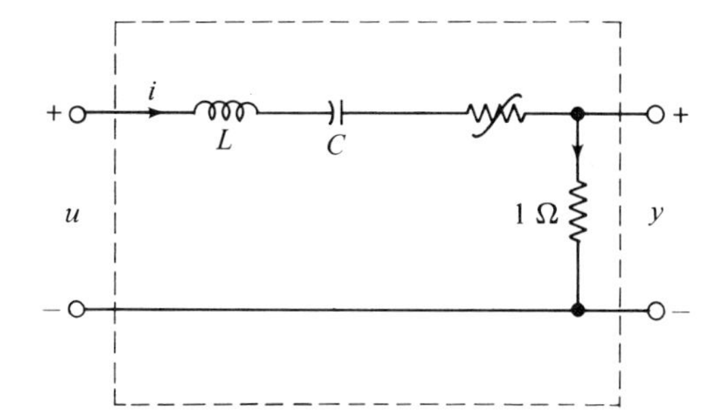

Figure 2.4-2 Example of a network whose input and output are related through an integro-differential operator.

by the network. In other words, $u(t)$ and $y(t)$ satisfying (2.4-2) are *admissible signal pairs* to the network.

In what follows we define the linearity of a network in terms of the operator $\mathfrak{N}$. Before doing so, however, we need to define homogeneity and additivity of an operator.

Definition 2.4-1 (Homogeneity). An integro-differential operator $\mathfrak{N}$ is called *homogeneous* if the following is true for all admissible **u** and **y**:

$$\text{if} \quad \mathfrak{N}(\mathbf{u}, \mathbf{y}) = 0 \quad \text{then} \quad \mathfrak{N}(\alpha\mathbf{u}, \alpha\mathbf{y}) = 0 \tag{2.4-3}$$

where α is *any* scalar constant.

This means that if the input of a homogeneous operator is multiplied by a constant α its output will also be multiplied by α. For example, the operator $\mathfrak{N}$ given by

$$2y(t) + \int_0^t y(\tau)\,d\tau + 2u(t) + \frac{du(t)}{dt} = 0 \tag{2.4-4}$$

is homogeneous (if all the initial conditions are zero), since if $u(t)$ is replaced by $\alpha u(t)$, to keep the equation valid, $y(t)$ must also be replaced by $\alpha y(t)$. In other words, if the pair (u, y) is an admissible input–output pair to (2.4-4), so is the pair $(\alpha u, \alpha y)$. On the other hand, the operator represented by (2.4-2) is not homogeneous since, due to the presence of the nonlinear term y^2, the input $\alpha u(t)$ does not result in the output of $\alpha y(t)$.

Foı simplicity, sometimes the homogeneity of an operator is represented by the following shorthand notation:

$$\text{if} \quad \mathbf{u}(t) \longrightarrow \mathbf{y}(t) \quad \text{then } \mathfrak{N} \text{ is homogeneous if} \quad \alpha\mathbf{u}(t) \longrightarrow \alpha\mathbf{y}(t)$$

Let us now define the additivity property of an operator.

Definition 2.4-2 (Additivity). Let **u** and $\hat{\mathbf{u}}$ be any two input vectors to the operator $\mathfrak{N}(\mathbf{u}, \mathbf{y})$; denote their corresponding outputs by **y** and $\hat{\mathbf{y}}$. The operator $\mathfrak{N}$ is then called *additive* if the input vector $\mathbf{u} + \hat{\mathbf{u}}$ will produce the output vector $\mathbf{y} + \hat{\mathbf{y}}$. Symbolically, this is written as

$$\text{if} \quad \mathbf{u} \longrightarrow \mathbf{y} \quad \text{and} \quad \hat{\mathbf{u}} \longrightarrow \hat{\mathbf{y}} \quad \text{then } \mathfrak{N} \text{ is additive if} \quad (\mathbf{u} + \hat{\mathbf{u}}) \longrightarrow (\mathbf{y} + \hat{\mathbf{y}})$$

For example, the operator given in (2.4-4) is additive. To show this, let $\hat{u}$ and $\hat{y}$ be an admissible signal pair; thus,

$$2\hat{y}(t) + \int_0^t \hat{y}(\tau)\,d\tau + 2\hat{u}(t) + \frac{d\hat{u}(t)}{dt} = 0$$

Adding this equation to (2.4-4) and arranging the resulting equation, we get

$$2[y(t) + \hat{y}(t)] + \int_0^t [y(\tau) + \hat{y}(\tau)]\,d\tau + 2[u(t) + \hat{u}(t)] + \frac{d}{dt}[u(t) + \hat{u}(t)] = 0$$

Comparing this equation with (2.4-4), it is obvious that the input $u(t) + \hat{u}(t)$ will produce the output $y(t) + \hat{y}(t)$. The operator given in (2.4-2), however, is not additive (show this).

If an operator satisfies both homogeneity and additivity properties, it is said to satisfy the *superposition principle.*

We are now ready to define a portwise linear network.

Definition 2.4-3 (Portwise Linear). Let the input–output relationship of an n-port network be given by the integro-differential operator $\mathfrak{N}(\mathbf{u}, \mathbf{y})$; the network is then called *portwise linear* if $\mathfrak{N}(\mathbf{u}, \mathbf{y})$ is both homogeneous *and* additive.

In other words, a network is called portwise linear if its input–output operator satisfies the superposition principle. If $\mathfrak{N}(\mathbf{u}, \mathbf{y})$ violates either the homogeneity or additivity condition, the network will be portwise nonlinear.

For example, the network shown in Fig. 2.4-2 is not portwise linear since its input–output relation, given by (2.4-2), does not satisfy either additivity or homogeneity conditions. However, the network whose input–output relation is given by (2.4-4) *is* portwise linear since it satisfies both homogeneity and additivity conditions.

Remark 2.4-1. According to the preceding definition of linearity, a portwise nonlinear network need not contain any nonlinear elements. Indeed, any initial conditions on the energy-storing elements or the presence of any independent sources will generally classify the network as portwise nonlinear. To shed some light on this fact, consider the circuit shown in Fig. 2.4-3, which is comprised of a single linear capacitor with initial voltage v_0. Let the input be the current source u and the output be the voltage across the capacitor. The input–output relationship of this circuit is

$$y(t) = \frac{1}{C}\int_0^t u(\tau)\,d\tau + v_0 \tag{2.4-5}$$

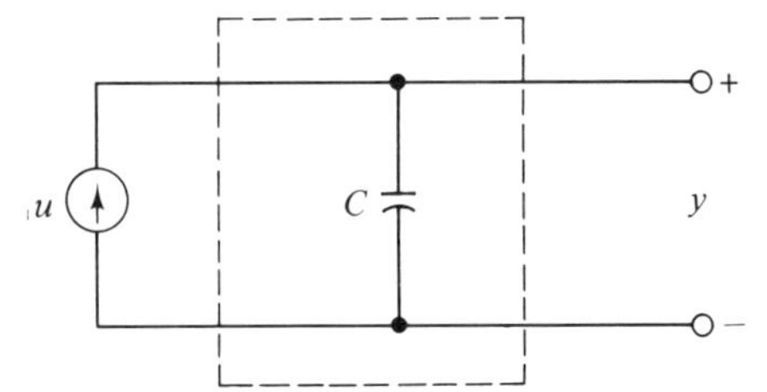

Figure 2.4-3 Circuit with nonzero initial condition that is not portwise linear.

This input–output relationship does not satisfy the superposition principle (why?). Consequently, the network is not portwise linear in the sense of Definition 2.4-3.

From what is discussed so far it should now be clear that networks comprised of linear elements are not necessarily portwise linear. A definite conclusion can, however, be drawn concerning the networks with zero initial conditions and no independent sources in their interior. This conclusion is summarized in the following theorem.

Theorem 2.4-1. An n-port network comprised of linear network elements with zero initial conditions and no independent sources is *portwise linear.*

Remark 2.4-2. Note that we have emphatically used the adjective portwise to stress the fact that we are only considering the port behavior of the networks. Theorem 2.4-1 suggests a more practical definition of linearity, one which excludes the effect of the initial conditions and independent sources available in the black box. In fact, from the *cause*-and-*effect* point of view, only the output corresponding to the applied input should be accounted for. For example, in (2.4-5) the output y is composed of two components; the first term is the contribution of the input and the second term is due to the initial conditions. If the initial conditions are assumed to be zero, the network is certainly portwise linear.

If the input–output behavior of the network is not our prime concern, the following conventional definition of linearity is used.

Definition 2.4-4 (Linear Network). A network comprised of linear network elements and independent sources is called a *linear network.*

Note the difference between the two definitions: one is *portwise linear*; the other is simply *linear*. Throughout the rest of this book we shall distinguish between these two definitions; if the input–output behavior of the network is of interest, the prefix "portwise" will be added to the adjective linear.

Let us now give some examples to illustrate the difference between these definitions.

Example 2.4-1. Consider the integrator circuit shown in Fig. 2.4-4. Assume that the initial voltage across the capacitor is zero. From what we discussed in Chapter 1, the voltage e is negligible in comparison to u and y; hence, the current through resistor R is almost the same as the current through the

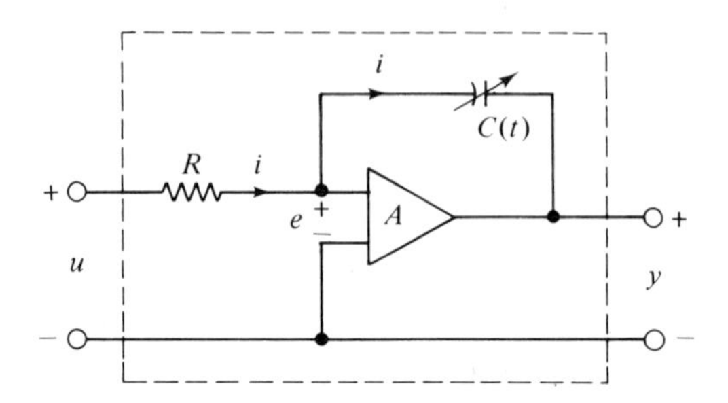

Figure 2.4-4 Integrator circuit with input u and output y.

capacitor; then the input–output relation of this two-port network is

$$\frac{1}{R}u(t) + \frac{d}{dt}[C(t)y(t)] = 0$$

or, equivalently,

$$\frac{1}{R}u(t) + y(t)\frac{d}{dt}C(t) + C(t)\frac{d}{dt}y(t) = 0$$

The operator as given clearly satisfies the homogeneity and additivity property; consequently, Definition 2.4-3 is satisfied and the network under consideration is portwise linear.

Since the operational amplifier is inherently nonlinear [see its v-i characteristic in Fig. 1.5-13(b)], Definition 2.4-4 is not satisfied. If we assume, however, that the operational amplifier operates in its linear range (i.e., if e is smaller than the saturating voltage), it can be considered as a linear element. Under this condition the network is linear in the sense of Definition 2.4-4.

Exercise 2.4-1. Consider the inductor-simulating circuit shown in Fig. 2.4-5. Let the initial voltage across the capacitor be equal to zero. Let the input be the voltage v_1 and the output be the current i_1. Show that this circuit is portwise linear. Is this network linear in the sense of Definition 2.4-4?

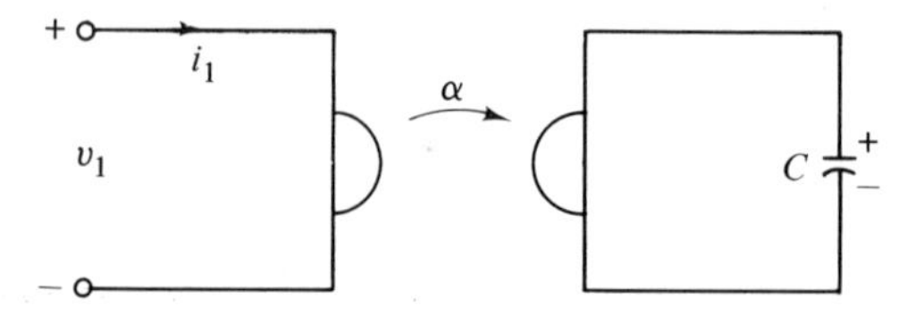

Figure 2.4-5 Inductor-simulating gyrator considered in Exercise 2.4-1.

Remark 2.4-3. It should be mentioned here that for a network to be linear in the sense of Definition 2.4-3 it is *not* necessary that all its elements be linear. Indeed, in the following counterexample we show that a particular network which contains nonlinear energy-storing elements will behave as a linear resistor as far as its port voltage and current are concerned.

Example 2.4-2. Consider the one-port network shown in Fig. 2.4-6. Let the input be the voltage source $v(t)$. Assume that the inductor is flux controlled

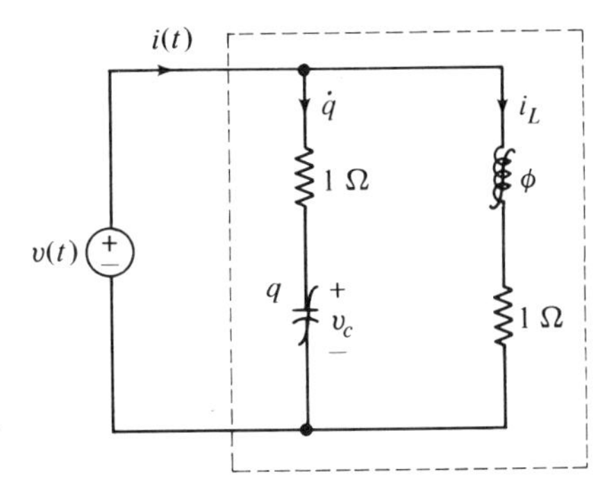

Figure 2.4-6 Portwise linear network with nonlinear elements.

and the capacitor is charge controlled with equations

$$i_L = f(\phi) \tag{2.4-6}$$

and

$$v_c = f(q) \tag{2.4-7}$$

where $f(\cdot)$ is the *same* function in both equations. To assure the existence and uniqueness of the solutions of the differential equations describing the network, assume that the derivative of $f(\cdot)$ is continuous. Furthermore, assume that the resistors are 1 Ω and all the initial conditions are equal to zero. We first show that this network is equivalent to a 1 Ω resistor.

The current through the capacitor is $(d/dt)q(t)$, and the voltage across the inductor is $(d/dt)\phi(t)$; then we have

$$v = v_c + 1\cdot\dot{q} \tag{2.4-8}$$

and

$$v = \dot{\phi} + 1\cdot i_L \tag{2.4-9}$$

where, for simplicity of notation, we have used a "dot" on the function to denote its derivative with respect to t; that is, $\dot{q} = (d/dt)q(t)$ and $\dot{\phi} = (d/dt)\phi(t)$. Replacing i_L and v_c from (2.4-6) and (2.4-7), we get

$$\dot{q} = -f(q) + v, \qquad q(t_0) = 0 \tag{2.4-10}$$

and

$$\dot{\phi} = -f(\phi) + v, \qquad \phi(t_0) = 0 \tag{2.4-11}$$

Since, by assumption, $f(\cdot)$ has continuous first derivative, then the solutions of equations (2.4-10) and (2.4-11) exist and are unique; furthermore, since $f(\cdot)$ is the same function in both equations we conclude that

$$q(t) = \phi(t) \quad \text{for all } t \geq t_0 \tag{2.4-12}$$

Now notice that the port current i is the sum of two currents $\dot{q}$ and i_L; that is,

$$i = \dot{q} + i_L \tag{2.4-13}$$

Replacing $\dot{q}$ from (2.4-8) and using (2.4-6) and (2.4-7), we can write

$$i = v - f(q) + f(\phi) \tag{2.4-14}$$

Then using (2.4-12), we obtain

$$v = i \tag{2.4-15}$$

Hence, the port voltage and the port current of the network under consideration are related by (2.4-15). Consequently, this network is equivalent to a 1 Ω resistor and is therefore portwise linear. A generalization of this example is given in the literature [3].

Exercise 2.4-2. Consider the network shown in Fig. 2.4-7. Let the input be the current source $i(t)$. Show that under the same assumptions as in Example 2.4-2, this network is linear and is equivalent to a 1 Ω resistor.

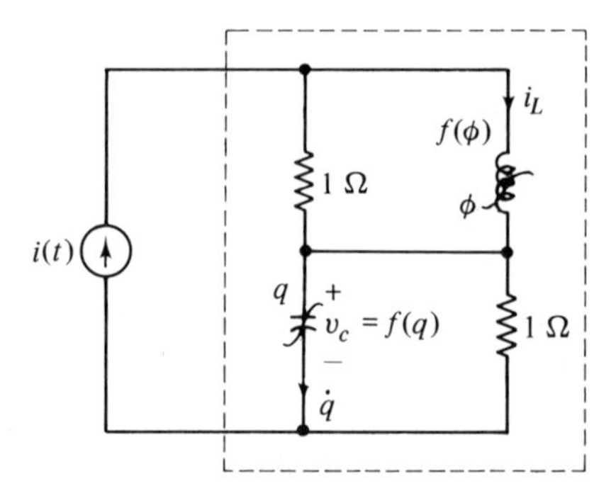

Figure 2.4-7 Network considered in Exercise 2.4-2.

Having discussed the definition of portwise linear networks, let us now study the definition of a nonlinear network. We just observed that a network with nonlinear elements may behave as a linear network as far as its ports are concerned; then what is a nonlinear network? In general, a network is said to be *portwise nonlinear* if its input–output relationship does not satisfy the homogeneity and/or additivity property. It is called *nonlinear* if it contains one or more nonlinear elements.

Example 2.4-3. Consider the *clipping circuit* shown in Fig. 2.4-8(a). The ideal diodes have either zero or infinite resistance. The input–output relationship for this circuit is shown in Fig. 2.4-8(b). Analytically, the input–

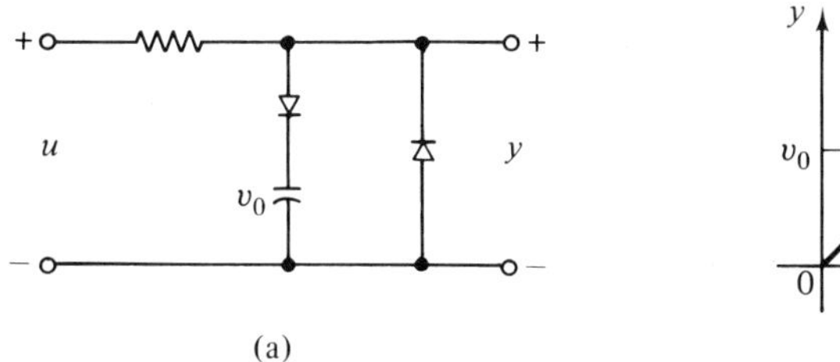

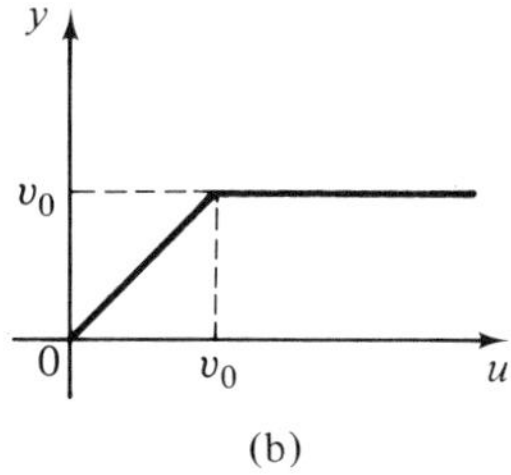

Figure 2.4-8 Clipping circuit and its input-output characteristic.

output relations can be represented as

$$y(t) = 0 \quad \text{for } u(t) \leq 0$$

$$y(t) = u(t) \quad \text{for } 0 \leq u(t) \leq v_0$$

$$y(t) = v_0 \quad \text{for } u(t) \geq v_0$$

These relationships clearly do not satisfy the homogeneity property [e.g., $u(t) = \frac{1}{2}v_0 \longrightarrow y(t) = \frac{1}{2}v_0$, but $u(t) = 2v_0 \not\rightarrow y(t) = 2v_0$]. Consequently, the network is portwise nonlinear. Furthermore, since it contains some nonlinear elements (ideal diodes), it is also nonlinear in the sense of Definition 2.4-4.

Example 2.4-4. Consider the circuit with coupled nonlinear inductors shown in Fig. 2.4-9. The ϕ-i relations of the inductors are

$$\phi_1 = 2i_1^2 + i_2^2$$

$$\phi_2 = i_1^2 + 2i_2^2$$

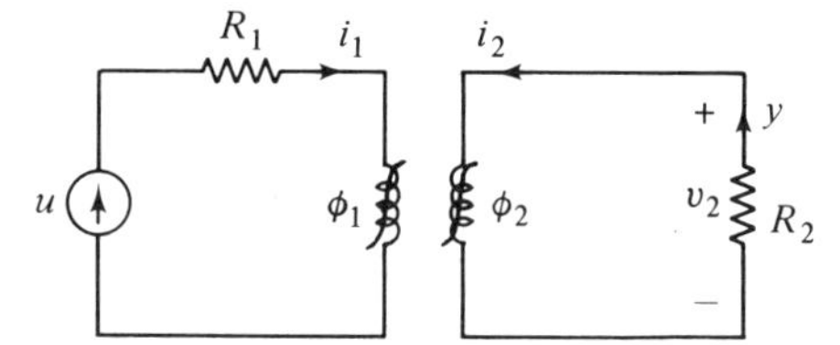

Figure 2.4-9 Circuit considered in Example 2.4-4.

Let the input be the current source i_1 and the output the current through the terminating resistor i_2. The input–output relationship can be obtained by writing the Kirchhoff voltage law in the second loop:

$$v_2 = \frac{d}{dt}\phi_2 = \frac{d}{dt}i_1^2 + 2\frac{d}{dt}i_2^2$$

Taking into account the v-i relation of the terminating resistor we get

$$R_2 i_2 + \frac{d}{dt}i_1^2 + 2\frac{d}{dt}i_2^2 = 0$$

Letting $i_1 = u$ and $i_2 = y$, this equation becomes

$$R_2 y + \frac{d}{dt}u^2 + 2\frac{d}{dt}y^2 = 0 \tag{2.4-16}$$

This is the operator representing the input–output relationship of the network under consideration. If $u \longrightarrow y$, it is clear that $\alpha u \not\rightarrow \alpha y$; hence, the homogeneity condition is not satisfied and the network is therefore portwise nonlinear. Due to the presence of nonlinear elements, it is also nonlinear in the sense of Definition 2.4-4.

2.5 TIME-VARYING VERSUS TIME-INVARIANT *n* PORTS

In this section we define the time-invariant and time-varying n-port networks in terms of their port behavior and compare these definitions with the conventional definition of time invariance. As in the previous section, we assume that the network under study is placed in a black box in the sense that our only access to it is through its terminals. It will turn out that, in some special cases, even if the network contains some time-varying elements, so far as its ports are concerned, it behaves as a time-invariant network.

There are three criteria for characterizing a network as time invariant. First, a network might be called time invariant if it contains no time-varying network elements; second, a network might be called time invariant if its input–output state equations can be represented as a set of first-order differential equations with constant coefficients. To verify the time invariance of a network by these definitions, one has to have access to the internal structure of the network under consideration. The third definition of time invariance is given in terms of its input–output relation. In contrast to the first two, the last definition does not require any knowledge of the topology or the element values of the network. In this section we shall show that the first definition implies the last two definitions in the sense that if a network is made up of time-invariant elements then its state equations are a set of differential equations with constant coefficients, and its input–output behavior is represented by a time-invariant operator. However, the converse is not true; we shall show that, under certain conditions, a network which satisfies the last two definitions of time invariance may contain some time-varying elements. Let us now proceed by defining a portwise time-invariant n-port network.

Consider a network with the input vector $\mathbf{u} = [u_1, u_2, \ldots, u_p]^T$ and the output vector $\mathbf{y} = [y_1, y_2, \ldots, y_q]^T$. As was shown in the previous section, if the network is comprised of lumped elements only, its input–output relation can be represented by an integro-differential operator $\mathfrak{N}(\mathbf{u}, \mathbf{y})$.

Intuitively, a network is said to be portwise time invariant if a time shift in an arbitrary input causes the same time shift in the corresponding output. That is,

$$\text{if} \quad \mathbf{u}(t) \longrightarrow \mathbf{y}(t) \quad \text{then} \quad \mathbf{u}(t-T) \longrightarrow \mathbf{y}(t-T) \quad \text{for all } T$$

For a single-input single-output network, this is illustrated in Fig. 2.5-1.

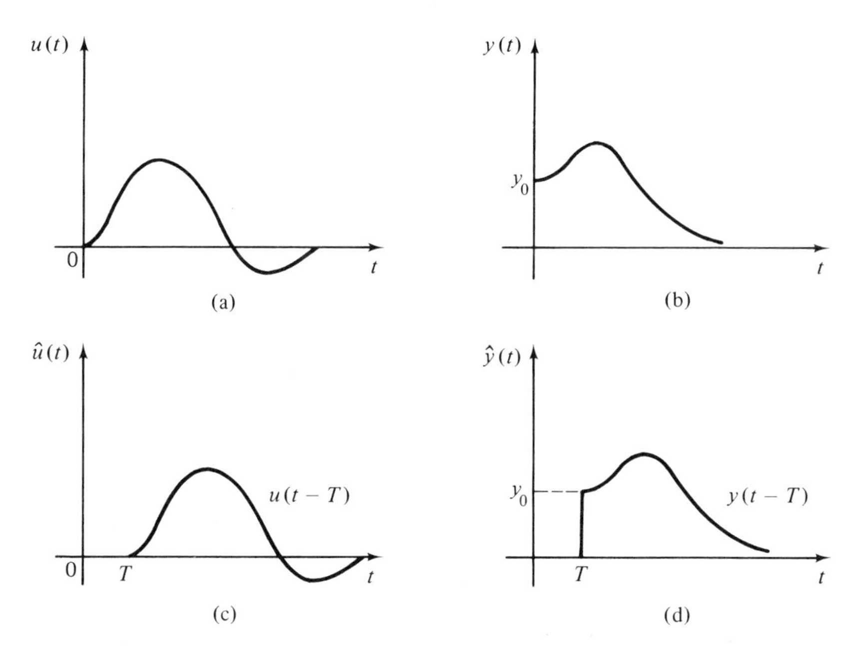

Figure 2.5-1 Input-output behavior of a portwise time-invariant network.

Before proceeding with the formal definition of time-invariant networks, we should point out that to come up with a useful definition we assume that the initial conditions of the network under study must be the same before any of the inputs $\mathbf{u}(t)$ or $\mathbf{u}(t-T)$ is applied [see Figs. 2.5-1(b) and (d)].

Definition 2.5-1 (Portwise Time Invariant). Let $[\mathbf{u}(t), \mathbf{y}(t)]$ and $[\hat{\mathbf{u}}(t), \hat{\mathbf{y}}(t)]$ be any two input–output pairs of a network. If $\hat{\mathbf{u}}(t) = \mathbf{u}(t-T)$ implies that $\hat{\mathbf{y}}(t) = \mathbf{y}(t-T)$ for all T, then the network is said to be *portwise time invariant*, provided that $\hat{\mathbf{y}}(T) = \mathbf{y}(0)$ to start with.

If a network is not portwise time invariant, it is called *portwise time varying*.

Example 2.5-1. Consider the simple circuit shown in Fig. 2.5-2. Let the input be the current source $u(t)$ and the output be the voltage across the capacitor

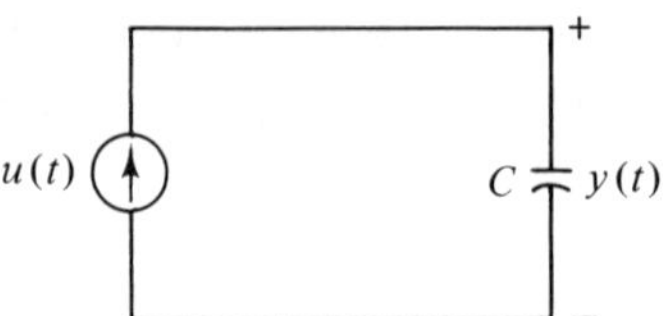

Figure 2.5-2 Portwise time-invariant network.

$y(t)$. The input–output relationship of the circuit is

$$y(t) = \begin{cases} \dfrac{1}{C}\displaystyle\int_0^t u(\tau)\,d\tau + y_0 & \text{for } t \geq 0 \\ 0 & \text{for } t < 0 \end{cases} \tag{2.5-1}$$

where y_0 denotes the initial voltage across the capacitor. To check the portwise time invariance of the network, let $\hat{u}(t)$ be another input defined by

$$\hat{u}(t) = \begin{cases} u(t - T) & \text{for } t \geq T \\ 0 & \text{for } t < 0 \end{cases} \tag{2.5-2}$$

Denote the corresponding output by $\hat{y}(t)$ (see Fig. 2.5-1). Since $\hat{u}(t)$ and $\hat{y}(t)$ are also admissible signal pairs, and since, according to (2.5-2), $\hat{u}(t) = 0$ for $t < T$, we can write

$$\hat{y}(t) = \begin{cases} \dfrac{1}{C}\displaystyle\int_T^t u(\tau)d\tau + y(T) & \text{for } t \geq T \\ 0 & \text{for } t < T \end{cases} \tag{2.5-3}$$

Writing $\hat{u}(t)$ in terms of $u(t)$ as prescribed in (2.5-2), and choosing the same initial value for $\hat{y}(t)$ as for $y(t)$ (i.e., letting $\hat{y}(T) = y_0$), we get

$$\hat{y}(t) = \begin{cases} \dfrac{1}{C}\displaystyle\int_T^t u(\tau - T)\,d\tau + y_0 & \text{for } t \geq T \\ 0 & \text{for } t < T \end{cases}$$

Letting $\sigma = \tau - T$, this relation becomes

$$\hat{y}(t) = \begin{cases} \dfrac{1}{C}\displaystyle\int_0^{t-T} u(\sigma)\,d\sigma + y_0 & \text{for } t \geq T \\ 0 & \text{for } t < T \end{cases}$$

Comparing this last equation with (2.5-1), we get $\hat{y}(t) = y(t - T)$ for $t \geq T$, and $\hat{y}(t) = 0$ for $t < T$ (see Fig. 2.5-1). The network under consideration is therefore portwise time invariant in the sense of Definition 2.5-1.

Remark 2.5-1. Note that the assumption that the initial values of the output must be the same in both cases (i.e., $\hat{\mathbf{y}}(T) = \mathbf{y}_0$) is crucial. If the initial condition in (2.5-3) had been chosen as $y_1 \neq y_0$, the corresponding output $\hat{y}(t)$

would not have been equal to $y(t - T)$ for $t \geq T$. Consequently, an inherently time-invariant network would have been classified as time varying.

The conclusion of Example 2.5-1 can be generalized to include any network that contains only time-invariant elements. This generalization is given in the form of the following theorem.

Theorem 2.5-1. Any n-port network containing no time-varying elements is portwise time invariant.

According to this theorem, if a given n-port network contains only time-invariant elements and voltage and current sources, there is no need for checking its input–output relation; the network is considered portwise time invariant.

Remark 2.5-2. The converse of the above theorem is not, however, true, since there are networks that contain some time-varying elements and yet are portwise time invariant. Once such example is given next.

Example 2.5-2. Consider the network shown in Fig. 2.5-3. Let the energy-storing elements be time varying and

$$L(t) = C(t) \quad \text{for all } t \tag{2.5-4}$$

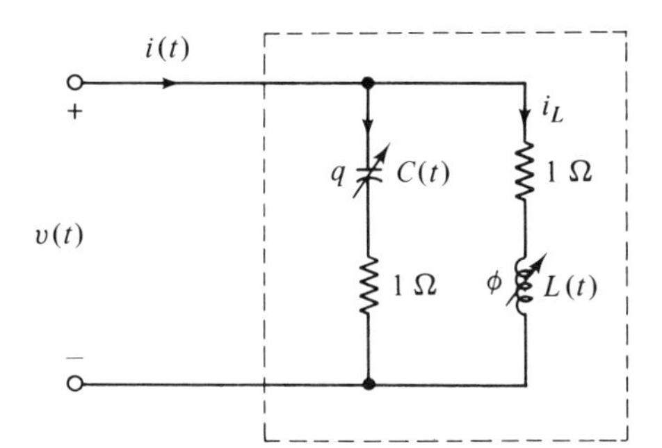

Figure 2.5-3 Network with time-varying elements that is portwise time-invariant.

Assume that the resistors are time invariant with resistance of 1 Ω and all the initial conditions are equal to zero. Then in a manner analogous to that of Example 2.4-2, we obtain

$$\dot{q} = -\frac{1}{C(t)}q + v, \qquad q(t_0) = 0 \tag{2.5-5}$$

and

$$\dot{\phi} = -\frac{1}{L(t)}\phi + v, \qquad \phi(t_0) = 0 \tag{2.5-6}$$

Then equations (2.5-4) through (2.5-6), and the fact that all the initial conditions are equal to zero, yield

$$\phi(t) = q(t) \quad \text{for all } t \tag{2.5-7}$$

Then $i(t)$ can be written as

$$i = \dot{q} + \frac{1}{L(t)}\phi(t) \tag{2.5-8}$$

Consequently, from (2.5-5) and (2.5-8) we get

$$v(t) = i(t)$$

That is, the network is equivalent to a 1 Ω resistor and hence is time invariant.

Exercise 2.5-1. Consider the network shown in Fig. 2.4-7. Assume that the capacitor and inductor are linear and time varying with capacitance and inductance $C(t)$ and $L(t)$. Furthermore, assume that

$$C(t) = L(t) \quad \text{for all } t$$

and all the initial conditions are equal to zero. Show that this network is portwise time invariant by first showing that it is equivalent to a 1 Ω resistor.

Exercise 2.5-2. Consider the one-port network shown in Fig. 2.5-4. Let all

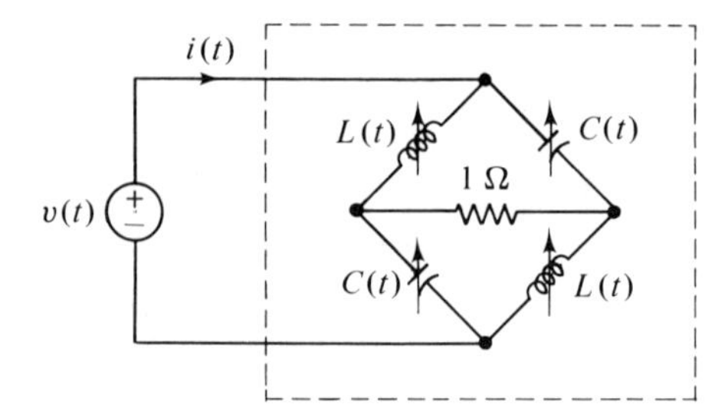

Figure 2.5-4

the initial conditions be equal to zero and assume that

$$L(t) = C(t) \quad \text{for all } t$$

Show that this network is time invariant by first showing that it is equivalent to a 1 Ω resistor.

As in the previous section, if the input–output behavior of the network is not the major concern, the following conventional definition of time invariance can be employed.

Definition 2.5-2 (Time-Invariant Network). A network is called *time invariant* if it contains no time-varying elements; it is called *time varying* if it contains one or more time-varying elements.

We shall show in Chapters 4 and 5 that the time-invariant networks just defined can be represented by a set of integro-differential equations with constant coefficients.

It is important to distinguish between the two definitions of time invariance as stated. A given network might be time invariant in the sense of Definition 2.5-1 and time varying in the sense of Definition 2.5-2; the networks given in Figs. 2.5-3 and 2.5-4 are examples of this case.

In concluding this section it should be mentioned that time-varying networks play an increasingly important role in modern engineering applications. Examples of such useful networks include parametric amplifiers, switching circuits, and filters with time-varying delay or bandwidth [6].

2.6 PASSIVE VERSUS ACTIVE n PORTS

The passivity and activity of network elements were discussed in Chapter 1; it was mentioned that the instantaneous power flowing into a network element is the product of the voltage across the element and the current through it at the instant of observation. It was also mentioned that the total energy absorbed or delivered by the element (depending on whether it was passive or active) from t_0 to t is the integral of the power over the interval $[t_0, t]$. A network element was defined to be passive if the energy delivered to it is nonnegative for all times t and for *all* admissible signal pairs $[v, i]$; accordingly, an element was said to be active if for *some* signal pair or for some interval $[t_0, t]$ the energy delivered to it is negative. In this section we generalize this idea and give a precise definition of passivity and activity of n-port networks. Throughout we assume that the network under consideration is in a black box and our only access to it is through its ports; that is, the only way we can verify the passivity or activity of a network is by examining its port behavior.

Let us consider an n-port network whose port voltages and port currents are

$$\mathbf{v}(t) = [v_1(t), \ldots, v_n(t)]^T \tag{2.6-1}$$

and

$$\mathbf{i}(t) = [i_1(t), \ldots, i_n(t)]^T \tag{2.6-2}$$

Definition 2.6-1 (Passive Networks). An n-port network is said to be *passive* if for all *real* admissible signal pairs $[\mathbf{v}, \mathbf{i}]$ the total energy delivered to its ports is nonnegative for all t; that is,

$$E(t) = \int_{-\infty}^{t} \mathbf{v}^T(\tau)\mathbf{i}(\tau)\, d\tau \geq 0 \quad \text{for all } t \geq -\infty \tag{2.6-3}$$

assuming that at $t = -\infty$ the network is relaxed; that is, $\mathbf{v}(-\infty) = \mathbf{0}$ and $\mathbf{i}(-\infty) = \mathbf{0}$.

Example 2.6-1. Consider the two-port gyrator shown in Fig. 2.6-1; let the gyration constant be α. Then the **v-i** relation of this two-port gyrator is

$$\begin{bmatrix} v_1 \\ v_2 \end{bmatrix} = \begin{bmatrix} 0 & -\alpha \\ \alpha & 0 \end{bmatrix} \cdot \begin{bmatrix} i_1 \\ i_2 \end{bmatrix} \tag{2.6-4}$$

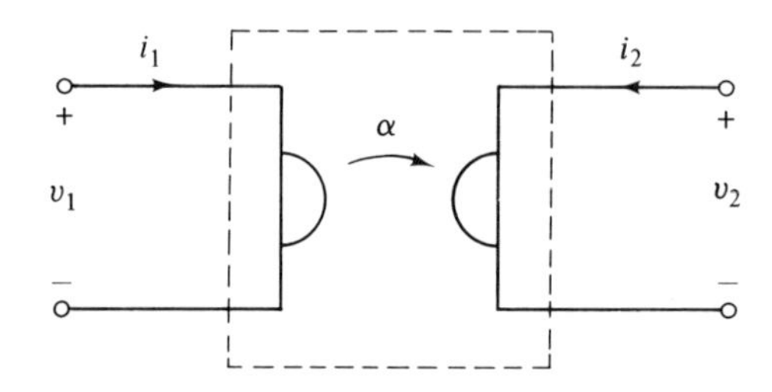

Figure 2.6-1 Gyrator as a passive two-port.

and the energy delivered to it from $-\infty$ to t is

$$E(t) = \int_{-\infty}^{t} [i_1(\tau) \quad i_2(\tau)] \begin{bmatrix} 0 & \alpha \\ -\alpha & 0 \end{bmatrix} \begin{bmatrix} i_1(\tau) \\ i_2(\tau) \end{bmatrix} d\tau$$

or, equivalently,

$$E(t) = \int_{-\infty}^{t} [i_1(\tau) \quad i_2(\tau)] \begin{bmatrix} \alpha i_2(\tau) \\ -\alpha i_1(\tau) \end{bmatrix} d\tau$$

Thus
$$E(t) = \int_{-\infty}^{t} [\alpha i_1(\tau) i_2(\tau) - \alpha i_2(\tau) i_1(\tau)]\, d\tau = 0 \tag{2.6-5}$$

Since the energy absorbed by a gyrator is never negative, the network under consideration is considered passive; moreover, since according to (2.6-5) the energy delivered or absorbed by this network is identically zero, the network is said to be *lossless*; more precisely,

Definition 2.6-2 (Lossless Networks). An n-port network is said to be *lossless* if

$$E \triangleq \int_{-\infty}^{\infty} \mathbf{v}^T(\tau)\mathbf{i}(\tau)\, d\tau = 0 \tag{2.6-6}$$

for all admissible signal pairs whose L_2 norm is bounded. This implies that $\mathbf{v}(t)$ and $\mathbf{i}(t)$ must satisfy the following conditions:

$$\int_{-\infty}^{\infty} \mathbf{v}^T(\tau)\mathbf{v}(\tau)\, d\tau < \infty \tag{2.6-7}$$

and
$$\int_{-\infty}^{\infty} \mathbf{i}^T(\tau)\mathbf{i}(\tau)\, d\tau < \infty \tag{2.6-8}$$

The losslessness of an n port network essentially means that for *any* bounded signal pair the *total* energy absorbed or delivered by the network at its ports is zero.

Exercise 2.6-1. Consider an ideal transformer whose symbolic representation is given in Fig. 2.6-2. The **v-i** relations of this transformer are

$$v_2 = \frac{1}{n} v_1 \tag{2.6-9}$$

and

$$i_2 = -n i_1 \tag{2.6-10}$$

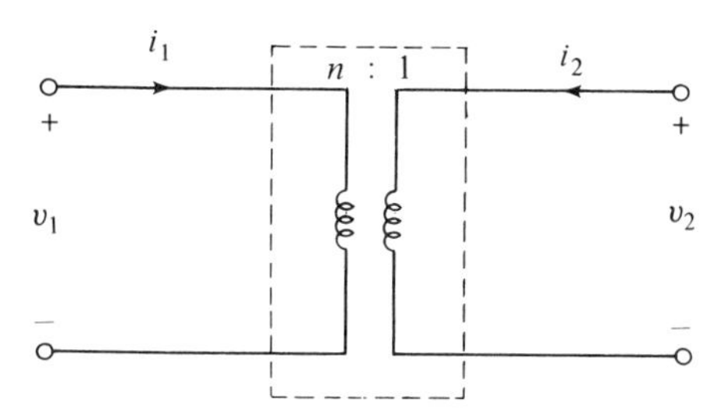

Figure 2.6-2 Symbolic representation of an ideal transformer.

Show that this ideal transformer is a lossless two-port network in the sense of Definition 2.6-2.

In Chapter 3 we use Tellegen's theorem to show that an n-port network made up of any interconnection of two-terminal lossless elements is lossless. The converse is not true, however; that is, a network made up of both active and passive elements might be passive in the sense of Definition 2.6-1 or lossless in the sense of Definition 2.6-2. The following counterexample illustrates this fact.

Example 2.6-2. Consider the one-port network shown in Fig. 2.4-6. Let the capacitor be time invariant with capacitance -1, that is,

$$q = -v_c \tag{2.6-11}$$

and the inductor be time invariant with inductance -1, that is,

$$\phi = -i_L \tag{2.6-12}$$

Then both the capacitor and the inductor are active (show this). However, as far as the input–output pairs $[v, i]$ are concerned, this one-port network is equivalent to a 1 Ω resistor; that is,

$$v = i$$

Hence, it is passive, since

$$E(t) = \int_{-\infty}^{t} i^2(\tau)\, d\tau \geq 0 \quad \text{for all } t > -\infty \tag{2.6-13}$$

Remark 2.6-1. So far in our definitions we have considered the real admissible signal pairs only. In many cases, however, complex signals such as $e^{j\omega t}$ are used. To incorporate such signals in the definition of passivity, we modify Definition 2.6-1 to read

Definition 2.6-3. An n-port network is said to be *passive* if for all complex admissible signal pairs $[\mathbf{v}, \mathbf{i}]$

$$E(t) = \text{Re}\left\{\int_{-\infty}^{t} \mathbf{v}^*(\tau)\mathbf{i}(\tau)\, d\tau\right\} \geq 0 \quad \text{for all } t \geq -\infty \qquad (2.6\text{-}14)$$

where * indicates the conjugate transpose of $\mathbf{v}(t)$ and Re{·} denotes the real part of a complex number.

Example 2.6-3. Consider a one-port network made up of a linear time-invariant capacitor with capacitance C. Let the voltage across it be

$$v(t) = v_1(t) + jv_2(t), \qquad j \triangleq \sqrt{-1}$$

Then the current through it is

$$i(t) = C[\dot{v}_1(t) + j\dot{v}_2(t)]$$

Hence,
$$E(t) = \text{Re}\left\{\int_{-\infty}^{t} C(v_1 - jv_2)(\dot{v}_1 + j\dot{v}_2)\, d\tau\right\}$$

Consequently, if we assume that $v_1(-\infty) = v_2(-\infty) = 0$, we obtain

$$E(t) = C\int_{-\infty}^{t} (v_1\dot{v}_1 + v_2\dot{v}_2)\, d\tau = \tfrac{1}{2}C[v_1^2(t) + v_2^2(t)] \geq 0$$

This holds for all $t \geq -\infty$; hence, the capacitor is passive. Notice that for an n-port network to be passive, equation (2.6-6) or (2.6-14) must hold for *all* admissible signal pairs, not just selected signals. This immediately suggests a definition for active networks, which is given next.

Definition 2.6-4 (Active Networks). An n-port network is said to be *active* if

$$E(t) = \int_{-\infty}^{t} \mathbf{v}^T(\tau)\mathbf{i}(\tau)\, d\tau < 0 \qquad (2.6\text{-}15)$$

for *some* admissible signal pairs or for some $t > -\infty$.

Example 2.6-4. Consider the one-port network shown in Fig. 2.6-3(a), which is made up of a nonlinear resistor whose i-v characteristic is shown in Fig. 2.6-3(b). The resistor is voltage controlled with i-v relation

$$i = f(v)$$

and the input is a 1-V battery, which is connected to the resistor at t_0. The

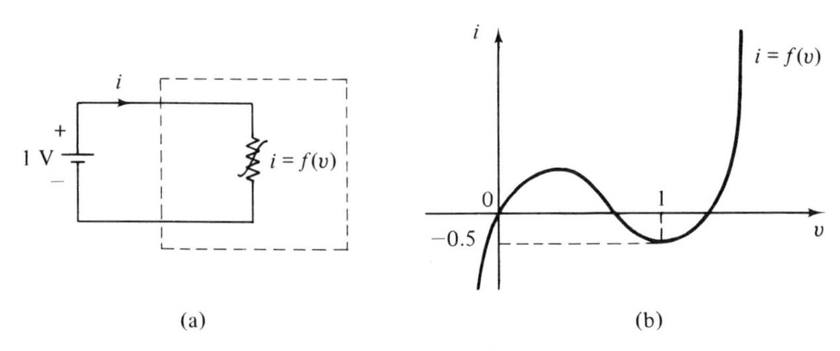

(a) (b)

Figure 2.6-3

energy delivered by the battery is

$$E(t) = \int_{-\infty}^{t} v(\tau)f(v(\tau))\,d\tau$$

$$E(t) = \int_{0}^{t} 1(-0.5)\,d\tau = -0.5t$$

which is negative for all $t > 0$. This means that the network actually delivers energy to the battery (battery is being charged). Hence, the resistor is active.

2.7 CAUSAL AND NONANTICIPATIVE NETWORKS

Causality and nonanticipativeness are among the most important properties of physical networks and systems—every physically realizable network and system must be nonanticipative and every linear passive network is causal. In this section we introduce a definition of causality and nonanticipativeness based on the input–output properties of the n-port networks. For this reason, we must specify the input (cause) and the output (effect) of the network under consideration. That is, for a given admissible signal pair $[\mathbf{v}\ \mathbf{i}]$, it is necessary to know which signal is the input and which the output. It will be shown later in this section that depending on whether the network is current excited–voltage response (i.e., $\mathbf{i}$ is the input and $\mathbf{v}$ is the output) or voltage excited–current response (i.e., $\mathbf{v}$ is the input and $\mathbf{i}$ is the output) the network can be considered causal or noncausal. To keep our discussion of causality and nonanticipativeness general, as usual, we shall denote the input of the network by $\mathbf{u}(t)$ and the corresponding output by $\mathbf{y}(t)$; that is, we denote the admissible signal pair of the network under consideration by $[\mathbf{u}, \mathbf{y}]$.

Intuitively speaking, a network is nonanticipative if the response (output) does not precede the excitation (input). To put this intuitive notion of nonanticipativeness in an analytical framework form, we introduce the following definition. Without any loss of generality we consider the relaxed networks, that is, those whose initial condition is zero; then we have

Definition 2.7-1 (Nonanticipative Networks). An n-port network with input $\mathbf{u}(t)$ and output $\mathbf{y}(t)$ is said to be *nonanticipative* if $\mathbf{u}(t) = \mathbf{0}$ implies that $\mathbf{y}(t) = \mathbf{0}$ for all $t < t_0$, for all t_0, and for all $[\mathbf{u}(t), \mathbf{y}(t)]$.

A network is anticipative if it is not nonanticipative. More precisely, an n-port network is said to be *anticipative* if for *some* input–output pair $[\mathbf{u}(t), \mathbf{y}(t)]$, if $\mathbf{u}(t) = 0$ for all $t < t_0$, then $\mathbf{y}(t) \neq 0$ for *some* $t < t_0$ and some t_0.

Every network discussed in this book is nonanticipative. Hypothetical examples of anticipative networks can be constructed; however, no anticipative network model can be physically realized. This is because in no physical system can the effect precede the cause.

The concept of causality can be defined in terms of nonanticipativeness and the uniqueness of the response for a given input. In contrast to the previous case, physical networks can be realized that are noncausal. Examples of such networks will be given later in this section. Let us now give a precise definition of causality in terms of the input–output properties of a network.

Definition 2.7-2. (Causality). An n-port network is said to be *causal* if for *all* admissible input–output pairs $[\mathbf{u}(t), \mathbf{y}(t)]$ and $[\hat{\mathbf{u}}(t), \hat{\mathbf{y}}(t)]$, if $\mathbf{u}(t) = \hat{\mathbf{u}}(t)$ for $t < t_1$, then $\mathbf{y}(t) = \hat{\mathbf{y}}(t)$ for $t \leq t_1$ and for any $t_1 > -\infty$.

That is, two identical excitations over any interval must give rise to identical responses over the same interval. We shall show through an example that this condition can be violated by physically realizable networks.

Accordingly, an n-port network is said to be *noncausal* if, for some admissible input–output pairs $[\mathbf{u}(t), \mathbf{y}(t)]$ and $[\hat{\mathbf{u}}(t), \hat{\mathbf{y}}(t)]$, $\mathbf{u}(t) = \hat{\mathbf{u}}(t)$ for all $t < t_1$ results in $\mathbf{y}(t) \neq \hat{\mathbf{y}}(t)$ for some $t < t_1$.

From this definition it is clear that two *sufficient* conditions for a network $\mathfrak{N}$ to be noncausal are (1) $\mathfrak{N}$ is anticipative, and (2) the output $\mathbf{y}(t)$ is not a unique function of its input $\mathbf{u}(t)$. The first condition is not satisfied by any physically realizable network; hence, noncausality of the physical networks is mainly due to the second condition, that is, the nonuniqueness of the response.

Example 2.7-1. Consider a one-port network consisting of a single tunnel diode whose v-i characteristic is shown in Fig. 2.7-1(b). We shall show that this network is causal under voltage excitation–current response, and is noncausal under current excitation–voltage response.

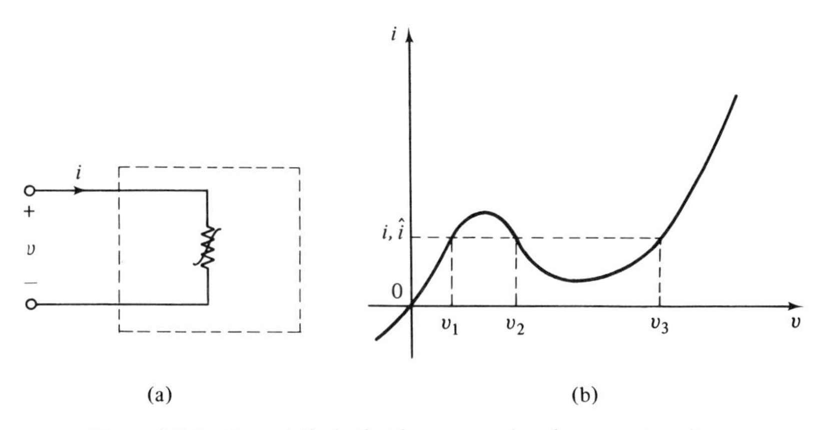

Figure 2.7-1 Tunnel diode that is noncausal under current excitation.

The i-v relation of this resistor is

$$i = f(v)$$

Then for two inputs $v(t)$ and $\hat{v}(t)$ such that

$$v(t) = \hat{v}(t) \quad \text{for all } t$$

the corresponding outputs $i(t)$ and $\hat{i}(t)$ are

$$i(t) = f(v(t)) = f(\hat{v}(t)) = \hat{i}(t) \quad \text{for all } t$$

Thus, two identical inputs produce identical outputs; the conditions of Definition 2.7-2 are satisfied; hence, the network is causal under voltage excitation–current response. Now let the input be a current source i; then to two identical current excitations $i = \hat{i}$ there correspond three different responses $v_1 \neq v_2 \neq v_3$ (see Fig. 2.7-1); hence, according to Definition 2.7-2, this resistor is noncausal under current excitation–voltage response. Notice that in either case the resistor is nonanticipative.

Example 2.7-2. Consider the two-port transformer discussed in Exercise 2.6-1. If the input is taken as the column vector

$$\mathbf{u}(t) = \begin{bmatrix} v_1 \\ i_1 \end{bmatrix}$$

and the output taken as the column vector

$$\mathbf{y}(t) = \begin{bmatrix} v_2 \\ i_2 \end{bmatrix}$$

then the transformer, taken as a two-port network, is causal, since the output $\mathbf{y}(t)$ is uniquely determined by the input $\mathbf{u}(t)$ through

$$\mathbf{y}(t) = \begin{bmatrix} \frac{1}{n} & 0 \\ 0 & -n \end{bmatrix} \mathbf{u}(t)$$

However, if the input $\mathbf{u}(t)$ is taken as the vector

$$\mathbf{u}(t) = \begin{bmatrix} i_1 \\ i_2 \end{bmatrix}$$

and the output taken as

$$\mathbf{y}(t) = \begin{bmatrix} v_1 \\ v_2 \end{bmatrix}$$

then the output $\mathbf{y}(t)$ is completely independent from the input $\mathbf{u}(t)$; hence, two identical inputs $\mathbf{u}(t)$ and $\hat{\mathbf{u}}(t)$ do not necessarily produce identical outputs. Therefore, the transformer under consideration is noncausal under current excitation–voltage response. In a similar manner we can show that it is also noncausal under voltage excitation–current response.

Remark 2.7-1. From the preceding discussion it is clear that before characterizing a network as causal or noncausal we must first specify the excitation and response of the network. In the special case of linear passive networks it can be shown that if the input to the n-port network under consideration is made up of independent voltage or current sources, then linearity and passivity imply causality.

2.8 STABLE AND UNSTABLE NETWORKS

The concept of stability is of considerable importance both in the analysis and the design of networks. There are a number of definitions of stability, each of which is suitable for a particular aspect of network behavior. In this section we shall consider one important type of stability, bounded-input bounded-output stability. Furthermore, since stability is a vast subject, in this chapter we discuss only the basic definitions and give some illustrative examples. Some other definitions of stability, such as Lyaponov stability, will be introduced in Chapter 11. Some necessary and sufficient conditions for stability of a given network will also be discussed in Chapter 11.

Before proceeding with the definition of stability of an n-port network, let us define the boundedness of a vector function. Generally speaking, a vector $\mathbf{x}(t)$ is called bounded if its norm remains finite for all t; more precisely,

Definition 2.8-1 (Boundedness). A vector function $\mathbf{x}(t)$ is said to be bounded if there exists a constant M such that

$$\|\mathbf{x}\| \leq M < \infty \tag{2.8-1}$$

where $\|\mathbf{x}\|$ is the norm of $\mathbf{x}(t)$ in the sense of Definition 2.3-3.

Let us now consider an n-port network with admissible signal pair $[\mathbf{v}(t), \mathbf{i}(t)]$. As in the previous sections, to study bounded-input bounded-output stability, we have to specify whether the network is voltage excited–current response, or vice versa. To be completely general, we denote the input vector by $\mathbf{u}(t)$ and the output vector by $\mathbf{y}(t)$. Then we have

Definition 2.8-2 (B.I.B.O. Stability). Consider an n-port network $\mathfrak{N}$ with the input vector $\mathbf{u}(t)$ and the output vector $\mathbf{v}(t)$; then $\mathfrak{N}$ is said to be *bounded-input bounded-output* stable (b.i.b.o. stable) if and only if for *any* bounded input $\mathbf{u}(t)$ the corresponding output $\mathbf{y}(t)$ is bounded; or equivalently,

A network $\mathfrak{N}$ is b.i.b.o. stable if for *any* input $\mathbf{u}(t)$ and some finite number M_1 there exists a finite number M_2 (which depends on M_1) such that

$$\|\mathbf{u}\| \leq M_1 < \infty \Longrightarrow \|\mathbf{y}\| \leq M_2 < \infty$$

A network is said to be b.i.b.o. *unstable* if it is not b.i.b.o. stable.

More precisely, an n-port network is said to be b.i.b.o. unstable if there exists a bounded input that produces an unbounded output.

Example 2.8-1. Consider the two-port gyrator discussed in Section 1.5; let $\mathbf{i}(t)$ be the input and $\mathbf{v}(t)$ be the output. Then, using (2.3-4), the l_p norm of $\mathbf{v}(t)$ is

$$|\mathbf{v}(t)|_p = (|v_1|^p + |v_2|^p)^{1/p} \tag{2.8-2}$$

By using (1.5-3) and (1.5-4), we get

$$|\mathbf{v}(t)|_p = \alpha(|i_1|^p + |i_2|^p)^{1/p} = \alpha|\mathbf{i}(t)|_p$$

From the definition of L_p given by (2.3-10), we can write

$$\|\mathbf{v}\|_p = \alpha\|\mathbf{i}\|_p \tag{2.8-3}$$

Now if the input is bounded, that is, if

$$\|\mathbf{i}\|_p \leq M_1 < \infty$$

then (2.8-3) implies that

$$\|\mathbf{v}\|_p \leq \alpha M_1 < \infty \quad \text{for } \|\mathbf{i}\|_p \leq M_1 < \infty$$

Hence, the gyrator is b.i.b.o. stable.

Example 2.8-2. Consider the network shown in Fig. 2.8-1. This network is made up of a linear time-invariant capacitor with capacitance C in series with a constant current source of 1 A. Let us consider the output to be the voltage across the capacitor and the input to be the current through the current source. Clearly, the input is bounded. The output, however, is given by

$$v_c(t) = \frac{1}{C}\int_0^t 1 \cdot dt = \frac{1}{C}t$$

Thus, as $t \longrightarrow \infty$, the output will go to infinity. As a result, the network is b.i.b.o. unstable.

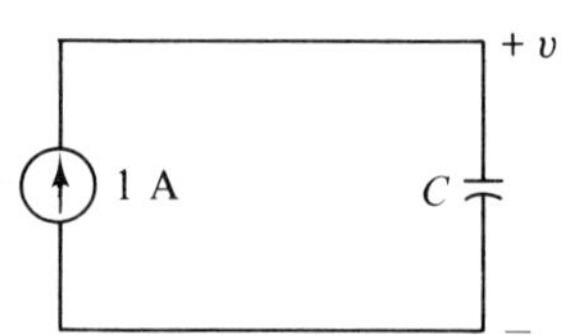

Figure 2.8-1 Simple capacitor circuit that is current excited-voltage response unstable.

Remark 2.8-1. In the stability studies of a network we usually deal with only two norms, L_2 and L_∞. The L_2 norm typically describes the behavior of the network in terms of the amount of the energy delivered to it. The L_∞ norm, on the other hand, describes the behavior of the network in terms of the amplitude of its admissible signal pair.

Exercise 2.8-1. Consider the one-port network shown in Fig. 2.8-2. Show that this network is b.i.b.o. unstable under voltage excitation–current response in the L_∞ sense. [Hint: Choose the particular input $v(t) = \sin t$ for which $\|v\|_\infty = 1$ and show that $\|i\|_\infty \longrightarrow \infty$ as $t \longrightarrow \infty$.]

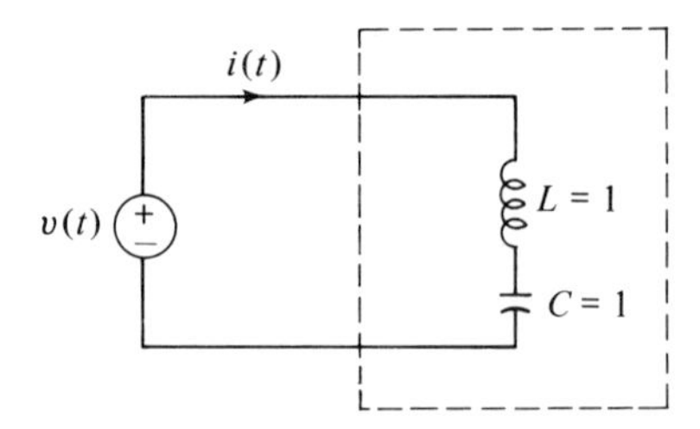

Figure 2.8-2

Exercise 2.8-2. Consider the *RLC* network shown in Fig. 2.8-3. Show that this network is b.i.b.o. stable in the L_∞ norm.

To obtain conditions for the stability of networks, we need to have complete knowledge of the differential equation describing the network; such conditions will be discussed in Chapter 11.

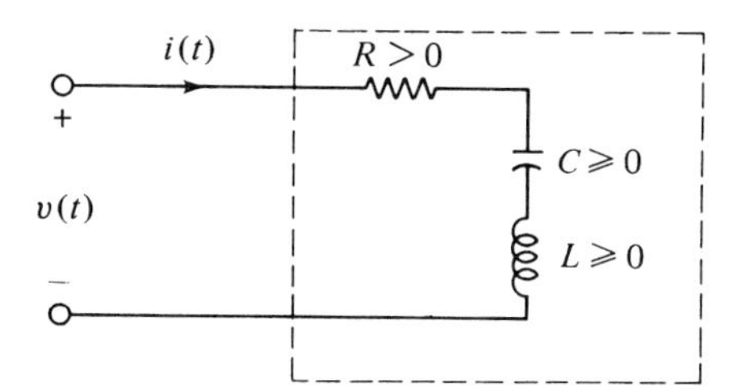

Figure 2.8-3

REFERENCES

[1] NEWCOMB, R. W., "On the Definition of a Network," *Proc. IEEE*, **53**, No. 5 (1965), 547–548.

[2] HOHN, F. E., *Elementary Matrix Algebra*, The Macmillan Company, New York, 1964.

[3] DESOER, C. A., and K. K. WONG, "Constant Resistance One-Ports Which Include Nonlinear Time-Varying Elements," *IEEE Trans. Circuit Theory*, **CT-13**, No. 4 (1966), 403–409.

[4] KUH, E. S., and R. A. ROHRER, *Theory of Linear Active Networks*, Holden-Day, Inc., San Francisco, 1967, pp. 42–48.

[5] KAPLAN, W., *Advanced Calculus*, Addison-Wesley Publishing Company, Inc., Reading, Mass., 1952.

[6] DESOER, C. A., and B. PEIKARI, "Design of Nonlinear Time-Invariant and Linear Time-Varying Networks," *IEEE Trans. Circuit Theory*, **CT-17**, No. 2 (1970), 233–240.

PROBLEMS

2.1 Find the l_1 and l_∞ norm of the following:

$$\mathbf{x} = [1 \quad -2 \quad 3]^T, \qquad \mathbf{A} = \begin{bmatrix} 1 & -5 & 6 \\ 0.5 & -2 & 0 \\ -1 & 0 & -1 \end{bmatrix}$$

2.2 Find the L_1, L_2, and L_∞ norm of the following vectors:

$$\mathbf{x}(t) = [e^{-t} \quad \sin t \quad te^{-t}]^T, \qquad t \geq 0$$

$$\mathbf{y}(t) = [e^{t} \quad 1 \quad e^{-t}], \qquad t \geq 0$$

2.3 Consider the network shown in Fig. P2.3(a). Let the input be the voltage source $u(t)$ and the output $y(t)$ be the current through the 1 Ω resistor. The i-v characteristic of the ideal diode is given in Fig. P2.3(b). Determine whether this network is portwise linear.

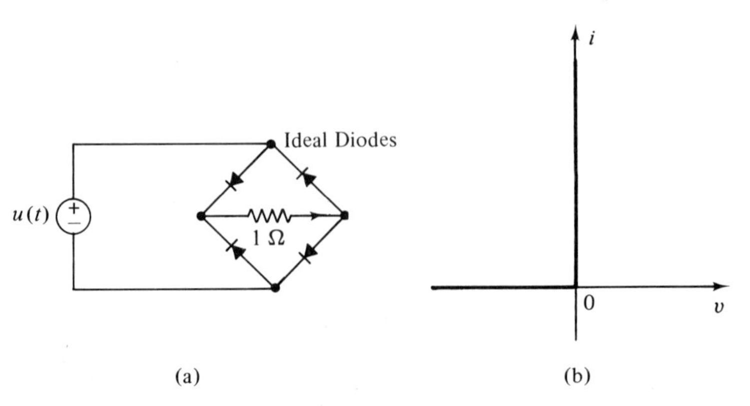

Figure P2.3

2.4 Consider the network shown in Fig. P2.4. The nonlinear resistor has an arbitrary v-i characteristic that goes through the origin. Show that this network is portwise linear.

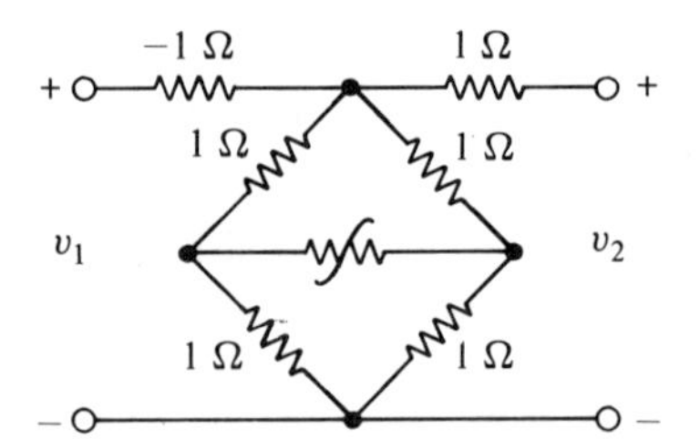

Figure P2.4

2.5 Determine whether the networks whose input $u(t)$ and output $y(t)$ are related by the following equations are linear or not:

(a) $y(t) + \dot{y}(t) = u(t) + u^2(t)$.

(b) $y(t) + \ddot{y}(t) = u(t) + \int_0^t u(\tau)\, d\tau$.

(c) $y(t) = 1 + u(t) \sin \omega t$.

2.6 Consider the network shown in Fig. P2.6. The initial condition on the capacitor is zero and the diodes are ideal. Show that this network is portwise linear.

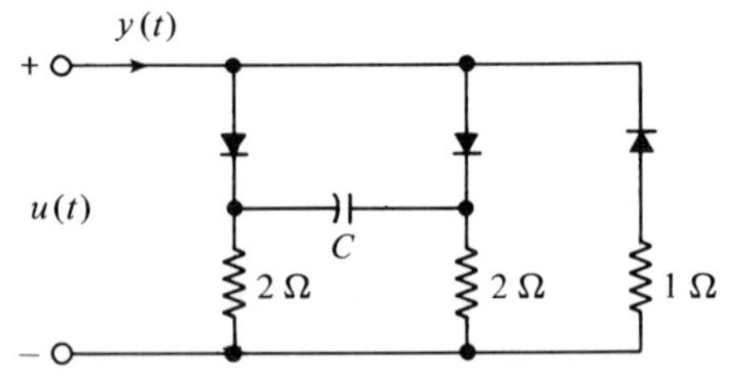

Figure P2.6

2.7 Determine whether the networks whose input $u(t)$ and output $y(t)$ are related by the following equations are time-invariant:

(a) $y(t) = tu(t) + 1$.

(b) $\dot{y}(t) + \ddot{y}(t) = u(t) + \dot{u}(t)$.

(c) $y(t) = t + \int_0^t u(\tau)\, d\tau$.

2.8 Show that the network shown in Fig. P2.8 is time invariant. Assume that

$$R_1(t) = 1 + 0.5 \sin \omega t$$
$$R_2(t) = 1 - 0.5 \sin \omega t$$

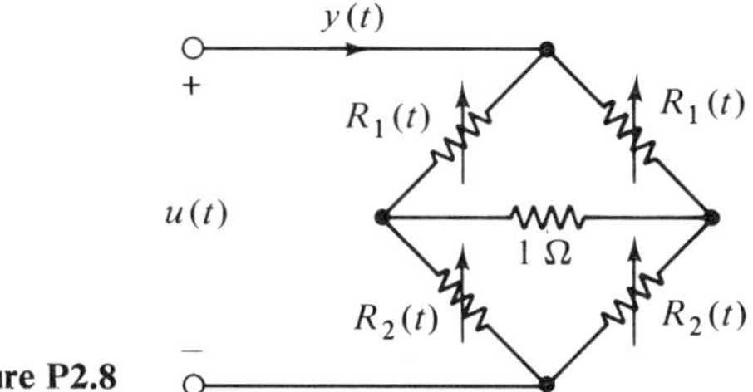

Figure P2.8

2.9 Show that the network shown in Fig. P2.9 is time invariant if $L(t) = C(t)$. (Hint: Use the conclusion of Example 2.5-2.)

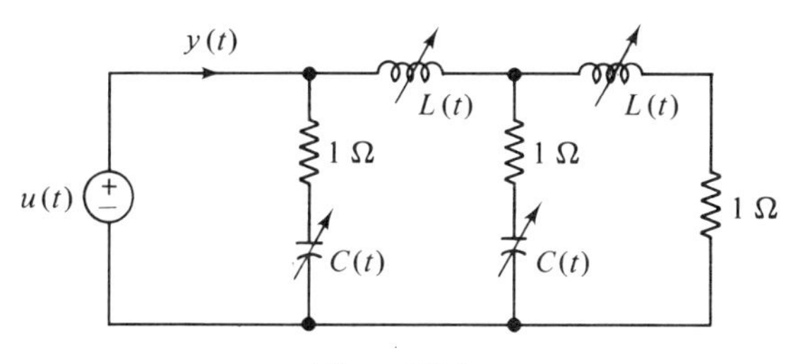

Figure P2.9

2.10 Determine whether the network shown in Fig. P2.4 is passive or active.

2.11 Show that the network shown in Fig. P2.6 is passive.

2.12 Consider the network shown in Fig. P2.12. Assume that the network is relaxed at $t = -\infty$.

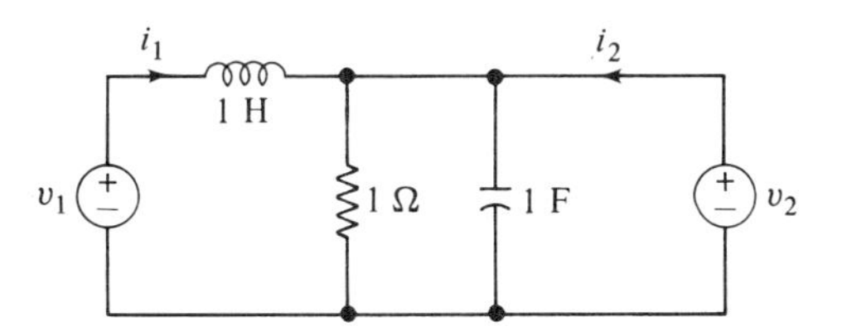

Figure P2.12

(a) Show that this network is passive.

(b) Now suppose that the time-invariant capacitor is replaced by a time-varying capacitor $C(t)$ with the requirement that $C(t) > 0$. Again assume that the network is relaxed at $-\infty$. Is it still true that the network is passive?

2.13 Show that the network shown in Fig. P2.9 is passive regardless of the form of $L(t)$ and $C(t)$.

2.14 Consider the two-port network shown in Fig. P2.14. Assume that the nonlinear resistor is current controlled with v-i relation

$$v_R = i_R^3$$

Determine whether this network is passive or active.

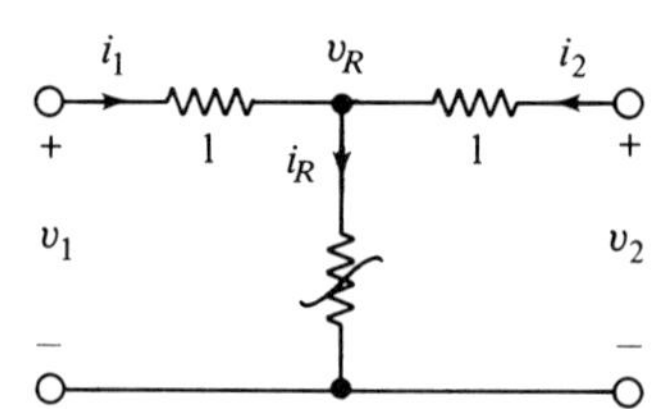

Figure P2.14

2.15 Repeat Problem 2.14, but this time assume that the v-i characteristic of the nonlinear resistor is

$$V_R = i_R - i_R^3$$

2.16 The input–output relation of a hypothetical network is

$$y(t) = u^2(t) + u(t + T), \qquad T > 0$$

Show that this network is anticipative and hence not physically realizable.

2.17 The v-i characteristic of a current-controlled resistor is

$$v = i - i^3$$

Show that this network is voltage input–current response noncausal.

2.18 Determine whether the network shown in Fig. P2.4 is stable or unstable.

2.19 The input–output relation of a network is

$$y(t) = e^{-(t-t_0)}y(t_0) + \int_{t_0}^{t} e^{-(t-\tau)}u(\tau)\,d\tau$$

Show that if $|y(t_0)|$ is finite and the L_1 norm of u is bounded then $|y(t)| < \infty$ for all t.

2.20 The input–output relation of a three-port resistive network is

$$\begin{bmatrix} v_1 \\ v_2 \\ v_3 \end{bmatrix} = \begin{bmatrix} -1 & -1 & 2 \\ -1 & 5 & 3 \\ 6 & -5 & -2 \end{bmatrix} \cdot \begin{bmatrix} i_1 \\ i_2 \\ i_3 \end{bmatrix}$$

Show that this network is stable in the L_∞ sense.

Network Graph Theory 3

3.1 NOTATIONS AND DEFINITIONS

In Chapter 2 we characterized a network in terms of its admissible signal pairs; throughout we assumed that the network is placed in a black box so that our only access to it was through its terminal pairs. In this chapter we assume that the *topology* or the inner structure of the network is also known and then implement some of the results of introductory graph theory to obtain general laws which apply to any network, regardless of its complexity or the nature of its elements. Among the basic results derived in this chapter are Kirchhoff's voltage and current laws; then, using these laws, we state and prove an extremely general result concerning the power delivered and absorbed by the network—Tellegen's theorem.

Let us first introduce some of the basic definitions most often used in network graph theory. Loosely speaking, the graph of a network is simply a diagram showing the interconnection of network elements; in this diagram every two-terminal network element is represented, regardless of its nature, by a line segment called a *branch*, and each end point of the element is denoted by a dot called a *node*. The collection of these branches and nodes is called the *graph* of the network under consideration. Before proceeding with the precise definition of some of the terms in graph theory, let us give a simple illustration. Consider the networks shown in Figs. 3.1-1(a) and (c). Let the arrows show the assumed direction of the current flow in the network; the corresponding *oriented* graphs are shown in Figs. 3.1-1(b) and (d). In Fig.

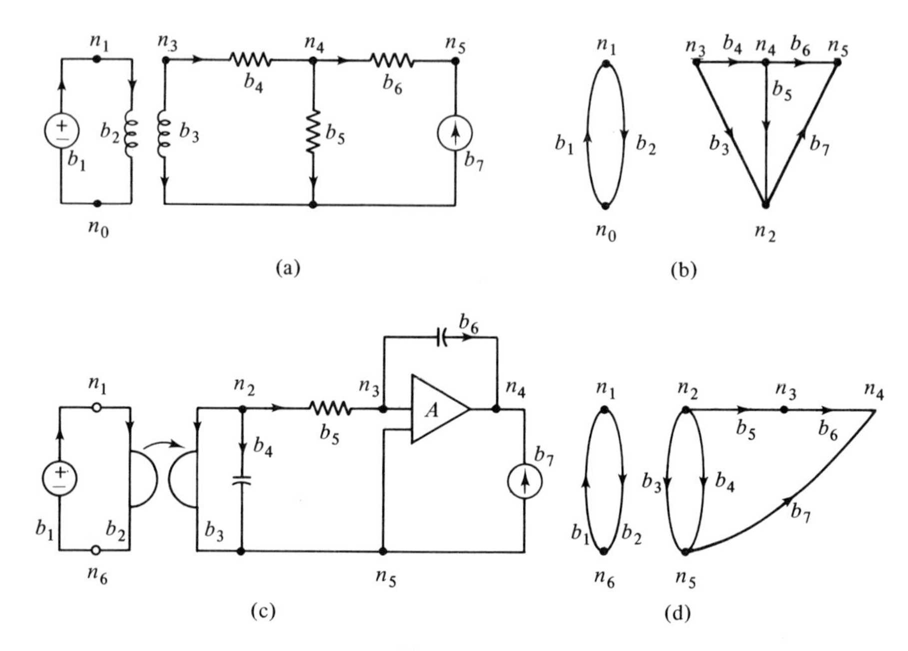

Figure 3.1-1

3.1-1(b) we say, for instance, that the branches b_4, b_5, and b_6 are *incident* to the node n_4 and the graph is comprised of two *disjoint* subgraphs. Note that the assumed directions of the current flow in the branches may or may not agree with the actual direction of the current flow. Let us now define some of the graph theory terms used in this and future chapters.

Definition 3.1-1 (Node). A *node*, n_i, is defined to be an end point of a line segment or an isolated point.

Definition 3.1-2 (Branch). A *branch*, b_k, is a line segment associated with two nodes n_i and n_j at its end points.

A branch sometimes is denoted by the pair $[n_i, n_j]_k$; this means that the branch b_k is incident to nodes n_i and n_j.

Definition 3.1-3 (Graph). A *graph*, G, is a collection of nodes and branches with the condition that branches intersect one another only at the nodes.

Denote the set of all branches in a graph by β, that is, $\beta \triangleq \{b_1, b_2, \ldots, b_B\}$, and the set of all the nodes by γ, that is, $\gamma \triangleq \{n_1, n_2, \ldots, n_{N+1}\}$, where B is the number of the branches and $N + 1$ is the number of the nodes in the graph; then the graph is sometimes denoted by $G(\gamma, \beta)$. If to each branch of a graph we assign a direction, the resulting graph is said to be *oriented* or *directed.* A graph $G_1(\gamma_1, \beta_1)$ is called a *subgraph* of $G(\gamma, \beta)$ if γ_1 is a subset of γ and β_1 is a subset of β; this implies that every branch of G_1 is a branch of G and every node of G_1 is a node of G. For example, consider the graph shown in Fig. 3.1-2; for this graph, $\gamma = \{n_1, n_2, n_3, n_4\}$ and $\beta = \{b_1, b_2, b_3, b_4, b_5, b_6\}$. A subgraph $G_1(\gamma_1, \beta_1)$ of this graph is, for instance, made up of $\gamma_1 = \{n_1, n_3, n_4\}$ and $\beta_1 = \{b_3, b_4, b_6\}$.

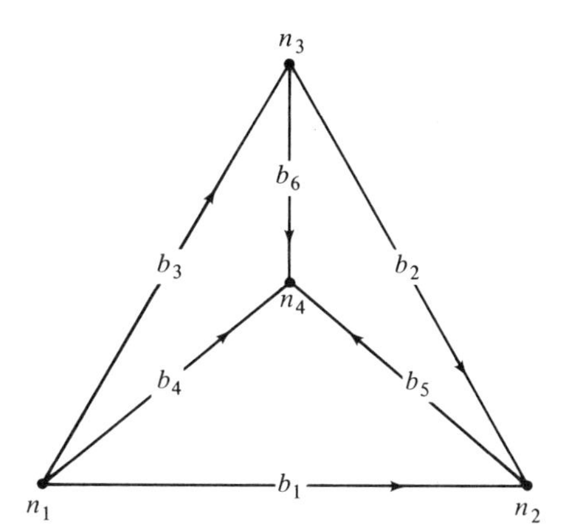

Figure 3.1-2 Connected graph with six branches and four nodes.

Definition 3.1-4 (Degree of a Node). The *degree* of a node is the number of the branches incident to it.

For example, in Fig. 3.1-2 all the nodes are of degree 3, and in Fig. 3.1-1(b) nodes n_3 and n_5 are of degree 2. A node of degree 2 is called a *simple* node and a node of degree 0 is called an *isolated* node.

Definition 3.1-5 (Path). A *path* of length m is a sequence of m *distinct* branches together with $m + 1$ *distinct* nodes such that every node is of degree 2 except the initial and terminal nodes, which are of degree 1.

In Fig. 3.1-2, branches (b_1, b_2, b_6) form a path; the corresponding nodes are n_1, n_2, n_3, n_4, where n_1 and n_4 are initial and terminal nodes.

Definition 3.1-6 (Loop). A *loop* of length m is a path such that its initial and terminal nodes coincide. A loop of length 1 is called a *self-loop*.

In Fig. 3.1-2, for example, branches (b_1, b_5, b_6, b_3) form a loop. To a given loop of a directed graph, we usually assign a *direction*—clockwise or counterclockwise.

Definition 3.1-7 (Connected Graph). A graph G is said to be *connected* if there is at least one path between any two of its nodes.

The graph shown in Fig. 3.1-2 is connected, whereas the graph shown in Fig. 3.1-1(b) is not.

Definition 3.1-8 (Tree). A *tree* T of a *connected graph* G is a connected subgraph of G with the following properties:

(a) T contains *all* the nodes of G.

(b) T does not contain any loop.

The branches of a tree are called *tree branches.* The collection of branches together with their corresponding nodes that are not in T form a subgraph of G, which is called the *cotree* of G corresponding to T; the branches of G that are not tree branches are called *links* or *chords.* In Fig. 3.1-2, for example, (b_3, b_6, b_2) form a tree; the corresponding cotree is made up of (b_1, b_4, b_5).

Definition 3.1-9 (Cutset). A *cutset* of a connected graph is a set of *minimum* number of branches that when deleted divide the graph into two separate subgraphs.

This implies that if any branch of the cutset is not deleted the graph remains connected. For example, in Fig. 3.1-2 branches (b_1, b_5, b_6, b_3) form a cutset, whereas the sets (b_1, b_5, b_6) and (b_2, b_5, b_2, b_3) are not cutsets, since the removal of the first set of branches does not divide the graph into two separate subgraphs, and the second set is not minimal.

Exercise 3.1-1. Find all the trees, loops, and cutsets of the graph shown in Fig. 3.1-2.

Exercise 3.1-2. Show that for a connected graph G there exists at least one tree. (Hint: If the given graph has any loops, remove a branch from each loop. This will leave the network connected and without loops.)

Exercise 3.1-3. Show that a tree T of a connected graph G with $N + 1$ nodes and B branches has exactly N branches and $B - N$ chords. (Hint: Use a mathematical induction to prove the first part, and apply the definition of a chord to establish the second part.)

Exercise 3.1-4. Show that a graph G is a tree if and only if there exists exactly one path between any two nodes of G [Hint: Use an indirect proof; if there is more than one path between any two nodes it forms a loop, and so on.]

Definition 3.1-10 (Fundamental Loop). A *fundamental loop* of a graph G with respect to a tree T is a loop that is made up of *one* chord and a unique set of branches of T.

Definition 3.1-11 (Fundamental Cutset). A *fundamental cutset* of a graph G with respect to a tree T is a cutset that is made up of *one* tree branch and a unique set of chords.

Example 3.1-1. Consider the graph shown in Fig. 3.1-3(a); a tree T of this graph is shown in Fig. 3.1-3(b) by solid lines. Fundamental loops corresponding to T are l_1, l_2, l_3, and l_4. The fundamental cutsets corresponding to T are c_1, c_2, c_3, c_4, and c_5.

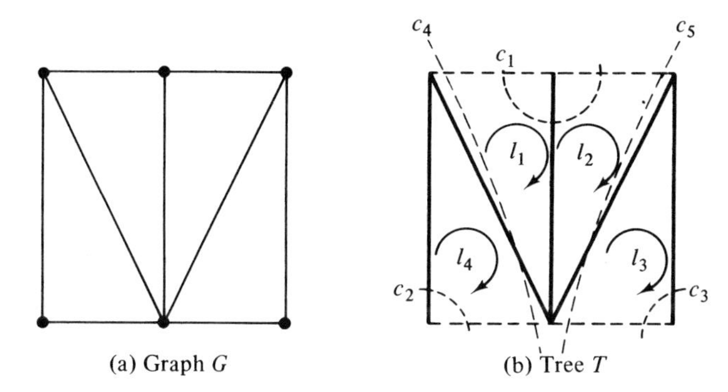

Figure 3.1-3 (*a*) Graph G and (*b*) the fundamental loops and cutsets corresponding to the tree T shown by solid line.

Next we state a useful result concerning the number of the fundamental cutsets and fundamental loops of a graph.

Theorem 3.1-1. Let G be a connected graph with $N + 1$ nodes and B branches. Then to each tree T of G there correspond N fundamental cutsets and $B - N$ fundamental loops.

Proof

According to Exercise 3.1-3, T has N branches and $B - N$ chords. Then, by Definitions 3.1-9 and 3.1-10, to each branch of T there corresponds a fundamental cutset and to each chord of T there corresponds a fundamental loop; hence, the conclusion of the theorem follows.

Example 3.1-2. Consider the graph shown in Fig. 3.1-4(a). There are 16 trees in this graph (show them!).

Consider a tree T of this graph [see Fig. 3.1-4(b)]. Each chord of this tree together with a unique set of tree branches forms a fundamental loop; for example, the chord b_4 together with branches b_1 and b_2 forms the fundamental loop l_2. Note that no other combination of the tree branches with b_4 forms a fundamental loop. Similarly, each tree branch with a unique set of chords forms a fundamental cutset; for instance, branch b_3 together with chords b_5 and b_6 forms the fundamental cutset c_3, and so on.

In the next two sections we use the basic definitions and results stated in this section to derive the matrix representation of a graph and a generalization of Kirchhoff's current and voltage laws.

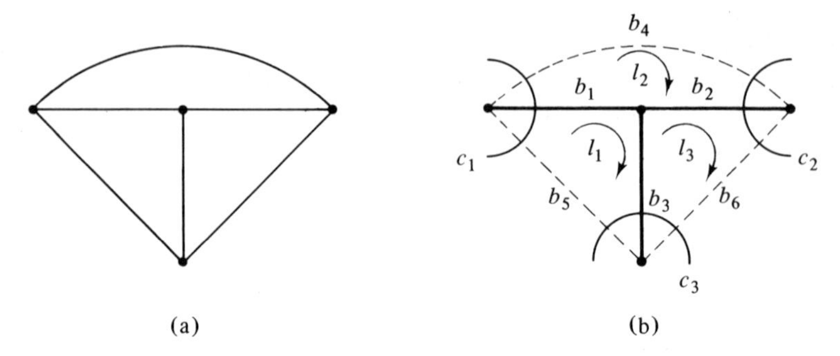

Figure 3.1-4 (a) Graph G; (b) fundamental cutsets and loops corresponding to a tree T of G, which is shown in solid lines.

3.2 INCIDENCE MATRIX AND KIRCHHOFF'S CURRENT LAW

In Section 3.1 we mentioned that the graph of a network is simply a diagram showing the topology and structure of the network. In this section we show how a graph can be represented by a matrix whose elements are $+1$, 0, or -1, and then proceed to give a matrix representation of Kirchhoff's current law. We show that all the information about a graph, such as its trees, loops, and cutsets, can be represented by appropriate matrices. This form of representation is extremely important if one has in mind the computer-aided analysis or design of networks, since it is much easier to store a matrix in a computer than to store a graph!

We start by defining the augmented incidence matrix. This matrix shows which branch is incident to which node. Suppose that the oriented graph under consideration consists of $N + 1$ nodes and B branches. Let us label these branches and nodes arbitrarily $b_1, b_2, \ldots, b_B$ and $n_1, n_2, \ldots, n_{N+1}$, respectively; then we have

Definition 3.2-1 (Augmented Incidence Matrix). The $(N + 1) \times B$ matrix $\mathbf{A}_a = (a_{kj})$ is said to be the *augmented incidence* matrix of a directed graph G with $N + 1$ nodes and B branches if

$$
\begin{aligned}
a_{kj} &= +1 \quad \text{when } b_j \text{ is incident to } n_k \text{ and is directed away from it} \\
&= -1 \quad \text{when } b_j \text{ is incident to } n_k \text{ and is directed toward it} \\
&= 0 \quad \text{when } b_j \text{ is not incident to } n_k
\end{aligned}
$$

As an illustration, consider the oriented graph shown in Fig. 3.2-1. The augmented incidence matrix is given by

$$
\mathbf{A}_a = \begin{array}{c} \\ n_1 \\ n_2 \\ n_3 \\ n_4 \\ n_5 \\ n_6 \end{array}
\begin{array}{c}
\begin{array}{ccccccccc} b_1 & b_2 & b_3 & b_4 & b_5 & b_6 & b_7 & b_8 & b_9 \end{array} \\
\begin{bmatrix}
-1 & 0 & 0 & 1 & 1 & 0 & 0 & 0 & 0 \\
1 & 1 & 0 & 0 & 0 & -1 & 0 & 0 & 0 \\
0 & -1 & -1 & 0 & 0 & 0 & -1 & -1 & 1 \\
0 & 0 & 1 & -1 & 0 & 0 & 0 & 0 & 0 \\
0 & 0 & 0 & 0 & -1 & 1 & 1 & 0 & 0 \\
0 & 0 & 0 & 0 & 0 & 0 & 0 & 1 & -1
\end{bmatrix}
\end{array}
\qquad (3.2\text{-}1)
$$

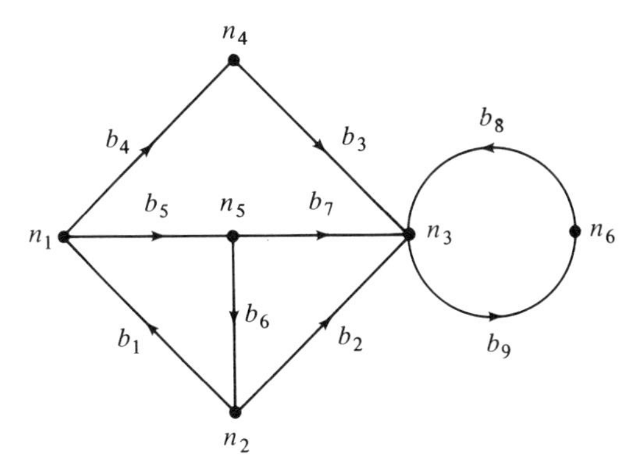

Figure 3.2-1

Remark 3.2-1. Since each branch of G is incident to exactly two nodes and since the arrow in each branch "leaves" one node and "enters" another, each column of $\mathbf{A}_a$ has exactly two nonzero elements, one of which is $+1$ and the other -1.

Notice that for *any* given graph we can obtain an incidence matrix using Definition 3.2-1. Conversely, given any incidence matrix we can draw its corresponding graph; simply indicate $N + 1$ nodes on a paper and connect these nodes according to Definition 3.2-1. Before we state a general theorem concerning the rank of $\mathbf{A}_a$, let us consider the augmented incidence matrix given in (3.2-1). If we add all the rows of this matrix to the last row, we obtain a row of zeros; this means that the rows of $\mathbf{A}_a$ are not linearly independent. Hence, the rank of $\mathbf{A}_a$ is less than $N + 1$. This is a consequence of the following general result.

Theorem 3.2-1. The rank of the augmented incidence matrix $\mathbf{A}_a$ of a connected graph G with $N + 1$ nodes and B branches is N.

Since $\mathbf{A}_a$ has $N + 1$ rows, this theorem implies that only N of these rows are linearly independent. Since we are primarily interested in the application of this theorem we omit its proof. A rigorous proof can be found in [1].

Aside from representing the topology of a lumped network, the augmented incidence matrix can be used to represent Kirchhoff's current law (KCL) in matrix form. Before discussing this topic in detail, let us state Kirchhoff's current law as it applies to the nodes of lumped networks.

Kirchhoff's Current Law. For any lumped electrical network the algebraic sum of all the currents entering or leaving a node is equal to zero.

This law is true for any lumped network and at any time; it does not depend on the nature of the elements—it only depends on the topology of the network. To put this law in analytical form, consider a network with B branches and $N + 1$ nodes. Assign an arbitrary direction to the currents in each branch, and designate this direction by an arrow. Furthermore, assign a *negative* sign to the currents whose direction points *toward* a node and a *positive* sign to the currents whose direction points *away* from a node. Now, denoting the current in branch b_j by i_j and applying KCL at node k, we get

$$\sum_{j=1}^{B} a_{kj} i_j = 0 \quad \text{for } k = 1, 2, \ldots, N + 1 \tag{3.2-2}$$

where a_{kj} is the same as in Definition 3.2-1. For example, applying (3.2-2) to nodes n_1 and n_2 of the network whose graph is given in Fig. 3.2-1, we get

$$n_1: \quad -i_1 + i_4 + i_5 = 0$$

$$n_2: \quad i_1 + i_2 - i_6 = 0$$

The KCL equations for the remaining nodes can be written in a similar manner. Now, if (3.2-2) is written in matrix form it reads

$$\mathbf{A}_a \mathbf{i}_b = \mathbf{0} \tag{3.2-3}$$

where the right-hand side of the equation is an $N + 1$ column vector whose elements are all zero, and

$$\mathbf{A}_a = (a_{kj}) \tag{3.2-4}$$

in which a_{kj} is the same as in Definition 3.2-1; $\mathbf{i}_b$ is a B-column vector given by

$$\mathbf{i}_b = [i_1, i_2, \ldots, i_B]^T \tag{3.2-5}$$

Equation (3.2-3) represents KCL for a network whose augmented incidence matrix $\mathbf{A}_a$ is given by (3.2-4). This equation holds for any lumped network whether or not its elements are linear, nonlinear, time varying, or time invariant.

Example 3.2-1. Consider the network whose graph is given in Fig. 3.2-1; then the matrix representation of KCL for this graph is given by

$$\begin{bmatrix} -1 & 0 & 0 & 1 & 1 & 0 & 0 & 0 & 0 \\ 1 & 1 & 0 & 0 & 0 & -1 & 0 & 0 & 0 \\ 0 & -1 & -1 & 0 & 0 & 0 & -1 & -1 & 1 \\ 0 & 0 & 1 & -1 & 0 & 0 & 0 & 0 & 0 \\ 0 & 0 & 0 & 0 & -1 & 1 & 1 & 0 & 0 \\ 0 & 0 & 0 & 0 & 0 & 0 & 0 & 1 & -1 \end{bmatrix} \begin{bmatrix} i_1 \\ i_2 \\ i_3 \\ i_4 \\ i_5 \\ i_6 \\ i_7 \\ i_8 \\ i_9 \end{bmatrix} = \begin{bmatrix} 0 \\ 0 \\ 0 \\ 0 \\ 0 \\ 0 \end{bmatrix} \tag{3.2-6}$$

According to Theorem 3.2-1, the rank of the augmented incidence matrix $\mathbf{A}_a$ is N; that is, among $N + 1$ equations represented by (3.2-3), only N equations are linearly independent. If, for example, we choose the first N equations in (3.2-6), the last equation does not convey any new information and hence can be deleted. Indeed, if we delete *any* one of the $N + 1$ equations in (3.2-3), the remaining N equations are linearly independent node equations. Therefore, (3.2-3) can be written in the reduced form

$$\mathbf{A}\mathbf{i}_b = \mathbf{0} \tag{3.2-7}$$

where $\mathbf{A}$ is the $N \times B$ incidence matrix defined next:

Definition 3.2-2 (Incidence Matrix). The *incidence* matrix $\mathbf{A}$ of a graph G with $N + 1$ nodes and B branches is an $N \times B$ matrix obtained from the augmented incidence matrix of G, $\mathbf{A}_a$, by deleting any one of its rows.

The node corresponding to the deleted row is called the *reference node* or *datum*.

Example 3.2-2. Consider the network of Fig. 3.2-2(a) and its corresponding oriented graph shown in Fig. 3.2-2(b). The augmented incidence matrix for this network is simply

$$\mathbf{A}_a = \begin{array}{c} \\ n_1 \\ n_2 \\ n_3 \\ n_4 \end{array} \begin{array}{c} \begin{matrix} b_1 & b_2 & b_3 & b_4 & b_5 \end{matrix} \\ \begin{bmatrix} -1 & 1 & 0 & 0 & 0 \\ 0 & -1 & -1 & 1 & 0 \\ 0 & 0 & 0 & -1 & -1 \\ 1 & 0 & 1 & 0 & 1 \end{bmatrix} \end{array}$$

If we now let n_4 be the datum, the corresponding incidence matrix can be obtained from $\mathbf{A}_a$ by deleting the last row:

$$\mathbf{A} = \begin{bmatrix} -1 & 1 & 0 & 0 & 0 \\ 0 & -1 & -1 & 1 & 0 \\ 0 & 0 & 0 & -1 & -1 \end{bmatrix}$$

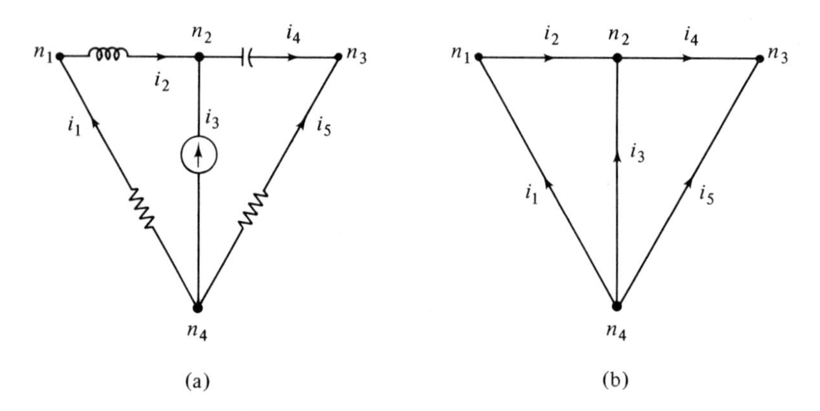

Figure 3.2-2

Then the linearly independent node equations of the network shown in Fig. 3.2-2, written in matrix form, are

$$\begin{bmatrix} -1 & 1 & 0 & 0 & 0 \\ 0 & -1 & -1 & 1 & 0 \\ 0 & 0 & 0 & -1 & -1 \end{bmatrix} \begin{bmatrix} i_1 \\ i_2 \\ i_3 \\ i_4 \\ i_5 \end{bmatrix} = \begin{bmatrix} 0 \\ 0 \\ 0 \end{bmatrix} \tag{3.2-8}$$

It should be mentioned that the application of KCL is not limited to the nodes of a lumped network. In fact, we can show that it applies to any cutset of a network. Before discussing this generalized application of KCL, let us define the augmented cutset matrix of a directed graph. Consider the graph of Example 3.2-2, redrawn in Fig. 3.2-3. Let us denote the cutsets of this graph by broken lines intersecting the branches of the graph that are in the corresponding cutset; in Fig. 3.2-3 we can designate six such cutsets—c_1 through c_6. Each cutset separates the graph into exactly two disjoint subgraphs. To assign an orientation to each cutset, choose an arbitrary direction for the net current flowing from one subgraph into another and denote this orientation by an arrow on the broken line representing the cutset. To state KCL for the cutsets of a graph, assign a *positive* sign to the currents in a branch whose direction is the same as the orientation of the cutset and a *negative* sign to the currents with opposite direction; then KCL for the cutsets of a network can be stated as follows:

For any lumped electrical network the algebraic sum of all the currents entering or leaving a cutset is equal to zero.

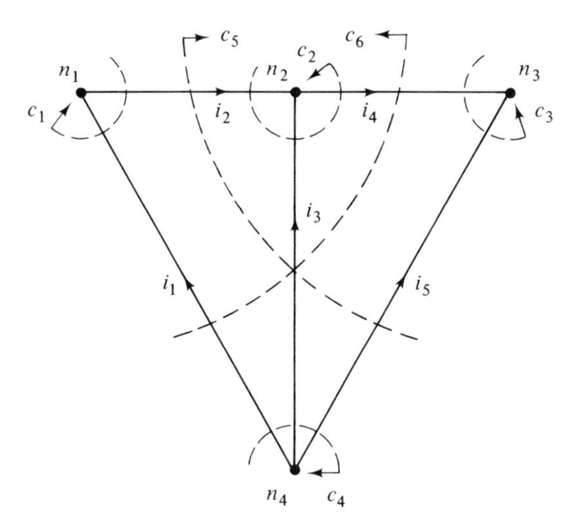

Figure 3.2-3 Cutsets of the graph of Fig. 3.2-2.

This law applied to the cutsets of the graph of Fig. 3.2-3 yields

$$
\begin{aligned}
c_1: &\quad i_1 - i_2 = 0 \\
c_2: &\quad i_2 + i_3 - i_4 = 0 \\
c_3: &\quad i_4 + i_5 = 0 \\
c_4: &\quad -i_1 - i_3 - i_5 = 0 \\
c_5: &\quad i_2 + i_3 + i_5 = 0 \\
c_6: &\quad i_1 + i_3 - i_4 = 0
\end{aligned}
\tag{3.2-9}
$$

These equations put in matrix form become

$$
\begin{array}{c}
\\ c_1 \\ c_2 \\ c_3 \\ c_4 \\ c_5 \\ c_6
\end{array}
\begin{array}{c}
\begin{array}{ccccc} b_1 & b_2 & b_3 & b_4 & b_5 \end{array} \\
\begin{bmatrix}
1 & -1 & 0 & 0 & 0 \\
1 & 0 & 1 & -1 & 0 \\
0 & 0 & 0 & 1 & 1 \\
-1 & 0 & -1 & 0 & -1 \\
0 & 1 & 1 & 0 & 1 \\
1 & 0 & 1 & -1 & 0
\end{bmatrix}
\end{array}
\begin{bmatrix} i_1 \\ i_2 \\ i_3 \\ i_4 \\ i_5 \end{bmatrix}
=
\begin{bmatrix} 0 \\ 0 \\ 0 \\ 0 \\ 0 \\ 0 \end{bmatrix}
\tag{3.2-10}
$$

or, in a more compact form,

$$\mathbf{Q}_a \mathbf{i}_b = \mathbf{0} \tag{3.2-11}$$

where $\mathbf{Q}_a$ is called the *augmented cutset* matrix of the graph under consideration; equations (3.2-11), therefore, are called cutset equations of the network. A quick look at (3.2-9) shows that only three equations are linearly independent; the last three equations can be obtained by linear combinations of the first three. In general, it can be shown that the rank of the augmented cutset matrix of a graph with $N+1$ nodes is exactly N. In contrast to the node equations, it is not generally possible to pick N linearly independent cutset equations arbitrarily. To obtain a set of linearly independent cutset equations, we have to make use of the notion of fundamental cutsets (see Definition 3.1-10) and the fundamental cutset matrix. This can be best illustrated by a specific example.

Example 3.2-3. Consider the graph shown in Fig. 3.2-4; a tree T of this graph is shown with solid lines and the corresponding chords are shown by broken lines. Since T has five branches, there are five fundamental cutsets; branch b_3, for example, together with chords b_7 and b_9 form the fundamental cutset c_3; the remaining fundamental cutsets are denoted by c_1, c_2, c_4, and c_5.

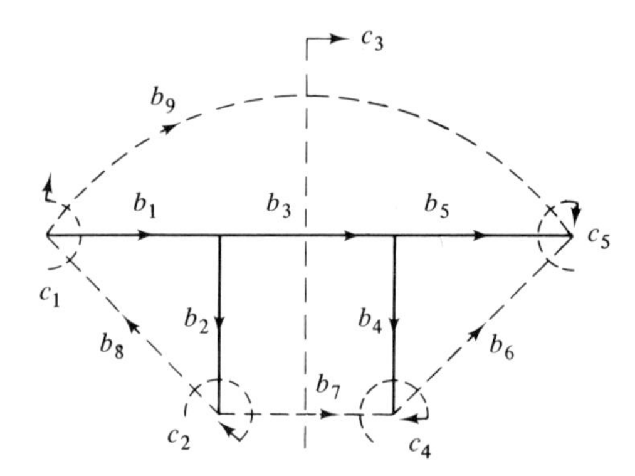

Figure 3.2-4 Fundamental cutsets of the graph G.

The basic property of the fundamental cutsets is that they yield linearly independent KCL equations. Since there are N fundamental cutsets in a graph with $N+1$ nodes, to obtain *all* the linearly independent cutset equations, it is sufficient to apply KCL to the fundamental cutsets only. For example, applying KCL to the fundamental cutsets of the graph shown in Fig. 3.2-4 and putting the results in matrix form, we get

$$\begin{array}{c} \\ c_1 \\ c_2 \\ c_3 \\ c_4 \\ c_5 \end{array}\!\!\begin{array}{c} \begin{array}{ccccccccc} b_1 & b_2 & b_3 & b_4 & b_5 & b_6 & b_7 & b_8 & b_9 \end{array} \\ \begin{bmatrix} 1 & 0 & 0 & 0 & 0 & 0 & 0 & -1 & 1 \\ 0 & 1 & 0 & 0 & 0 & 0 & -1 & -1 & 0 \\ 0 & 0 & 1 & 0 & 0 & 0 & 1 & 0 & 1 \\ 0 & 0 & 0 & 1 & 0 & -1 & 1 & 0 & 0 \\ 0 & 0 & 0 & 0 & 1 & 1 & 0 & 0 & 1 \end{bmatrix} \end{array} \begin{bmatrix} i_1 \\ i_2 \\ i_3 \\ i_4 \\ i_5 \\ i_6 \\ i_7 \\ i_8 \\ i_9 \end{bmatrix} = \begin{bmatrix} 0 \\ 0 \\ 0 \\ 0 \\ 0 \end{bmatrix} \qquad (3.2\text{-}12)$$

or, equivalently,

$$\mathbf{Q}_f \mathbf{i}_b = \mathbf{0} \tag{3.2-13}$$

where $\mathbf{Q}_f$ denotes the 5×9 matrix given in (3.2-12). To formalize the procedure for obtaining the fundamental cutset matrix, we first label the branches of the corresponding graph in the following manner:

Consider a graph G with $N + 1$ nodes and B branches. Choose a tree T of G and label the branches of T by b_1 through b_N (T has exactly N branches; see Exercise 3.1-3) and label the remaining branches (i.e., chords of G with respect to T) b_{N+1} through b_B. Also, denote the fundamental cutsets corresponding to tree branches $b_1, b_2, \ldots, b_N$ by $c_1, c_2, \ldots, c_N$, respectively, and let the orientation of c_k be the same as the tree branch b_k for $k = 1, 2, \ldots, N$. Then we have

Definition 3.2-3 (Fundamental Cutset Matrix). The *fundamental cutset* matrix $\mathbf{Q}_f$ of a graph G with $N + 1$ nodes and B branches corresponding to a tree T is an $N \times B$ matrix

$$\mathbf{Q}_f = (q_{kj}) \tag{3.2-14}$$

where

$$\begin{aligned} q_{kj} &= 1 && \text{when } b_j \text{ is in } c_k \text{ and has the same orientation} \\ &= -1 && \text{when } b_j \text{ is in } c_k \text{ and has the opposite orientation} \\ &= 0 && \text{when } b_j \text{ is not in } c_k \end{aligned}$$

Let us now state and prove an important theorem concerning the rank of the fundamental cutset matrix just defined.

Theorem 3.2-2. The rank of the fundamental cutset matrix of a graph with $N + 1$ nodes is N.

Proof

According to the procedure just outlined, since b_1 through b_N are tree branches and since the orientation of the cutsets is the same as the orientation of the tree branches, the fundamental cutset matrix $\mathbf{Q}_f$ can be partitioned in the form

$$\mathbf{Q}_f = [\mathbf{1} \,\vdots\, \mathbf{F}] \tag{3.2-15}$$

where $\mathbf{1}$ is an $N \times N$ identity matrix corresponding to the tree branches, and F is an $N \times (B - N)$ matrix that corresponds to the chords of T. From (3.2-15) it is clear that $\mathbf{Q}_f$ is of rank N, since it has an $N \times N$ nonsingular submatrix $\mathbf{1}$.

An illustration of the partitioning of $\mathbf{Q}_f$ is given in (3.2-12); the 5×5 identity matrix corresponds to the tree branches b_1 through b_5.

3.3 LOOP MATRIX AND KIRCHHOFF'S VOLTAGE LAW

In this section we show that for a network with $N + 1$ nodes and B branches another set of $B - N$ linearly independent equations can be obtained by considering Kirchhoff's voltage law (KVL). We start by introducing the augmented loop matrix, which represents all the loops in a graph, and then proceed to obtain a set of linearly independent loop equations.

Consider an oriented graph G with B branches and L loops; let us label the branches arbitrarily $b_1, b_2, \ldots, b_B$ and the loops $l_1, l_2, \ldots, l_L$. Let us also assign arbitrary orientation (clockwise or counterclockwise) to loops l_1 through l_L; then

Definition 3.3-1 (Augmented Loop Matrix). An $L \times B$ matrix $\mathbf{B}_a = (b_{ij})$ is said to be the *augmented loop matrix* of an oriented graph G with B branches and L loops if

$$
\begin{aligned}
b_{kj} &= 1 && \text{when branch } b_j \text{ is in loop } l_k \text{ and has the same orientation} \\
&= -1 && \text{when branch } b_j \text{ is in loop } l_k \text{ and has the opposite orientation} \\
&= 0 && \text{when branch } b_j \text{ is not in loop } l_k
\end{aligned}
$$

Example 3.3-1. Consider the oriented graph G shown in Fig. 3.3-1; in this graph $N + 1 = 5$, $B = 7$, and $L = 4$. Label the branches b_1 through b_7 and the loops l_1 through l_4; choose arbitrary directions for the loops and indicate them by arrows as shown. The augmented loop matrix is given in equation (3.3-1).

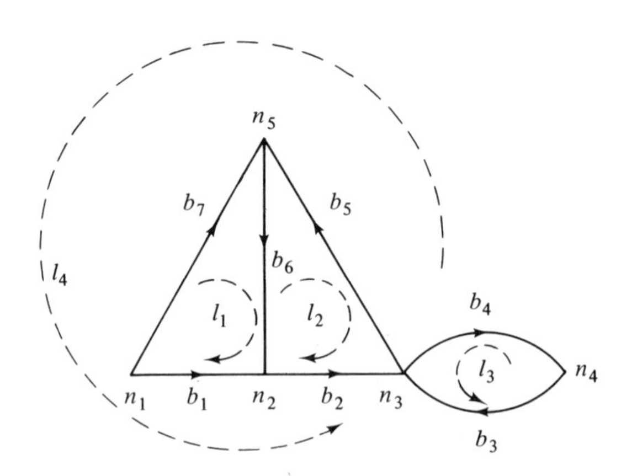

Figure 3.3-1

$$\mathbf{B}_a = \begin{array}{c} \\ l_1 \\ l_2 \\ l_3 \\ l_4 \end{array} \begin{array}{c} \begin{array}{ccccccc} b_1 & b_2 & b_3 & b_4 & b_5 & b_6 & b_7 \end{array} \\ \begin{bmatrix} -1 & 0 & 0 & 0 & 0 & 1 & 1 \\ 0 & -1 & 0 & 0 & -1 & -1 & 0 \\ 0 & 0 & -1 & -1 & 0 & 0 & 0 \\ 1 & 1 & 0 & 0 & 1 & 0 & -1 \end{bmatrix} \end{array} \tag{3.3-1}$$

By simple elementary row and column operations, we can show that the rank of $\mathbf{B}_a$ given in (3.3-1) is exactly 3. In general, we can state the following theorem concerning the rank of the augmented loop matrix of a graph. A proof of this theorem can be found in [2].

Theorem 3.3-1. The rank of the augmented loop matrix of a graph G with $N + 1$ nodes and B branches is exactly $B - N$.

The augmented loop matrix of a graph can also be used to represent KVL applied to the loops of the graph under consideration. To show this, we must first give the precise statement of KVL. Consider a lumped network with $N + 1$ nodes and B branches. Let us label the branches b_1 through b_B and use the sign convention given in Fig. 1.1-2 and equation (1.1-10) to assign the polarity of each branch. Denote the *voltage drop* across branch b_j by v_j, and indicate the direction of the voltage drop by an arrow. Let us also label the loops l_1 through l_L and assign arbitrary directions (clockwise or counterclockwise) to each loop. Then if the direction of the voltage drop across a branch is the same as the direction of the loop, assign a positive sign to the voltage drop; otherwise, assign a negative sign. With this sign convention we can state the following.

Kirchhoff's Voltage Law (KVL). For any lumped electrical network the algebraic sum of the voltage drops in any loop is equal to zero.

Kirchhoff's voltage law is completely general and applies to any lumped circuit regardless of the nature of its elements. Putting KVL in analytic form, it reads

$$\sum_{j=1}^{B_k} b_{kj} v_j = 0, \qquad k = 1, 2, \ldots, L \tag{3.3-2}$$

where b_{kj} is the same as in Definition 3.3-1 and B_k is the number of branches in loop l_k.

Example 3.3-2. Consider the network shown in Fig. 3.3-2; for this network $B = 5$, $N + 1 = 4$, and $L = 3$. Then from (3.3-2) we can write

$$\begin{aligned} l_1: &\quad v_1 - v_4 - v_5 = 0 \\ l_2: &\quad v_2 + v_3 - v_4 = 0 \\ l_3: &\quad v_1 - v_2 - v_3 - v_5 = 0 \end{aligned} \tag{3.3-3}$$

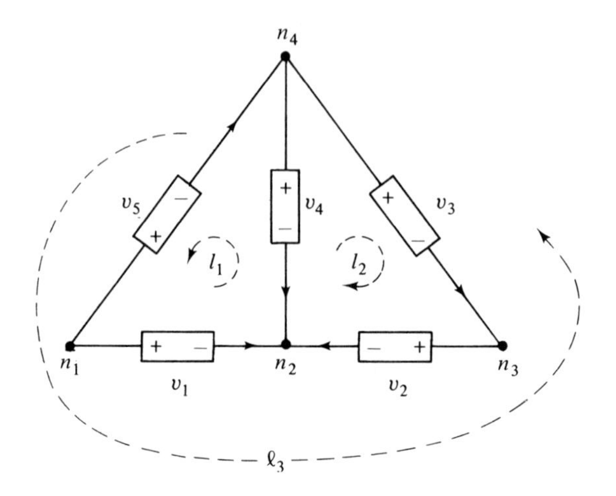

Figure 3.3-2

or, equivalently,

$$\begin{bmatrix} 1 & 0 & 0 & -1 & -1 \\ 0 & 1 & 1 & -1 & 0 \\ 1 & -1 & -1 & 0 & -1 \end{bmatrix} \begin{bmatrix} v_1 \\ v_2 \\ v_3 \\ v_4 \\ v_5 \end{bmatrix} = \begin{bmatrix} 0 \\ 0 \\ 0 \end{bmatrix} \tag{3.3-4}$$

If we write (3.3-2) in matrix form, we get

$$\mathbf{B}_a \mathbf{v}_b = \mathbf{0} \tag{3.3-5}$$

where

$$\mathbf{B}_a = (b_{kj}) \tag{3.3-6}$$

and b_{kj} is the same as in Definition 3.3-1, and $\mathbf{v}_b$ is a B-column vector defined by

$$\mathbf{v}_b = [v_1, v_2, \dots, v_B]^T \tag{3.3-7}$$

According to Theorem 3.3-1, the rank of $\mathbf{B}_a$ is $B - N$; hence, in equation (3.3-4) the rank of the matrix in the left-hand side is $5 - 3 = 2$. This can be verified by observing that the last row can be obtained by subtracting the second row from the first. From the conclusion of Theorem 3.3-1 it is therefore clear that among all the loop equations of a network only $B - N$ are linearly independent.

To give a systematic procedure for obtaining linearly independent loop equations, we must make use of the fundamental loops defined in Definition 3.1-9. In this section we show that to each set of fundamental loops there corresponds a fundamental loop matrix with $B - N$ linearly independent rows. We first show this for a simple example and then proceed with the general case.

Example 3.3-3. Consider the lumped network whose graph is given in Fig. 3.3-3. Choose a tree T of this graph (shown in Fig. 3.3-3 by solid lines) and indicate the direction of the voltage drop in each branch by an arrow.

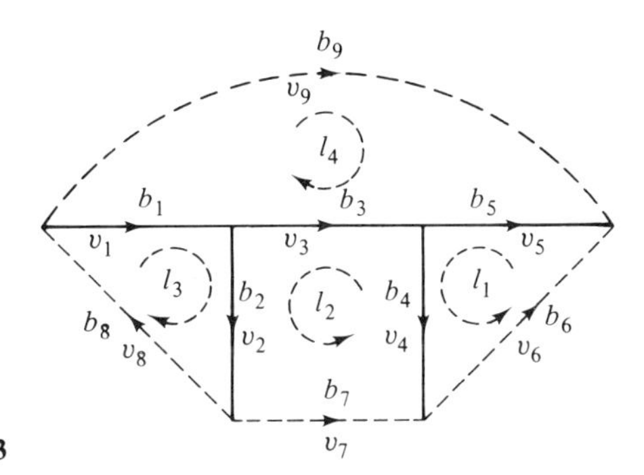

Figure 3.3-3

According to Definition 3.1-9, to each chord of T there corresponds a fundamental loop; denote these fundamental loops by l_1, l_2, l_3, and l_4, and apply KVL to them. The resulting equations, put in matrix form, will be

$$\begin{array}{c} \\ l_1 \\ l_2 \\ l_3 \\ l_4 \end{array} \begin{array}{c} \begin{array}{ccccccccc} b_1 & b_2 & b_3 & b_4 & b_5 & b_6 & b_7 & b_8 & b_9 \end{array} \\ \begin{bmatrix} 0 & 0 & 0 & 1 & -1 & 1 & 0 & 0 & 0 \\ 0 & 1 & -1 & -1 & 0 & 0 & 1 & 0 & 0 \\ 1 & 1 & 0 & 0 & 0 & 0 & 0 & 1 & 0 \\ -1 & 0 & -1 & 0 & -1 & 0 & 0 & 0 & 1 \end{bmatrix} \end{array} \begin{bmatrix} v_1 \\ v_2 \\ v_3 \\ v_4 \\ v_5 \\ v_6 \\ v_7 \\ v_8 \\ v_9 \end{bmatrix} = \begin{bmatrix} 0 \\ 0 \\ 0 \\ 0 \end{bmatrix} \tag{3.3-8}$$

The loop matrix in this equation is clearly of rank 4, since it has a 4×4 submatrix with nonzero determinant (the identity submatrix). To give a systematic procedure for writing $B - N$ linearly independent loop equations for a graph G with $N + 1$ nodes and B branches, we first define the fundamental loop matrix corresponding to a tree T of G.

Consider an oriented graph G with $N + 1$ nodes and B branches. Choose a tree T of G and label its branches $b_1, b_2, \ldots, b_N$ (according to Exercise 3.1-3, T has exactly N branches) and the chords by $b_{N+1}, \ldots, b_B$. By Definition 3.1-9, to each chord of T there corresponds a fundamental loop; denote the fundamental loop corresponding to b_{N+k} by l_k, $k = 1, 2, \ldots, B - N$, and choose the orientation of l_k to be the same as that of b_{N+k}. Then we have

Definition 3.3-2 (Fundamental Loop Matrix). The *fundamental loop matrix* $\mathbf{B}_f$ of an oriented graph G with $N + 1$ nodes and B branches corresponding to a tree T is a $(B - N) \times B$ matrix

$$\mathbf{B}_f = (b_{kj}) \tag{3.3-9}$$

where

$$\begin{aligned} b_{kj} &= 1 && \text{when } b_j \text{ is in the fundamental loop } l_k \text{ and has the same orientation} \\ &= -1 && \text{when } b_j \text{ is in the fundamental loop } l_k \text{ and has opposite orientation} \\ &= 0 && \text{when } b_j \text{ is not in the fundamental loop } l_k \end{aligned}$$

According to the procedure just discussed, the resulting fundamental loop matrix can be partitioned as

$$\mathbf{B}_f = [-\mathbf{F}^T \vdots \mathbf{1}] \tag{3.3-10}$$

where $\mathbf{1}$ is a $(B - N) \times (B - N)$ identity matrix, which corresponds to the chords of T, and $\mathbf{F}$ is the same matrix as in (3.2-15).

Exercise 3.3-1. Find the fundamental cutset matrix of the graph shown in Fig. 3.3-3 with respect to the tree shown in solid lines and verify (3.3-10). Compare your result with the 4×9 matrix given in (3.3-8).

The following theorem gives the rank of the fundamental loop matrix of any connected graph.

Theorem 3.3-2. The rank of the fundamental loop matrix of a connected graph with $N + 1$ nodes and B branches is $B - N$.

Proof

It follows immediately from (3.3-10).

From the conclusion of this theorem and equation (3.3-5), we can now give an algorithm for obtaining all the linearly independent loop equations of a network by first obtaining the fundamental loop matrix. Consider a connected network $\mathfrak{N}$ with $N + 1$ nodes and B branches.

Step 1. Draw the graph of $\mathfrak{N}$ and choose a tree T; assign arbitrary orientation to all branches of this graph.

Step 2. Designate the fundamental loops corresponding to T and assign arbitrary orientation to them.

Step 3. Apply KVL to each fundamental loop and put the resulting equations in matrix form:

$$\mathbf{B}_f \mathbf{v}_b = \mathbf{0} \tag{3.3-11}$$

where B_f is given in Definition 3.3-2 and $\mathbf{v}_b$ is the B-column vector given in (3.3-7). Example 3.3-3 is an illustration of this procedure.

It should be mentioned that it is not necessary to label the branches or the chords of the chosen tree in any specific order. The particular numbering of the chords developed previously was for the purpose of putting B_f in the form (3.3-10) to facilitate the proof of Theorem 3.3-2. Furthermore, the particular representation of the fundamental loop matrix given in (3.3-10) and the fundamental cutset matrix given in (3.2-15) are extremely useful in the systematic derivations of the state equations, which we discuss in Chapter 5. Let us now work out another example of this procedure.

Example 3.3-4. Consider the network whose graph is given in Fig. 3.3-4. In this graph, $N + 1 = 6$ and $B = 9$; hence, the rank of the fundamental

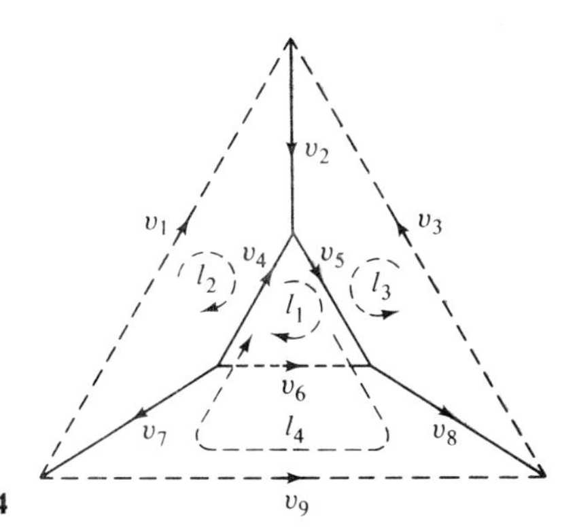

Figure 3.3-4

loop matrix is $B - N = 9 - 5 = 4$. A tree T of G is shown in solid lines and its corresponding fundamental loops are denoted by l_1, l_2, l_3, and l_4. Let the branch voltage vector be given by

$$\mathbf{v}_b = [v_1, v_2, \ldots, v_9]^T \tag{3.3-12}$$

Then applying KVL to the fundamental loops and putting the result in matrix form, we get

$$\mathbf{B}_f\mathbf{v}_b = \mathbf{0} \tag{3.3-13}$$

where $\mathbf{B}_f$ is given by

$$\mathbf{B}_f = \begin{bmatrix} 0 & 0 & 0 & 1 & 1 & -1 & 0 & 0 & 0 \\ 1 & 1 & 0 & -1 & 0 & 0 & 1 & 0 & 0 \\ 0 & 1 & 1 & 0 & 1 & 0 & 0 & 1 & 0 \\ 0 & 0 & 0 & 1 & 1 & 0 & -1 & 1 & -1 \end{bmatrix}$$

Mesh Equations. Next we outline an alternative method of writing a set of linearly independent KVL equations for a special class of networks, planar networks. We first define planar networks and then proceed with the description of the method.

Definition 3.3-3 (Planar Network). A network is said to be *planar* if its graph can be drawn on a plane in such a way that no two branches intersect except at the nodes.

A network that is not planar is called *nonplanar*. All the graphs considered so far in this chapter are planar; two examples of nonplanar graphs are given in Fig. 3.3-5. These two graphs are called basic Kuratowski nonplanar graphs. In a planar graph we can consider a particular form of loops called *meshes* or *windows*; more precisely,

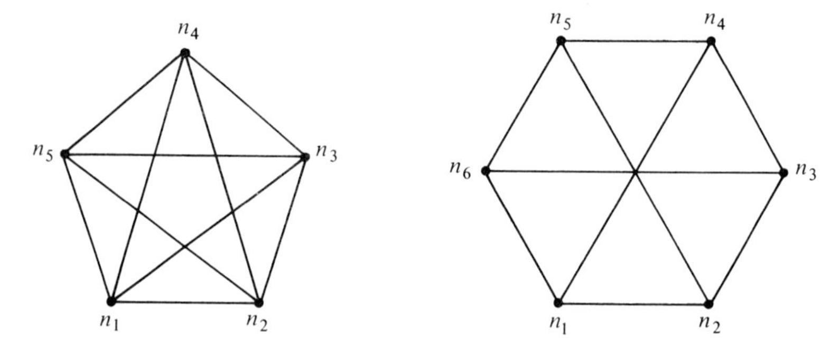

Figure 3.3-5 Basic Kuratowski nonplanar graphs.

Definition 3.3-4 (Mesh). A loop of a connected planar graph is called a *mesh* if it does not contain any branches in its interior.

For example, the planar graph shown in Fig. 3.3-6 has three meshes, m_1, m_2, and m_3; the graph shown in Fig. 3.3-7(a) has four meshes since it can be redrawn as in Fig. 3.3-7(b).

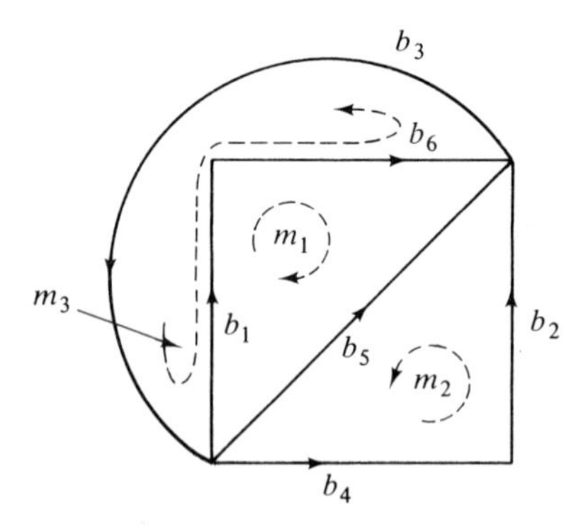

Figure 3.3-6

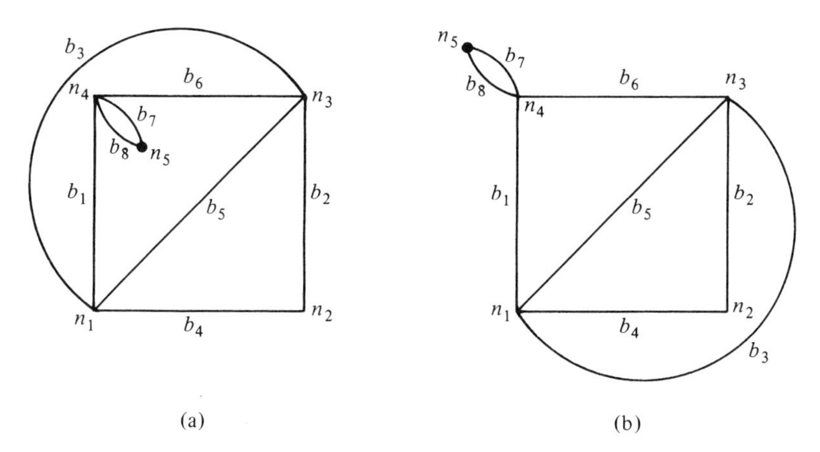

Figure 3.3-7

Next we state an important theorem concerning the number of the meshes in a planar network. A proof of this theorem can be constructed by mathematical induction.

Theorem 3.3-3. A connected planar network with $N + 1$ nodes and B branches has exactly $B - N$ meshes.

Exercise 3.3-2. Use a mathematical induction to prove Theorem 3.3-3.

To illustrate how meshes can be used to obtain a set of $B - N$ linearly independent KVL equations, consider the graph shown in Fig. 3.3-6. Let the voltage drop in branch b_k be denoted by v_k, and let its polarity be indicated by the corresponding arrows. Then applying KVL to m_1, m_2, and m_3 and putting the result in matrix form, we get

$$\begin{array}{c} \\ m_1 \\ m_2 \\ m_3 \end{array}\begin{array}{c} \begin{array}{cccccc} b_1 & b_2 & b_3 & b_4 & b_5 & b_6 \end{array} \\ \begin{bmatrix} 1 & 0 & 0 & 0 & -1 & 1 \\ 0 & 1 & 0 & 1 & -1 & 0 \\ 1 & 0 & 1 & 0 & 0 & 1 \end{bmatrix} \end{array} \begin{bmatrix} v_1 \\ v_2 \\ v_3 \\ v_4 \\ v_5 \\ v_6 \end{bmatrix} = \begin{bmatrix} 0 \\ 0 \\ 0 \end{bmatrix} \tag{3.3-14}$$

or, in a more compact form,

$$\mathbf{M}\mathbf{v}_b = \mathbf{0} \tag{3.3-15}$$

where $\mathbf{M}$ is called the *mesh matrix* of the graph under consideration. By subtracting the first row of $\mathbf{M}$ from its third row, we can see that the rank of $\mathbf{M}$ given in (3.3-14) is equal to 3. Let us now give a general procedure for writing $B - N$ linearly independent KVL equations for a planar network by defining the mesh matrix $\mathbf{M}$.

Consider a connected planar graph G with $N + 1$ nodes and B branches. According to Theorem 3.3-3, we can specify $B - N$ meshes for this graph; label these by m_1 through m_{B-N} and assign arbitrary orientation to each (clockwise or counterclockwise); then we have

Definition 3.3-5 (Mesh Matrix). The *mesh matrix* $\mathbf{M}$ of a connected planar graph G with $N + 1$ nodes and B branches is a $(B - N) \times B$ matrix

$$\mathbf{M} = (m_{kj}) \tag{3.3-16}$$

where

$$\begin{aligned} m_{kj} &= 1 && \text{when } b_j \text{ is in mesh } m_k \text{ and their orientations coincide} \\ &= -1 && \text{when } b_j \text{ is in mesh } m_k \text{ and their orientations do not coincide} \\ &= 0 && \text{when } b_j \text{ is not in mesh } m_k \end{aligned}$$

Then we have the following result concerning the rank of $\mathbf{M}$.

Theorem 3.3-4. The rank of the mesh matrix of a connected planar graph with $N + 1$ nodes and B branches is $B - N$.

The proof of this theorem is similar to the proof of Theorem 3.3-2 and is left as an exercise for the reader.

3.4 INTERRELATIONSHIP BETWEEN MATRICES OF A GRAPH

In this section we derive some useful relationships between $\mathbf{A}$, $\mathbf{B}_f$, and $\mathbf{Q}_f$; more precisely, we show that

$$\mathbf{A}\mathbf{B}_f^T = \mathbf{0} \tag{3.4-1}$$

and

$$\mathbf{B}_f\mathbf{Q}_f^T = \mathbf{0} \tag{3.4-2}$$

where $\mathbf{A}$, $\mathbf{B}_f$, and $\mathbf{Q}_f$ are the incidence matrix, fundamental loop matrix, and fundamental cutset matrix defined in the previous sections. Before proving these relations, let us give a simple illustration of them.

Consider the graph G shown in Fig. 3.4-1; take n_4 as the datum and assume the orientation is given by the corresponding arrows. Then by Definition 3.2-2 we can write

$$\mathbf{A} = \begin{array}{c} \\ n_1 \\ n_2 \\ n_3 \end{array} \begin{array}{c} \begin{array}{cccccc} b_1 & b_2 & b_3 & b_4 & b_5 & b_6 \end{array} \\ \begin{bmatrix} -1 & 0 & 0 & 1 & 1 & 0 \\ 0 & 1 & 0 & 0 & -1 & 1 \\ 0 & 0 & 1 & -1 & 0 & -1 \end{bmatrix} \end{array} \tag{3.4-3}$$

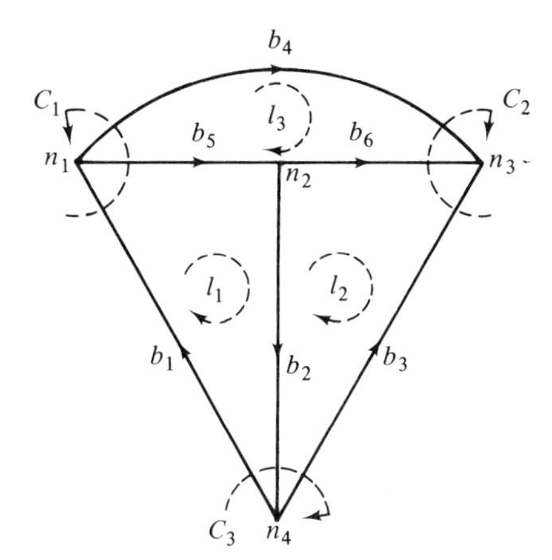

Figure 3.4-1

Next, choose a tree T consisting of b_2, b_5, and b_6 and denote the corresponding fundamental loops l_1, l_2, and l_3; then, from Definition 3.3-2, we can write the corresponding fundamental matrix as

$$\mathbf{B}_f = \begin{matrix} l_1 \\ l_2 \\ l_3 \end{matrix} \overset{\begin{matrix} b_1 & b_2 & b_3 & b_4 & b_5 & b_6 \end{matrix}}{\begin{bmatrix} 1 & 1 & 0 & 0 & 1 & 0 \\ 0 & -1 & 1 & 0 & 0 & 1 \\ 0 & 0 & 0 & 1 & -1 & -1 \end{bmatrix}} \tag{3.4-4}$$

Postmultiplying (3.4-3) by $\mathbf{B}_f^T$ given in (3.4-4), we get

$$\overbrace{\begin{bmatrix} -1 & 0 & 0 & 1 & 1 & 0 \\ 0 & 1 & 0 & 0 & -1 & 1 \\ 0 & 0 & 1 & -1 & 0 & -1 \end{bmatrix}}^{\mathbf{A}} \overbrace{\begin{bmatrix} 1 & 0 & 0 \\ 1 & -1 & 0 \\ 0 & 1 & 0 \\ 0 & 0 & 1 \\ 1 & 0 & -1 \\ 0 & 1 & -1 \end{bmatrix}}^{\mathbf{B}_f^T} = \begin{bmatrix} 0 & 0 & 0 \\ 0 & 0 & 0 \\ 0 & 0 & 0 \end{bmatrix}$$

That is, equation (3.4-1) holds for this particular example. In general, considering a graph with $N + 1$ nodes and B branches, we can state the following:

Theorem 3.4-1. Let **A** and $\mathbf{B}_f$ denote the incidence and fundamental loop matrices of a graph G; then

$$\mathbf{A}\mathbf{B}_f^T = \mathbf{0} \tag{3.4-5}$$

and

$$\mathbf{B}_f\mathbf{A}^T = \mathbf{0} \tag{3.4-6}$$

Proof

It is sufficient to prove (3.4-5) only, since (3.4-6) can be obtained simply by transposing (3.4-5). To show (3.4-5), recall that $\mathbf{A}$ is an $N \times B$ matrix and $\mathbf{B}_f^T$ is a $B \times (B - N)$; then the product

$$\mathbf{C} = \mathbf{A}\mathbf{B}_f^T$$

is an $N \times (B - N)$ matrix. We want to show that all the elements of $\mathbf{C}$ are zero. Denote the rows of $\mathbf{A}$ by $n_1, n_2, \ldots, n_N$ and its columns by $b_1, b_2, \ldots, b_B$. Also, denote the rows of $\mathbf{B}_f^T$ by $b_1, b_2, \ldots, b_B$ and its columns by $l_1, l_2, \ldots, l_{B-N}$; then

$$c_{ij} = \sum_{k=1}^{B} a_{ik} b_{jk}, \qquad \begin{array}{l} i = 1, 2, \ldots, N \\ j = 1, 2, \ldots, B - N \end{array} \tag{3.4-7}$$

or

$$\mathbf{C} = \begin{array}{c} n_1 \\ \cdot \\ \cdot \\ \cdot \\ n_i \\ \cdot \\ \cdot \\ \cdot \\ n_N \end{array} \begin{bmatrix} a_{11} & & \cdots & & a_{1B} \\ & & & & \\ & & & & \\ & & & & \\ a_{i1} & \cdots & a_{ik} & \cdots & a_{iB} \\ & & & & \\ & & & & \\ & & & & \\ a_{N1} & & \cdots & & a_{NB} \end{bmatrix} \overset{\begin{array}{ccccc} l_1 & \cdots & l_j & \cdots & l_{B-N} \end{array}}{\begin{bmatrix} & & b_{j1} & & \\ & & \cdot & & \\ & & \cdot & & \\ \cdot & & \cdot & & \cdot \\ \cdot & & b_{jk} & & \cdot \\ \cdot & & \cdot & & \cdot \\ & & \cdot & & \\ & & \cdot & & \\ & & b_{jB} & & \end{bmatrix}} \tag{3.4-8}$$

That is, the ijth element of $\mathbf{C}$ is obtained by multiplying column l_j of $\mathbf{B}_f^T$ to the row n_i of $\mathbf{A}$. We consider two possible cases and in each case show that $c_{ij} = 0$.

Case 1. The node n_i is not in the loop l_j. Then none of the branches of l_j are incident to n_i; that is, corresponding to any nonzero element in the column l_j there will be a zero element in the row n_i; hence, $c_{ij} = 0$.

Case 2. The node n_i is in the loop l_j; then exactly two branches of the loop l_j are incident to n_i (see Fig. 3.4-2). If these two branches are oriented as in Fig. 3.4-2(a), the elements of the column l_j corresponding to these branches are both $+1$, and the elements of the row n_j corresponding to these branches are $+1$ and -1; hence, the algebraic sum of these add to zero and again $c_{ij} = 0$. If the branches are oriented as in Fig. 3.4-2(b), then the elements of the column l_j corresponding to these branches are $+1$ and -1, and the elements of the row n_j corresponding to these branches are both -1; consequently, $c_{ij} = 0$. This reasoning holds for $i = 1, 2, \ldots, N$ and for $j = 1, 2, \ldots, B - N$; hence, the conclusion of the theorem follows.

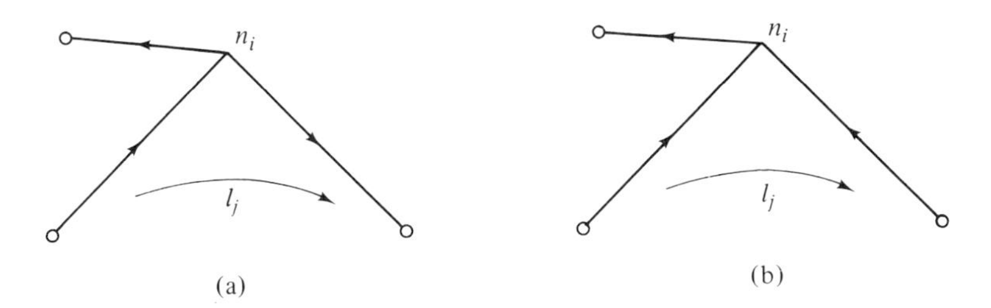

Figure 3.4-2

The result of this theorem can easily be generalized to apply to the augmented incidence matrix $\mathbf{A}_a$ and augmented loop matrix $\mathbf{B}_a$; more precisely,

Theorem 3.4-2. Consider a graph G with $N + 1$ nodes, B branches, and L loops. Let $\mathbf{A}_a$ and $\mathbf{B}_a$ be augmented incidence and loop matrices defined by Definitions 3.2-1 and 3.3-1, respectively. Then

$$\mathbf{A}_a\mathbf{B}_a^T = \mathbf{0} \tag{3.4-9}$$

and

$$\mathbf{B}_a\mathbf{A}_a^T = \mathbf{0} \tag{3.4-10}$$

Exercise 3.4-1. Prove Theorem 3.4-2. (Hint: Follow the proof of Theorem 3.4-1.)

Next we turn to the relationship between $\mathbf{B}_f$ and $\mathbf{Q}_f$. Let us first find the fundamental cutset matrix for the graph of Fig. 3.4-1 and show that (3.4-2) holds for this particular example. The matrix $\mathbf{Q}_f$ can be written down by considering the fundamental cutsets corresponding to tree T (composed of branches b_2, b_5, and b_6) of Fig. 3.4-1.

$$\mathbf{Q}_f = \begin{array}{c} \\ c_1 \\ c_2 \\ c_3 \end{array}\!\!\begin{array}{c} \begin{array}{cccccc} b_1 & b_2 & b_3 & b_4 & b_5 & b_6 \end{array} \\ \begin{bmatrix} 1 & 0 & 0 & -1 & -1 & 0 \\ 0 & 0 & -1 & 1 & 0 & 1 \\ -1 & 1 & 1 & 0 & 0 & 0 \end{bmatrix} \end{array}$$

Then we have

$$\underbrace{\begin{bmatrix} 1 & 0 & 0 & -1 & -1 & 0 \\ 0 & 0 & -1 & 1 & 0 & 1 \\ -1 & 1 & 1 & 0 & 0 & 0 \end{bmatrix}}_{\mathbf{Q}_f} \underbrace{\begin{bmatrix} 1 & 0 & 0 \\ 1 & -1 & 0 \\ 0 & 1 & 0 \\ 0 & 0 & 1 \\ 1 & 0 & -1 \\ 0 & 1 & -1 \end{bmatrix}}_{\mathbf{B}_f^T} = \begin{bmatrix} 0 & 0 & 0 \\ 0 & 0 & 0 \\ 0 & 0 & 0 \end{bmatrix}$$

In general, consider a graph G with $N + 1$ nodes and B branches. Choose a tree T of this graph and denote the fundamental cutsets and branches of G with respect to T by $c_1, c_2, \ldots, c_N$ and $b_1, b_2, \ldots, b_B$, respectively; then we have

Theorem 3.4-3. Let $\mathbf{B}_f$ and $\mathbf{Q}_f$ represent the fundamental loop matrix and fundamental cutset matrix of a graph G with respect to a tree T; then

$$\mathbf{B}_f\mathbf{Q}_f^T = \mathbf{0} \tag{3.4-11}$$

and

$$\mathbf{Q}_f\mathbf{B}_f^T = \mathbf{0} \tag{3.4-12}$$

Exercise 3.4-2. Give a detailed proof of Theorem 3.4-3. [Hint: Use equations (3.2-15) and (3.3-10).]

3.5 TELLEGEN'S THEOREM AND ITS APPLICATION

In this section we state and prove Tellegen's theorem, which undoubtedly is one of the most general theorems in network theory; it applies to any network made up of lumped two-terminal elements, regardless of their nature. Before proceeding with the statement of the theorem, let us establish an important relationship between the branch voltage vector $\mathbf{v}_b$ and the node voltage vector $\mathbf{v}_n$ of a network.

Relationship Between Branch and Node Voltages. Consider a network with $N + 1$ nodes and B branches. Denote the node voltages with respect to an arbitrary reference by $v_{n_1}, v_{n_2}, \ldots, v_{n_N}, v_{n_{N+1}}$ and the branch voltages by $v_{b_1}, v_{b_2}, \ldots, v_{b_B}$. Then each branch voltage v_{b_j} can be written in terms of two node voltages, say, v_{n_i} and v_{n_k}. For example, for the simple network whose

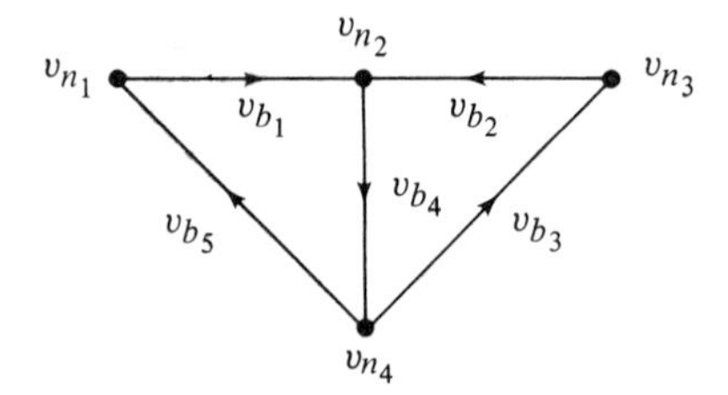

Figure 3.5-1

graph is shown in Fig. 3.5-1 we can write

$$v_{b_1} = v_{n_1} - v_{n_2}$$

$$v_{b_2} = v_{n_3} - v_{n_2}$$

$$v_{b_3} = v_{n_4} - v_{n_3}$$

$$v_{b_4} = v_{n_2} - v_{n_4}$$

$$v_{b_5} = v_{n_4} - v_{n_1}$$

In general, we can write

$$\begin{bmatrix} v_{b_1} \\ v_{b_2} \\ \cdot \\ \cdot \\ \cdot \\ v_{b_B} \end{bmatrix} = \begin{bmatrix} p_{11} & \cdots & p_{1(N+1)} \\ \cdot & & \cdot \\ \cdot & & \cdot \\ \cdot & & \cdot \\ p_{B1} & \cdots & p_{B(N+1)} \end{bmatrix} \begin{bmatrix} v_{n_1} \\ v_{n_2} \\ \cdot \\ \cdot \\ \cdot \\ v_{n_{N+1}} \end{bmatrix} \tag{3.5-1}$$

where p_{jk} is defined by

$$\begin{aligned} p_{jk} &= 1 && \text{when } b_j \text{ is incident to } n_k \text{ and is directed away from it} \\ &= -1 && \text{when } b_j \text{ is incident to } n_k \text{ and is directed toward it} \\ &= 0 && \text{when } b_j \text{ is not incident to } n_k \end{aligned}$$

Let

$$\mathbf{v}_b \triangleq (v_{b_1}, v_{b_2}, \ldots, v_{b_B})^T \tag{3.5-2}$$

and

$$\hat{\mathbf{v}}_n \triangleq (v_{n_1}, v_{n_2}, \ldots, v_{n_{N+1}})^T \tag{3.5-3}$$

Then (3.5-1) can be written as

$$\mathbf{v}_b = \mathbf{P}\hat{\mathbf{v}}_n \tag{3.5-4}$$

where $\mathbf{P}$ is the $B \times (N + 1)$ matrix whose jkth element p_{jk} is as given previously. Comparing p_{jk} with a_{kj} given in Definition 3.2-1, we immediately recognize that

$$p_{jk} = a_{kj} \quad \text{for all } j \text{ and } k \tag{3.5-5}$$

Hence, the matrix $\mathbf{P}$ given in (3.5-4) is the transpose of the familiar augmented incidence matrix $\mathbf{A}_a$; consequently, (3.5-4) can be written as

$$\mathbf{v}_b = \mathbf{A}_a^T\hat{\mathbf{v}}_n \tag{3.5-6}$$

This equation gives a relation between the node voltage vector $\hat{\mathbf{v}}_n$ and the branch voltage vector $\mathbf{v}_b$.

Let us now choose an arbitrary node as the datum and measure the node voltages with respect to this datum. For simplicity of the notations, choose n_{N+1} to be the datum and let $v_{n_{N+1}} = 0$. Then the last element of $\hat{\mathbf{v}}_n$ is zero and can be deleted together with the last column of $\mathbf{A}_a^T$; more precisely, if we let

$$\mathbf{v}_n = (v_{n_1}, v_{n_2}, \ldots, v_{n_N})^T \tag{3.5-7}$$

then (3.5-6) can be written as

$$\mathbf{v}_b = \mathbf{A}^T\mathbf{v}_n \tag{3.5-8}$$

where $\mathbf{A}$ is the incidence matrix of the network under consideration as defined in Definition 3.2-2. We can now state and prove the following.

Theorem 3.5-1 (Tellegen). Consider a lumped network $\mathfrak{N}$ consisting of $N+1$ nodes and B branches. Let $\mathbf{v}_b$ and $\mathbf{i}_b$ be given by

$$\mathbf{v}_b = (v_{b_1}, v_{b_2}, \ldots, v_{b_B})^T$$

and

$$\mathbf{i}_b = (i_{b_1}, i_{b_2}, \ldots, i_{b_B})^T$$

where v_{b_k} and i_{b_k} denote the voltage across and current through the branch b_k for $k = 1, 2, \ldots, B$. Then

$$\mathbf{v}_b^T \mathbf{i}_b = 0 \tag{3.5-9}$$

or, equivalently,

$$\sum_{k=1}^{B} v_{b_k} i_{b_k} = 0 \tag{3.5-10}$$

Proof

Transposing both sides of (3.5-8), we get

$$\mathbf{v}_b^T = \mathbf{v}_n^T \mathbf{A} \tag{3.5-11}$$

Postmultiplying (3.5-11) by $\mathbf{i}_b$ yields

$$\mathbf{v}_b^T \mathbf{i}_b = \mathbf{v}_n^T \mathbf{A} \mathbf{i}_b \tag{3.5-12}$$

But from (3.2-7) we have

$$\mathbf{A}\mathbf{i}_b = \mathbf{0} \tag{3.5-13}$$

Hence, from (3.5-13) and (3.5-12) we can write

$$\mathbf{v}_b^T(t)\mathbf{i}_b(t) = 0 \quad \text{for all } t \tag{3.5-14}$$

which proves the theorem.

The physical interpretation of Tellegen's theorem is the conservation of power. According to (3.5-14), the sum of the powers delivered to or absorbed by *all* the branches of a given lumped network is equal to zero. That is, the power delivered by the active elements of a network is completely absorbed by the passive elements at each instant of time. The mathematical implications of this theorem, however, are much broader. In the following we introduce an important generalization of equation (3.5-14).

Theorem 3.5-2. Consider two networks $\mathfrak{N}$ and $\hat{\mathfrak{N}}$ made up of lumped two-terminal elements whose graphs are identical. Denote their corresponding branch voltage and current vectors $\mathbf{v}_b$, $\mathbf{i}_b$ and $\hat{\mathbf{v}}_b$, $\hat{\mathbf{i}}_b$; then

$$\mathbf{v}_b^T(t)\hat{\mathbf{i}}_b(t) = 0 \quad \text{for all } t \tag{3.5-15}$$

and

$$\hat{\mathbf{v}}_b^T(t)\mathbf{i}_b(t) = 0 \quad \text{for all } t \tag{3.5-16}$$

Proof

For the network $\mathfrak{N}$ we can write

$$\mathbf{v}_b(t) = \mathbf{A}^T \mathbf{v}_n(t)$$

or, equivalently,

$$\mathbf{v}_b^T(t) = \mathbf{v}_n^T(t)\mathbf{A} \tag{3.5-17}$$

where $\mathbf{A}$ denotes the incidence matrix of $\mathfrak{N}$. Postmultiplying both sides of (3.5-17) by $\hat{\mathbf{i}}_b(t)$, we get

$$\mathbf{v}_b^T(t)\hat{\mathbf{i}}_b(t) = \mathbf{v}_n^T(t)\mathbf{A}\hat{\mathbf{i}}_b(t) \tag{3.5-18}$$

Let $\hat{\mathbf{A}}$ denote the incidence matrix of $\hat{\mathfrak{N}}$; then, according to (3.2-7), we have

$$\hat{\mathbf{A}}\hat{\mathbf{i}}_b(t) = \mathbf{0} \tag{3.5-19}$$

Since, by hypothesis of the theorem, the graphs of the two networks are identical, we have

$$\hat{\mathbf{A}} = \mathbf{A} \tag{3.5-20}$$

Hence, from (3.5-20), (3.5-19), and (3.5-18), we obtain

$$\mathbf{v}_b^T(t)\hat{\mathbf{i}}_b(t) = 0 \quad \text{for all } t$$

This proves the first part of the theorem; the second part can also be shown in an analogous manner. This theorem is sometimes referred to as Tellegen's quasi-power theorem [4].

Remark 3.5-1. Note that the conclusion of Theorem 3.5-2 cannot be interpreted as a conservation of power; it is merely a mathematical relationship that exists between the branch voltages of one circuit and the branch currents of another circuit with the same topology. There is no constraint on the element characterization of either network. For example, consider the two networks shown in Fig. 3.5-2. Denote the branch voltages of one network by $v_1, v_2, \ldots, v_6$ and the currents of the other network by $\hat{i}_1, \hat{i}_2, \ldots, \hat{i}_6$. Then

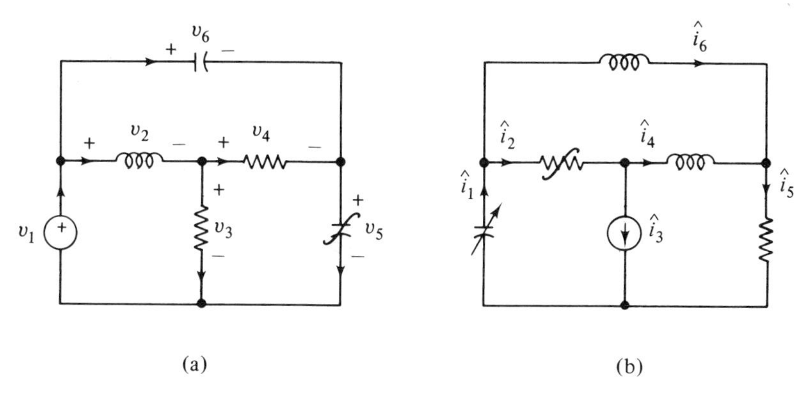

Figure 3.5-2 Two different networks with the same topology; voltages of one network are related to the currents of the other through equation (3.5-15).

Theorem 3.5-2 implies that

$$v_1\hat{i}_1 + v_2\hat{i}_2 + v_3\hat{i}_3 + v_4\hat{i}_4 + v_5\hat{i}_5 + v_6\hat{i}_6 = 0$$

regardless of what the element types and values are.

Exercise 3.5-1. Let $\mathbf{v}_b(t)$ and $\mathbf{i}_b(t)$ denote the branch voltage and current vectors of a network made up of lumped two-terminal elements; show that

$$\mathbf{v}_b^T(t)\mathbf{i}_b(t-T) = 0 \quad \text{for all } t \text{ and for all } T \tag{3.5-21}$$

Exercise 3.5-2. Use Theorem 3.5-1 to show that a one-port network comprised of passive two-terminal elements is passive.

We should point out that Tellegen's theorem has been known, in one form or another, since 1883. Tellegen was, however, the first who recognized its generality and numerous applications. An interesting account of the history of this theorem and its various forms is given in [4].

Before concluding this chapter, let us establish a useful relationship between the branch current vector $\mathbf{i}_b$ and the mesh current vector $\mathbf{i}_m$ of a planar graph.

Relationship Between Branch and Mesh Currents. Consider a *planar* network with $N + 1$ nodes and B branches. According to Theorem 3.3-3, there are exactly $B - N$ meshes in this network. Denote these meshes by $m_1, m_2, \ldots, m_{B-N}$ and assign an arbitrary direction to each mesh. Let us also assign a fictitious current i_{m_k} to each mesh; these fictitious mesh currents circulate in the branches of their corresponding meshes and are assumed to have the same orientation as their respective meshes. The branch currents can then be expressed in terms of these mesh currents. For example, consider the

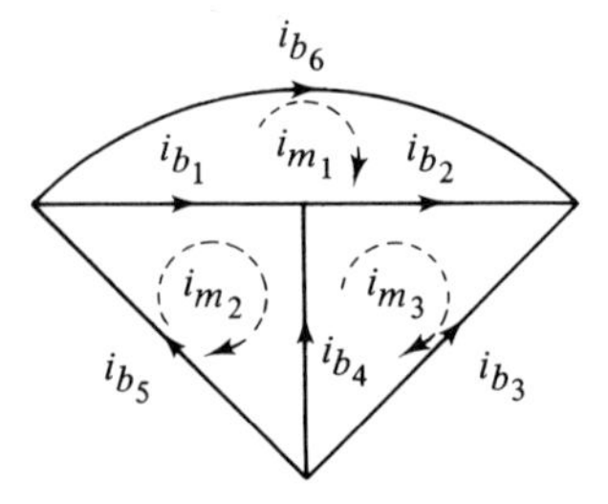

Figure 3.5-3

network whose graph is shown in Fig. 3.5-3. We can write

$$i_{b_1} = i_{m_2} - i_{m_1}$$
$$i_{b_2} = i_{m_3} - i_{m_1}$$
$$i_{b_3} = -i_{m_3}$$

$$i_{b_4} = i_{m_3} - i_{m_2}$$
$$i_{b_5} = i_{m_2}$$
$$i_{b_6} = i_{m_1}$$

In general, we can write each branch current as a linear combination of the mesh currents; that is,

$$\begin{bmatrix} i_{b_1} \\ i_{b_2} \\ \cdot \\ \cdot \\ \cdot \\ i_{b_B} \end{bmatrix} = \begin{bmatrix} p_{11} & \cdots & p_{1(B-N)} \\ p_{21} & \cdots & p_{2(B-N)} \\ \cdot & & \cdot \\ \cdot & & \cdot \\ \cdot & & \cdot \\ p_{B1} & \cdots & p_{B(B-N)} \end{bmatrix} \begin{bmatrix} i_{m_1} \\ i_{m_2} \\ \cdot \\ \cdot \\ \cdot \\ i_{m_{(B-N)}} \end{bmatrix} \tag{3.5-22}$$

where p_{jk} is defined by

$$\begin{aligned} p_{jk} &= 1 \quad \text{when } b_j \text{ is in } m_k \text{ and is oriented in the same direction} \\ &= -1 \quad \text{when } b_j \text{ is in } m_k \text{ and is oriented in the opposite direction} \\ &= 0 \quad \text{when } b_j \text{ is not in } m_k \end{aligned}$$

If we now denote

$$\mathbf{i}_b = (i_{b_1}, i_{b_2}, \ldots, i_{b_B})^T \tag{3.5-23}$$

$$\mathbf{i}_m = (i_{m_1}, i_{m_2}, \ldots, i_{m_{(B-N)}})^T \tag{3.5-24}$$

and

$$\mathbf{P} = (p_{ij}) \tag{3.5-25}$$

then

$$\mathbf{i}_b = \mathbf{P}\mathbf{i}_m \tag{3.5-26}$$

where $\mathbf{P}$ is the $B \times (B - N)$ matrix whose jkth element is as just defined. Comparing p_{jk} with m_{jk} given in Definition 3.3-5, we get

$$p_{jk} = m_{kj} \tag{3.5-27}$$

Hence, (3.5-26) can be written as

$$\mathbf{i}_b = \mathbf{M}^T\mathbf{i}_m \tag{3.5-28}$$

where $\mathbf{M}$ is given in (3.3-16).

Exercise 3.5-3. Use equations (3.3-15) and (3.5-28) to prove Tellegen's theorem for a planar network.

Let us next generalize equation (3.5-28) to include nonplanar networks also.

Relationship Between Branch and Loop Currents. Consider a connected graph G (planar or nonplanar) with $N + 1$ nodes and B branches. Choose a tree T for this graph and assign arbitrary orientations for the tree branches and chords. There are exactly $B - N$ chords in G corresponding to T. Recall that each chord together with a set of unique tree branches forms a fundamental loop. Assign fictitious currents to each fundamental loop; each fictitious loop current is assumed to circulate in its corresponding fundamental loop and have the same direction as its respective chord. As in the case of mesh currents, the branch currents of G can then be expressed in terms of these loop currents.

In general, each branch current i_{b_k} can be written as a linear combination of the loop currents; let

$$\mathbf{i}_l = (i_{l_1}, i_{l_2}, \ldots, i_{l_{(B-N)}})^T \tag{3.5-29}$$

Then, in a similar manner as in the case of planar networks, it can be shown that

$$\mathbf{i}_b = \mathbf{B}_f^T \mathbf{i}_l \tag{3.5-30}$$

where $\mathbf{i}_b$ is a B-column vector whose elements are branch currents of G, $\mathbf{B}_f$ is the fundamental loop matrix given in Definition 3.3-2, and $\mathbf{i}_l$ is a $(B - N)$-column vector whose elements are the loop currents just discussed.

Equation (3.5-30) is a useful relation in the sense that to obtain B branch currents one can first solve for $B - N$ loop currents and then use (3.5-30) to obtain the branch currents. In many cases this procedure greatly reduces the number of the integro-differential equations that must be solved for obtaining the branch currents and voltages of a network. This point will be further discussed in Chapter 4.

REFERENCES

[1] SESHU, S., and M. B. REED, *Linear Graphs and Electrical Networks*, Addison-Wesley Publishing Company, Inc., Reading, Mass., 1961.

[2] CHAN, S. P., *Introductory Topological Analysis of Electrical Networks*, Holt, Rinehart and Winston, Inc., New York, 1969.

[3] BERGE, C., *Theory of Graphs and Its Applications*, John Wiley & Sons, Inc., New York, 1962.

[4] PENFIELD, P., JR., R. SPENCE, and S. DUNKER, *Tellegen's Theorem and Electrical Networks*, The MIT Press, Cambridge, Mass., 1970.

PROBLEMS

3.1 Prove that a subgraph G', which contains all the nodes of a connected graph G, is a tree of G if and only if there exists exactly one path between every two nodes of G'.

3.2 Prove that every tree of a connected graph with $N + 1$ nodes contains exactly N branches.

3.3 Consider the planar graph shown in Fig. P3.3.
(a) Find all the trees of this graph (there are 16).
(b) Find all the cutsets of this graph.
(c) Find all the loops of this graph.

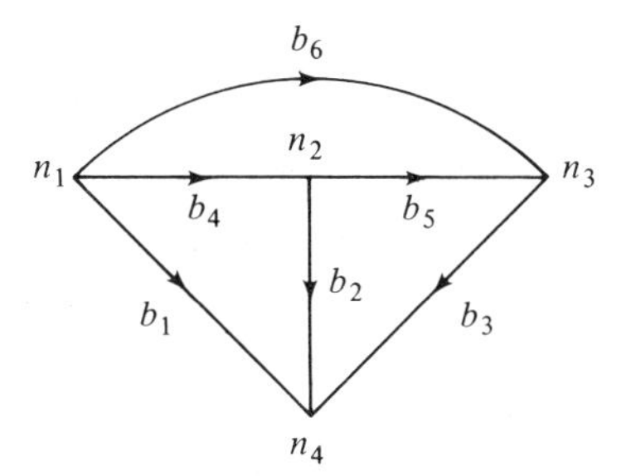

Figure P3.3

3.4 Consider the graph shown in Fig. P3.3. Denote the tree made up of b_6, b_5, and b_2 by T.
(a) Find all the fundamental loops of this graph corresponding to T.
(b) Find all the fundamental cutsets of this graph corresponding to T.
Check your answer with the conclusion of Theorem 3.1-1.

3.5 Write the augmented incidence matrix of the graph shown in Fig. P3.3; check the conclusion of Remark 3.2-1.

3.6 Show that the augmented incidence matrix obtained in Problem 3.5 has rank 3.

3.7 Write down the augmented incidence matrix of the graph shown in Fig. P3.7.

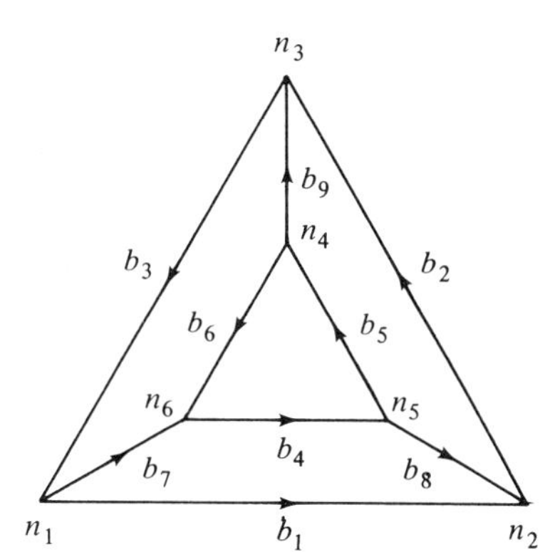

Figure P3.7

3.8 Write the linearly independent KCL equations for the network whose graph is shown in Fig. P3.3; choose n_4 as the datum.

3.9 Take n_1 as the datum for the network whose graph is given in Fig. P3.7 and use the result of Problem 3.7 to obtain all the linearly independent KCL equations of the network.

3.10 Write all the linearly independent KCL equations for the network of Fig. 3.1-1(a).

3.11 The incidence matrix of a network is given by

$$\mathbf{A} = \begin{bmatrix} 0 & -1 & -1 & 1 & 0 \\ 0 & 0 & 0 & -1 & -1 \\ 1 & 0 & 1 & 0 & 1 \end{bmatrix}$$

Draw the graph of this network.

3.12 It can be shown [2] that the total number of the trees n_T in a graph is given by

$$n_T = \det (\mathbf{A}_a \mathbf{A}_a^T)$$

where A_a is the augmented incidence matrix of the graph. Verify this relation for the graph of Fig. P3.3.

3.13 Write the augmented cutset matrix of Fig. P3.3.

3.14 Consider the graph shown in Fig. P3.3. Write down the fundamental cutset matrix of this graph with respect to the tree that is made up of branches b_5, b_2, and b_1.

3.15 Write down the fundamental cutset matrix of the graph of Fig. P3.7 corresponding to the tree whose branches are b_9, b_5, b_8, b_1, and b_7.

3.16 Consider the graph shown in Fig. P3.16. Choose a tree of this graph and label its branches in such a way that the fundamental cutset matrix corresponding to this tree can be partitioned as in equation (3.2-15). Write down the corresponding fundamental matrix.

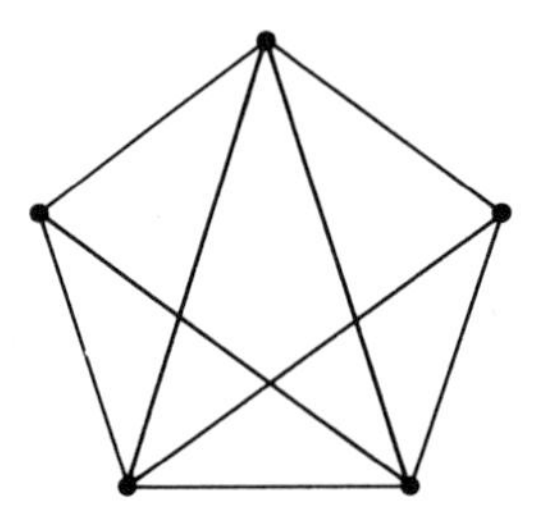

Figure P3.16

3.17 Write down the augmented loop matrix of the network shown in Fig. P3.7.

3.18 Show that the rank of the augmented loop matrix of the graph shown in Fig. P3.3 is 3.

3.19 Obtain the fundamental loop matrix of the graph shown in Fig. P3.7 and show that its rank is 4; choose a tree whose branches are b_9, b_5, b_8, b_1, and b_7.

3.20 Write down all the linearly independent KVL equations for the network whose graph is given in Fig. P3.3.

3.21 Choose a tree of the graph shown in Fig. P3.7 and use its corresponding fundamental loop matrix to write all the linearly independent KVL equations.

3.22 Consider the graph shown in Fig. P3.16. Use the same tree chosen in Problem 3.16 to obtain the fundamental loop matrix of this graph. Show that if the fundamental cutset matrix of this graph obtained in Problem 3.16 is in the form $\mathbf{Q}_f = [\mathbf{1} \vdots \mathbf{F}]$ the corresponding fundamental loop matrix is $\mathbf{B}_f = [-\mathbf{F}^T \vdots \mathbf{1}]$.

3.23 Choose appropriate meshes for the planar graph of Fig. P3.3 and obtain its mesh matrix **M**. Show that **M** is of rank 3.

3.24 Write down the mesh matrix of the planar graph of Fig. P3.7 and obtain all the linearly independent mesh equations.

3.25 Show that the graph of Fig. P3.16 is nonplanar.

3.26 Use the results of Problems 3.7 and 3.19 to show that $\mathbf{A}\mathbf{B}_f^T = \mathbf{0}$ for the graph shown in Fig. P3.7.

3.27 Use the results of Problems 3.15 and 3.19 to show that $\mathbf{B}_f\mathbf{Q}_f^T = \mathbf{0}$ for the graph of Fig. P3.7.

3.28 Consider the resistive black-box network shown in Fig. P3.28. For $R_2 = 1\ \Omega$, if $v_1 = 6$ V, then $i_1 = 1$ A and $v_2 = 1$ V. For $R_2 = 2\ \Omega$, if $v_1 = 5$ V, then $i_1 = 1$ A and $v_2 =$ unknown. Find the unknown voltage v_2 in the second case. (Hint: Use the conclusion of Theorem 3.5-2.)

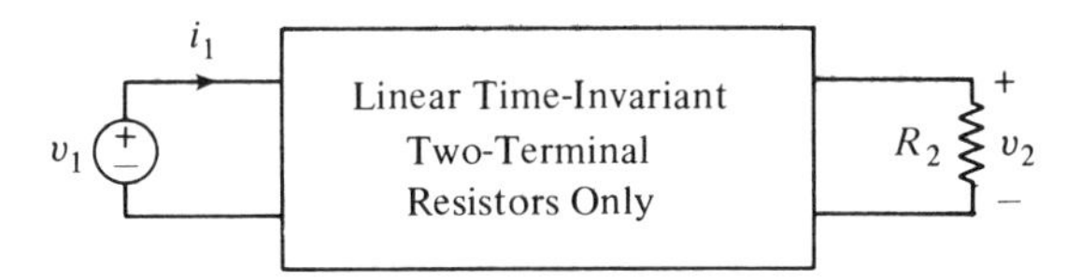

Figure P3.28

Analysis of Linear Time-Invariant Networks 4

4.1 INTRODUCTION

In Chapter 3 we introduced a systematic procedure for obtaining N linearly independent KCL equations and $B - N$ linearly independent KVL equations for a network with $N + 1$ nodes and B branches:

$$\text{KCL:} \qquad \mathbf{A}\mathbf{i}_b = \mathbf{0} \tag{4.1-1}$$

$$\text{KVL:} \qquad \mathbf{B}_f\mathbf{v}_b = \mathbf{0} \tag{4.1-2}$$

These provide B equations in $2B$ unknowns (B branch voltages and B branch currents). Equations (4.1-1) and (4.1-2) depend only on the topology of the network and not the nature of its elements. In this chapter we use the branch voltage–current relations of a special class of networks, the networks that are comprised of linear time-invariant elements, to obtain another set of B linearly independent equations. These branch Voltage–Current Relations (VCR) together with KCL and KVL equations provide $2B$ equations in $2B$ unknowns that can be solved for the branch current vector $\mathbf{i}_b$ and the branch voltage vector $\mathbf{v}_b$. This method, however, will later be abandoned in favor of more practical methods: nodal analysis and loop analysis. In these latter methods, we introduce auxiliary network variables, such as node voltages and loop currents to reduce the number of the integro-differential equations describing the network to N, in the case of nodal analysis, and to $B - N$ in the case of loop analysis. Since the networks under study in this chapter are linear and time invariant, we use the Laplace-transform method to formulate various analysis procedures. The steady-state sinusoidal analysis is discussed in Section 4.6, and the final section is devoted to a general discussion on network functions. It is assumed that the reader has some knowledge of the definition and properties of the Laplace transform. A brief discussion on the Laplace transform is given in the Appendix.

4.2 DIRECT ANALYSIS METHODS

Consider a network made up of linear time-invariant two-terminal elements and independent voltage and current sources. To simplify the notations, let the initial voltages on the capacitors be represented as voltage sources in series with the capacitors and the initial currents through the inductors be represented as current sources in parallel with the inductors. To illustrate that this can always be done, consider the linear time-invariant capacitor with capacitance C shown in Fig. 4.2-1(a); let the initial voltage on

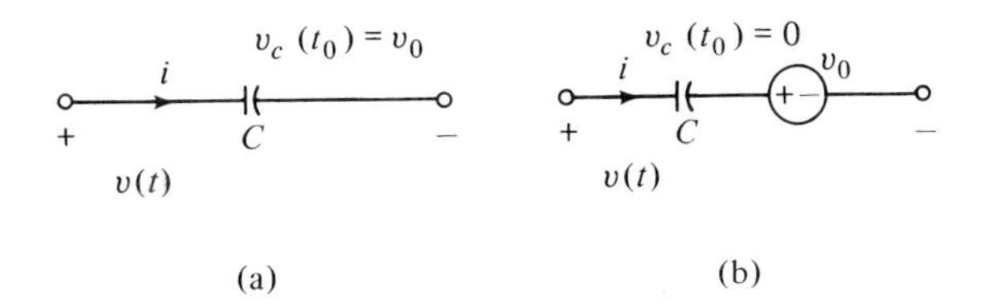

Figure 4.2-1 Two equivalent capacitor branches.

this capacitor be v_o; then the voltage across the branch is given by

$$v(t) = \frac{1}{C}\int_{t_0}^{t} i(\tau)\,d\tau + v_0 \tag{4.2-1}$$

Consequently, this capacitor is equivalent to a capacitor with capacitance C, whose initial voltage is equal to zero, in series with a voltage source v_0 [Fig. 4.2-1(b)]. Similarly, a linear time-invariant inductor with initial current i_0 is equivalent to an inductor with the same inductance whose initial current is zero and is in parallel with a current source i_0 [see Fig. 4.2-2(a) and (b)]. The current through the branch in either case is given by

$$i(t) = \frac{1}{L}\int_{t_0}^{t} v(\tau)\,d\tau + i_0 \tag{4.2-2}$$

To keep the analysis general, assume that a typical branch, say b_k, of the network under consideration is given as in Fig. 4.2-3, where from the preceding discussion we can assume that all the initial conditions are included in the independent sources. The branch current i_{b_k} is then given as the sum of the

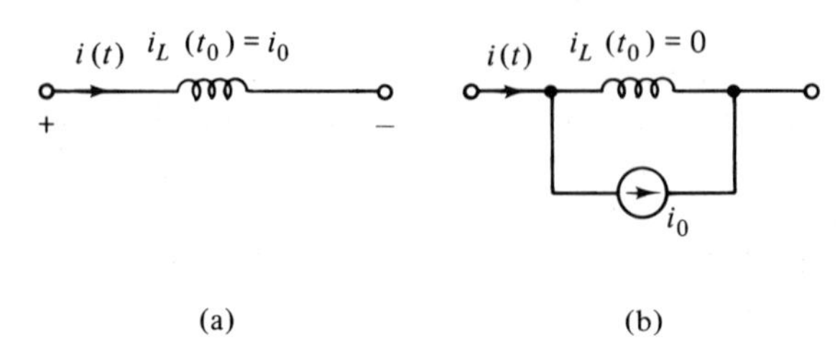

Figure 4.2-2 Two equivalent inductor branches.

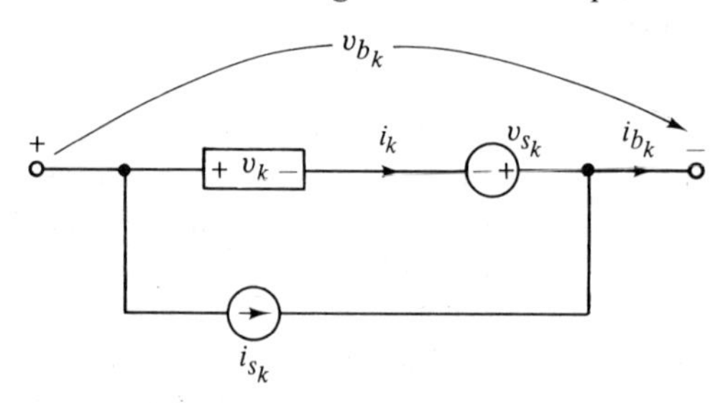

Figure 4.2-3 General branch of a linear time-invariant network.

current source i_{s_k} and the *element current* i_k; that is,

$$i_{b_k} = i_k + i_{s_k}, \qquad k = 1, 2, \ldots, B \tag{4.2-3}$$

and the branch voltage v_{b_k} is the algebraic sum of the voltage drop across the voltage source v_{s_k} and the voltage drop across the element v_k:

$$v_{b_k} = v_k - v_{s_k}, \qquad k = 1, 2, \ldots, B \tag{4.2-4}$$

More concisely, let

$$\mathbf{i}_b = (i_{b_1}, i_{b_2}, \ldots, i_{b_B})^T \tag{4.2-5}$$

$$\mathbf{i}_s = (i_{s_1}, i_{s_2}, \ldots, i_{s_B})^T \tag{4.2-6}$$

and

$$\mathbf{i} = (i_1, i_2, \ldots, i_B)^T \tag{4.2-7}$$

Then we have

$$\mathbf{i}_b = \mathbf{i} + \mathbf{i}_s \tag{4.2-8}$$

Similarly, let

$$\mathbf{v}_b = (v_{b_1}, v_{b_2}, \ldots, v_{b_B})^T \tag{4.2-9}$$

$$\mathbf{v}_s = (v_{s_1}, v_{s_2}, \ldots, v_{s_B})^T \tag{4.2-10}$$

and

$$\mathbf{v} = (v_1, v_2, \ldots, v_B)^T \tag{4.2-11}$$

Then

$$\mathbf{v}_b = \mathbf{v} - \mathbf{v}_s \tag{4.2-12}$$

The *element voltage* v_k is related to the element current i_k through the following equations:

1. If b_k is a resistor with resistance R_k,

$$v_k = R_k i_k \tag{4.2-13}$$

2. If b_k is a capacitor with capacitance C_k,

$$v_k(t) = \frac{1}{C_k}\int_{t_0}^{t} i_k(\tau)\,d\tau \tag{4.2-14}$$

By taking the Laplace transform of equation (4.2-14), it can be written as

$$v_k(s) = \frac{1}{sC_k} i_k(s) = \frac{1}{s} S_k i_k(s) \tag{4.2-15}$$

where $s = \sigma + j\omega$ is the Laplace variable; $v_k(s)$ and $i_k(s)$ are the Laplace transforms of $v_k(t)$ and $i_k(t)$, respectively (see the Appendix). Also, $S_k = 1/C_k$ is the elastance of the kth capacitor.

3. If b_k is an inductor with self-inductance L_k and mutual inductance M_{kj}, then

$$v_k = L_k \frac{d}{dt} i_k + \sum_{\substack{j=1 \\ j \neq k}}^{B} M_{kj} \frac{d}{dt} i_j \tag{4.2-16}$$

Again, taking the Laplace transform of both sides of this equation, it can be written as

$$v_k(s) = sL_k i_k(s) + \sum_{\substack{j=1 \\ j \neq k}}^{B} sM_{kj} i_j(s) \tag{4.2-17}$$

With these conventions we are now ready to introduce a systematic method for obtaining the branch voltage and current vectors $\mathbf{v}_b(t)$ and $\mathbf{i}_b(t)$ for a given network.

Impedance Matrix Method. For a given network, the element voltage $\mathbf{v}$ can be written as

$$\mathbf{v}(s) = \mathbf{Z}(s)\mathbf{i}(s) \tag{4.2-18}$$

where $\mathbf{Z}(s)$ is the *element impedance matrix* of the network under study and is defined by

$$\mathbf{Z}(s) = \mathbf{R} + \frac{1}{s}\mathbf{S} + s\mathbf{L} \tag{4.2-19}$$

so that $\mathbf{R}$ is a diagonal $B \times B$ resistance matrix whose kth diagonal element is R_k, $\mathbf{S}$ is a diagonal $B \times B$ elastance matrix whose diagonal element $S_k = 1/C_k$ is the elastance of the kth branch, and $\mathbf{L}$ is a $B \times B$ matrix whose kth diagonal element is the self-inductance L_k and its jkth off-diagonal element is the mutual inductance M_{jk}. Equation (4.2-18) represents the *element voltage–current relationship* for all the branches of the network; $Z(s)$ can be determined by knowing the element type and values of each branch. From (4.2-18), (4.2-12), and (4.2-8), we get the *branch voltage–current relationship* (VCR):

$$\mathbf{v}_b(s) = \mathbf{Z}(s)\mathbf{i}_b(s) - \mathbf{v}_s(s) - \mathbf{Z}(s)\mathbf{i}_s(s) \tag{4.2-20}$$

or equivalently,

$$\mathbf{v}_b(s) = \mathbf{Z}(s)\mathbf{i}_b(s) - [\mathbf{v}_s(s) + \hat{\mathbf{v}}_s(s)]$$

where $\hat{\mathbf{v}}_s(s)$ is defined by

$$\hat{\mathbf{v}}_s(s) = \mathbf{Z}(s)\mathbf{i}_s(s) \tag{4.2-21}$$

Equation (4.2-20) represents B equations relating the branch voltage vector $\mathbf{v}_b$ to the branch current vector $\mathbf{i}_b$; for this reason it is called the branch voltage–current relation (VCR). Hence, equations (4.2-20), (4.1-1), and (4.1-2) together form $2B$ linearly independent equations in $2B$ unknowns that can be solved for all the branch voltages and branch currents $\mathbf{i}_b(s)$ and $\mathbf{v}_b(s)$. To put these equations in a more compact form, replace $\mathbf{v}_b$ in (4.1-2) from (4.2-20) to get

$$\mathbf{B}_f\mathbf{Z}(s)\mathbf{i}_b = \mathbf{B}_f\mathbf{Z}(s)\mathbf{i}_s + \mathbf{B}_f\mathbf{v}_s \tag{4.2-22}$$

Equation (4.2-22) represents a set of $B - N$ linearly independent equations in B unknowns; combining these with N equations given in (4.1-1), we get

$$\begin{bmatrix} \mathbf{B}_f\mathbf{Z}(s) \\ \hdashline \mathbf{A} \end{bmatrix} \mathbf{i}_b(s) = \begin{bmatrix} \mathbf{B}_f\mathbf{Z}(s) \\ \hdashline \mathbf{0} \end{bmatrix} \mathbf{i}_s(s) + \begin{bmatrix} \mathbf{B}_f \\ \hdashline \mathbf{0} \end{bmatrix} \mathbf{v}_s(s) \tag{4.2-23}$$

where the dotted line indicates matrix partitioning. If the coefficient of $\mathbf{i}_b(s)$ is nonsingular, this equation yields

$$\mathbf{i}_b(s) = \begin{bmatrix} \mathbf{B}_f\mathbf{Z}(s) \\ \hdashline \mathbf{A} \end{bmatrix}^{-1} \begin{bmatrix} \mathbf{B}_f\mathbf{Z}(s) \\ \hdashline \mathbf{0} \end{bmatrix} \mathbf{i}_s(s) + \begin{bmatrix} \mathbf{B}_f\mathbf{Z}(s) \\ \hdashline \mathbf{A} \end{bmatrix}^{-1} \begin{bmatrix} \mathbf{B}_f \\ \hdashline \mathbf{0} \end{bmatrix} \mathbf{v}_s(s) \tag{4.2-24}$$

Since the right-hand side of this equation is completely specified, the branch current vector $\mathbf{i}_b(t)$ can be obtained by taking the inverse Laplace transform of both sides of the equation. Having found $\mathbf{i}_b(s)$ from (4.2-24), the branch voltage vector $\mathbf{v}_b(s)$ can then be easily obtained from (4.2-20).

Let us now summarize the steps required in obtaining $\mathbf{v}_b(t)$ and $\mathbf{i}_b(t)$ for a general *RLC* network by the impedance matrix method.

Step 1. Draw the graph of the network, select one of the nodes as the datum, and choose an arbitrary tree T.

Step 2. Write down the incidence matrix $\mathbf{A}$ and the fundamental loop matrix $\mathbf{B}_f$ corresponding to T.

Step 3. From the element VCR write down the element impedance matrix $\mathbf{Z}(s)$.

Step 4. From the network write down the voltage and current source vectors $\mathbf{v}_s(s)$ and $\mathbf{i}_s(s)$.

Step 5. Use equation (4.2-24) to obtain $\mathbf{i}_b(s)$; take the inverse Laplace transform to get $\mathbf{i}_b(t)$.

Step 6. Use equations (4.2-20) and (4.2-24) to compute $\mathbf{v}_b(s)$; take the inverse Laplace transform to get $\mathbf{v}_b(t)$.

Before working out an example to illustrate this procedure, let us consider a special situation that is frequently encountered in practical cases.

Remark 4.2-1 (Source Transformation). If some of the branches of the network under consideration are made up of independent sources alone, the outlined method of analysis is still valid. In fact, the presence of source-only branches reduces the number of the equations that must be solved for the branch voltages and branch currents.

Let us first assume that one of the branches of the network under consideration is made up of a voltage source only. To be more specific, consider Fig. 4.2-4(a), which shows the portion of a given network that has a voltage source-only branch, say b_k. Let the voltage source at b_k be v_{s_k}. If we now remove this voltage source and, instead, add a voltage source v_{s_k} to all the other branches that are incident to *one* of the terminals of b_k, the resulting network will be equivalent to the original network. In this process the branch b_k is then replaced by a short circuit and hence eliminated. Performing this source transformation on the network of Fig. 4.2-4(a) results in the network shown in Fig. 4.2-4(b). Other voltage source-only branches can also be eliminated in the same manner.

Let us now assume that the branch b_k consists of a current source i_{s_k} only.

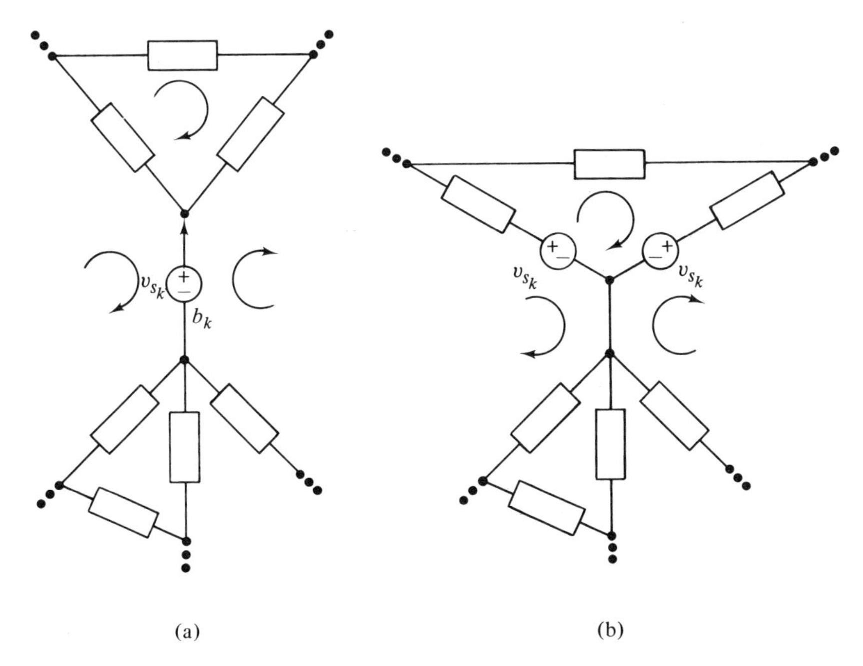

Figure 4.2-4 Voltage source-only branch in (a) is replaced by a short circuit by redrawing the network as in (b).

For example, consider the portion of a given network that is shown in Fig. 4.2-5(a). The branch b_k is made up of the independent current source i_{s_k} alone.

Let us remove the branch b_k altogether and, instead, add a current source i_{s_k} parallel to each branch of any *one* loop of which b_k is a branch. The resulting network will be equivalent to the original network. In this process the branch b_k is replaced by an open circuit. Applying this source transformation to the network shown in Fig. 4.2-5(a), the network shown in Fig. 4.2-5(b) will result.

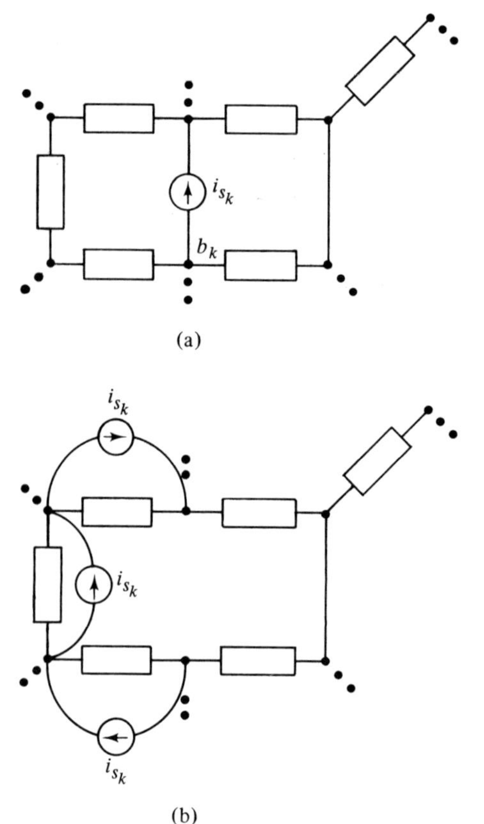

Figure 4.2-5 Current source-only branch in (a) is eliminated by redrawing the network as in (b).

Exercise 4.2-1. Show that the networks in Figs. 4.2-4(b) and 4.2-5(b) are equivalent to the networks shown in Figs. 4.2-4(a) and 4.2-5(a), respectively. (Hint: Show that KCL and KVL in both networks are the same.)

Let us now work out an example to illustrate the analysis technique discussed so far.

Example 4.2-1. Consider the network shown in Fig. 4.2-6. Assume that $L_5 L_6 \neq M^2$, where M is the mutual inductance between inductors L_5 and L_6.

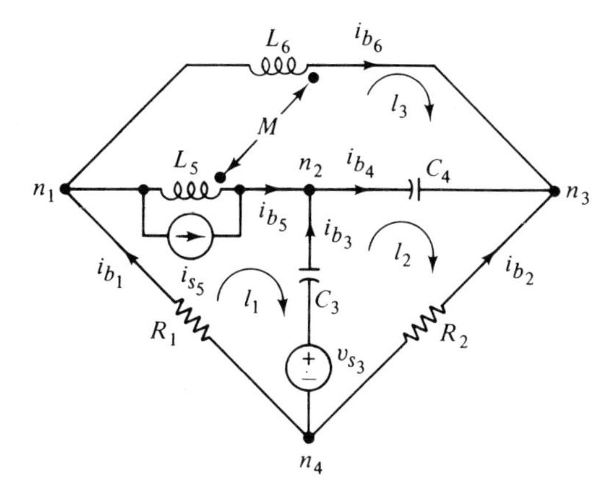

Figure 4.2-6 Network considered in Example 4.2-1.

All the elements are assumed to be linear and time invariant. The problem is to find a set of linearly independent equations that can be solved for all the branch currents and all the branch voltages.

Solution

Taking n_4 as the datum, the incidence matrix $\mathbf{A}$ is

$$\mathbf{A} = \begin{matrix} n_1 \\ n_2 \\ n_3 \end{matrix} \overset{\begin{matrix} R_1 & R_2 & C_3 & C_4 & L_5 & L_6 \end{matrix}}{\begin{bmatrix} -1 & 0 & 0 & 0 & 1 & 1 \\ 0 & 0 & -1 & 1 & -1 & 0 \\ 0 & -1 & 0 & -1 & 0 & -1 \end{bmatrix}} \tag{4.2-25}$$

Furthermore, if we choose a tree T consisting of branches C_3, C_4, and L_5 and assume that its corresponding fundamental loops l_1, l_2, and l_3 are all clockwise oriented, the fundamental loop matrix $\mathbf{B}_f$ will be

$$\mathbf{B}_f = \begin{matrix} l_1 \\ l_2 \\ l_3 \end{matrix} \overset{\begin{matrix} R_1 & R_2 & C_3 & C_4 & L_5 & L_6 \end{matrix}}{\begin{bmatrix} 1 & 0 & -1 & 0 & 1 & 0 \\ 0 & -1 & 1 & 1 & 0 & 0 \\ 0 & 0 & 0 & -1 & -1 & 1 \end{bmatrix}} \tag{4.2-26}$$

The branch currents and element currents are related through (4.2-8); in this case $\mathbf{i}_s(s)$ is

$$\mathbf{i}_s(s) = [0 \quad 0 \quad 0 \quad 0 \quad i_{s_5} \quad 0]^T \tag{4.2-27}$$

Also, the branch voltages and element voltages are related through (4.2-12); for the example in hand, $\mathbf{v}_s(s)$ is

$$\mathbf{v}_s(s) = [0 \quad 0 \quad \mathbf{v}_{s_3} \quad 0 \quad 0 \quad 0]^T \tag{4.2-28}$$

The element impedance matrix $\mathbf{Z}(s)$ defined by (4.2-19) can then be constructed by observing that $\mathbf{R}$ is a 6×6 matrix all of whose elements are zero except the first and second diagonal terms, which are R_1 and R_2; the matrix

$(1/s)\mathbf{S}$ is also a 6×6 matrix all of whose elements are zero except the third and fourth diagonal terms, which are $1/sC_3$ and $1/sC_4$, respectively. The matrix $s\mathbf{L}$ can also be specified in a similar manner; adding these together, we get

$$\mathbf{Z}(s) = \begin{bmatrix} R_1 & 0 & 0 & 0 & 0 & 0 \\ 0 & R_2 & 0 & 0 & 0 & 0 \\ 0 & 0 & \dfrac{1}{sC_3} & 0 & 0 & 0 \\ 0 & 0 & 0 & \dfrac{1}{sC_4} & 0 & 0 \\ 0 & 0 & 0 & 0 & sL_5 & sM \\ 0 & 0 & 0 & 0 & sM & sL_6 \end{bmatrix} \tag{4.2-29}$$

Then $\mathbf{B}_f\mathbf{Z}(s)$ required in (4.2-22) can be obtained by postmultiplying (4.2-26) by $\mathbf{Z}(s)$ obtained in (4.2-29):

$$\mathbf{B}_f\mathbf{Z}(s) = \begin{bmatrix} R_1 & 0 & -\dfrac{1}{sC_3} & 0 & sL_5 & sM \\ 0 & -R_2 & \dfrac{1}{sC_3} & \dfrac{1}{sC_4} & 0 & 0 \\ 0 & 0 & 0 & -\dfrac{1}{sC_4} & s(M-L_5) & s(L_6-M) \end{bmatrix} \tag{4.2-30}$$

which upon augmentation by $\mathbf{A}$ results in

$$\begin{bmatrix} \mathbf{B}_f\mathbf{Z}(s) \\ \hline \mathbf{A} \end{bmatrix} = \begin{bmatrix} R_1 & 0 & -\dfrac{1}{sC_3} & 0 & sL_5 & sM \\ 0 & -R_2 & \dfrac{1}{sC_3} & \dfrac{1}{sC_4} & 0 & 0 \\ 0 & 0 & 0 & -\dfrac{1}{sC_4} & s(M-L_5) & s(L_6-M) \\ \hline -1 & 0 & 0 & 0 & 1 & 1 \\ 0 & 0 & -1 & 1 & -1 & 0 \\ 0 & -1 & 0 & -1 & 0 & -1 \end{bmatrix} \tag{4.2-31}$$

Consequently, all the $B \times B$ matrices in (4.2-23) are determined, and we have obtained six linearly independent equations in six unknowns that can be solved for i_{b_1} through i_{b_6} as in (4.2-24). The branch voltages v_{b_1} through v_{b_6} can then be found by solving (4.2-20). Let us now introduce an alternative approach to obtaining the branch voltages and currents.

Admittance Matrix Method. The branch voltage-current relations can also be derived in terms of the *element admittance matrix* $\mathbf{Y}(s)$. Consider the branch b_k of a given network:

1. If b_k is a resistor with conductance G_k, we can write

$$i_k = G_k v_k \tag{4.2-32}$$

2. If b_k is a capacitor with capacitance C_k,

$$i_k = sC_k v_k \tag{4.2-33}$$

3. To obtain the VCR for the inductive branches, assume that the inductance matrix $\mathbf{L}$ can be written as

$$\mathbf{L} = \begin{bmatrix} \mathbf{0} & \mathbf{0} \\ \mathbf{0} & \hat{\mathbf{L}} \end{bmatrix}_{B \times B}$$

where $\hat{\mathbf{L}}$ is an $m \times m$ submatrix of $\mathbf{L}$, and m is the number of the inductors in the network under study; hence,

$$\hat{\mathbf{L}} = \begin{bmatrix} L_1 & M_{12} & \cdots & M_{1m} \\ M_{21} & L_2 & \cdots & M_{2m} \\ \vdots & \vdots & & \vdots \\ M_{m1} & M_{m2} & \cdots & L_{mm} \end{bmatrix}$$

where L_k is the self-inductance of the branch b_k, and M_{ij} is the mutual inductance between branches b_i and b_j. If $\hat{\mathbf{L}}$ is nonsingular, the *reciprocal inductance* matrix $\mathbf{\Gamma}$ can be defined by

$$\mathbf{\Gamma} = \begin{bmatrix} \mathbf{0} & \mathbf{0} \\ \mathbf{0} & \hat{\mathbf{\Gamma}} \end{bmatrix}_{B \times B} \tag{4.2-34}$$

where $\hat{\mathbf{\Gamma}}$ is the inverse of $\hat{\mathbf{L}}$; that is,

$$\hat{\mathbf{\Gamma}} = \begin{bmatrix} \mathbf{\Gamma}_1 & W_{12} & \cdots & W_{1m} \\ W_{21} & \mathbf{\Gamma}_2 & \cdots & W_{2m} \\ \vdots & \vdots & & \vdots \\ W_{m1} & W_{m2} & \cdots & \mathbf{\Gamma}_{mm} \end{bmatrix} = \begin{bmatrix} L_1 & M_{12} & \cdots & M_{1m} \\ M_{21} & L_2 & \cdots & M_{2m} \\ \vdots & \vdots & & \vdots \\ M_{m1} & M_{m2} & \cdots & L_{mm} \end{bmatrix}^{-1} \tag{4.2-35}$$

in which $\mathbf{\Gamma}_k$ is called the reciprocal self-inductance of b_k, and W_{ij} is called the reciprocal mutual inductance between branches b_i and b_j. Consequently, the current in the kth inductive branch is

$$i_k = \frac{1}{s}\left(\mathbf{\Gamma}_k v_k + \sum_{\substack{j=1 \\ j \neq k}}^{m} W_{kj} v_j\right)$$

Thus, using the preceding conventions, we can write

$$\mathbf{i}(s) = \mathbf{Y}(s)\mathbf{v}(s) \tag{4.2-36}$$

in which the admittance matrix $\mathbf{Y}(s)$ is defined by

$$\mathbf{Y}(s) = \mathbf{G} + s\mathbf{C} + \frac{1}{s}\mathbf{\Gamma} \tag{4.2-37}$$

where $\mathbf{G}$ is the $B \times B$ diagonal conductance matrix whose kth diagonal element is G_k, $\mathbf{C}$ is the $B \times B$ capacitance matrix, and $\mathbf{\Gamma}$ is the $B \times B$ reciprocal inductance matrix defined in (4.2-34).

Using (4.2-36), (4.2-8), and (4.2-12), the corresponding VCR can be written as

$$\mathbf{i}_b(s) = \mathbf{Y}(s)\mathbf{v}_b(s) + \mathbf{i}_s(s) + \mathbf{Y}(s)\mathbf{v}_s(s) \tag{4.2-38}$$

This relation together with (4.1-1) yields

$$\mathbf{AY}(s)\mathbf{v}_b(s) = -\mathbf{AY}(s)\mathbf{v}_s(s) - \mathbf{Ai}_s(s) \tag{4.2-39}$$

Equation (4.2-39) represents N linearly independent equations in B unknowns, which together with (4.1-2) will give

$$\begin{bmatrix} \mathbf{AY}(s) \\ \hdashline \mathbf{B}_f \end{bmatrix} \mathbf{v}_b(s) = -\begin{bmatrix} \mathbf{AY}(s) \\ \hdashline \mathbf{0} \end{bmatrix} \mathbf{v}_s(s) - \begin{bmatrix} \mathbf{A} \\ \hdashline \mathbf{0} \end{bmatrix} \mathbf{i}_s(s) \tag{4.2-40}$$

or equivalently

$$\mathbf{v}_b(s) = -\begin{bmatrix} \mathbf{AY}(s) \\ \hdashline \mathbf{B}_f \end{bmatrix}^{-1} \begin{bmatrix} \mathbf{AY}(s) \\ \hdashline \mathbf{0} \end{bmatrix} \mathbf{v}_s(s) - \begin{bmatrix} \mathbf{AY}(s) \\ \hdashline \mathbf{B}_f \end{bmatrix}^{-1} \begin{bmatrix} \mathbf{A} \\ \hdashline \mathbf{0} \end{bmatrix} \mathbf{i}_s(s) \tag{4.2-41}$$

Consequently, given the admittance matrix $\mathbf{Y}(s)$, (4.2-41) can be solved for $\mathbf{v}_b(s)$, and $\mathbf{i}_b$ can be obtained from (4.2-38).

As in the previous case, let us summarize the important steps in the admittance matrix method just discussed:

Step 1. Draw the graph of the network under consideration; choose a datum and an arbitrary tree T.

Step 2. Write down the incidence matrix $\mathbf{A}$ and the fundamental loop matrix $\mathbf{B}_f$ corresponding to tree T.

Step 3. Obtain the admittance matrix $\mathbf{Y}(s)$ from the network VCR (express these relations in terms of the complex variable s by using the Laplace transform).

Step 4. Write down the source voltage and current vectors $\mathbf{v}_s(s)$ and $\mathbf{i}_s(s)$.

Step 5. Use equation (4.2-41) to compute $\mathbf{v}_b(s)$ and take the inverse Laplace transform to get $\mathbf{v}_b(t)$.

Step 6. Use equation (4.2-38) together with $\mathbf{v}_b(s)$ obtained in step 5 to get $\mathbf{i}_b(s)$.

Let us now work out a simple example to illustrate the method. In this example we assume that the network under consideration contains a gyrator; it will be shown that the presence of a gyrator will not alter the general formulations as outlined.

Example 4.2-2. Consider the gyrator circuit shown in Fig. 4.2-7(a). The corresponding graph is given in Fig. 4.2-7(b). The gyration constant of the gyrator is denoted by α and the conductances of the resistors are represented by G_3 and G_4. To use (4.2-41), we need to find $\mathbf{A}$, $\mathbf{B}_f$, $\mathbf{Y}(s)$, $\mathbf{v}_s(s)$, and $\mathbf{i}_s(s)$.

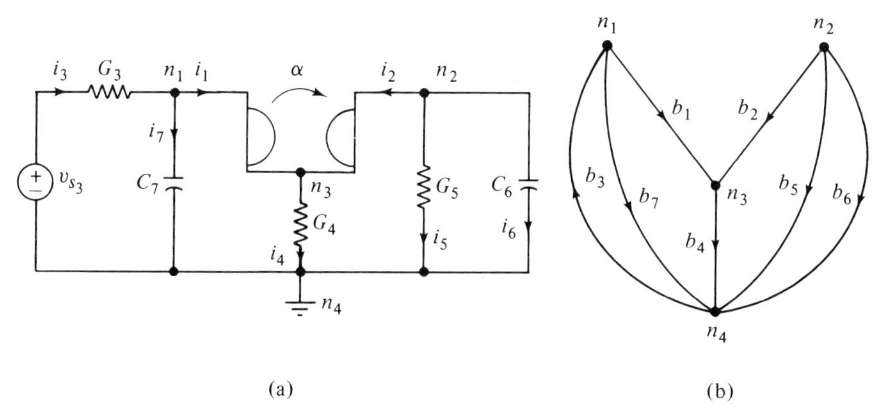

Figure 4.2-7 Gyrator network and its graph considered in Example 4.2-2.

Take n_4 as the datum; the $\mathbf{A}$ matrix can therefore be written as

$$\mathbf{A} = \begin{matrix} n_1 \\ n_2 \\ n_3 \end{matrix} \overset{\begin{matrix} b_1 & b_2 & b_3 & b_4 & b_5 & b_6 & b_7 \end{matrix}}{\begin{bmatrix} 1 & 0 & -1 & 0 & 0 & 0 & 1 \\ 0 & 1 & 0 & 0 & 1 & 1 & 0 \\ -1 & -1 & 0 & 1 & 0 & 0 & 0 \end{bmatrix}} \tag{4.2-42}$$

To write down $\mathbf{B}_f$, let us choose a tree T consisting of the branches b_1, b_2, and b_3. Choosing the orientation of the fundamental loops (l_1, l_2, l_3, l_4) the same as the orientation of the corresponding chords (b_4, b_5, b_6, b_7), the fundamental loop matrix will become

$$\mathbf{B}_f = \begin{matrix} l_1 \\ l_2 \\ l_3 \\ l_4 \end{matrix} \overset{\begin{matrix} b_1 & b_2 & b_3 & b_4 & b_5 & b_6 & b_7 \end{matrix}}{\begin{bmatrix} 1 & 0 & 1 & 1 & 0 & 0 & 0 \\ 1 & -1 & 1 & 0 & 1 & 0 & 0 \\ 1 & -1 & 1 & 0 & 0 & 1 & 0 \\ 0 & 0 & 1 & 0 & 0 & 0 & 1 \end{bmatrix}} \tag{4.2-43}$$

From the branch VCR the branch admittance matrix $\mathbf{Y}(s)$ can be written as

$$\mathbf{Y}(s) = \begin{bmatrix} 0 & \frac{1}{\alpha} & 0 & 0 & 0 & 0 & 0 \\ -\frac{1}{\alpha} & 0 & 0 & 0 & 0 & 0 & 0 \\ 0 & 0 & G_3 & 0 & 0 & 0 & 0 \\ 0 & 0 & 0 & G_4 & 0 & 0 & 0 \\ 0 & 0 & 0 & 0 & G_5 & 0 & 0 \\ 0 & 0 & 0 & 0 & 0 & sC_6 & 0 \\ 0 & 0 & 0 & 0 & 0 & 0 & sC_7 \end{bmatrix} \tag{4.2-44}$$

(Recall that the v-i relations of a gyrator are $v_1 = -\alpha i_2$ and $v_2 = \alpha i_1$.) Since there is only one voltage source in the network, v_{s_3}, the source vectors become

$$\mathbf{v}_s(s) = [0 \quad 0 \quad v_{s_3} \quad 0 \quad 0 \quad 0 \quad 0]^T \tag{4.2-45}$$

and

$$\mathbf{i}_s(s) = \mathbf{0} \tag{4.2-46}$$

For this particular example, the equation (4.2-41) reduces to

$$\mathbf{v}_b(s) = -\begin{bmatrix} \mathbf{AY}(s) \\ \text{-----} \\ \mathbf{B}_f \end{bmatrix}^{-1} \begin{bmatrix} \mathbf{AY}(s) \\ \text{-----} \\ \mathbf{0} \end{bmatrix} \mathbf{v}_s(s)$$

Utilizing the corresponding matrices, this becomes

$$\mathbf{v}_b(s) = \begin{bmatrix} 0 & \frac{1}{\alpha} & -G_3 & 0 & 0 & 0 & sC_7 \\ -\frac{1}{\alpha} & 0 & 0 & 0 & G_5 & sC_6 & 0 \\ \frac{1}{\alpha} & -\frac{1}{\alpha} & 0 & G_4 & 0 & 0 & 0 \\ 1 & 0 & 1 & 1 & 0 & 0 & 0 \\ 1 & -1 & 1 & 0 & 1 & 0 & 0 \\ 1 & -1 & 1 & 0 & 0 & 1 & 0 \\ 0 & 0 & 1 & 0 & 0 & 0 & 1 \end{bmatrix}^{-1} \begin{bmatrix} -G_3 v_{s_3}(s) \\ 0 \\ 0 \\ 0 \\ 0 \\ 0 \\ 0 \end{bmatrix}$$

Carrying out the matrix inversion and multiplication, the Laplace transform of the branch voltage vector $\mathbf{v}_b(s)$ can be determined, which upon taking the inverse Laplace transform yields $\mathbf{v}_b(t)$. The branch current vector $\mathbf{i}_b(s)$ can also be obtained from (4.2-38).

Remark 4.2-2. In the analysis methods discussed we have excluded the presence of dependent sources, ideal transformers, and operational amplifiers. The reason is that these network elements do not have an admittance or impedance representation; for example, the v-i relationship of an ideal transformer is given by $v_2 = nv_1$ and $i_2 = -(1/n)i_1$. That is, the voltages v_1 and v_2 are independent of the currents i_2 and i_1; consequently, neither the admittance matrix $\mathbf{Y}(s)$ nor the impedance matrix $\mathbf{Z}(s)$ is defined for networks containing ideal transformers. This difficulty can be remedied by representing such elements via dependent sources. We take up the analysis of such networks in Section 4.5.

A different approach for obtaining a set of linearly independent equations that can be solved for $\mathbf{i}_b$ and $\mathbf{v}_b$ is to introduce node voltages as auxiliary variables. This method is discussed in the next section.

4.3 NODAL ANALYSIS

Consider a network with $N + 1$ nodes and B branches; if the number of the nodes is considerably less than the number of the branches, the nodal analysis discussed in this section becomes more efficient than other methods. In the nodal analysis we use the node voltages as auxiliary variables and obtain a set of N linear independent equations in N unknowns; this, in contrast to $2B$ independent equations in $2B$ unknowns obtained earlier, provides considerable savings in effort and computation time if N is much smaller than $2B$.

Let us choose the $(N + 1)$st node as the datum and denote the node-to-datum voltages by $v_{n_1}, v_{n_2}, \ldots, v_{n_N}$, and define the node-to-datum voltage vector $\mathbf{v}_n$ by

$$\mathbf{v}_n = (v_{n_1}, v_{n_2}, \ldots, v_{n_N})^T \tag{4.3-1}$$

Then the branch voltage vector $\mathbf{v}_b$ and the node voltage vector $\mathbf{v}_n$ are related by (3.5-8), which we repeat for convenience:

$$\mathbf{v}_b = \mathbf{A}^T\mathbf{v}_n \tag{4.3-2}$$

Here our objective is to obtain a set of N linearly independent equations that can be solved for $\mathbf{v}_n$. To do so, let us replace $\mathbf{v}_b$ from (4.3-2) into (4.2-12) to obtain

$$\mathbf{v} - \mathbf{v}_s = \mathbf{A}^T\mathbf{v}_n \tag{4.3-3}$$

Also, replace $\mathbf{i}_b$ from (4.2-8) into (4.1-1) to get

$$\mathbf{Ai} = -\mathbf{Ai}_s \tag{4.3-4}$$

But $\mathbf{i}$ can be expressed in terms of $\mathbf{v}$ through the admittance matrix $\mathbf{Y}(s)$; hence, from (4.3-4) and (4.2-36), we have

$$\mathbf{AY}(s)\mathbf{v} = -\mathbf{Ai}_s(s) \tag{4.3-5}$$

Replacing $\mathbf{v}$ in (4.3-5) from (4.3-3), we get

$$\mathbf{AY}(s)\mathbf{A}^T\mathbf{v}_n(s) = -\mathbf{AY}(s)\mathbf{v}_s(s) - \mathbf{Ai}_s(s)$$

or, equivalently,

$$\mathbf{Y}_n(s)\mathbf{v}_n(s) = -\mathbf{A}\mathbf{Y}(s)\mathbf{v}_s(s) - \mathbf{A}\mathbf{i}_s(s) \tag{4.3-6}$$

where $\mathbf{Y}_n(s)$ is the $N \times N$ *node admittance matrix* defined by

$$\mathbf{Y}_n(s) \triangleq \mathbf{A}\mathbf{Y}(s)\mathbf{A}^T \tag{4.3-7}$$

Upon multiplication of both sides of (4.3-6) by $\mathbf{Y}_n^{-1}(s)$ (if it exists), we get

$$\mathbf{v}_n(s) = -\mathbf{Y}_n^{-1}(s)\mathbf{A}\mathbf{Y}(s)\mathbf{v}_s(s) - \mathbf{Y}_n^{-1}(s)\mathbf{A}\mathbf{i}_s(s) \tag{4.3-8}$$

Since $\mathbf{v}_s(s)$ and $\mathbf{i}_s(s)$ are given voltage and current sources, (4.3-8) represents a set of N linearly independent equations in $\mathbf{v}_n(s)$, which has a unique solution. Having solved (4.3-8) for $\mathbf{v}_n(s)$, the branch voltage vector $\mathbf{v}_b$ can be obtained from (4.3-2) and then the branch current vector $\mathbf{i}_b$ is readily found using (4.2-38).

Example 4.3-1. Consider the network shown in Fig. 4.2-6. Let us use nodal analysis to obtain a set of linearly independent equations that can be solved for the node voltages v_{n_1}, v_{n_2}, and v_{n_3}. For this network, the conductance matrix $\mathbf{G}$ is a 6×6 matrix all of whose elements are zero except the first and second diagonal terms, which are $1/R_1$ and $1/R_2$, respectively. The capacitance matrix $\mathbf{C}$ is a 6×6 matrix all of whose elements are zero except the third and fourth diagonal elements, which are C_3 and C_4, respectively. The reciprocal inductance matrix $\mathbf{\Gamma}$ can be obtained by inverting the inductance submatrix $\hat{L}$,

$$\hat{\mathbf{L}} = \begin{bmatrix} L_5 & M \\ M & L_6 \end{bmatrix} \tag{4.3-9}$$

and using (4.2-35).

Then the element admittance matrix $\mathbf{Y}(s)$ can be written as

$$\mathbf{Y}(s) = \begin{bmatrix} \frac{1}{R_1} & 0 & 0 & 0 & 0 & 0 \\ 0 & \frac{1}{R_2} & 0 & 0 & 0 & 0 \\ 0 & 0 & sC_3 & 0 & 0 & 0 \\ 0 & 0 & 0 & sC_4 & 0 & 0 \\ 0 & 0 & 0 & 0 & \frac{L_6}{s\Delta} & -\frac{M}{s\Delta} \\ 0 & 0 & 0 & 0 & -\frac{M}{s\Delta} & \frac{L_5}{s\Delta} \end{bmatrix} \tag{4.3-10}$$

where $$\Delta = L_5 L_6 - M^2$$

The incidence matrix $\mathbf{A}$ for this network was found in (4.2-25); using (4.3-7), we get

$$\mathbf{Y}_n(s) = \begin{bmatrix} \dfrac{1}{R_1} + \dfrac{L_5 + L_6 - 2M}{s\Delta} & \dfrac{M - L_6}{s\Delta} & \dfrac{M - L_5}{s\Delta} \\ \dfrac{M - L_6}{s\Delta} & s(C_3 + C_4) + \dfrac{L_6}{s\Delta} & -sC_4 - \dfrac{M}{s\Delta} \\ \dfrac{M - L_5}{s\Delta} & -sC_4 - \dfrac{M}{s\Delta} & \dfrac{1}{R_2} + sC_4 + \dfrac{L_5}{s\Delta} \end{bmatrix} \tag{4.3-11}$$

Also, from (4.2-25), (4.2-27), (4.2-28), (4.3-6), and (4.3-10) we obtain

$$\mathbf{Y}_n(s)\mathbf{v}_n(s) = -\mathbf{AY}(s)\mathbf{v}_s - \mathbf{Ai}_s = \begin{bmatrix} -i_{s_5} \\ -sC_3 v_{s_3} + i_{s_5} \\ 0 \end{bmatrix} \tag{4.3-12}$$

Finally, (4.3-11) and (4.3-12) yield

$$\begin{bmatrix} v_{n_1}(s) \\ v_{n_2}(s) \\ v_{n_3}(s) \end{bmatrix} = \begin{bmatrix} \dfrac{1}{R_1} + \dfrac{L_5 + L_6 - 2M}{s\Delta} & \dfrac{M - L_5}{s\Delta} & \dfrac{M - L_5}{s\Delta} \\ \dfrac{M - L_6}{s\Delta} & s(C_3 + C_4) + \dfrac{L_6}{s\Delta} & -sC_4 - \dfrac{M}{s\Delta} \\ \dfrac{M - L_5}{s\Delta} & -sC_4 - \dfrac{M}{s\Delta} & \dfrac{1}{R_2} + sC_4 + \dfrac{L_5}{s\Delta} \end{bmatrix}^{-1} \times \begin{bmatrix} -i_{s_5} \\ -sC_3 v_{s_3} + i_{s_5} \\ 0 \end{bmatrix}$$

This is a set of three linearly independent equations in v_{n_1}, v_{n_2}, and v_{n_3}. The branch voltages and currents can then be found from (4.3-2) and (4.2-38).

Let us now summarize the steps required for a complete nodal analysis of a given network.

Step 1. Choose an arbitrary node as the datum and write down the incidence matrix $\mathbf{A}$ from the network graph.

Step 2. Find the element admittance matrix $\mathbf{Y}(s)$ from the branch VCR.

Step 3. Use $\mathbf{A}$, obtained in step 1, to calculate $\mathbf{Y}_n(s)$ from:

$$\mathbf{Y}_n(s) = \mathbf{AY}(s)\mathbf{A}^T.$$

Step 4. Obtain the voltage and current source vectors $\mathbf{v}_s(s)$ and $\mathbf{i}_s(s)$ from the network.

Step 5. Use $\mathbf{Y}_n(s)$, $\mathbf{Y}(s)$, $\mathbf{v}(s)$, and $\mathbf{i}_s(s)$ to compute $\mathbf{v}_n(s)$ from

$$\mathbf{v}_n(s) = -\mathbf{Y}_n^{-1}(s)\mathbf{AY}(s)\mathbf{v}_s(s) - \mathbf{Y}_n^{-1}(s)\mathbf{Ai}_s(s)$$

Step 6. Use $\mathbf{v}_b(s) = \mathbf{A}^T\mathbf{v}_n(s)$ and (4.2-38) to compute branch voltage and current vectors $\mathbf{v}_b(s)$ and $\mathbf{i}_b(s)$.

Example 4.3-2. Let us apply the nodal analysis to the gyrator network of Fig. 4.2-7. The incidence matrix $\mathbf{A}$ and element admittance matrix $\mathbf{Y}(s)$ for this network are given in equations (4.2-42) and (4.2-44), respectively. The node admittance matrix is therefore

$$\mathbf{Y}_n(s) = \mathbf{A}\mathbf{Y}(s)\mathbf{A}^T = \begin{bmatrix} G_3 + sC_7 & \dfrac{1}{\alpha} & -\dfrac{1}{\alpha} \\ -\dfrac{1}{\alpha} & G_5 + sC_6 & \dfrac{1}{\alpha} \\ \dfrac{1}{\alpha} & \dfrac{1}{\alpha} & G_4 \end{bmatrix}$$

Using $\mathbf{v}_s(s)$ and $\mathbf{i}_s(s)$ from equations (4.2-45) and (4.2-46), the node voltage vector $\mathbf{v}_n(s)$ can be written as

$$\mathbf{v}_n(s) = \begin{bmatrix} G_3 + sC_7 & \dfrac{1}{\alpha} & -\dfrac{1}{\alpha} \\ -\dfrac{1}{\alpha} & G_5 + sC_6 & \dfrac{1}{\alpha} \\ \dfrac{1}{\alpha} & -\dfrac{1}{\alpha} & G_4 \end{bmatrix}^{-1} \begin{bmatrix} -G_3 v_{s_3} \\ 0 \\ 0 \end{bmatrix}$$

Carrying through the matrix inversion and multiplication, we get

$$\mathbf{v}_n(s) = \frac{-G_3 v_{s_3}(s)}{\Delta(s)} \begin{bmatrix} G_4 G_5 + sC_6 G_4 + \dfrac{1}{\alpha^2} \\ \dfrac{G_4}{\alpha} + \dfrac{1}{\alpha^2} \\ \dfrac{1}{\alpha^2} + \dfrac{G_5 + sC_6}{\alpha} \end{bmatrix}$$

where $\Delta(s)$ denotes the determinant of $\mathbf{Y}_n(s)$.

4.4 LOOP ANALYSIS

An alternative set of auxiliary network variables is the fictitious loop or mesh currents discussed in Chapter 3. The method that makes use of such variables is called *loop* or *mesh analysis.* In this section, we introduce a detailed discussion of loop analysis, which is a generalization of the mesh analysis and can be used to study the behavior of nonplanar as well as planar networks. This method is particularly useful for the analysis of networks in which the number of the fundamental loops is considerably less than the number of the nodes. This is because in loop analysis the number of simultaneous equations describing the network is the same as the number of its fundamental loops.

Consider a network with $N + 1$ nodes and B branches; choose an arbitrary tree T for this network and indicate its corresponding fundamental loops. According to Theorem 3.1-1, there are exactly $B - N$ fundamental loops; denote these by $l_1, l_2, \ldots, l_{(B-N)}$; assign an arbitrary orientation to each loop and denote the corresponding loop currents by $i_{l_1}, \ldots, i_{l_{(B-N)}}$. The loop current vector $\mathbf{i}_l$ can then be defined as

$$\mathbf{i}_l = (i_{l_1}, i_{l_2}, \ldots, i_{l_{(B-N)}})^T \tag{4.4-1}$$

Recall that the branch current vector $\mathbf{i}_b$ can be expressed in terms of $\mathbf{i}_l$ by [see equation (3.5-30)]

$$\mathbf{i}_b = \mathbf{B}_f^T \mathbf{i}_l \tag{4.4-2}$$

where $\mathbf{B}_f$ is the fundamental loop matrix given in Definition 3.3-2. Replacing $\mathbf{i}_b$ in (4.4-2) from $\mathbf{i} = \mathbf{i}_b - \mathbf{i}_s$, we get

$$\mathbf{i} = \mathbf{B}_f^T \mathbf{i}_l - \mathbf{i}_s \tag{4.4-3}$$

Also, replacing $\mathbf{v}_b$ from $\mathbf{v}_b = \mathbf{v} - \mathbf{v}_s$ in the KVL equation $\mathbf{B}_f \mathbf{v}_b = \mathbf{0}$, we obtain

$$\mathbf{B}_f \mathbf{v} = \mathbf{B}_f \mathbf{v}_s \tag{4.4-4}$$

But, the element voltage vector $\mathbf{v}$ is related to the element current vector $\mathbf{i}$ through the impedance matrix $\mathbf{Z}(s)$; that is,

$$\mathbf{v} = \mathbf{Z}(s)\mathbf{i} \tag{4.4-5}$$

Then (4.4-4) and (4.4-5) yield

$$\mathbf{B}_f \mathbf{Z}(s)\mathbf{i} = \mathbf{B}_f \mathbf{v}_s \tag{4.4-6}$$

Furthermore, replacing $\mathbf{i}$ from (4.4-3) in (4.4-6), we get

$$\mathbf{B}_f \mathbf{Z}(s) \mathbf{B}_f^T \mathbf{i}_l = \mathbf{B}_f \mathbf{Z}(s) \mathbf{i}_s + \mathbf{B}_f \mathbf{v}_s \tag{4.4-7}$$

To simplify the notations, let us use the $(B - N) \times (B - N)$ *loop impedance matrix* $\mathbf{Z}_l(s)$ defined by

$$\mathbf{Z}_l(s) = \mathbf{B}_f \mathbf{Z}(s) \mathbf{B}_f^T \tag{4.4-8}$$

Then (4.4-7) and (4.4-8) give

$$\mathbf{Z}_l(s)\mathbf{i}_l = \mathbf{B}_f \mathbf{Z}(s)\mathbf{i}_s + \mathbf{B}_f \mathbf{v}_s \tag{4.4-9}$$

Equation (4.4-9) represents a set of $B - N$ linearly independent simultaneous equations in $B - N$ unknowns that can be solved for the loop currents i_{l_1} through $i_{l_{(B-N)}}$. Indeed, if $\mathbf{Z}_l(s)$ is nonsingular, upon multiplication on both sides by $\mathbf{Z}_l^{-1}(s)$, equation (4.4-9) can be written as

$$\mathbf{i}_l(s) = \mathbf{Z}_l^{-1}(s)\mathbf{B}_f \mathbf{Z}(s)\mathbf{i}_s(s) + \mathbf{Z}_l^{-1}(s)\mathbf{B}_f \mathbf{v}_s(s) \tag{4.4-10}$$

Having solved for $\mathbf{i}_l$, the vector $\mathbf{i}_b$ can be obtained from (4.4-2) and, consequently, $\mathbf{v}_b$ may be calculated using (4.2-23).

Let us now illustrate the application of this method through an example.

Example 4.4-1. Consider the network shown in Fig. 4.4-1; use loop analysis to obtain a set of linearly independent equations that can be solved for the loop currents l_1 and l_2.

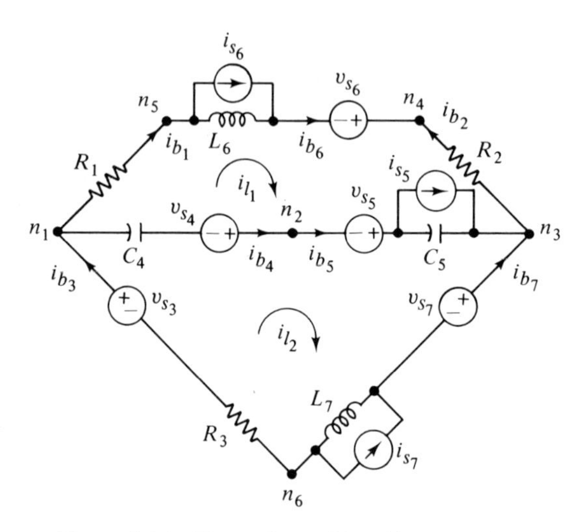

Figure 4.4-1 Network considered in Example 4.4-1.

Solution

For this network $N = 5$ and $B = 7$; hence, $B - N = 2$; then loop analysis is the simplest method in the sense that it yields fewer simultaneous equations that must be solved. Let us choose a tree T consisting of b_6, b_1, b_4, b_5, and b_3; the corresponding chords are b_2 and b_7. There are two fundamental loops l_1 and l_2, as shown in the figure; denote their corresponding currents i_{l_1} and i_{l_2} and assign clockwise orientation to both loops. Notice that since the network under consideration is planar, we can use either mesh or loop analysis. To keep the analysis general, we use the latter one; then we have

$$\mathbf{i}_l = \begin{bmatrix} i_{l_1} \\ i_{l_2} \end{bmatrix} \tag{4.4-11}$$

and

$$\mathbf{B}_f = \begin{array}{c} \\ l_1 \\ l_2 \end{array} \begin{array}{c} \begin{array}{ccccccc} b_1 & b_2 & b_3 & b_4 & b_5 & b_6 & b_7 \end{array} \\ \begin{bmatrix} 1 & -1 & 0 & -1 & -1 & 1 & 0 \\ 0 & 0 & 1 & 1 & 1 & 0 & -1 \end{bmatrix} \end{array} \tag{4.4-12}$$

Also,

$$\mathbf{i}_s = [0 \quad 0 \quad 0 \quad 0 \quad i_{s_5} \quad i_{s_6} \quad i_{s_7}]^T \tag{4.4-13}$$

and

$$\mathbf{v}_s = [0 \quad 0 \quad v_{s_3} \quad v_{s_4} \quad v_{s_5} \quad v_{s_6} \quad v_{s_7}]^T \tag{4.4-14}$$

Since there are no mutual inductances, the element impedance matrix is a diagonal and can be obtained from the figure:

$$\mathbf{Z}(s) = \operatorname{diag}\begin{bmatrix} R_1 & R_2 & R_3 & \frac{1}{sC_4} & \frac{1}{sC_5} & sL_6 & sL_7 \end{bmatrix} \tag{4.4-15}$$

Then using (4.4-8), we get

$$\mathbf{Z}_l(s) = \begin{bmatrix} R_1 + R_2 + \frac{1}{sC_4} + \frac{1}{sC_5} + sL_6 & -\frac{1}{sC_4} - \frac{1}{sC_5} \\ -\frac{1}{sC_4} - \frac{1}{sC_5} & R_3 + \frac{1}{sC_4} + \frac{1}{sC_5} + sL_7 \end{bmatrix} \tag{4.4-16}$$

Finally, from equation (4.4-10) we obtain

$$\begin{bmatrix} i_{l_1}(s) \\ i_{l_2}(s) \end{bmatrix} = \begin{bmatrix} R_1 + R_2 + \frac{1}{sC_4} + \frac{1}{sC_5} + sL_6 & -\frac{1}{sC_4} - \frac{1}{sC_5} \\ -\frac{1}{sC_4} - \frac{1}{sC_5} & R_3 + \frac{1}{sC_4} + \frac{1}{sC_5} + sL_7 \end{bmatrix}^{-1}$$

$$\times \begin{bmatrix} si_{s_6} - \frac{i_{s_5}}{sC_5} - v_{s_4} + v_{s_5} + v_{s_6} \\ -si_{s_6}L_7 - \frac{i_{s_5}}{sC_6} + v_{s_3} + v_{s_4} + v_{s_5} - v_{s_7} \end{bmatrix}$$

The loop currents $i_{l_1}(t)$ and $i_{l_2}(t)$ can then be determined by carrying out the matrix inversion and multiplication and taking the inverse Laplace transform of both sides of the equation.

Example 4.4-2. Consider the bridged gyrator network shown in Fig. 4.4-2(a) and its corresponding graph in part (b). It is required to obtain a set of simultaneous equations that can be solved for the loop currents i_{l_1}, i_{l_2}, and i_{l_3}. As

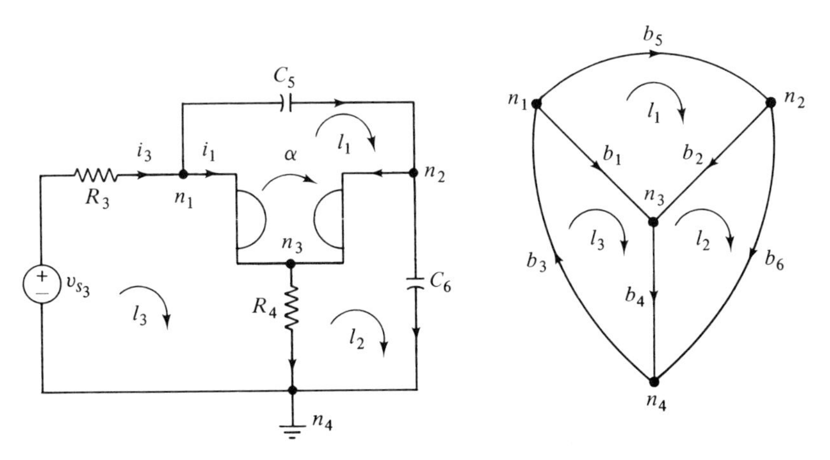

Figure 4.4-2 Network considered in Example 4.4-2 and its corresponding graph.

a numerical example, choose $\alpha = 1$, $R_4 = 1\,\Omega$, $R_3 = 2\,\Omega$, $C_5 = \frac{1}{2}$ F, and $C_6 = \frac{1}{3}$ F, and let the voltage source be a unit step function. Compute the voltage across the capacitor C_6 for all $t \geq 0$.

Solution

Choose a tree consisting of branches b_1, b_2, and b_4. The corresponding fundamental loop matrix $\mathbf{B}_f$ is therefore

$$\mathbf{B}_f = \begin{matrix} l_1 \\ l_2 \\ l_3 \end{matrix} \overset{\begin{matrix} b_1 & b_2 & b_3 & b_4 & b_5 & b_6 \end{matrix}}{\begin{bmatrix} -1 & 1 & 0 & 0 & 1 & 0 \\ 0 & -1 & 0 & -1 & 0 & 1 \\ 1 & 0 & 1 & 1 & 0 & 0 \end{bmatrix}}$$

Furthermore, the element impedance matrix $\mathbf{Z}(s)$ can be directly obtained from the branch VCR:

$$\mathbf{Z}(s) = \begin{bmatrix} 0 & -\alpha & 0 & 0 & 0 & 0 \\ \alpha & 0 & 0 & 0 & 0 & 0 \\ 0 & 0 & R_3 & 0 & 0 & 0 \\ 0 & 0 & 0 & R_4 & 0 & 0 \\ 0 & 0 & 0 & 0 & \dfrac{1}{sC_5} & 0 \\ 0 & 0 & 0 & 0 & 0 & \dfrac{1}{sC_6} \end{bmatrix}$$

The loop impedance matrix $\mathbf{Z}_l(s)$ can then be computed as

$$\mathbf{Z}_l(s) = \mathbf{B}_f \mathbf{Z}(s) \mathbf{B}_f^T = \begin{bmatrix} \dfrac{1}{sC_5} & -\alpha & \alpha \\ \alpha & R_4 + \dfrac{1}{sC_6} & -\alpha - R_4 \\ -\alpha & \alpha - R_4 & R_3 + R_4 \end{bmatrix}$$

The voltage and current source vectors are easily determined from the figure:

$$\mathbf{v}_s(s) = [0 \quad 0 \quad v_{s_3}(s) \quad 0 \quad 0 \quad 0]^T$$

and

$$\mathbf{i}_s(s) = \mathbf{0}$$

Thus, equation (4.4-10) reduces to

$$\mathbf{i}_l(s) = \begin{bmatrix} \dfrac{1}{sC_5} & -\alpha & \alpha \\ \alpha & R_4 + \dfrac{1}{sC_6} & -\alpha - R_4 \\ -\alpha & \alpha - R_4 & R_3 + R_4 \end{bmatrix}^{-1} \begin{bmatrix} 0 \\ 0 \\ v_{s_3}(s) \end{bmatrix}$$

Inserting the numerical values of the network elements, $\mathbf{i}_l(s)$ becomes

$$\mathbf{i}_l(s) = \begin{bmatrix} \frac{2}{s} & -1 & 1 \\ 1 & 1+\frac{3}{s} & -2 \\ -1 & 0 & 3 \end{bmatrix}^{-1} \begin{bmatrix} 0 \\ 0 \\ \frac{1}{s} \end{bmatrix}$$

Carrying through the matrix inversion and multiplication, we get

$$\mathbf{i}_l(s) = \begin{bmatrix} \dfrac{s-3}{2s^2+9s+18} \\ \dfrac{4+s}{2s^2+9s+18} \\ \dfrac{s^2+2s+6}{s(2s^2+9s+18)} \end{bmatrix}$$

Now the voltage across the terminating capacitor $v_c(s)$ can be determined from

$$v_c(s) = \frac{1}{sC_6} i_{l_2}(s)$$

Thus,

$$v_c(s) = \frac{12+3s}{s(2s^2+9s+18)}$$

Using a partial fraction expansion, this can be written as

$$v_c(s) = \frac{2}{3}\left[\frac{1}{s} - \frac{s+\frac{9}{4}}{\left(s+\frac{9}{4}\right)^2 + \left(\frac{3\sqrt{7}}{4}\right)^2}\right]$$

Finally, taking the inverse Laplace transform from both sides of this equation yields

$$v_c(t) = \frac{2}{3} - \frac{2}{3}e^{-(9/4)t}\cos\frac{3\sqrt{7}}{4}t \quad \text{for } t \geq 0$$

Let us now give a summary of the steps involved in this loop analysis.

Step 1. Choose a tree T and find its corresponding fundamental loop matrix $\mathbf{B}_f$.

Step 2. Write down the element impedance matrix $\mathbf{Z}(s)$ using the branch VCR.

Step 3. Compute the loop impedance matrix $\mathbf{Z}_l(s)$ from $\mathbf{Z}_l(s) = \mathbf{B}_f\mathbf{Z}(s)\mathbf{B}_f^T$.

Step 4. Write down the voltage source vector $\mathbf{v}_s(s)$ and the current source vector $\mathbf{i}_s(s)$ from the network.

Step 5. Use the results of steps 2, 3, and 4 to write the desired loop equations as

$$\mathbf{i}_l(s) = \mathbf{Z}_l^{-1}(s)\mathbf{B}_f\mathbf{Z}(s)\mathbf{i}_s(s) + \mathbf{Z}_l^{-1}(s)\mathbf{B}_f\mathbf{v}_s(s)$$

Step 6. Carry out the matrix inversion and multiplication and take the inverse Laplace transform to obtain $\mathbf{i}_l(t)$.

4.5 ANALYSIS OF NETWORKS CONTAINING DEPENDENT SOURCES

So far in this chapter we have assumed that the network under study does not have any dependent sources. In many practical applications, however, the presence of dependent sources is quite common. Networks containing transistors, operational amplifiers, ideal transformers, vacuum tubes, negative convertors, and so on, often can be modeled as linear time-invariant networks with dependent voltage or current sources. Table 4.5-1 shows some of the typical network elements that can be modeled with dependent sources. In this section we develop a general method for the loop analysis of such networks. The nodal analysis can then be obtained in a similar manner. At first, we assume that all the voltage sources are voltage controlled and all the current sources are current controlled. Then we show that current controlled voltage sources and voltage-controlled current sources can also be handled with a simple transformation.

The general procedure for the analysis of such networks can be best understood if we start with an example.

Example 4.5-1. Consider the network shown in Fig. 4.5-1 in which all the dependent current sources are current controlled and all the dependent voltage sources are voltage controlled. The scalar constants α_1, μ_1, α_3, and μ_5 are assumed to be positive. The branch voltage vector v_b can be written as

$$\mathbf{v}_b = -\mathbf{v}_s + \mathbf{v} - \mathbf{P}\mathbf{v} \tag{4.5-1}$$

where $\mathbf{v}_s$ represents the independent voltage sources, $\mathbf{v}$ is the element voltage vector, and $\mathbf{P}\mathbf{v}$ represents the dependent voltage-controlled voltage sources; $\mathbf{P}$ is a $B \times B$ matrix. For the example in hand, we have

$$\mathbf{v}_s = [0 \quad 0 \quad v_{s_3} \quad v_{s_4} \quad 0]^T$$

and

$$\mathbf{P}\mathbf{v} = [\mu_5 v_5 \quad 0 \quad 0 \quad 0 \quad \mu_1 v_1]^T$$

Then

$$\mathbf{P} = \begin{bmatrix} 0 & 0 & 0 & 0 & \mu_5 \\ 0 & 0 & 0 & 0 & 0 \\ 0 & 0 & 0 & 0 & 0 \\ 0 & 0 & 0 & 0 & 0 \\ \mu_1 & 0 & 0 & 0 & 0 \end{bmatrix}$$

Table 4.5-1

Network Element	Model
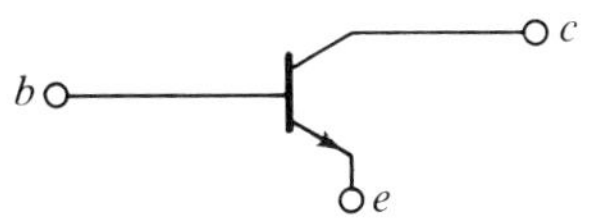 b, c, e Common-Emitter Transistor	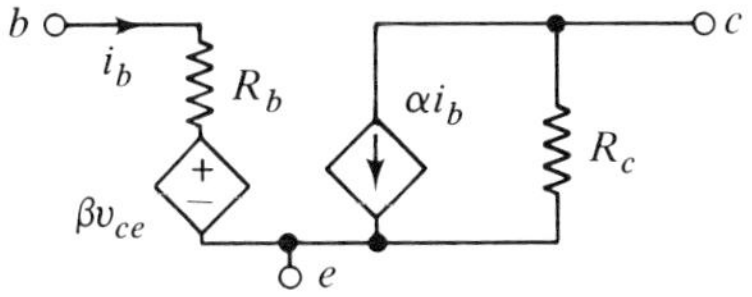 b, c, i_b, R_b, αi_b, R_c, βv_{ce}, e Hybrid π Model
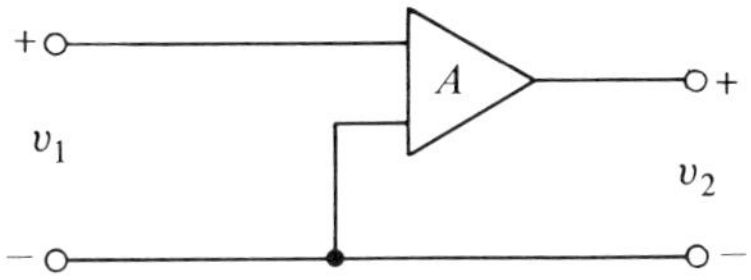 $+$, v_1, $-$, A, $+$, v_2, $-$ Open-Loop Operational Amplifier	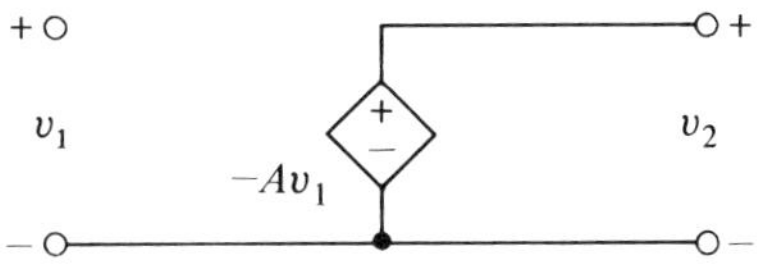 $+$, v_1, $-$, $-Av_1$, $+$, v_2, $-$
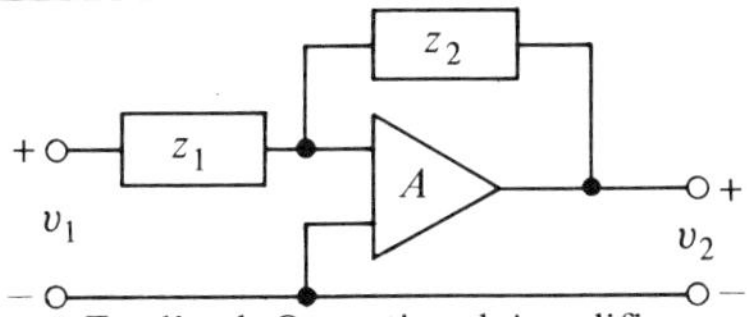 z_2, z_1, $+$, v_1, $-$, A, $+$, v_2, $-$ Feedback Operational Amplifier	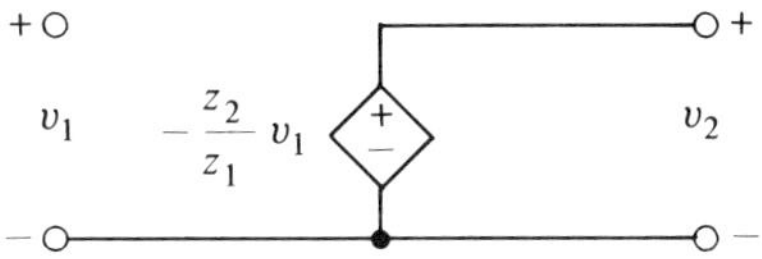 $+$, v_1, $-$, $-\frac{z_2}{z_1}v_1$, $+$, v_2, $-$
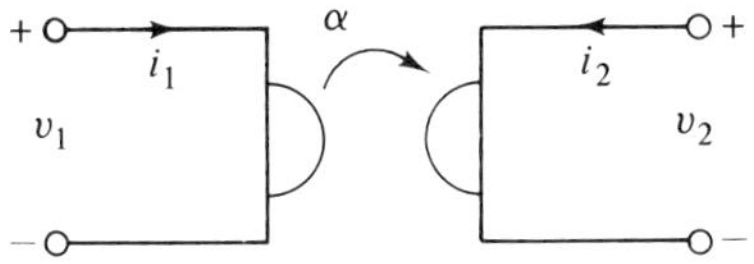 $+$, i_1, v_1, $-$, α, i_2, $+$, v_2, $-$ Gyrator	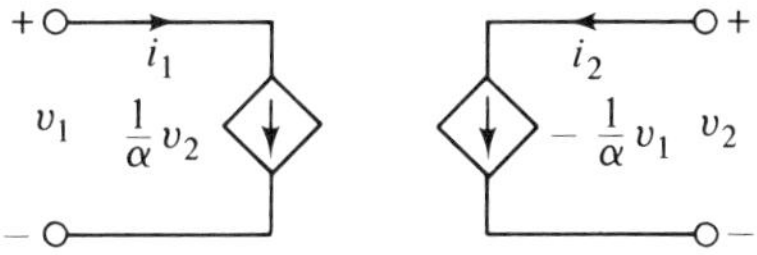 $+$, i_1, v_1, $-$, $\frac{1}{\alpha}v_2$, $-\frac{1}{\alpha}v_1$, i_2, $+$, v_2, $-$ One of Two Possible Two-Source Models
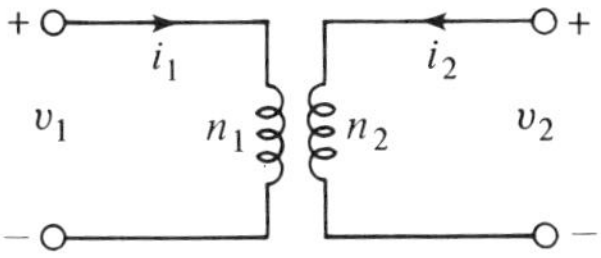 $+$, i_1, v_1, $-$, n_1, n_2, i_2, $+$, v_2, $-$ Ideal Transformer	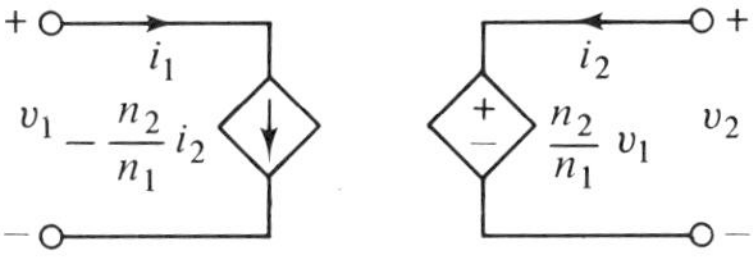 $+$, i_1, v_1, $-$, $-\frac{n_2}{n_1}i_2$, $\frac{n_2}{n_1}v_1$, i_2, $+$, v_2, $-$ One of Two Possible Two-Source Models
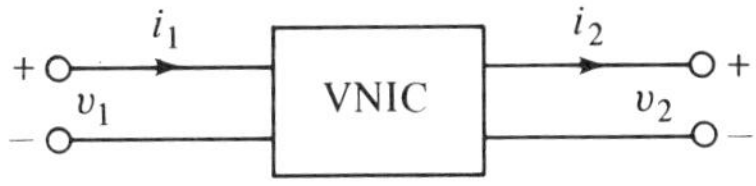 i_1, $+$, v_1, $-$, VNIC, i_2, $+$, v_2, $-$ Voltage-Inversion Negative-Impedance Convertor	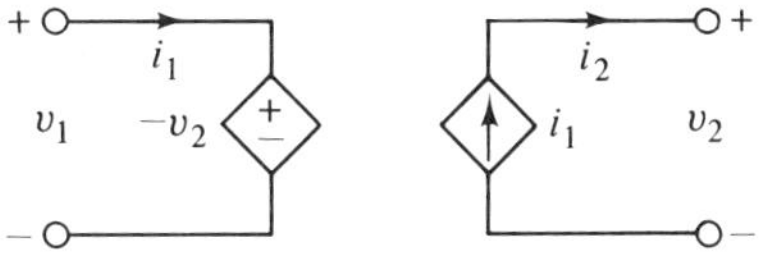 $+$, i_1, v_1, $-$, $-v_2$, i_1, i_2, $+$, v_2, $-$ One of Two Possible Two-Source Models
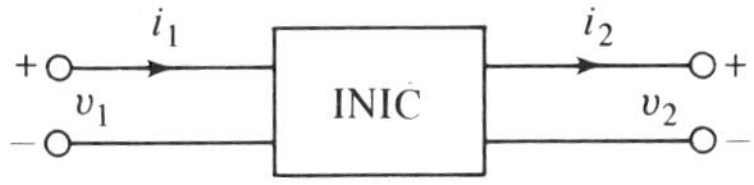 i_1, $+$, v_1, $-$, INIC, i_2, $+$, v_2, $-$ Current-Inversion Negative-Impedance Convertor	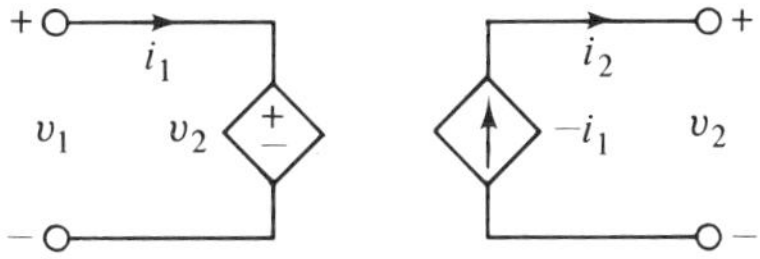 $+$, i_1, v_1, $-$, v_2, $-i_1$, i_2, $+$, v_2, $-$ One of Two Possible Two-Source Models

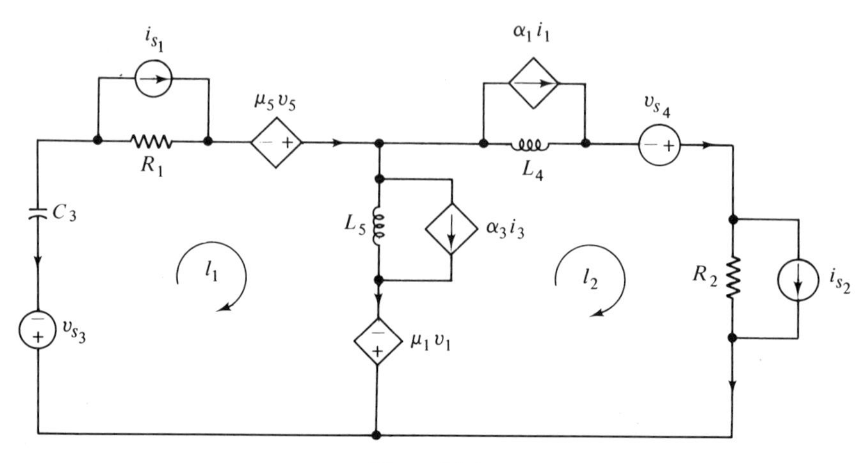

Figure 4.5-1

The branch current vector $\mathbf{i}_b$ can be written as

$$\mathbf{i}_b = \mathbf{i}_s + \mathbf{i} + \mathbf{Q}\mathbf{i} \tag{4.5-2}$$

where $\mathbf{i}_s$ represents the independent current sources, i is the element current vector, and $\mathbf{Qi}$ represents the dependent current-controlled current sources; $\mathbf{Q}$ is a $B \times B$ constant matrix. In this example we have

$$\mathbf{i}_s = [i_{s_1} \quad i_{s_2} \quad 0 \quad 0 \quad 0]^T$$

and

$$\mathbf{i} = [0 \quad 0 \quad 0 \quad \alpha_1 i_1 \quad \alpha_3 i_3]^T$$

Then

$$\mathbf{Q} = \begin{bmatrix} 0 & 0 & 0 & 0 & 0 \\ 0 & 0 & 0 & 0 & 0 \\ 0 & 0 & 0 & 0 & 0 \\ \alpha_1 & 0 & 0 & 0 & 0 \\ 0 & 0 & \alpha_3 & 0 & 0 \end{bmatrix}$$

Let us now derive a general formulation of the loop analysis of networks containing dependent sources.

Rewrite (4.5-1) and (4.5-2) as

$$\mathbf{v}_b = -\mathbf{v}_s + (\mathbf{1} - \mathbf{P})\mathbf{v} \tag{4.5-3}$$

$$\mathbf{i}_b = \mathbf{i}_s + (\mathbf{1} + \mathbf{Q})\mathbf{i} \tag{4.5-4}$$

where $\mathbf{1}$ represents the $B \times B$ identity matrix.

From KVL, we have $\mathbf{B}_f\mathbf{v}_b = \mathbf{0}$, which together with (4.5-3) yields

$$\mathbf{B}_f(\mathbf{1} - \mathbf{P})\mathbf{v} = \mathbf{B}_f\mathbf{v}_s \tag{4.5-5}$$

Recall from (3.5-30) that the branch currents can be written in terms of loop currents by

$$\mathbf{i}_b = \mathbf{B}_f^T \mathbf{i}_l$$

Then, using (4.5-4), we get

$$(\mathbf{1} + \mathbf{Q})\mathbf{i} = \mathbf{B}_f^T \mathbf{i}_l - \mathbf{i}_s \tag{4.5-6}$$

If we now assume that the matrix $(\mathbf{1} + \mathbf{Q})$ is nonsingular, we can write (4.5-6) as

$$\mathbf{i} = (\mathbf{1} + \mathbf{Q})^{-1}\mathbf{B}_f^T \mathbf{i}_l - (\mathbf{1} + \mathbf{Q})^{-1}\mathbf{i}_s \tag{4.5-7}$$

Furthermore, the element voltage vector $\mathbf{v}$ is related to the element current vector $\mathbf{i}$ through the impedance matrix; that is,

$$\mathbf{v} = \mathbf{Z}(s)\mathbf{i}$$

Then (4.5-5) can be written

$$\mathbf{B}_f(\mathbf{1} - \mathbf{P})\mathbf{Z}(s)\mathbf{i} = \mathbf{B}_f \mathbf{v}_s \tag{4.5-8}$$

Replacing $\mathbf{i}$ from (4.5-7) in (4.5-8), we obtain

$$\mathbf{B}_f(\mathbf{1} - \mathbf{P})\mathbf{Z}(s)(\mathbf{1} + \mathbf{Q})^{-1}\mathbf{B}_f^T \mathbf{i}_l = \mathbf{B}_f[(\mathbf{1} - \mathbf{P})\mathbf{Z}(s)(\mathbf{1} + \mathbf{Q})^{-1}\mathbf{i}_s + \mathbf{v}_s]$$

or, simply,

$$\mathbf{i}_l(s) = \mathbf{Z}_l^{-1}(s)\mathbf{B}_f[(\mathbf{1} - \mathbf{P})\mathbf{Z}(s)(\mathbf{1} + \mathbf{Q})^{-1}\mathbf{i}_s(s) + \mathbf{v}_s(s)] \tag{4.5-9}$$

where we have used the following shorthand notations:

$$\mathbf{Z}_l(s) \triangleq \mathbf{B}_f(\mathbf{1} - \mathbf{P})\mathbf{Z}(s)(\mathbf{1} + \mathbf{Q})^{-1}\mathbf{B}_f^T \tag{4.5-10}$$

Equation (4.5-9) is the desired loop equation; $\mathbf{Z}_l(s)$ is the loop impedance matrix and can be obtained using (4.5-10). Note that if there are no dependent sources in the network, the matrices $\mathbf{P}$ and $\mathbf{Q}$ will be zero and (4.5-9) reduces to (4.4-10).

For the example of Fig. 4.5-1, we have

$$\mathbf{Z}(s) = \operatorname{diag}\begin{bmatrix} R_1 & R_2 & \dfrac{1}{sC_3} & sL_4 & vL_5 \end{bmatrix}$$

and

$$(\mathbf{1} + \mathbf{Q})^{-1} = \begin{bmatrix} 1 & 0 & 0 & 0 & 0 \\ 0 & 1 & 0 & 0 & 0 \\ 0 & 0 & 1 & 0 & 0 \\ -\alpha_1 & 0 & 0 & 1 & 0 \\ 0 & 0 & -\alpha_3 & 0 & 1 \end{bmatrix}$$

Then

$$(\mathbf{1} - \mathbf{P})\mathbf{Z}(s)(\mathbf{1} + \mathbf{O})^{-1} = \begin{bmatrix} R_1 & 0 & \alpha_3\mu_5 sL_5 & 0 & -\mu_5 sL_5 \\ 0 & R_2 & 0 & 0 & 0 \\ 0 & 0 & \dfrac{1}{sC_3} & 0 & 0 \\ -s\alpha_1 L_4 & 0 & 0 & sL_4 & 0 \\ -\mu_1 R_1 & 0 & -\alpha_3 sL_5 & 0 & sL_5 \end{bmatrix}$$

Also, from the network under study we can write

$$\mathbf{B}_f = \begin{bmatrix} 1 & 0 & -1 & 0 & 1 \\ 0 & 1 & 0 & 1 & -1 \end{bmatrix}$$

Then

$$\mathbf{Z}_l(s) = \begin{bmatrix} (1-\mu_1)R_1 + \dfrac{1}{sC_3} + (1+\alpha_3)(1-\mu)sL_5 & -(1+\mu_5)sL_5 \\ -\alpha_1 sL_4 + \mu_1 R_1 - (1+\alpha_3)sL_5 & R_2 + sL_4 + sL_5 \end{bmatrix}$$

The loop current vector $i_l(s)$ can then be determined using (4.5-9). As was mentioned earlier, the nodal analysis of networks containing dependent sources can also be developed along the same lines.

Exercise 4.5-1. Use a nodal analysis to obtain a set of independent equations that can be solved for the node voltage vector $\mathbf{v}_n(s)$.

Ans: $\mathbf{v}_n(s) = -\mathbf{Y}_n^{-1}(s)\mathbf{A}(\mathbf{1}+\mathbf{Q})\mathbf{Y}(s)(\mathbf{1}-\mathbf{P})^{-1}\mathbf{v}_s(s) - \mathbf{Y}_n^{-1}(s)\mathbf{A}\mathbf{i}_s(s)$ (4.5-11)

where the node impedance matrix $\mathbf{Y}_n(s)$ is given by

$$\mathbf{Y}_n(s) = \mathbf{A}(\mathbf{1}+\mathbf{Q})\mathbf{Y}(s)(\mathbf{1}-\mathbf{P})^{-1}\mathbf{A}^T \tag{4.5-12}$$

Remark 4.5-1. In the derivation of loop equations we assumed that the matrix $(\mathbf{1}+\mathbf{Q})$ is nonsingular; in nodal analysis we must assume that $(\mathbf{1}-\mathbf{P})$ is nonsingular. Then the first step toward nodal or loop analysis is to check the determinants of $(\mathbf{1}-\mathbf{P})$ and $(\mathbf{1}+\mathbf{Q})$ matrices and then choose the appropriate method.

Remark 4.5-2. So far in this section we have assumed that all the dependent voltage sources are voltage controlled and all the current sources are current-controlled. A glance at Table 4.5-1, however, shows that some of the most useful network elements are modeled by voltage-controlled current sources and/or current-controlled voltage sources. To include this latter group of dependent sources in our analysis, we first perform the source transformation discussed earlier in this chapter to put all the voltage sources in series and all the current sources in parallel with a nonzero impedance. A typical current-controlled voltage source will then appear in the branch b_k of a given network as in Fig. 4.5-2(a); the voltage across the dependent voltage source depends on the i_j current through the branch b_j. This current-controlled voltage source, however, can be transformed to a current-controlled current source in *parallel* with the same branch impedance $z_k(s)$. This resulting branch is shown in Fig. 4.5-2(b). It is an easy exercise to show that these two branches are equivalent. Indeed, the VCR of either branch is given by

$$i_k(s) = \frac{1}{z_k(s)}v_k(s) - \frac{\alpha}{z_k(s)}i_j(s) \tag{4.5-13}$$

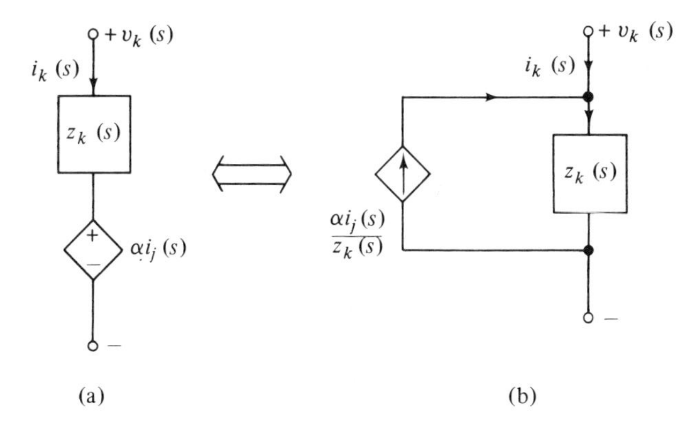

Figure 4.5-2 Current-controlled voltage source of (a) can be transformed to the current-controlled current source of (b).

The final result is, therefore, a current-controlled current source, which can be treated with the method just discussed.

Voltage-controlled current sources can also be dealt with in a similar manner. Consider the branch b_k of a network consisting of a voltage-controlled current source in parallel with the impedance $z_k(s)$ [see Fig. 4.5-3(a)]. The

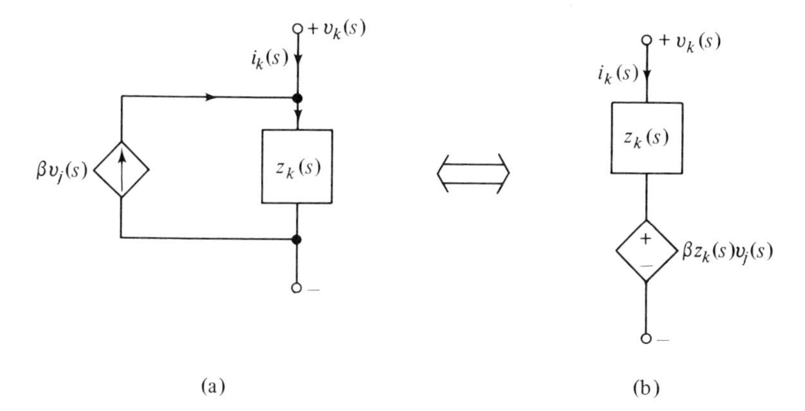

Figure 4.5-3 Voltage-controlled current source of (a) can be transformed to the voltage-controlled voltage source of (b).

current through the dependent current source depends on the voltage v_j across the branch b_j. This source can be transformed to a current-controlled current source in series with the same impedance z_k; the resulting branch is shown in Fig. 4.5-3(b).

The VCR of either branch is given by

$$v_k(s) = z_k(s)i_k(s) + \beta z_k(s)v_j(s) \tag{4.5-14}$$

Having performed the transformation of dependent sources, either loop analysis [equation (4.5-9)] or nodal analysis [equation (4.5-11)] can be used. Let us now conclude this section with an example.

Example 4.5-2. Consider the operational amplifier circuit shown in Fig. 4.5-4. Let us perform a nodal analysis on this circuit to find the node voltage v_1 and v_2. Using the third entry from Table 4.5-1, this circuit can be modeled as in Fig. 4.5-5(a); the voltage across the dependent voltage source is $-sC_6R_7v_2$.

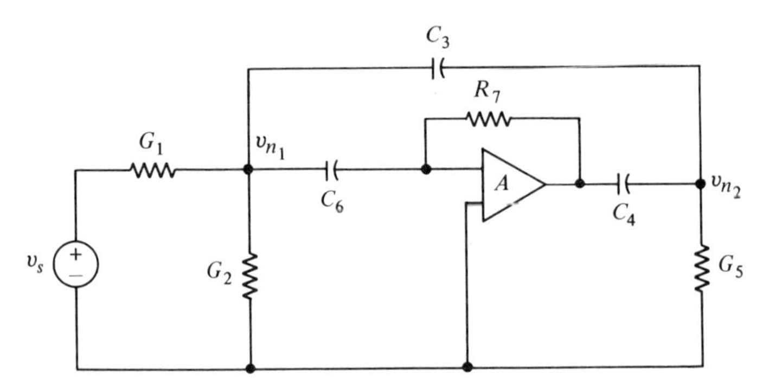

Figure 4.5-4 Operational amplifier circuit considered in Example 4.5-2.

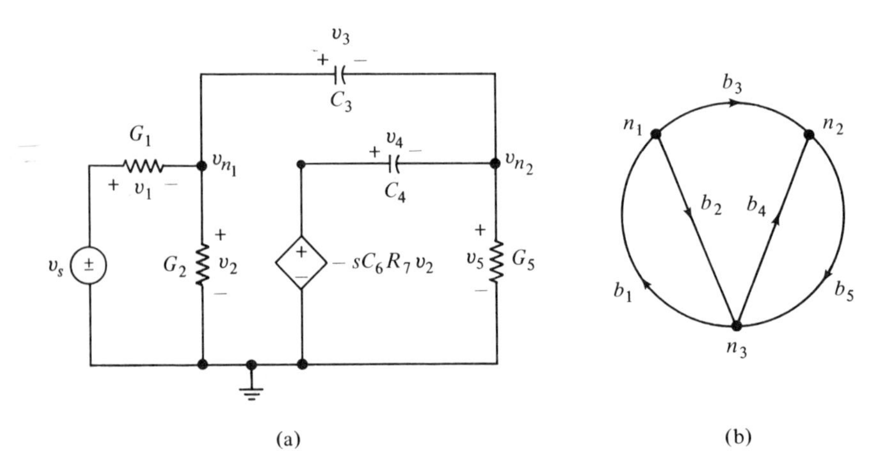

Figure 4.5-5 (a) Model of the circuit shown in Fig. 4.5-4, and (b) its graph.

The graph of the resulting circuit is shown in Fig. 4.5-5(b). To be able to use (4.5-11) and (4.5-12) for a nodal analysis, we need to obtain **A**, **Q**, **P**, $\mathbf{v}_s(s)$, $\mathbf{i}_s(s)$, and $\mathbf{Y}(s)$. Taking n_3 as the datum, the incidence matrix **A** is

$$\mathbf{A} = \begin{matrix} & b_1 & b_2 & b_3 & b_4 & b_5 \\ n_1 & -1 & 1 & 1 & 0 & 0 \\ n_2 & 0 & 0 & -1 & -1 & 1 \end{matrix}$$

Since there are no dependent current sources in the network, the **Q** matrix is identically zero. Furthermore, since there is only one voltage-controlled source present, the **P** matrix is a 5×5 matrix with all the entries equal to zero except p_{42}, which is equal to $-sC_6R_7$; that is,

$$\mathbf{P} = \begin{bmatrix} 0 & 0 & 0 & 0 & 0 \\ 0 & 0 & 0 & 0 & 0 \\ 0 & 0 & 0 & 0 & 0 \\ 0 & -sC_6R_7 & 0 & 0 & 0 \\ 0 & 0 & 0 & 0 & 0 \end{bmatrix}$$

and $\mathbf{Q} = \mathbf{0}$

From the circuit it is clear that

$$\mathbf{i}_s(s) = \mathbf{0}, \qquad \mathbf{v}_s(s) = [v_s(s) \quad 0 \quad 0 \quad 0 \quad 0]^T$$

and $\mathbf{Y}(s) = \text{diag}\{G_1 \quad G_2 \quad sC_3 \quad sC_4 \quad G_5\}$

Using (4.5-12), the node impedance matrix becomes

$$\mathbf{Y}_n(s) = \begin{bmatrix} G_1 + G_2 + sC_3 & -sC_3 \\ -sC_3 + s^2C_4C_6R_7 & sC_3 + sC_4 + G_5 \end{bmatrix}$$

Using $\mathbf{Y}_n(s)$ in (4.5-11) yields

$$\mathbf{v}_n(s) = -\begin{bmatrix} G_1 + G_2 + sC_3 & -sC_3 \\ -sC_3 + s^2C_4C_6R_7 & sC_3 + sC_4 + G_5 \end{bmatrix}^{-1} \begin{bmatrix} -G_1 \\ 0 \end{bmatrix} v_s(s)$$

As a numerical example, let $G_1 = G_2 = G_5 = 1$, $C_3 = C_4 = C_6 = 1$, $R_7 = 1$, and $v_s(t)$ be a unit step function [i.e., $v_s(s) = 1/s$]; then

$$\mathbf{v}_n(s) = -\begin{bmatrix} 2 + s & -s \\ -s + s^2 & 2s + 1 \end{bmatrix}^{-1} \begin{bmatrix} -\dfrac{1}{s} \\ 0 \end{bmatrix} = \begin{bmatrix} \dfrac{1 + 2s}{s(s^3 + s^2 + 5s + 2)} \\ \dfrac{-s + 1}{s^3 + s^2 + 5s + 2} \end{bmatrix}$$

which upon taking the inverse Laplace transform yields $\mathbf{v}_n(t)$.

Before proceeding to the next section, let us point out that the general formulations developed in this and the previous sections serve mainly as a systematic procedure for attacking large-scale network analysis problems. Furthermore, these procedures constitute the basic foundations of computer-aided network analysis programs. In many simple cases, however, shortcuts can be used, particularly if the network under study has but a few elements. Details of such techniques are discussed in several introductory texts [3–7].

4.6 SINUSOIDAL STEADY-STATE ANALYSIS

So far we have assumed that the network under consideration is excited by arbitrary voltage and/or current sources $v_s(t)$ and $i_s(t)$. In many electrical networks, such as power systems and communication networks, however, these sources are primarily of sinusoidal nature. A special formulation of the network response to such sources, therefore, is of considerable importance. Since the sinusoidal steady-state analysis of networks is extensively covered in introductory texts [3, 7], we briefly mention the basic idea behind this procedure in conjunction with the nodal and loop analysis developed in the previous sections.

Let us start by assuming that a typical independent source, say the voltage source $v_k(t)$, is given as

$$v_k(t) = |V_k| \cos(\omega_0 t + \phi_k) \tag{4.6-1}$$

where $|V_k|$, ω_0, and ϕ_k are real constant numbers representing the *amplitude*, *angular frequency*, and the *phase* of the sinusoidal source $v_k(t)$. To simplify the manipulations, we use *phasor* notation; more specifically, we define the phasor V_k to be a complex number given by

$$V_k = |V_k| e^{j\phi_k} \tag{4.6-2}$$

The input signal $v_k(t)$ can then be written in terms of the phasor V_k:

$$v_k(t) = \text{Re}\{V_k e^{j\omega_0 t}\} \tag{4.6-3}$$

In words, $v_k(t)$ is equal to the real part of $V_k e^{j\omega_0 t}$. Note that the phase of the signal, ϕ_k, is included in the phasor V_k. The validity of equation (4.6-3) can be checked by using the Euler equation

$$e^{j\theta} = \cos\theta + j\sin\theta \tag{4.6-4}$$

If the source vectors $\mathbf{v}_s(t)$ and $\mathbf{i}_s(t)$ are composed of sinusoidal functions with arbitrary amplitudes and phases but *the same frequency* ω_0, they can be written as

$$\hat{\mathbf{v}}_s(t) = \text{Re}\{\boldsymbol{V} e^{j\omega_0 t}\} \tag{4.6-5}$$

and

$$\hat{\mathbf{i}}_s(t) = \text{Re}\{\boldsymbol{I} e^{j\omega_0 t}\} \tag{4.6-6}$$

where $\boldsymbol{V}$ and $\boldsymbol{I}$ are column vectors whose kth elements V_k and I_k are voltage and current phasors, respectively. Note that we have deviated from our usual notation and have represented the vector phasors in boldface, italic capital letters to emphasis their phasor nature.

Since the network under consideration is linear and time invariant, to simplify the manipulations we can first find the response of the network of the complex exponential excitations:

$$\mathbf{v}_s(t) = \boldsymbol{V} e^{j\omega_0 t} \tag{4.6-7}$$

$$\mathbf{i}_s(t) = \boldsymbol{I} e^{j\omega_0 t} \tag{4.6-8}$$

and then take the real part of the corresponding complex response.

With this assumption, let us obtain the sinusoidal response of a linear time-invariant network using a nodal analysis. Recall that for given voltage and current source vectors $\mathbf{v}_s(s)$ and $\mathbf{i}_s(s)$ the complex node voltage vector $\mathbf{v}_n(s)$ can be obtained [see equation (4.3-6)] by solving

$$\mathbf{Y}_n(s)\mathbf{v}_n(s) = -\mathbf{A}\mathbf{i}_s(s) - \mathbf{A}\mathbf{Y}(s)\mathbf{v}_s(s) \tag{4.6-9}$$

where $\mathbf{A}$ is the incidence matrix of the network, $\mathbf{Y}(s)$ is the element admittance matrix, and $\mathbf{Y}_n(s) = \mathbf{A}^T\mathbf{Y}(s)\mathbf{A}^T$. Taking the Laplace transform of $\mathbf{v}_s(t)$ and $\mathbf{i}_s(t)$ given in (4.6-7) and (4.6-8), we get

$$\mathbf{v}_s(s) = \boldsymbol{V}\frac{1}{s - j\omega_0} \tag{4.6-10}$$

and

$$\mathbf{i}_s(s) = \boldsymbol{I}\frac{1}{s - j\omega_0} \tag{4.6-11}$$

Using these relations, equation (4.6-9) can be written as

$$\mathbf{v}_n(s) = -\mathbf{Y}_n^{-1}(s)[\mathbf{A}\boldsymbol{I} + \mathbf{A}\mathbf{Y}(s)\boldsymbol{V}]\frac{1}{s - j\omega_0} \tag{4.6-12}$$

Ordinarily, one can compute the right-hand side of equation (4.6-12) and then take the inverse Laplace transform to obtain $\mathbf{v}_n(t)$ for all $t \geq 0$. In the present case, however, we are only interested in the steady-state response of the network. We would, therefore, like to utilize the sinusoidal nature of the sources to obtain a simple method for computing $\mathbf{v}_n(t)$ as t approaches ∞. This is done in the following.

Steady-State Response. To obtain the steady-state response, let us use a partial fraction expansion (see the Appendix) to write (4.6-12) in the form

$$\mathbf{v}_n(s) = \sum_{k=1}^{n} \frac{1}{s - p_k}\hat{\mathbf{R}}_k + \frac{1}{s - j\omega_0}\mathbf{R}(j\omega_0) \tag{4.6-13}$$

where $\hat{\mathbf{R}}_k$ is the residue vector of $-\mathbf{Y}_n^{-1}(s)[\mathbf{A}\boldsymbol{I} + \mathbf{A}\mathbf{Y}(s)\boldsymbol{V}]$ at its kth pole, $s = p_k$, and $\mathbf{R}(j\omega_0)$ is the residue vector corresponding to the pole of $\mathbf{v}_n(s)$ at $s = j\omega_0$. Note that both $\hat{\mathbf{R}}_k$ and $\mathbf{R}(j\omega_0)$ are n-column vectors with complex elements.

To achieve a sinusoidal steady-state, the network under consideration must satisfy the following two crucial assumptions; the reason for them becomes clear shortly.

Assumption 1. $-\mathbf{Y}_n^{-1}(s)[\mathbf{A}\boldsymbol{I} + \mathbf{A}\mathbf{Y}(s)\boldsymbol{V}]$ must have no poles at $j\omega_0$; that is, we must have

$$p_k \neq j\omega_0 \quad \text{for all } k = 1, 2, \ldots, n$$

where p_k is the kth pole of $\mathbf{v}_n(s)$ given in (4.6-12).

Assumption 2. The real part of the complex pole p_k must be strictly negative; that is,

$$\text{Re}\{p_k\} < 0 \quad \text{for all } k = 1, 2, \ldots, n$$

In practice, any linear network comprised of *positive* resistors, inductors, and capacitors satisfies Assumption 2. The first assumption, however, should be checked for each individual network, since p_k depends on the topology as well as the element values of the network under consideration.

Taking the inverse Laplace transform of (4.6-13), we obtain

$$\mathbf{v}_n(t) = \sum_{k=1}^{n} \hat{\mathbf{R}}_k e^{p_k t} + \mathbf{R}(j\omega_0)e^{j\omega_0 t} \quad \text{for } t \geq 0 \tag{4.6-14}$$

This is the response of the network for all $t \geq 0$; to obtain the steady-state response, we must let $t \longrightarrow \infty$. In doing so, since by Assumption 2 $\text{Re}\{p_k\} < 0$, we get

$$\lim_{t \to \infty} e^{p_k t} = 0 \quad \text{for } k = 1, 2, \ldots, n \tag{4.6-15}$$

Hence, as $t \longrightarrow \infty$ the only nonzero term in the right-hand side of (4.6-14) will be $\mathbf{R}(j\omega_0)e^{j\omega_0 t}$. This, in fact, is the desired complex steady-state response of the network:

$$\mathbf{v}_n^{ss}(t) = \lim_{t \to \infty} \mathbf{v}_n(t) = \mathbf{R}(j\omega_0)e^{j\omega_0 t} \tag{4.6-16}$$

The computation of the steady-state response, therefore, reduces to finding $\mathbf{R}(j\omega_0)$, the residue vector of $-\mathbf{Y}_n^{-1}(s)[\mathbf{A}\boldsymbol{I} + \mathbf{A}\mathbf{Y}(s)\boldsymbol{V}]$ at $s = j\omega_0$. This can be determined by multiplying both sides of (4.6-13) by $s - j\omega_0$ and letting $s \longrightarrow j\omega_0$, which yields

$$\mathbf{R}(j\omega_0) = \lim_{s \to j\omega_0} (s - j\omega_0)\mathbf{v}_n(s) \tag{4.6-17}$$

[The summation term in the right-hand side of (4.6-13) will vanish, since, by Assumption 1, $p_k \neq j\omega_0$.] Now, replacing $\mathbf{v}_n(s)$ in (4.6-17) from (4.6-12) yields

$$\mathbf{R}(j\omega_0) = -\mathbf{Y}_n^{-1}(j\omega_0)[\mathbf{A}\boldsymbol{I} + \mathbf{A}\mathbf{Y}(j\omega_0)\boldsymbol{V}] \tag{4.6-18}$$

The steady-state response of the network to the exponential excitations $\boldsymbol{V}e^{j\omega_0 t}$ *and* $\boldsymbol{I}e^{j\omega_0 t}$ *is therefore given by*

$$\mathbf{v}_n^{ss}(t) = -\mathbf{Y}_n^{-1}(j\omega_0)[\mathbf{A}\boldsymbol{I} + \mathbf{A}\mathbf{Y}(j\omega_0)\boldsymbol{V}]e^{j\omega_0 t} \tag{4.6-19}$$

Furthermore, the steady-state response $\hat{\mathbf{v}}_n^{ss}(t)$ corresponding to the excitations $\text{Re}\{\boldsymbol{V}e^{j\omega_0 t}\}$ and $\text{Re}\{\boldsymbol{I}e^{j\omega_0 t}\}$ can be obtained by taking the real part of (4.6-19);

that is,

$$\hat{\mathbf{v}}_n^{ss}(t) = \text{Re}\{-\mathbf{Y}_n^{-1}(j\omega_0)[\mathbf{A}\boldsymbol{I} + \mathbf{A}\mathbf{Y}(j\omega_0)\boldsymbol{V}]e^{j\omega_0 t}\} \tag{4.6-20}$$

The steps involved in sinusoidal steady-state nodal analysis can then be summarized as follows:

Step 1. Replace all the capacitors and inductors with their equivalent impedances $1/j\omega_0 C_k$ and $j\omega_0 L_k$.

Step 2. Put all the sinusoidal sources in the exponential form $\boldsymbol{V}e^{j\omega_0 t}$ and $\boldsymbol{I}e^{j\omega_0 t}$; compute $\boldsymbol{V}$ and $\boldsymbol{I}$.

Step 3. Write down the element impedance matrix $\mathbf{Y}(j\omega_0)$ and compute $\mathbf{Y}_n(j\omega_0)$ from $\mathbf{Y}_n(j\omega_0) = \mathbf{A}\mathbf{Y}(j\omega_0)\mathbf{A}^T$.

Step 4. Compute $\mathbf{Y}_n^{-1}(j\omega_0)[\mathbf{A}\boldsymbol{I} + \mathbf{A}\mathbf{Y}(j\omega_0)\boldsymbol{V}]$. The steady-state response is therefore $\hat{\mathbf{v}}_n^{ss}(t) = \text{Re}\{-\mathbf{Y}_n^{-1}(j\omega_0)[\mathbf{A}\boldsymbol{I} + \mathbf{A}\mathbf{Y}(j\omega_0)\boldsymbol{V}]e^{j\omega_0 t}]\}$.

Note that, although in the preceding derivations we only used nodal analysis, loop analysis may also be used to obtain similar results. The details are left for the reader as an exercise.

Remark 4.6-1. In the beginning of this section we made two important assumptions concerning the existence of sinusoidal steady-state response of linear time-invariant networks. We can now discuss the reason behind such assumptions. Let us consider the second assumption first, that is, $\text{Re}\{p_k\} < 0$. Suppose that, for some p_k, $\text{Re}\{p_k\} > 0$; then the corresponding term in the time domain, $r_k e^{p_k t}$, is an exponential function with positive exponent and hence will grow unbounded as $t \longrightarrow \infty$. Consequently, the network under consideration will be unstable and steady-state response will have no meaning. If $\text{Re}\{p_k\} = 0$, the corresponding terms in the time domain will be $r_k e^{j\omega_k t}$, where ω_k denotes the imaginary part of p_k. In this case the response is not a simple sinusoidal one with angular frequency ω_0; rather it is a combination of two or more sinusoids with different frequencies. Hence, the analysis discussed in this section does not apply.

The other assumption was that $p_k \neq j\omega_0$ for $k = 1, 2, \ldots, n$. To see the necessity of this assumption for the existence of sinusoidal steady-state response, let $p_k = j\omega_0$ for some k. Then the partial fraction expansion of (4.6-13) will contain terms like $[1/(s - j\omega_0)^2]\mathbf{R}_k$, which in the time domain corresponds to $\mathbf{R}_k t e^{j\omega_0 t}$. Hence, as $t \longrightarrow \infty$, this term will go to infinity, and consequently the network becomes unstable.

Remark 4.6-2. It should be emphasized that the sinusoidal steady-state analysis discussed here applies only if the network is excited with a *single* angular frequency ω_0. If, however, there is more than one frequency present, we can use the superposition principle discussed in Chapter 2 to obtain the response of the network to each frequency and then combine the results to get the total solution.

Let us now work out an example to illustrate the procedure discussed.

Example 4.6-1. Consider the bridged T network shown in Fig. 4.6-1. This network is excited by two independent sources with the same angular frequency ω_0. Let us obtain the node voltages $v_1(t)$, $v_2(t)$, and $v_3(t)$ in the steady state.

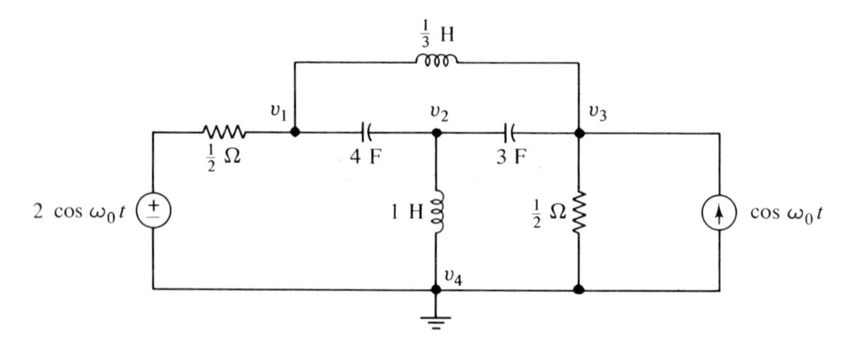

Figure 4.6-1 Linear time-invariant network considered in Example 4.6-1.

Solution

Since both sources have the same angular frequency, using steps 1 through 3, the transformed network can be drawn as in Fig. 4.6-2. We have also used the source transformation discussed in Remark 4.5-2 to transform the voltage source to a current source. This will simplify the computations to a large

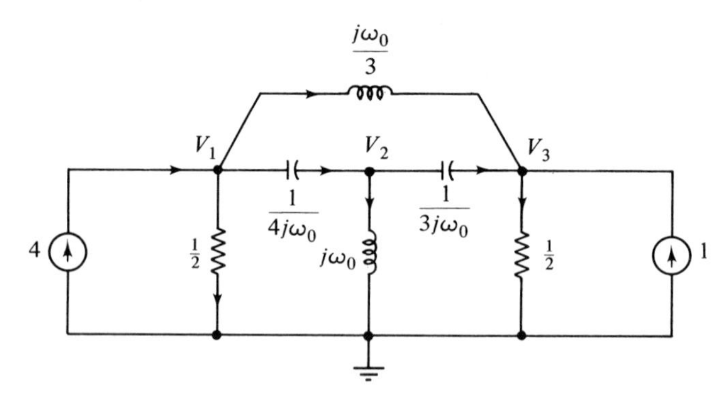

Figure 4.6-2 Steady-state equivalent of the network shown in Fig. 4.6-1.

degree, since in (4.6-12) $\boldsymbol{V}$ will now be equal to zero. From (4.6-19) it is clear that we need to obtain $\mathbf{A}$, $\boldsymbol{I}$, $\boldsymbol{V}$, and $\mathbf{Y}(j\omega_0)$. From Fig. 4.6-2 we get

$$\mathbf{A} = \begin{array}{c} \\ n_1 \\ n_2 \\ n_3 \end{array}\begin{array}{c} \begin{array}{cccccc} b_1 & b_2 & b_3 & b_4 & b_5 & b_6 \end{array} \\ \begin{bmatrix} 1 & 1 & 0 & 0 & 0 & 1 \\ 0 & -1 & 1 & 1 & 0 & 0 \\ 0 & 0 & 0 & -1 & 1 & -1 \end{bmatrix} \end{array}$$

and

$$\mathbf{Y}(j\omega_0) = \operatorname{diag}\left\{2 \quad 4j\omega_0 \quad \frac{1}{j\omega_0} \quad 3j\omega_0 \quad 2 \quad \frac{3}{j\omega_0}\right\}$$

Also,

$$\boldsymbol{V} = \mathbf{0}$$

and

$$\boldsymbol{I} = [-4 \quad 0 \quad 0 \quad 0 \quad -1 \quad 0]^T$$

Computing $\mathbf{Y}_n(j\omega_0)$ from $\mathbf{Y}_n(j\omega_0) = \mathbf{A}\mathbf{Y}(j\omega_0)\mathbf{A}^T$ and using (4.6-18), we get

$$\mathbf{R}(j\omega_0) = -\begin{bmatrix} 2 + 4j\omega_0 + \dfrac{3}{j\omega_0} & -4j\omega_0 & -\dfrac{3}{j\omega_0} \\ -4j\omega_0 & 7j\omega_0 + \dfrac{1}{j\omega_0} & -3j\omega_0 \\ -\dfrac{3}{j\omega_0} & -3j\omega_0 & 2 + 3j\omega_0 + \dfrac{3}{j\omega_0} \end{bmatrix}^{-1} \begin{bmatrix} -4 \\ 0 \\ -1 \end{bmatrix}$$

As a numerical example, let $\omega_0 = 1$ radian per second; then using the preceding equation in (4.6-19) yields the complex response

$$\mathbf{v}_n^{ss}(t) = \begin{bmatrix} 1.38 + 0.71j \\ 1.36 + 0.46j \\ 0.88 - 0.03j \end{bmatrix} e^{jt}$$

This is the response (node voltage v_1, v_2, and v_3) corresponding to the complex voltage source $2e^{j\omega_0 t}$ and the current source $e^{j\omega_0 t}$. The node voltages corresponding to the real inputs [$v_s(t) = 2\cos\omega_0 t$ and $i_s(t) = \cos\omega_0 t$] can be obtained by taking the real part of $\mathbf{v}_n^{ss}(t)$:

$$\hat{v}_1^{ss}(t) = 1.38\cos t - 0.71\sin t, \qquad \hat{v}_2^{ss}(t) = 1.36\cos t - 0.46\sin t,$$

$$\hat{v}_3^{ss}(t) = 0.88\cos t + 0.03\sin t$$

4.7 NETWORK FUNCTIONS

So far in this chapter we have discussed three general methods for the analysis of linear time-invariant networks: (1) direct application of KVL, KCL, and VCR, (2) nodal analysis, and (3) loop analysis. These methods are completely general and can be used to analyze any network comprised of linear time-invariant elements and voltage and current sources. In many practical applications, however, the network under consideration can be modeled as a multiport network in which the sources are considered as the inputs to the network and the responses are the voltages across and/or the currents through some of the terminals. In this case, if the *network functions* are known, the response of the network to a set of inputs can be obtained without resorting to the general (and sometimes lengthy) analysis methods discussed in the previous sections. This, in effect, is a shortcut, which in many cases saves much time and effort in computing the network response. In this section we define several useful network functions and in each case discuss their basic properties. Throughout we assume that the network under consideration is comprised of linear time-invariant elements so that the Laplace transform can be used in representing various network functions. Let us start by defining the open-circuit impedance matrix.

Open-Circuit Impedance Matrix. The open-circuit impedance matrix is a network function relating the Laplace transform of the port currents to the Laplace transform of the port voltages of a network when all the initial conditions and internal independent sources are put equal to zero. More specifically, consider the linear time-invariant n-port network shown in Fig. 4.7-1.

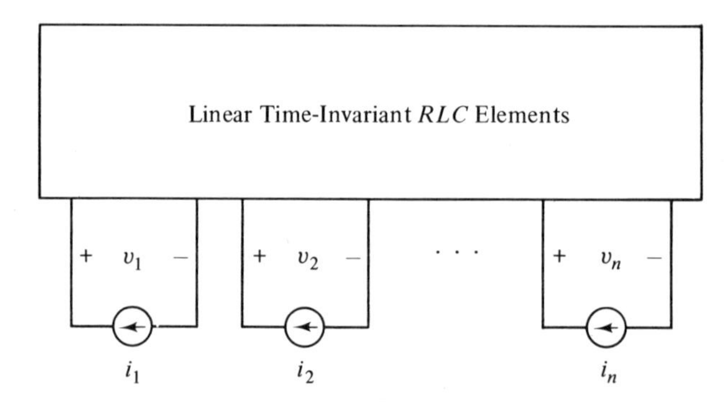

Figure 4.7-1 Current excited–voltage response multiterminal network.

Assume that the black box contains no independent sources of any kind and all the initial conditions are zero (dependent voltage or current sources may be present). Let the *port voltage* vector be denoted by

$$\mathbf{v}(s) = [v_1(s), v_2(s), \ldots, v_n(s)]^T \tag{4.7-1}$$

and the *port current* vector be denoted by

$$\mathbf{i}(s) = [i_1(s), i_2(s), \ldots, i_n(s)]^T \tag{4.7-2}$$

Since in the time domain $\mathbf{v}(t)$ and $\mathbf{i}(t)$ are related through a set of linear integro-differential equations, in the frequency domain each element of $\mathbf{v}(s)$ can be expressed as a linear combination of the elements of $\mathbf{i}(s)$. More specifically, the vector $\mathbf{v}(s)$ can be written as

$$\mathbf{v}(s) = \mathbf{Z}_{oc}(s)\mathbf{i}(s) \tag{4.7-3}$$

where $\mathbf{Z}_{oc}(s)$ is an $n \times n$ matrix whose elements are functions of the complex variable s. For reasons to be made clear shortly, $\mathbf{Z}_{oc}(s)$ is called the *driving-point impedance* or the *open-circuit impedance matrix* of the network under consideration. Quite clearly, given $\mathbf{Z}_{oc}(s)$ for a network whose input vector is $\mathbf{i}(s)$, the response vector $\mathbf{v}(t)$ can be obtained by taking the inverse Laplace transform of (4.7-3). Consequently, to obtain the port voltages, there is no need to go through any of the analysis techniques discussed in the previous sections; it is sufficient to obtain $\mathbf{Z}_{oc}(s)$ and use $\mathbf{v}(s) = \mathbf{Z}_{oc}(s)\mathbf{i}(s)$. There are several methods of obtaining $\mathbf{Z}_{oc}(s)$, one of which is discussed next.

Let us rewrite (4.7-3) in a more explicit form:

$$\begin{bmatrix} v_1(s) \\ v_2(s) \\ \cdot \\ \cdot \\ \cdot \\ v_n(s) \end{bmatrix} = \begin{bmatrix} z_{11}(s) & z_{12}(s) & \dots & z_{1n}(s) \\ z_{21}(s) & z_{22}(s) & \dots & z_{2n}(s) \\ \cdot & & & \\ \cdot & & & \\ \cdot & & & \\ z_{n1}(s) & z_{n2}(s) & \dots & z_{nn}(s) \end{bmatrix} \begin{bmatrix} i_1(s) \\ i_2(s) \\ \cdot \\ \cdot \\ \cdot \\ i_n(s) \end{bmatrix} \tag{4.7-4}$$

To obtain $z_{jk}(s)$, let us open circuit all the ports except the kth port; that is, choose $\mathbf{i}(s)$ to be

$$\mathbf{i}(s) = [0, 0, \dots, 0, i_k(s), 0, \dots, 0]^T$$

Then from (4.7-4) we can write

$$v_j(s) = z_{jk}(s)i_k(s)$$

Hence, $$z_{jk}(s) = \frac{v_j(s)}{i_k(s)}; \text{ all current sources zero except } i_k \tag{4.7-5}$$

Consequently, by open circuiting all the current sources except i_k and finding $v_j(s)$, the jkth element of $\mathbf{Z}_{oc}(s)$ can be obtained from equation (4.7-5). Repeating this process for all j and k, the matrix $\mathbf{Z}_{oc}(s)$ will be completely specified. An example of this procedure will be worked out later. Let us first define another important network function.

Short-Circuit Admittance Matrix. As in the previous case, the short-circuit admittance matrix is a network function relating the Laplace transform of the port currents to the Laplace transform of the port voltages. But in this case, the port currents are expressed in terms of the port voltages. If the n-port network under consideration is excited by independent voltage sources, the port currents can readily be obtained by using the short-circuit admittance matrix defined next. Consider the linear time-invariant network shown in Fig. 4.7-2. Assume that the black box contains no independent sources, and initial capacitor voltages and inductor currents are zero. Let us adopt the notations of (4.7-1) and (4.7-2). Then the Laplace transform of the current vector $\mathbf{i}(s)$ is related to the Laplace transform of voltage source vector $\mathbf{v}(s)$ through an $n \times n$ matrix $\mathbf{Y}_{sc}(s)$:

$$\mathbf{i}(s) = \mathbf{Y}_{sc}(s)\mathbf{v}(s) \tag{4.7-6}$$

$\mathbf{Y}_{sc}(s)$ is called the *driving-point admittance matrix* or the *short-circuit admittance matrix* of the network under consideration. The elements of $\mathbf{Y}_{sc}(s)$ are sometimes called the *y parameters* of the network.

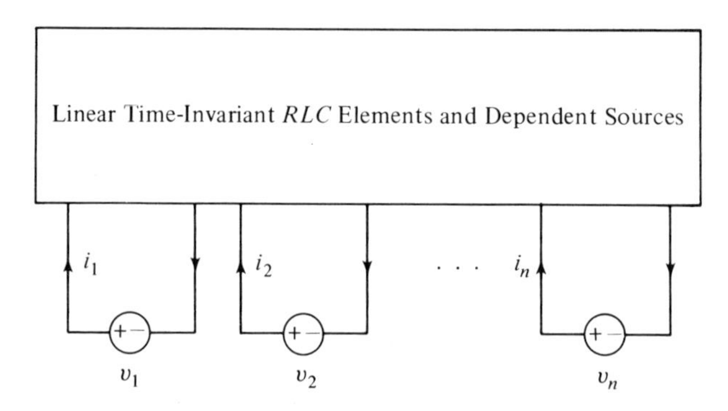

Figure 4.7-2 Voltage excited–current response n-port network.

Given $\mathbf{Y}_{sc}(s)$ and the input vector $\mathbf{v}(t)$, if all the initial conditions are zero, the port current vector $\mathbf{i}(t)$ can be obtained by taking the inverse Laplace transform of (4.7-6).

The elements of $\mathbf{Y}_{sc}(s)$ can be obtained in a similar manner as those of $\mathbf{Z}_{oc}(s)$. From (4.7-6) we can write

$$\begin{bmatrix} i_1(s) \\ i_2(s) \\ \cdot \\ \cdot \\ \cdot \\ i_n(s) \end{bmatrix} = \begin{bmatrix} y_{11}(s) & y_{12}(s) & \dots & y_{1n}(s) \\ y_{21}(s) & y_{22}(s) & \dots & y_{2n}(s) \\ \cdot & & & \\ \cdot & & & \\ \cdot & & & \\ y_{n1}(s) & y_{n2}(s) & \dots & y_{nn}(s) \end{bmatrix} \begin{bmatrix} v_1(s) \\ v_2(s) \\ \cdot \\ \cdot \\ \cdot \\ v_n(s) \end{bmatrix} \tag{4.7-7}$$

If we now *short circuit* all the voltage sources of the network shown in Fig. 4.7-2 except v_k, that is, if we choose

$$\mathbf{v}_s(s) = [0, 0, \dots, v_k(s), \dots, 0]^T$$

the current through the jth port is then given by

$$i_j(s) = y_{jk}(s)v_k(s) \tag{4.7-8}$$

Consequently, the jkth element of $\mathbf{Y}_{sc}(s)$ can be obtained from

$$y_{jk}(s) = \frac{i_j(s)}{v_k(s)}; \text{ all voltage sources zero except } v_k(s) \tag{4.7-9}$$

Note that if the open-circuit impedance matrix $\mathbf{Z}_{oc}(s)$ of a network is given and if $\mathbf{Z}_{sc}^{-1}(s)$ exists, then clearly

$$\mathbf{Y}_{sc}(s) = \mathbf{Z}_{oc}^{-1}(s) \tag{4.7-10}$$

It is not, however, difficult to give examples of networks for which either $\mathbf{Y}_{sc}(s)$ or $\mathbf{Z}_{oc}(s)$ is not defined. Besides, if the dimension of the matrix $\mathbf{Z}_{oc}(s)$ is large, obtaining $\mathbf{Z}_{oc}^{-1}(s)$ could be a formidable task. Therefore, in such cases it might be easier to obtain $\mathbf{Y}_{sc}(s)$ independently.

Example 4.7-1. Consider the linear time-invariant three-port network shown in Fig. 4.7-3. Obtain $\mathbf{Y}_{sc}(s)$ and $\mathbf{Z}_{oc}(s)$ of this network.

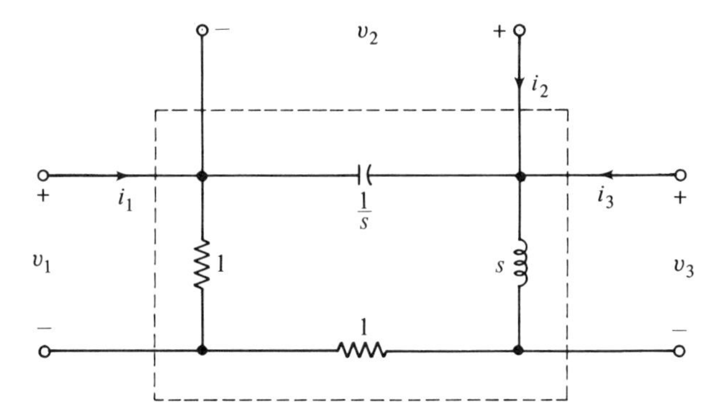

Figure 4.7-3 Three-port network considered in Example 4.7-1.

Solution

Short circuiting ports 2 and 3 and applying a voltage source $v_1(s)$ to port 1, the network under study can be represented as in Fig. 4.7-4. Using (4.7-9), we can then obtain $y_{11}(s)$, $y_{21}(s)$, and $y_{31}(s)$ as follows:

$$y_{11}(s) = \frac{i_1(s)}{v_1(s)} = 2, \qquad y_{21}(s) = \frac{i_2(s)}{v_1(s)} = 1, \qquad y_{31}(s) = \frac{i_3(s)}{v_1(s)} = -1$$

Repeating this process for ports 2 and 3, the remaining elements of $\mathbf{Y}_{sc}(s)$ can

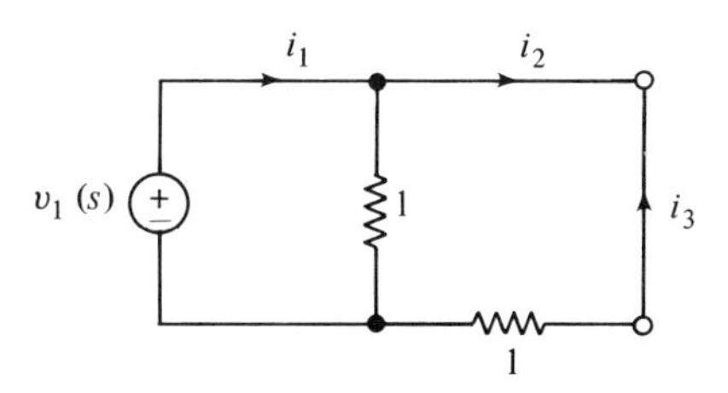

Figure 4.7-4 Network used for computing y_{11}, y_{21}, and y_{31}.

also be found; thus,

$$\mathbf{Y}_{sc}(s) = \begin{bmatrix} 2 & 1 & -1 \\ 1 & s+1 & -1 \\ -1 & -1 & 1+\dfrac{1}{s} \end{bmatrix}$$

To obtain $\mathbf{Z}_{oc}(s)$, we can then invert $\mathbf{Y}_{sc}(s)$ just obtained. Since in this case det $\mathbf{Y}_{sc}(s) = 2 + (1/s) + s \neq 0$ for almost all s, inverting $\mathbf{Y}_{sc}(s)$ yields

$$\mathbf{Z}_{oc}(s) = \frac{1}{2 + \dfrac{1}{s} + s} \begin{bmatrix} s+1+\dfrac{1}{s} & & s \\ -\dfrac{1}{s} & 1+\dfrac{2}{s} & 1 \\ s & 1 & 1+2s \end{bmatrix}$$

Exercise 4.7-1. Obtain $\mathbf{Z}_{os}(s)$ for the network shown in Fig. 4.7-3 by using (4.7-5) and check your result with $\mathbf{Z}_{oc}(s)$ just obtained.

Remark 4.7-1. Although $\mathbf{Z}_{oc}(s)$ and $\mathbf{Y}_{sc}(s)$ are defined under the assumption that the network under study does not contain any independent sources in its interior, these matrices can still be used to obtain the response of the network even if it contains some internal independent sources. Since the network is linear, it satisfies the superposition principle discussed in Chapter 2. Then under current excitation–voltage response the voltage vector $\mathbf{v}(s)$ for such networks can be written as the sum of two terms:

$$\mathbf{v}(s) = \mathbf{Z}_{oc}(s)\mathbf{i}(s) + \hat{\mathbf{v}}(s) \tag{4.7-11}$$

where $\mathbf{Z}_{oc}(s)\mathbf{i}(s)$ is the response of the network when all the internal sources are put equal to zero (i.e., internal voltage sources are short circuited and the current sources are open circuited). The second term, $\hat{\mathbf{v}}(s)$, is the response of the network due to the internal sources alone [when all the ports are open circuited, i.e., when $\mathbf{i}(s) = \mathbf{0}$].

The initial conditions in the circuit can also be taken care of in a similar manner. In fact, initial conditions may first be represented as independent sources (see Figs. 4.2-1 and 4.2-2), and then (4.7-11) can be used to obtain the total response.

If the open-circuit admittance matrix is used, the corresponding relation is

$$\mathbf{i}(s) = \mathbf{Y}_{sc}(s)\mathbf{v}(s) + \hat{\mathbf{i}}(s) \tag{4.7-12}$$

where $\mathbf{Y}_{sc}(s)\mathbf{v}(s)$ is the response of the network to external voltage sources alone and $\hat{\mathbf{i}}(s)$ is the response of the network due to internal sources.

Remark 4.7-2. If the network under consideration contains linear reciprocal elements such as resistors, inductors, capacitors, and ideal transformers only, the corresponding $\mathbf{Z}_{oc}(s)$ and $\mathbf{Y}_{sc}(s)$ matrices will be symmetric. On the other hand, if it contains some nonreciprocal elements such as gyrators or dependent sources, these matrices are not generally symmetric. Let us now work out another example to illustrate how dependent sources can be taken care of.

Example 4.7-2. Consider the transistor amplifier whose small-signal equivalent circuit is shown in Fig. 4.7-5. It is desired to obtain the short-circuit

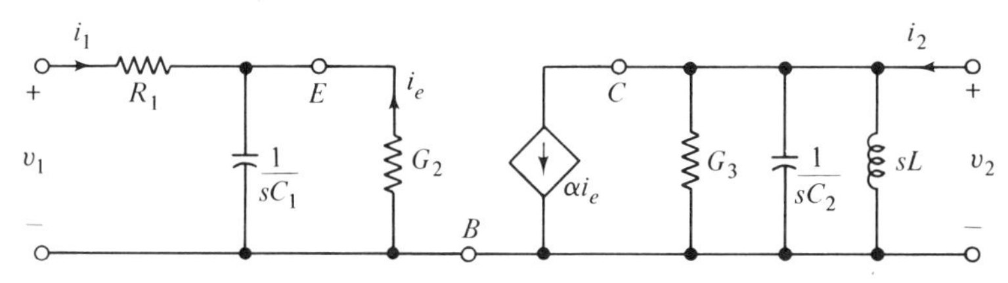

Figure 4.7-5 Transistor amplifier circuit for which $\mathbf{y}_{sc}(s)$ is to be computed.

admittance matrix of this network. This network can be considered as a two-port network; then its y parameters can be obtained from

$$y_{11} = \left.\frac{i_1}{v_1}\right|_{v_2=0}, \qquad y_{21} = \left.\frac{i_2}{v_1}\right|_{v_2=0}, \qquad y_{22} = \left.\frac{i_2}{v_2}\right|_{v_1=0}, \qquad y_{12} = \left.\frac{i_1}{v_2}\right|_{v_1=0}$$

Thus, short circuiting port 2, the VCR of port 1 becomes

$$v_1 = \left(R_1 + \frac{1}{G_2 + sC_1}\right)i_1$$

and

$$i_2 = \alpha i_e = \frac{-\alpha G_2}{sC_1 + G_2} i_1$$

Consequently,

$$y_{11}(s) = \frac{G_2 + sC_1}{1 + R_1(G_2 + sC_1)}$$

and

$$y_{21}(s) = \frac{-\alpha G_2}{1 + R_1(G_2 + sC_1)}$$

Similarly, short circuiting port 1, the VCR of port 2 becomes

$$i_e = 0, \qquad i_1 = 0, \qquad i_2 = \left(G_3 + sC_2 + \frac{1}{sL}\right)v_2$$

Hence,

$$y_{22}(s) = G_3 + sC_2 + \frac{1}{sL} \quad \text{and} \quad y_{12} = 0$$

The corresponding short-circuit impedance matrix is therefore

$$\mathbf{Y}_{sc}(s) = \begin{bmatrix} \dfrac{G_2 + sC_1}{1 + R_1(G_2 + sC_1)} & 0 \\ \dfrac{-\alpha G_2}{1 + R_1(G_2 + sC_1)} & G_3 + sC_2 + \dfrac{1}{sL} \end{bmatrix}$$

Transfer-Function Matrix. In defining the open-circuit impedance matrix $\mathbf{Z}_{oc}(s)$ and the short-circuit admittance matrix $\mathbf{Y}_{sc}(s)$, we imposed two conditions on the ports of the network under study. First, the excitations and the responses must correspond to the same terminal, and, second, the excitations or the responses must not be of the same kind. For example, in defining $\mathbf{Z}_{oc}(s)$ the excitations are the port currents and the responses are the voltages across the same ports. However, there are many useful electrical networks that do not satisfy these conditions. For example, in voltage amplifiers both the excitation and response are voltages across different terminals of the amplifier. Furthermore, in some electronic devices some of the inputs may be voltage sources and others may be current sources. In such cases the network may be represented by a transfer function or a transfer-function matrix.

Consider the multiport network shown in Fig. 4.7-6. To be completely

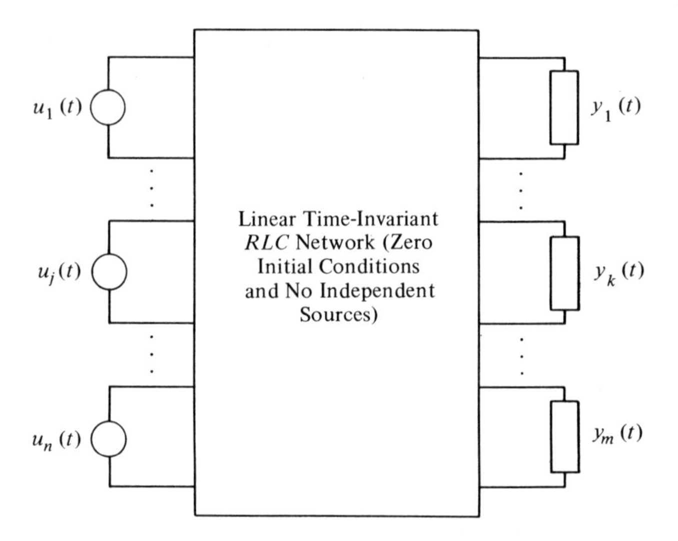

Figure 4.7-6 Multiport network used in defining the transfer-function matrix.

general, let us denote the inputs by $u_1(t), u_2(t), \ldots, u_n(t)$, where $u_j(t)$ may represent a voltage or a current source. Similarly, denote the outputs by $y_1(t), y_2(t), \ldots, y_m(t)$, where $y_k(t)$ can be either the voltage across or the current through the kth output port. Thus, the input vector $\mathbf{u}(t)$ is an n-column vector given by

$$\mathbf{u}(t) = [u_1(t), u_2(t), \ldots, u_n(t)]^T \tag{4.7-13}$$

and the output vector $\mathbf{y}(t)$ is an m-column vector defined by

$$\mathbf{y}(t) = [y_1(t), y_2(t), \ldots, y_m(t)]^T \tag{4.7-14}$$

We define the *transfer-function matrix* to be an $m \times n$ matrix $\mathbf{H}(s)$ relating the Laplace transform of the output vector $\mathbf{y}(s)$ to the Laplace transform of the input vector $\mathbf{u}(s)$ when all the initial conditions of the network are put equal to zero. Thus, we have

$$\mathbf{y}(s) = \mathbf{H}(s)\mathbf{u}(s) \tag{4.7-15}$$

If both $\mathbf{u}(t)$ and $\mathbf{y}(t)$ represent voltages, $\mathbf{H}(s)$ is called the *voltage transfer-function matrix;* if $u(t)$ and $y(t)$ represent currents, $\mathbf{H}(s)$ is called the *current transfer-function matrix.* If $\mathbf{u}(t)$ and $\mathbf{y}(t)$ contain both voltages and currents, $\mathbf{H}(s)$ is called the *hybrid transfer-function matrix.* In the special case of a single input and a single output (i.e., when $n = m = 1$), $\mathbf{H}(s)$ is scalar and is simply called the *transfer function.*

The elements of $\mathbf{H}(s)$ can be determined in a similar manner to those of $\mathbf{Z}_{oc}(s)$ and $\mathbf{Y}_{sc}(s)$. Let us rewrite (4.7-15) as

$$\begin{bmatrix} y_1(s) \\ y_2(s) \\ \cdot \\ \cdot \\ \cdot \\ y_k(s) \\ \cdot \\ \cdot \\ \cdot \\ y_m(s) \end{bmatrix} = \begin{bmatrix} h_{11}(s) & h_{12}(s) & \cdots & h_{1j}(s) & \cdots & h_{1n}(s) \\ h_{21}(s) & h_{22}(s) & \cdots & h_{2j}(s) & \cdots & h_{2n}(s) \\ \cdot & \cdot & & \cdot & & \cdot \\ \cdot & \cdot & & \cdot & & \cdot \\ \cdot & \cdot & & \cdot & & \cdot \\ h_{k1}(s) & h_{k2}(s) & \cdots & h_{kj}(s) & \cdots & h_{kn}(s) \\ \cdot & \cdot & & \cdot & & \cdot \\ \cdot & \cdot & & \cdot & & \cdot \\ \cdot & \cdot & & \cdot & & \cdot \\ h_{m1}(s) & h_{m2}(s) & \cdots & h_{mj}(s) & \cdots & h_{mn}(s) \end{bmatrix} \begin{bmatrix} u_1(s) \\ u_2(s) \\ \cdot \\ \cdot \\ \cdot \\ u_j(s) \\ \cdot \\ \cdot \\ \cdot \\ u_n(s) \end{bmatrix} \tag{4.7-16}$$

Quite clearly, we can then write

$$h_{kj}(s) = \frac{y_k(s)}{u_j(s)}; \quad \text{when all the inputs except } u_j \text{ are put equal to zero} \tag{4.7-17}$$

Note that in defining $\mathbf{H}(s)$ we have assumed that the initial conditions are equal to zero and no independent sources exist in the interior of the network. Dependent sources, however, may exist in the network under study. Let us now work out a simple example to see how dependent sources can be handled.

Example 4.7-3. Consider the network shown in Fig. 4.7-7(a). This network is extensively used as a low-pass filter in communication networks and is used to "pass" or amplify the signal with low frequencies and reject the signals with high frequencies. The voltage-controlled voltage source equivalent network of this filter is shown in Fig. 4.7-7(b) (see Table 4.5-1). Let us obtain the voltage transfer function of this network. To express $v_2(s)$ in terms of $v_1(s)$, one can use the nodal analysis discussed earlier. However, since the network under consideration is quite simple, we can apply KCL equations to nodes

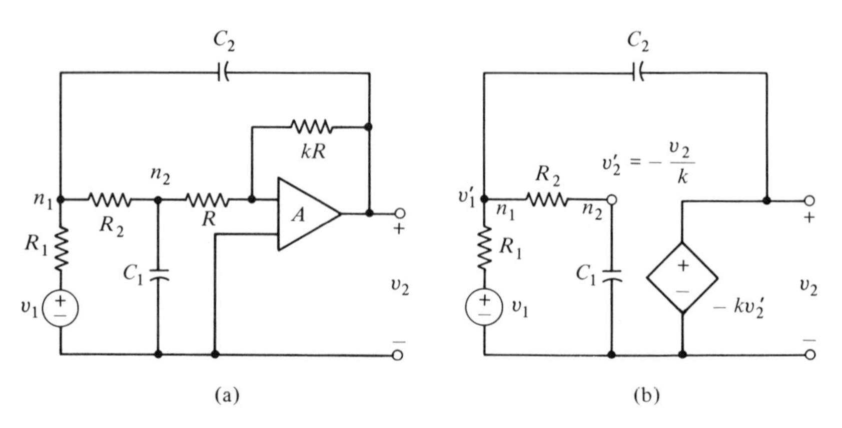

Figure 4.7-7 Low-pass filter used in Example 4.7-3 and its equivalent network.

n_1 and n_2 to get

$$(v_1 - v_1')\frac{1}{R_1} = \left(v_1' + \frac{v_2}{k}\right)\frac{1}{R_2} + (v_1' - v_2)sC_2$$

and
$$\left(v_1' + \frac{v_2}{k}\right)\frac{1}{R_2} = -\frac{sC_1v_2}{k}$$

Eliminating v_1' from the equations, we get

$$v_2(s) = \frac{-k}{s^2C_1C_2R_1R_2 + (R_1C_2k + C_1R_1 + C_2R_1 + C_1R_2)s + 1}v_1(s)$$

Choosing $R_1 = 1\ \Omega$, $R_2 = 10\ \Omega$, $C_1 = 0.1$ F, $C_2 = 1$ F, and $k = 1.1$, the transfer function $H(s)$ becomes

$$H(s) = \frac{-1.1}{s^2 + 3.2s + 1}$$

Example 4.7-4. Consider the RC circuit shown in Fig. 4.7-8. This network has two inputs, the voltage source u_1 and the current source u_2. The outputs

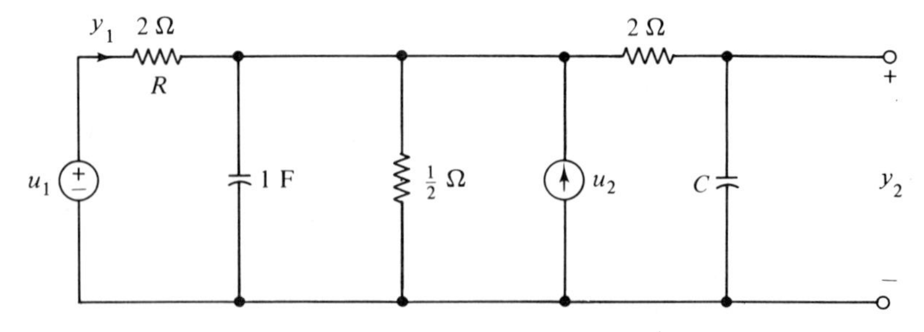

Figure 4.7-8 Multiinput-multioutput network of Example 4.7-4.

y_1 and y_2 are the current through the resistor R and the voltage across the capacitor C, respectively. The transfer function matrix $\mathbf{H}(s)$ is therefore a 2×2 matrix defined by

$$\begin{bmatrix} y_1(s) \\ y_2(s) \end{bmatrix} = \begin{bmatrix} h_{11}(s) & h_{12}(s) \\ h_{21}(s) & h_{22}(s) \end{bmatrix} \begin{bmatrix} u_1(s) \\ u_2(s) \end{bmatrix}$$

where $h_{ij}(s)$ can be obtained from

$$h_{11}(s) = \left.\frac{y_1}{u_1}\right|_{u_2=0}, \quad h_{12}(s) = \left.\frac{y_1}{u_2}\right|_{u_1=0}, \quad h_{21}(s) = \left.\frac{y_2}{u_1}\right|_{u_2=0}, \quad h_{22}(s) = \left.\frac{y_2}{u_2}\right|_{u_1=0}$$

For the network under consideration, these yield

$$\mathbf{H}(s) = \frac{1}{4s^2 + 14s + 5} \begin{bmatrix} 2s^2 + 6s + 2 & -2s - 1 \\ -2s - 1 & 2 \end{bmatrix}$$

Two-Port Parameters. Since two-port networks play an important role in communication systems, special attention has been given to their representation. Depending on the application of such networks, they can be represented by their y parameters, z parameters, transmission parameters, or hybrid parameters. All these representations are, of course, special cases of the transfer function matrix just discussed. Nevertheless, some of the most important two-port parameters are redefined in the first column of Table 4.7-1. The remaining columns of this table are used to convert one set of parameters to others; the matrices on each row in the table are equivalent. The corresponding voltage and currents are as shown in Fig. 4.7-9. It should be mentioned that it is not always possible to convert one set of parameters to others. This is because for some networks some of the parameters may not be defined. An example of this situation is given in the following.

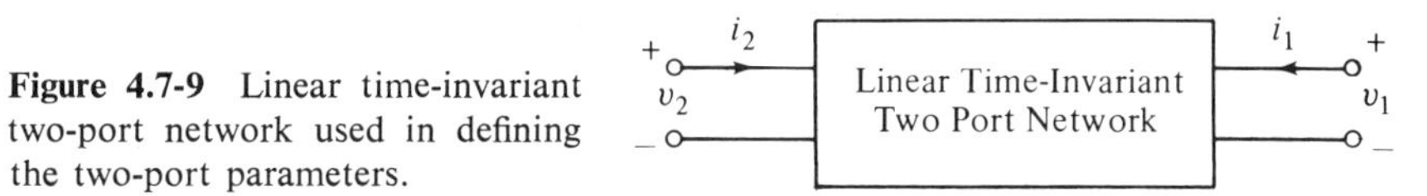

Figure 4.7-9 Linear time-invariant two-port network used in defining the two-port parameters.

Exercise 4.7-2. Consider the two-port networks shown in Fig. 4.7-10(a) and (b). Show that for (a) the y parameters and for (b) the z parameters are not defined.

As was mentioned earlier, each of the two-port parameters defined in Table 4.7-1 has a useful application in the interconnection of two-port networks; for example, if the two-ports are connected in tandem, as in Fig. 4.7-11, the transmission parameters are used to represent the overall network function in terms of individual networks. Thus,

$$\begin{bmatrix} v_1 \\ i_1 \end{bmatrix} = \begin{bmatrix} A_1 & B_1 \\ C_1 & D_1 \end{bmatrix} \begin{bmatrix} v_2' \\ -i_2' \end{bmatrix} \quad \text{and} \quad \begin{bmatrix} v_1' \\ i_1' \end{bmatrix} = \begin{bmatrix} A_2 & B_2 \\ C_2 & D_2 \end{bmatrix} \begin{bmatrix} v_2 \\ -i_2 \end{bmatrix}$$

Table 4.7-1

Definition	*z parameters*	*y parameters*	*ABCD parameters*	*h parameters*	*g parameters*
Open-circuit or z parameters $\begin{bmatrix} v_1 \\ v_2 \end{bmatrix} = \begin{bmatrix} z_{11} & z_{12} \\ z_{21} & z_{22} \end{bmatrix} \begin{bmatrix} i_1 \\ i_2 \end{bmatrix}$	$\begin{matrix} z_{11} & z_{12} \\ z_{21} & z_{22} \end{matrix}$	$\begin{matrix} \frac{y_{22}}{\Delta y} & \frac{-y_{12}}{\Delta y} \\ \frac{-y_{21}}{\Delta y} & \frac{y_{11}}{\Delta y} \end{matrix}$	$\begin{matrix} \frac{A}{C} & \frac{\Delta T}{C} \\ \frac{1}{C} & \frac{D}{C} \end{matrix}$	$\begin{matrix} \frac{\Delta h}{h_{22}} & \frac{h_{12}}{h_{22}} \\ \frac{-h_{21}}{h_{22}} & \frac{1}{h_{22}} \end{matrix}$	$\begin{matrix} \frac{1}{g_{11}} & \frac{-g_{12}}{g_{11}} \\ \frac{g_{21}}{g_{11}} & \frac{\Delta g}{g_{11}} \end{matrix}$
Short-circuit or y parameters $\begin{bmatrix} i_1 \\ i_2 \end{bmatrix} = \begin{bmatrix} y_{11} & y_{12} \\ y_{21} & y_{22} \end{bmatrix} \begin{bmatrix} v_1 \\ v_2 \end{bmatrix}$	$\begin{matrix} \frac{z_{22}}{\Delta z} & \frac{-z_{12}}{\Delta z} \\ \frac{-z_{21}}{\Delta z} & \frac{z_{11}}{\Delta z} \end{matrix}$	$\begin{matrix} y_{11} & y_{12} \\ y_{21} & y_{11} \end{matrix}$	$\begin{matrix} \frac{D}{B} & \frac{-\Delta T}{B} \\ \frac{-1}{B} & \frac{A}{B} \end{matrix}$	$\begin{matrix} \frac{1}{h_{11}} & \frac{-h_{12}}{h_{11}} \\ \frac{h_{21}}{h_{11}} & \frac{\Delta h}{h_{11}} \end{matrix}$	$\begin{matrix} \frac{\Delta g}{g_{22}} & \frac{g_{12}}{g_{22}} \\ \frac{-g_{21}}{g_{22}} & \frac{1}{g_{22}} \end{matrix}$
Transmission or *ABCD* parameters $\begin{bmatrix} v_1 \\ i_1 \end{bmatrix} = \begin{bmatrix} A & B \\ C & D \end{bmatrix} \begin{bmatrix} v_2 \\ -i_2 \end{bmatrix}$	$\begin{matrix} \frac{z_{11}}{z_{21}} & \frac{\Delta z}{z_{21}} \\ \frac{1}{z_{21}} & \frac{z_{22}}{z_{21}} \end{matrix}$	$\begin{matrix} \frac{-y_{22}}{y_{21}} & \frac{1}{-y_{21}} \\ \frac{-\Delta y}{y_{21}} & \frac{-y_{11}}{y_{21}} \end{matrix}$	$\begin{matrix} A & B \\ C & D \end{matrix}$	$\begin{matrix} \frac{-\Delta h}{h_{21}} & \frac{h_{11}}{h_{21}} \\ \frac{-h_{22}}{h_{21}} & \frac{-1}{h_{21}} \end{matrix}$	$\begin{matrix} \frac{1}{g_{21}} & \frac{g_{22}}{g_{21}} \\ \frac{g_{11}}{g_{21}} & \frac{\Delta g}{g_{21}} \end{matrix}$
Hybrid h parameters $\begin{bmatrix} v_1 \\ i_2 \end{bmatrix} = \begin{bmatrix} h_{11} & h_{12} \\ h_{21} & h_{22} \end{bmatrix} \begin{bmatrix} i_1 \\ v_2 \end{bmatrix}$	$\begin{matrix} \frac{\Delta z}{z_{22}} & \frac{z_{12}}{z_{22}} \\ \frac{-z_{21}}{z_{22}} & \frac{1}{z_{22}} \end{matrix}$	$\begin{matrix} \frac{1}{y_{11}} & \frac{-y_{12}}{y_{11}} \\ \frac{y_{21}}{y_{11}} & \frac{\Delta y}{y_{11}} \end{matrix}$	$\begin{matrix} \frac{B}{D} & \frac{\Delta T}{D} \\ \frac{-1}{D} & \frac{C}{D} \end{matrix}$	$\begin{matrix} h_{11} & h_{12} \\ h_{21} & h_{22} \end{matrix}$	$\begin{matrix} \frac{g_{22}}{\Delta g} & \frac{-g_{12}}{\Delta g} \\ \frac{-g_{21}}{\Delta g} & \frac{g_{11}}{\Delta g} \end{matrix}$
Hybrid g parameters $\begin{bmatrix} i_1 \\ v_2 \end{bmatrix} = \begin{bmatrix} g_{11} & g_{12} \\ g_{21} & g_{22} \end{bmatrix} \begin{bmatrix} v_1 \\ i_2 \end{bmatrix}$	$\begin{matrix} \frac{1}{z_{11}} & \frac{-z_{12}}{z_{11}} \\ \frac{z_{21}}{z_{11}} & \frac{\Delta z}{z_{11}} \end{matrix}$	$\begin{matrix} \frac{\Delta y}{y_{22}} & \frac{y_{12}}{y_{22}} \\ \frac{-y_{21}}{y_{22}} & \frac{1}{y_{22}} \end{matrix}$	$\begin{matrix} \frac{C}{A} & \frac{-\Delta T}{A} \\ \frac{1}{A} & \frac{B}{A} \end{matrix}$	$\begin{matrix} \frac{h_{22}}{\Delta h} & \frac{-h_{12}}{\Delta h} \\ \frac{-h_{21}}{\Delta h} & \frac{h_{11}}{\Delta h} \end{matrix}$	$\begin{matrix} g_{11} & g_{12} \\ g_{21} & g_{22} \end{matrix}$

$\Delta z = z_{11}z_{22} - z_{21}z_{12},\ \Delta y = y_{11}y_{22} - y_{21}y_{12},\ \Delta T = AD - BC,\ \Delta h = h_{11}h_{22} - h_{12}h_{21},\ \Delta g = g_{11}g_{22} - g_{12}g_{21}$

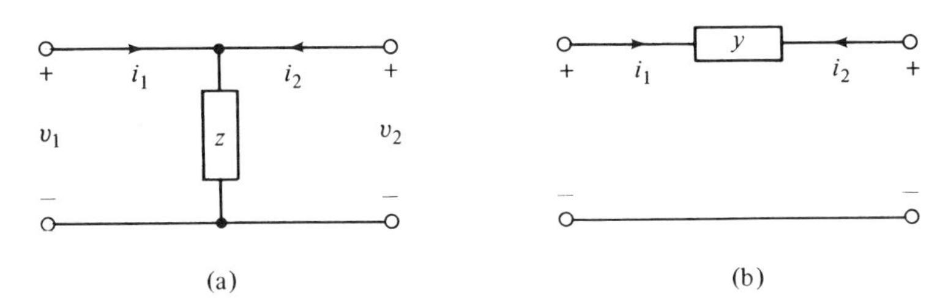

Figure 4.7-10 (a) Two-port network whose y parameters are not defined; (b) two-port network whose z parameters are not defined.

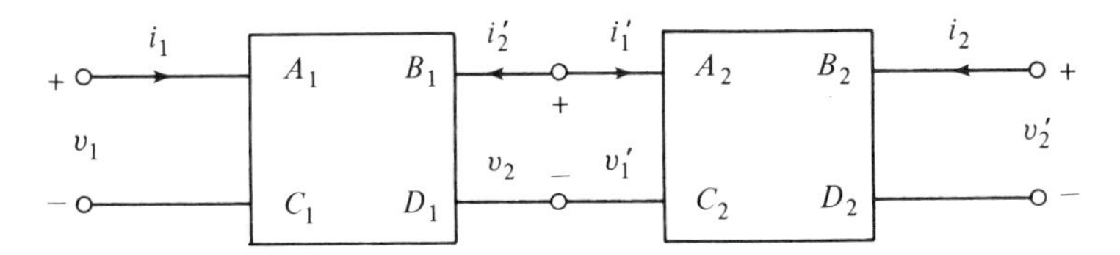

Figure 4.7-11 Tandem connection of two two-ports.

From the figure it is clear that $-i_2' = i_1'$ and $v_2' = v_1'$; then we can combine these equations to get

$$\begin{bmatrix} v_1 \\ i_1 \end{bmatrix} = \begin{bmatrix} A_1 & B_1 \\ C_1 & D_1 \end{bmatrix} \begin{bmatrix} A_2 & B_2 \\ C_2 & D_2 \end{bmatrix} \begin{bmatrix} v_2 \\ -i_2 \end{bmatrix}$$

That is, the transmission matrix of the overall network is the product of the transmission matrices of the individual networks. Two-port networks can also be connected in series, parallel, series parallel, and so on (see Problem P4.23).

Scattering Parameters. An alternative method of representing a linear time-invariant network is through its scattering parameters. This form of representation is particularly useful in microwave communication networks and transmission lines, since at very high frequencies it is practically impossible to measure conventional network parameters such as z, y, or hybrid parameters. Furthermore, networks for which z and/or y parameters do not exist (such as ideal transformers) can be best represented by scattering parameters.

Scattering parameters can be thought of as a generalization of the reflection coefficient in transmission lines. Therefore, to better understand the basic idea behind the scattering parameters, let us first consider the simple transmission line shown in Fig. 4.7-12. This line is used to transmit the signal v_s to the load z_l. The voltage source v_s is assumed to have an internal resistance R_s, which is shown in series with the source. At each point x on the line the voltage can be considered as the sum of two voltages, the *incident* voltage

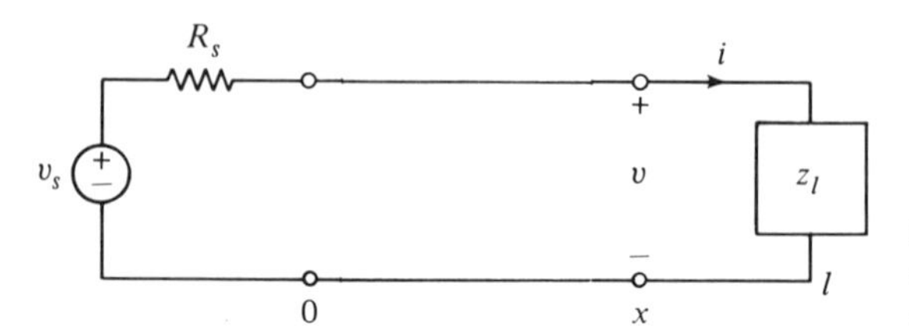

Figure 4.7-12 Transmission line used in definition of reflection coefficient.

v_i and the *reflected* voltage v_r:

$$v = v_i + v_r \tag{4.7-18}$$

Similarly, the current i can be thought of as the difference between the incident current i_i and the reflected current i_r:

$$i = i_i - i_r \tag{4.7-19}$$

The currents i_i and i_r are related to the line's characteristic impedance z_0 by $v_i = z_0 i_i$ and $v_r = z_0 i_r$. Equation (4.7-19) can therefore be written as

$$i = \frac{1}{z_0}(v_i - v_r) \tag{4.7-20}$$

From (4.7-18) and (4.7-20) we can then get

$$v_r = \tfrac{1}{2}(v - z_0 i) \tag{4.7-21}$$

and

$$v_i = \tfrac{1}{2}(v + z_0 i) \tag{4.7-22}$$

The *reflection coefficient* ρ is then defined to be

$$\rho = \frac{v_r}{v_i} = \frac{v - z_0 i}{v + z_0 i}$$

To see the significance of the reflection coefficient, let us obtain the instantaneous power $p(t)$ delivered to the load impedance:

$$p(t) = v(t)i(t) = \frac{1}{z_0}(v_i + v_r)(v_i - v_r) = \frac{1}{z_0}(v_i^2 - v_r^2)$$

To maximize the power, it is sufficient to minimize v_r and, consequently, ρ. Hence, the smaller the reflection coefficient the higher will be the power transferred to the load impedance. For a detailed derivation of the these relations, any elementary text in electromagnetic field theory can be consulted [8]. In analogy to the preceding example, the input voltage and current of the lumped element network of Fig. 4.7-13 can be decomposed into incident and reflected quantities:

$$v = v_i + v_r \tag{4.7-23}$$

$$i = i_i - i_r \tag{4.7-24}$$

Furthermore, from the network we can write

$$v_s = v + Ri \tag{4.7-25}$$

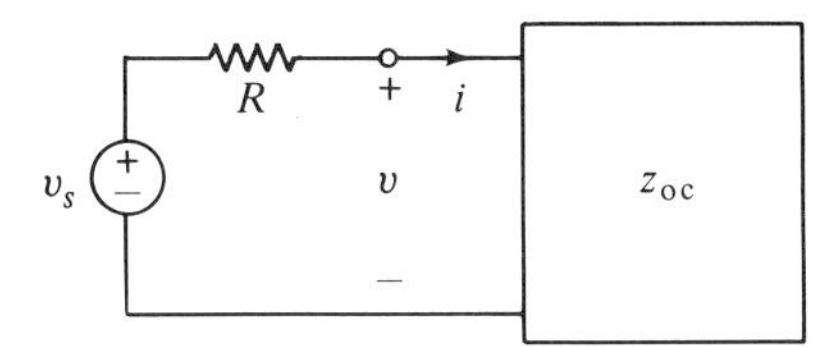

Figure 4.7-13 One-port lumped network used in defining scattering parameter.

where R is the internal resistance of the voltage source. In contrast to the distributed case, the choice of v_i, v_r, i_i, and i_r is arbitrary in the lumped network under consideration; as long as equations (4.7-23) and (4.7-24) are satisfied, it is unimportant how we decompose v into v_i and v_r or how we decompose i into i_i and i_r. However, to obtain a simple formulation for the corresponding scattering parameter, we choose v_i to be

$$v_i = \tfrac{1}{2} v_s \tag{4.7-26}$$

and

$$i_i = \frac{1}{2R} v_s \tag{4.7-27}$$

Let us digress for a moment to explain the physical significance of this choice of v_i and i_i. If the network is *matched*, that is, if the open-circuit impedance of the network z_{oc} is the same as the source resistance R ($z_{oc} = R$), then the port voltage v is the same as the incident voltage v_i:

$$v = v_i \quad \text{and} \quad i = i_i$$

which in comparing with (4.7-23) implies that $v_r = 0$ and $i_r = 0$, which in turn implies that the power transferred to the load is maximum. Let us now return to defining the scattering parameter for the network under study.

From (4.7-23), (4.7-25), and (4.7-26) we get

$$v_r = \tfrac{1}{2}(v - Ri) \tag{4.7-28}$$

and

$$v_i = \tfrac{1}{2}(v + Ri) \tag{4.7-29}$$

Similarly, from (4.7-24), (4.7-25), and (4.7-27) we obtain

$$i_r = \frac{1}{2R}(v - Ri)$$

and

$$i_i = \frac{1}{2R}(v + Ri)$$

The scattering parameter of the one-port network under study is then defined as the ratio of the reflected wave to the incident wave:

$$S = \frac{v_r}{v_i} = \frac{i_r}{i_i} = \frac{v - Ri}{v + Ri} \tag{4.7-30}$$

Now since

$$v = z_{oc} i \quad \text{or} \quad i = y_{sc} v$$

the scattering parameter can be written as

$$S = \frac{z_{oc} - R}{z_{oc} + R} = \frac{1 - Ry_{sc}}{1 + Ry_{sc}} \tag{4.7-31}$$

Thus, the scattering parameter of a one-port network can be determined from its open-circuit impedance or its short-circuit admittance. Before defining the scattering matrix of an n-port network, let us work out an illustrative example.

Example 4.7-5. Consider the one-port network of Fig. 4.7-14(a). The open-circuit impedance of this circuit is clearly

$$z_{oc} = sL + \frac{R}{1 + sRC}$$

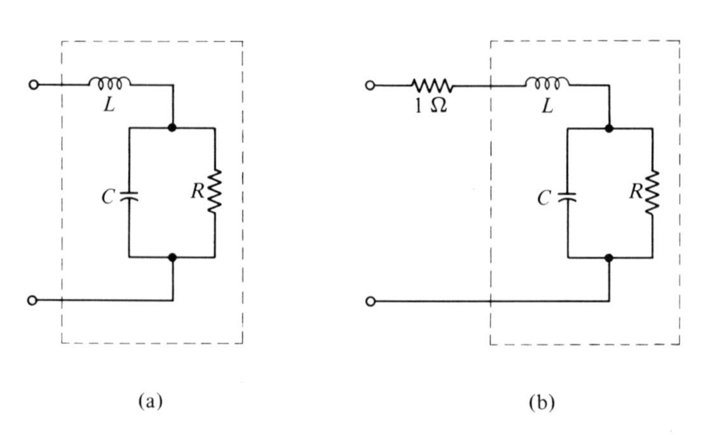

Figure 4.7-14 One-port *RLC* network of Example 4.7-5.

To put the network in the general form of Fig. 4.7-13, let us augment it by adding an arbitrary resistance, say 1 **Ω**, to its input terminal. The corresponding scattering parameter can then be obtained using (4.7-31):

$$S = \frac{LCRs^2 + (L - RC)s + R - 1}{LCRs^2 + (L + RC)s + R + 1}$$

Let us now define the scattering matrix of an n-port network and derive its relationship to the open-circuit impedance matrix. Consider the n-port network of Fig. 4.7-15. The resistor R_k is the internal resistance of the voltage source v_{s_k}. Let

$$\mathbf{v} = [v_1 \quad v_2 \quad \ldots \quad v_n]^T$$

$$\mathbf{i} = [i_1 \quad i_2 \quad \ldots \quad i_n]^T$$

$$\mathbf{v}_s = [v_{s_1} \quad v_{s_2} \quad \ldots \quad v_{s_n}]^T$$

and

$$\mathbf{R} = \text{diag}\,\{R_1, R_2, \ldots, R_n\} \tag{4.7-32}$$

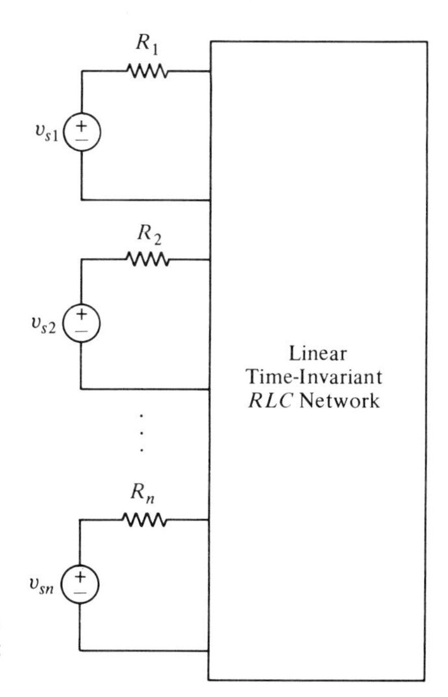

Figure 4.7-15 The n-port network shown in defining the scattering matrix.

As in the scalar case, the voltage and the current vectors can be decomposed into incident and reflected waves:

$$\mathbf{v} = \mathbf{v}_i + \mathbf{v}_r \tag{4.7-33}$$

and

$$\mathbf{i} = \mathbf{i}_i - \mathbf{i}_r \tag{4.7-34}$$

Also, from the figure we can write

$$\mathbf{v}_s = \mathbf{v} + \mathbf{R}\mathbf{i} \tag{4.7-35}$$

where $\mathbf{R}$ is the diagonal matrix defined in (4.7-32). As in the scalar case, let

$$\mathbf{v}_i = \tfrac{1}{2}\mathbf{v}_s \tag{4.7-36}$$

and

$$\mathbf{i}_i = \tfrac{1}{2}\mathbf{R}^{-1}\mathbf{v}_s \tag{4.7-37}$$

Then from (4.7-33), (4.7-35), and (4.7-36) we can easily get

$$\mathbf{v}_r = \tfrac{1}{2}(\mathbf{v} - \mathbf{R}\mathbf{i}) \tag{4.7-38}$$

and

$$\mathbf{v}_i = \tfrac{1}{2}(\mathbf{v} + \mathbf{R}\mathbf{i}) \tag{4.7-39}$$

Similarly, from (4.7-34), (4.7-35), and (4.7-37) we obtain

$$\mathbf{i}_r = \tfrac{1}{2}\mathbf{R}^{-1}(\mathbf{v} - \mathbf{R}\mathbf{i}) \tag{4.7-40}$$

and

$$\mathbf{i}_i = \tfrac{1}{2}\mathbf{R}^{-1}(\mathbf{v} + \mathbf{R}\mathbf{i}) \tag{4.7-41}$$

We are now in a position to define the scattering matrices for the network under study.

The *voltage scattering matrix* $\mathbf{S}^v$ of the n-port of Fig. 4.7-15 is an $n \times n$ matrix relating the Laplace transform of the reflected voltages to the Laplace transform of the incident voltages; that is,

$$\mathbf{v}_r(s) = \mathbf{S}^v \mathbf{v}_i(s) \tag{4.7-42}$$

From this relation and the expression for $\mathbf{v}_r$ and $\mathbf{v}_i$ given in (4.7-38) and (4.7-39), we get

$$\mathbf{v} - \mathbf{Ri} = \mathbf{S}^v(\mathbf{v} + \mathbf{Ri}) \tag{4.7-43}$$

But $\mathbf{v}$ is related to $\mathbf{i}$ through the network's open-circuit impedance matrix:

$$\mathbf{v} = \mathbf{Z}_{oc}\mathbf{i}$$

Hence, (4.7-43) can be written as

$$(\mathbf{Z}_{oc} - \mathbf{R})\mathbf{i} = \mathbf{S}^v(\mathbf{Z}_{oc} + \mathbf{R})\mathbf{i}$$

This equation holds for *all* $\mathbf{i}$; consequently, we must have

$$(\mathbf{Z}_{oc} - \mathbf{R}) = \mathbf{S}^v(\mathbf{Z}_{oc} + \mathbf{R})$$

or, equivalently,

$$\mathbf{S}^v = (\mathbf{Z}_{oc} - \mathbf{R})(\mathbf{Z}_{oc} + \mathbf{R})^{-1} \tag{4.7-44}$$

If the short-circuit admittance matrix is used, the resulting expression would be

$$\mathbf{S}^v = (\mathbf{1} - \mathbf{RY}_{sc})(\mathbf{1} + \mathbf{RY}_{sc})^{-1} \tag{4.7-45}$$

The reflected and incident currents are also related through another scattering matrix called the *current scattering matrix*. More specifically, the current scattering matrix $\mathbf{S}^i$ of the n-port of Fig. 4.7-15 is an $n \times n$ matrix relating the Laplace transform of the reflected currents to the Laplace transform of the incident currents; that is,

$$\mathbf{i}_r(s) = \mathbf{S}^i \mathbf{i}_i(s) \tag{4.7-46}$$

Using $\mathbf{i}_r$ and $\mathbf{i}_i$ of (4.7-40) and (4.7-41) in (4.4-46) results in

$$\mathbf{S}^i = \mathbf{R}^{-1}(\mathbf{Z}_{oc} - \mathbf{R})(\mathbf{Z}_{oc} + \mathbf{R})^{-1}\mathbf{R} \tag{4.7-47}$$

or, equivalently,

$$\mathbf{S}^i = (\mathbf{R}^{-1}\mathbf{Z}_{oc} - \mathbf{1})(\mathbf{R}^{-1}\mathbf{Z}_{oc} + \mathbf{1})^{-1} \tag{4.7-48}$$

If the short-circuit admittance matrix is used, we get

$$\mathbf{S}^i = (\mathbf{R}^{-1} - \mathbf{Y}_{sc})(\mathbf{R}^{-1} + \mathbf{Y}_{sc})^{-1} \tag{4.7-49}$$

Consequently, knowing either $\mathbf{Y}_{sc}$ or $\mathbf{Z}_{oc}$, the scattering matrix of the network under study can be determined using these relations.

Note that the relationship between $\mathbf{S}^v$ and $\mathbf{S}^i$ can be obtained by comparing (4.7-44) and (4.7-47):

$$\mathbf{S}^i = \mathbf{R}^{-1}\mathbf{S}^v\mathbf{R} \tag{4.7-50}$$

Thus, if the network under consideration is augmented by 1 Ω resistors (i.e., $\mathbf{R} = \mathbf{1}$), we have

$$\mathbf{S} = \mathbf{S}^i = \mathbf{S}^v = (\mathbf{Z}_{oc} - \mathbf{1})(\mathbf{Z}_{oc} + \mathbf{1})^{-1}$$

Let us now work out an example to illustrate the application of these relations.

Example 4.7-6. Consider the two-port network shown in Fig. 4.7-16. The open-circuit impedance matrix and the source resistance matix are, respectively,

$$\mathbf{Z}_{oc} = \begin{bmatrix} 1+s & 1 \\ 1 & 1+s \end{bmatrix}, \qquad \mathbf{R} = \begin{bmatrix} 1 & 0 \\ 0 & 1 \end{bmatrix}$$

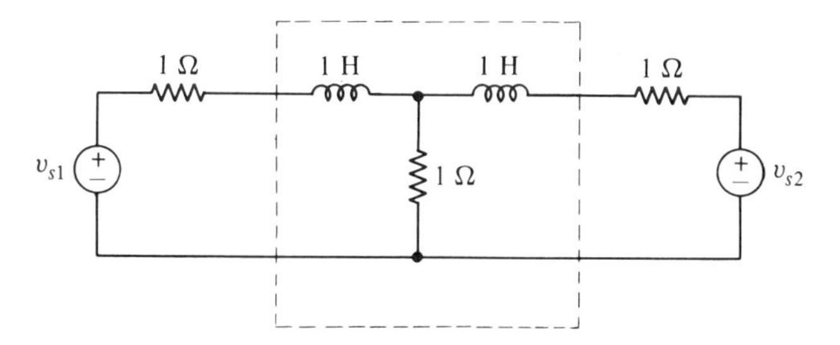

Figure 4.7-16 Two-port network used in Example 4.7-6.

Using (4.7-44), the resulting scattering matrix becomes

$$\mathbf{S}^i = \mathbf{S}^v = \mathbf{S} = \frac{1}{s^2+4s+3}\begin{bmatrix} s^2+2s-1 & 2 \\ 2 & s^2+2s-1 \end{bmatrix}$$

It should be mentioned that this discussion is merely an introduction to the scattering parameters. For more details and application the reader may consult [1, 2].

REFERENCES

[1] KUH, E. S., and R. A. ROHRER, *Theory of Linear Active Networks*, Holden-Day, Inc., San Francisco, 1967.

[2] BALABANIAN, N., T. A. BICKART, and S. SESHU, *Electrical Network Theory*, John Wiley & Sons, Inc., New York, 1969.

[3] GUPTA, S. C., J. W. BAYLESS, and B. PEIKARI, *Circuit Analysis: With Computer Usage in Problem Solving*, International Textbook Company—College Division, Scranton, Pa., 1972.

[4] DESOER, C. A., and E. S., KUH, *Basic Circuit Theory*, McGraw-Hill Book Company, New York, 1970.

[5] CHIRLIAN, P. M., *Basic Network Theory*, McGraw-Hill Book Company, New York, 1969.

[6] HUELSMAN, L. P., *Basic Circuit Theory with Digital Computers*, Prentice-Hall, Inc., Englewood Cliffs, N.J., 1972.

[7] WING, O., *Circuit Theory with Computer Methods*, Holt, Rinehart and Winston, Inc., New York, 1972.

[8] RAMO, S., J. R. WHINERY, and T. VANDUZER, *Fields and Waves in Communication Electronics*, John Wiley & Sons, Inc., New York, 1965.

[9] KARNI, S., *Intermediate Network Analysis*, Allyn and Bacon, Inc., Boston, 1971.

PROBLEMS

4.1 Consider the network shown in Fig. P4.1; the elements are in ohms, farads, and henries.

Write down a set of linearly independent equations that can be solved for the Laplace transform of the element voltages and currents using (1) element impedance matrix and (2) element admittance matrix.

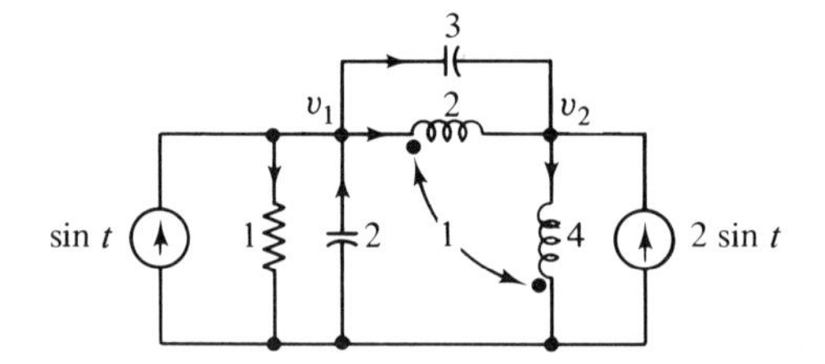

Figure P4.1

4.2 Repeat Problem 4.1 for the network shown in Fig. P4.2.

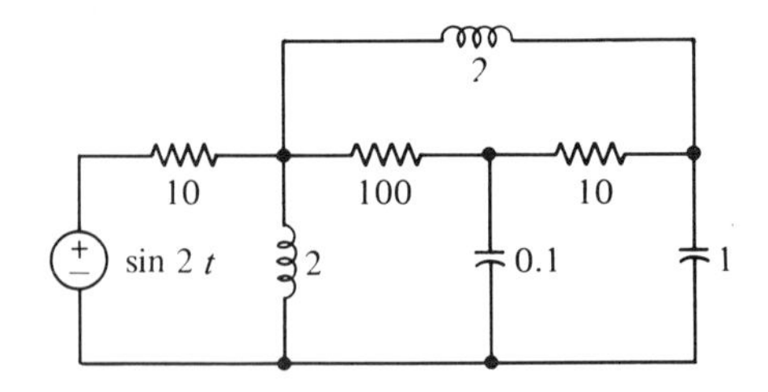

Figure P4.2

4.3 Consider the network shown in Fig. P4.3. Let the resistance of each resistor be 1 Ω, the inductance of each inductor be 1H, and the capacitance of each capacitor be 1F.

(a) Perform a source transformation to eliminate the branch containing the voltage source v_3.

(b) Obtain the incident matrix **A**, the element admittance matrix $\mathbf{Y}(s)$, and the node admittance matrix $\mathbf{Y}_n(s)$.

(c) Perform a nodal analysis on this network to obtain a set of linearly independent equations that can be solved for the Laplace transform of all the node voltages.

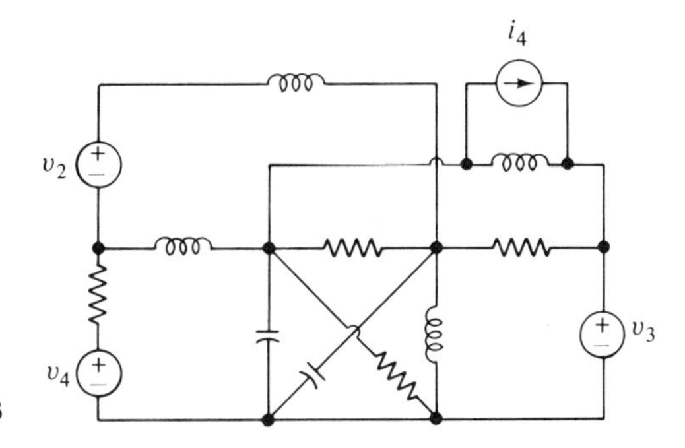

Figure P4.3

4.4 (a) Obtain the fundamental loop matrix $\mathbf{B}_f$, element impedance matrix $\mathbf{Z}(s)$, and the loop impedance matrix of the network shown in Fig. P4.3.

(b) Perform a loop analysis to obtain a set of linearly independent equations that can be solved for the Laplace transform of the loop currents.

(c) Compare the loop analysis of part (b) to the nodal analysis of Problem 4.3. Which one is simpler to solve?

4.5 Consider the gyrator network shown in Fig. P4.5. Perform a nodal analysis on this network.

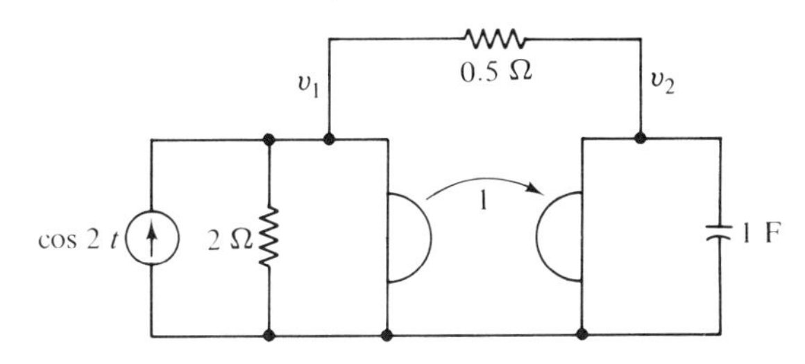

Figure P4.5

4.6 Perform a loop analysis on the network shown in Fig. P4.5.

4.7 Consider the network shown in Fig. P4.7 (see next page).

(a) Obtain the incidence matrix $\mathbf{A}$, the element admittance matrix $\mathbf{Y}(s)$, and the node admittance matrix $\mathbf{Y}_n(s)$ for this network.

(b) Perform a nodal analysis on this network and obtain a set of independent equations that can be solved for $v_1(s)$ and $v_2(s)$.

(c) Let $R_1 = R_2 = R_3 = 2\,\Omega$, $L_4 = L_5 = 1$ H, and $M = 0.5$; then find $v_1(t)$ and $v_2(t)$ for $t \geq 0$. Assume that all the initial conditions are zero.

4.8 (a) Obtain the fundamental loop equation $\mathbf{B}_f$, the element impedance matrix $\mathbf{Z}(s)$, and the loop impedance matrix $\mathbf{Z}_l(s)$ for the network of Fig. P4.7.

(b) Perform a loop analysis on this network.

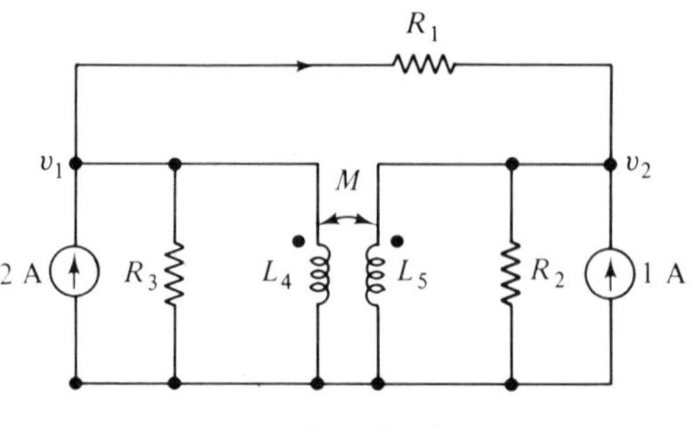

Figure P4.7

4.9 Consider the filter shown in Fig. P4.9.
(a) Represent the operational amplifier by a dependent source.
(b) Perform a nodal analysis on this network to obtain $v_2(t)$.

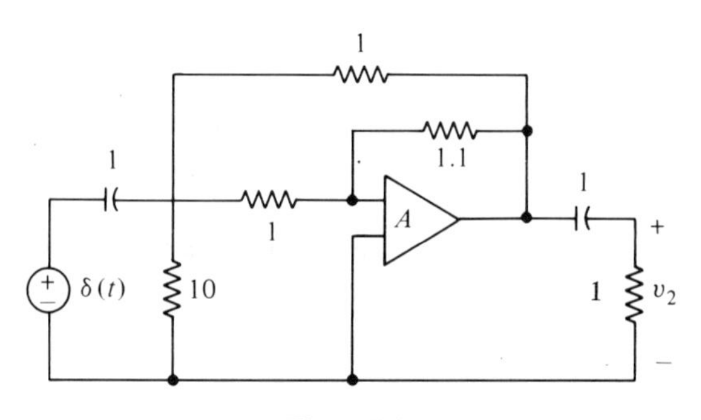

Figure P4.9

4.10 Consider the network shown in Fig. P4.10. Perform a loop analysis on this network.

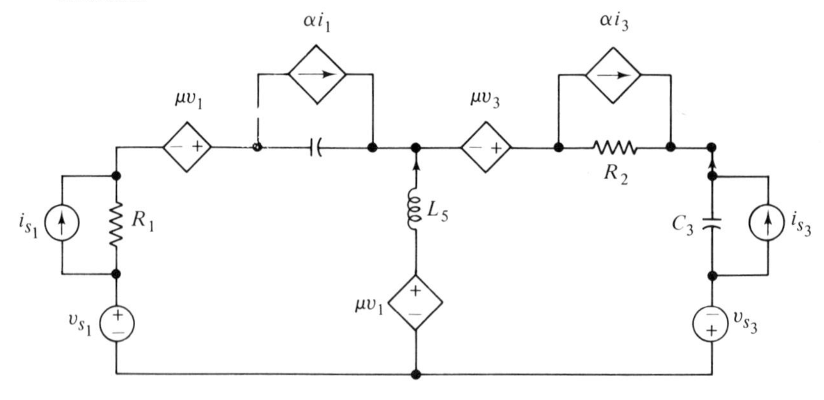

Figure P4.10

4.11 Perform a nodal analysis on the network shown in Fig. P4.10.

4.12 Consider the network shown in Fig. P4.12. Let $i_{s_1} = i_{s_3} =$ unit step function and assume that all the initial conditions are zero. Perform a nodal analysis on this network.

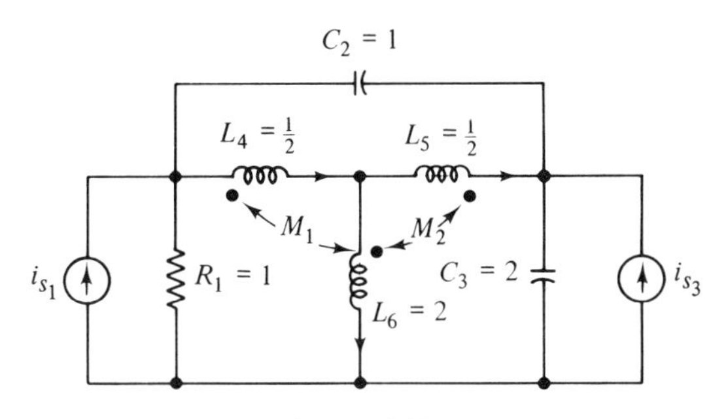

Figure P4.12

4.13 Perform a loop analysis on the network shown in Fig. P4.12. Assume that the initial conditions and sources are as in Problem 4.12.

4.14 Use the steady state nodal analysis to find the voltages v_1 and v_2 in Fig. P4.1.

4.15 Repeat Problem 4.14 for the network of Fig. P4.5.

4.16 Obtain the open-circuit impedance matrix of the network shown in Fig. P4.16.

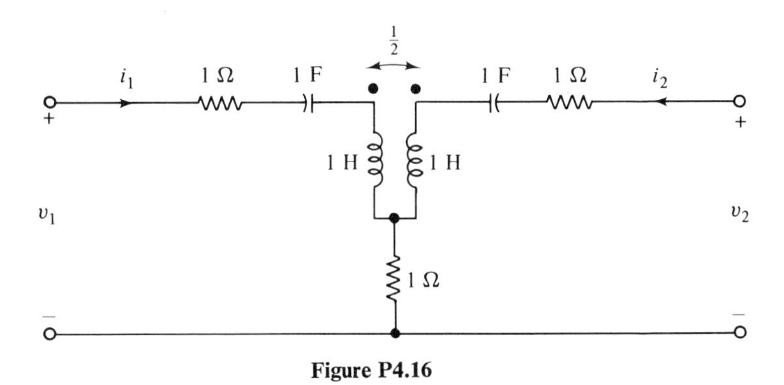

Figure P4.16

4.17 Find the open-circuit impedance matrix of the three-port network shown in Fig. P4.17. (see next page).

4.18 Find the short-circuit admittance matrix of the network shown in Fig. P4.16.

4.19 Obtain the short-circuit admittance matrix of the network shown in Fig. P4.17. (see next page).

4.20 Consider the transistor amplifier model shown in Fig. P4.20. (see next page). Find the z parameters, y parameters, and hybrid parameters for this network.

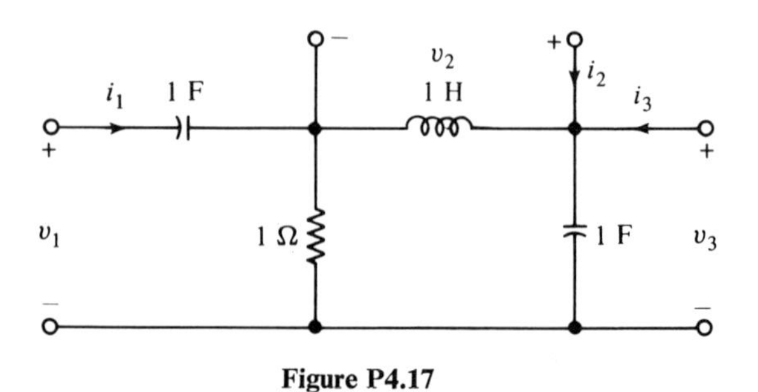

Figure P4.17

4.21 Obtain the voltage transfer function of the network in Fig. P4.20.

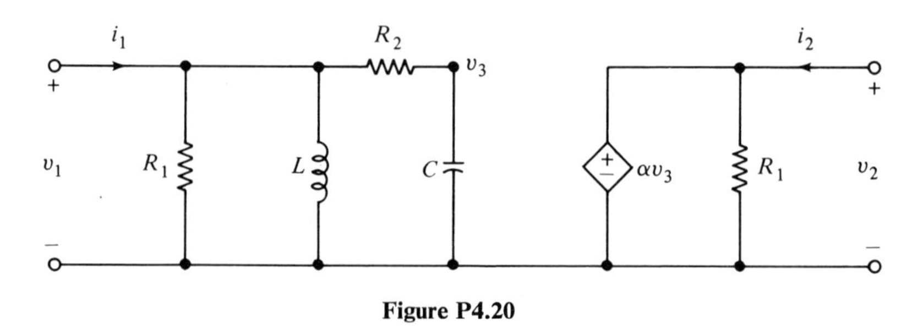

Figure P4.20

4.22 Verify the second column in Table 4.7-1.

4.23 Consider the parallel and series connection of the two-port networks shown in Fig. P4.23. Show that in (a) the short-circuit admittance matrix of the interconnected network is the sum of the individual short-circuit admittance matrices, and, also, that in (b) the open-circuit impedance matrix of the interconnected network is the sum of the individual open-circuit impedance matrices. Explain the conditions under which these interconnections are valid.

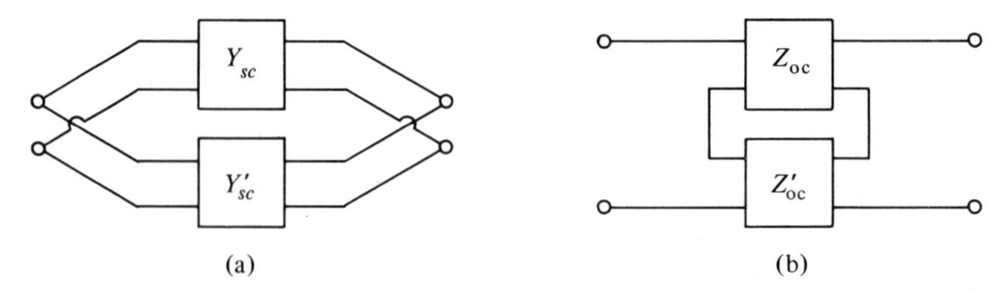

Figure P4.23

4.24 Find the voltage transfer function of the network of Fig. P4.9.

4.25 Obtain the voltage scattering matrix of the network shown in Fig. P4.16.

4.26 Obtain the current scattering matrix of the network of Fig. P4.17.

5 State-Variable Representation of Networks

5.1 INTRODUCTION

In Chapter 4 we used the Laplace transform technique to obtain an algebraic relationship between the Laplace transform of the input and the Laplace transform of the output of a given linear time-invariant network. In the single-input single-output case this relationship was represented as

$$y(s) = H(s)u(s) \tag{5.1-1}$$

where $u(s)$ and $y(s)$ are the transforms of the input and the response, respectively, and $H(s)$ is the transfer function of the network. If the network under consideration is comprised of lumped elements only, then $H(s)$ is given as the quotient of two rational polynomials in the complex variable s; that is,

$$H(s) = \frac{N(s)}{D(s)} \tag{5.1-2}$$

As was shown in Chapter 4, to obtain $y(t)$ for a given $u(t)$ when the roots of $D(s)$ are known, we can use a partial fraction expansion of $H(s)u(s)$ and then take the inverse Laplace transform. An alternative procedure is to introduce a set of auxiliary variables called the *state variables* (or *phase variables*) to put (5.1-1) in the form of a set of simultaneous first-order differential equations. The advantage of representing the equations in this form is threefold. First, there are numerous techniques readily available, both analytical and numerical, for solving a set of first-order differential equations. Second, first-order differential equations are much easier to program on digital computers than transfer functions; there are abundant computer routines available for this purpose. Third, the existing techniques in systems theory can be used to determine stability, controllability, and observability of a network if its describing equation is in the form of a set of first-order differential equations.

To show how (5.1-1) can be put in the form of a set of simultaneous differential equations, let us combine it with (5.1-2) to get

$$\frac{y(s)}{u(s)} = \frac{N(s)}{D(s)} \tag{5.1-3}$$

Then, using an auxiliary variable $x(s)$, equation (5.1-3) can be broken up into two equations:

$$u(s) = D(s)x(s) \tag{5.1-4}$$

and

$$y(s) = N(s)x(s) \tag{5.1-5}$$

The polynomials $D(s)$ and $N(s)$ are typically in the following forms:

$$D(s) = a_n s^n + a_{n-1}s^{n-1} + \cdots + a_1 s + a_0 \tag{5.1-6}$$

and

$$N(s) = b_m s^m + b_{m-1}s^{m-1} + \cdots + b_1 s + b_0 \tag{5.1-7}$$

For simplicity, we assume that the degree of $N(s)$ is less than or equal to the degree of $D(s)$; that is, $m \leq n$. Then (5.1-4) and (5.1-5) may be written as

$$u(s) = (a_n s^n + a_{n-1}s^{n-1} + \cdots + a_1 s + a_0)x(s)$$

$$y(s) = (b_m s^m + b_{m-1}s^{m-1} + \cdots + b_1 s + b_0)x(s)$$

which upon transformation to the time domain yield

$$u(t) = a_n x^{(n)}(t) + a_{n-1}x^{(n-1)}(t) + \cdots + a_1 x^{(1)}(t) + a_0 x(t) \tag{5.1-8}$$

and

$$y(t) = b_m x^{(m)}(t) + b_{m-1}x^{(m-1)}(t) + \cdots + b_1 x^{(1)}(t) + b_0 x(t) \tag{5.1-9}$$

where $x^{(k)}$ denotes the kth derivative of $x(t)$. We are now ready to choose a set of n auxiliary variables $x_1(t)$ through $x_n(t)$ for representing equation (5.1-1). Let

$$x_1 = x,\ x_2 = x^{(1)},\ x_3 = x^{(2)},\ \ldots,\ x_n = x^{(n-1)} \tag{5.1-10}$$

Then using (5.1-10) with (5.1-8), we get

$$\begin{aligned}
\dot{x}_1(t) &= x_2(t) \\
\dot{x}_2(t) &= x_3(t) \\
&\ \vdots \\
\dot{x}_{n-1}(t) &= x_n(t) \\
\dot{x}_n(t) &= \frac{-1}{a_n}[a_0 x_1(t) + a_1 x_2(t) + \cdots + a_{n-1}x_n(t)] + \frac{1}{a_n}u(t)
\end{aligned} \tag{5.1-11}$$

where for simplicity of notation we have denoted the first derivative of $x_k(t)$ by $\dot{x}_k(t)$.

Similarly, using (5.1-9) and (5.1-10), we obtain

$$y(t) = b_0 x_1(t) + b_1 x_2(t) + b_2 x_3(t) + \cdots + b_m x_{m+1}(t) \tag{5.1-12}$$

In the matrix form, equations (5.1-11) and (5.1-12) will be

$$\begin{bmatrix} \dot{x}_1 \\ \dot{x}_2 \\ \dot{x}_3 \\ \cdot \\ \cdot \\ \cdot \\ \dot{x}_n \end{bmatrix} = \begin{bmatrix} 0 & 1 & 0 & & \cdots & 0 \\ 0 & 0 & 1 & 0 & \cdots & 0 \\ 0 & 0 & 0 & 1 & \cdots & 0 \\ \cdot & & & & & \cdot \\ \cdot & & & & & \cdot \\ \cdot & & & & & \cdot \\ \dfrac{-a_0}{a_n} & \dfrac{-a_1}{a_n} & \dfrac{-a_2}{a_n} & & \cdots & \dfrac{-a_{n-1}}{a_n} \end{bmatrix} \begin{bmatrix} x_1 \\ x_2 \\ x_3 \\ \cdot \\ \cdot \\ \cdot \\ x_n \end{bmatrix} + \begin{bmatrix} 0 \\ 0 \\ 0 \\ \cdot \\ \cdot \\ \cdot \\ \dfrac{1}{a_n} \end{bmatrix} u(t) \tag{5.1-13}$$

and

$$y(t) = [b_0 \quad b_1 \quad b_2 \cdots b_m \quad 0 \quad 0 \cdots 0] \begin{bmatrix} x_1 \\ x_2 \\ x_3 \\ \cdot \\ \cdot \\ \cdot \\ x_n \end{bmatrix} \tag{5.1-14}$$

Or, in a more compact form,

$$\dot{\mathbf{x}}(t) = \mathbf{A}\mathbf{x}(t) + \mathbf{b}u(t) \tag{5.1-15}$$

$$y(t) = \mathbf{c}^T \mathbf{x}(t) \tag{5.1-16}$$

where $\mathbf{x}(t)$ is an n-column vector called the *state vector* and $\mathbf{A}$ is the $n \times n$ constant matrix in (5.1-13) whose elements are given in terms of $a_0, a_1, \ldots, a_n$. The column vectors $\mathbf{b}$ and $\mathbf{c}$ are also constant vectors whose elements are completely specified in terms of the coefficients of $D(s)$ and $N(s)$. These equations are equivalent to (5.1-1). In this form, however, $y(t)$ is related to $u(t)$ through the auxiliary variables $x_1, x_2, \ldots, x_n$. For a given input the vector differential equation (5.1-15) can be solved for $\mathbf{x}(t)$; the network response $y(t)$ can then be easily obtained from (5.1-16). In Chapters 6 and 7 we discuss the solution of such equations in detail. It turns out that in practice it is often easier to solve the state equations in the time domain than to take the inverse Laplace transform of (5.1-1).

Let us now work out a simple example to illustrate this procedure.

Example 5.1-1. The input–output relationship of a linear time-invariant network is given as follows.

$$y(s) = \frac{5s^2 + 2s + 3}{s^3 + 2s^2 + 4s + 1} u(s)$$

Obtain its input–output state equation.

Solution

In this example $n = 3$ and $m = 2$; then the state variables are $x_1(t)$, $x_2(t)$, and $x_3(t)$. Using (5.1-13) and (5.1-14), the input–output state equations will be

$$\begin{bmatrix} \dot{x}_1 \\ \dot{x}_2 \\ \dot{x}_3 \end{bmatrix} = \begin{bmatrix} 0 & 1 & 0 \\ 0 & 0 & 1 \\ -1 & -4 & -2 \end{bmatrix} \begin{bmatrix} x_1 \\ x_2 \\ x_3 \end{bmatrix} + \begin{bmatrix} 0 \\ 0 \\ 1 \end{bmatrix} u(t)$$

$$y(t) = [3 \quad 2 \quad 5] \begin{bmatrix} x_1 \\ x_2 \\ x_3 \end{bmatrix}$$

If the network under consideration is represented by a matrix transfer function, a similar approach can be used to obtain its state equations [17]. However, instead of elaborating on this technique, we concentrate on obtaining the state equations directly from the network. In the remaining sections of this chapter we introduce a systematic method for obtaining the auxiliary variables (to be called state variables) from the network variables, such as capacitor voltages, inductor currents, and so on, and employ KVL, KCL, and VCR to write the dynamical equations of the network in terms of these variables. The basic property of the state variables is that they yield a set of first-order differential equations that completely describe the network behavior. One other important advantage of using the state variables over other auxiliary variables (such as loop currents or node voltages) is that the techniques developed for writing the state equations of linear time-invariant networks can easily be generalized to include nonlinear and time-varying networks as well. For this reason, the state variable method is, in many cases, preferred over the previously mentioned methods of analysis.

Two types of network variables usually qualify as the state variables for lumped *RLC* networks; first, voltages across the capacitors and currents through the inductors, and, second, charges across the capacitors and fluxes through the inductors. The latter class of variables is particularly suitable for the state variable of linear time-varying networks. For nonlinear networks, however, the choice of the state variables depends on the nature of the element characteristics.

In the remainder of this chapter we first give an informal discussion concerning the selection of the state variables and the writing of state equations of simple networks (linear, nonlinear, time invariant, and time varying) by inspection, and then introduce a systematic method of writing the state equations of a network that makes use of the graph-theoretic results obtained in Chapter 3.

5.2 PRELIMINARY CONSIDERATIONS

In this section we give a method for choosing the state variable and writing the state equations of simple *RLC* networks by inspection. We start by first defining the following useful terms:

Definition 5.2-1 (Capacitor-Only Loop). A loop is called a *capacitor-only loop* if it is made up of capacitors only (two or more) and possibly some independent voltage sources.

Definition 5.2-2 (Inductor-Only Cutset). A cutset is called an *inductor-only cutset* if it is made up of inductors only (two or more) and possibly some current sources.

Since one of the properties of the state variables is that their corresponding state equations represent the dynamical behavior of the network under study, it seems natural to choose the state variables to be either the charges or voltages on the capacitors and the fluxes or currents through the inductors. This choice of variables guarantees that the resulting dynamical equations are a set of first-order differential equations. Furthermore, if the network under consideration does not have any capacitor-only loops or inductor-only cutsets, the resulting state equations are linearly independent and can be solved for capacitor voltages or charges and inductor currents or fluxes. Let us now outline the steps required for writing the state equations of a network by inspection and then proceed with some examples.

Basic Steps Involved in Writing the State Equations of Simple Networks

Step 1. If there are no capacitor-only loops and no inductor-only cutsets, choose either

1. the capacitor voltages and inductor currents *or*
2. the capacitor charges and inductor fluxes as the state variables.

Step 2. Write down the linearly independent KVL and KCL equations.

Step 3. Use the branch VCR to express all the network variables that are not the state variables in terms of the state variables chosen in step 1.

Step 4. Use the equations obtained in step 3 to eliminate all the variables that are not the chosen state variables in the equations obtained in step 2.

The resulting equations contain only the state variables and their derivatives. Then transferring all the derivatives to the left-hand side, the resulting equations will be the desired state equations in the *normal form.* As was mentioned earlier, the procedure outlined is useful primarily for simple networks. For more complicated networks one has to resort to a more systematic procedure; one such method will be discussed later in this chapter. Let us now consider some simple examples. We postpone the derivation of the state equations of the networks that contain capacitor-only loops or inductor-only cutsets until Section 5.5.

Example 5.2-1 Linear Time-Invariant Network. Consider the *RLC* network shown in Fig. 5.2-1. Let all the elements be linear and time invariant. Find the linearly independent state equations for this network.

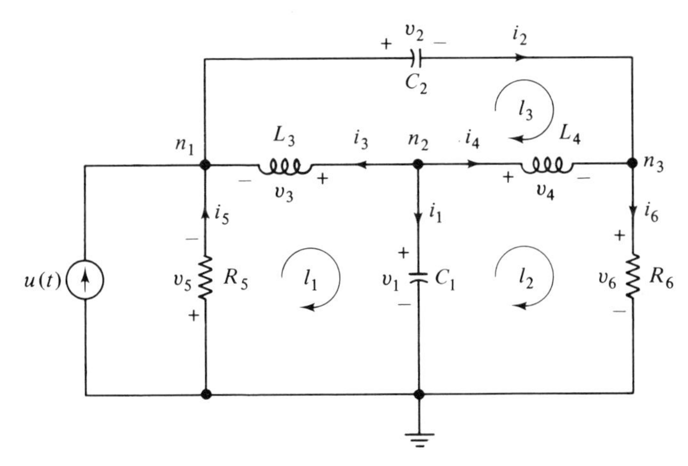

Figure 5.2-1 Linear time-invariant network whose state variables are chosen to be the inductor currents and the capacitor voltages.

Solution

Since there are no capacitor-only loops and no inductor-only cutsets, the outlined procedure can be employed:

Step 1. Choose the state variables to be the capacitor voltages and inductor currents (this set of variables is suitable if the network under study is linear and time invariant). That is, choose

$$x_1(t) = v_1(t) \tag{5.2-1}$$

$$x_2(t) = v_2(t) \tag{5.2-2}$$

$$x_3(t) = i_3(t) \tag{5.2-3}$$

$$x_4(t) = i_4(t) \tag{5.2-4}$$

where x_k denotes the kth state variable.

Step 2. Use the notations of step 1 to write the independent KCL and KVL equations:

$$\text{KCL} \quad n_1: \quad i_5 = i_2 - x_3 - u \tag{5.2-5}$$

$$n_2: \quad x_3 = -i_1 - x_4 \tag{5.2-6}$$

$$n_3: \quad i_6 = x_4 + i_2 \tag{5.2-7}$$

$$\text{KVL} \quad l_1: \quad x_1 = v_3 - v_5 \tag{5.2-8}$$

$$l_2: \quad x_1 = v_4 + v_6 \tag{5.2-9}$$

$$l_3: \quad x_2 = v_4 - v_3 \tag{5.2-10}$$

Step 3. Write the branch VCR to express i_1, i_2, v_3, v_4, v_5, v_6, i_5, and i_6 in terms of x_1 through x_4:

$$\text{VCR} \quad i_1 = C_1\dot{x}_1 \tag{5.2-11}$$

$$i_2 = C_2\dot{x}_2 \tag{5.2-12}$$

$$v_3 = L_3\dot{x}_3 \tag{5.2-13}$$

$$v_4 = L_4\dot{x}_4 \tag{5.2-14}$$

$$v_5 = R_5 i_5 \tag{5.2-15}$$

$$v_6 = R_6 i_6 \tag{5.2-16}$$

Step 4. Replacing (5.2-11) through (5.2-16) into KCL and KVL obtained in step 2, and eliminating i_6 and i_5, we get

$$\dot{x}_1 = -\frac{1}{C_1}x_3 - \frac{1}{C_1}x_4$$

$$R_6C_2\dot{x}_2 + L_4\dot{x}_4 = x_1 - R_6x_4$$

$$R_5C_2\dot{x}_2 - L_3\dot{x}_3 = -x_1 + R_5x_3 + R_5u$$

$$-L_3\dot{x}_3 + L_4\dot{x}_4 = x_2$$

Solving these equations for $\dot{x}_1$, $\dot{x}_2$, $\dot{x}_3$, and $\dot{x}_4$, and putting the result in the matrix form, we obtain

$$\begin{bmatrix} \dot{x}_1 \\ \dot{x}_2 \\ \dot{x}_3 \\ \dot{x}_4 \end{bmatrix} = \frac{1}{R_5 + R_6}\begin{bmatrix} 0 & 0 & -\dfrac{R_5 + R_6}{C_1} & -\dfrac{R_5 + R_6}{C_1} \\ 0 & -\dfrac{1}{C_2} & \dfrac{R_5}{C_2} & -\dfrac{R_6}{C_2} \\ \dfrac{R_5 + R_6}{L_3} & -\dfrac{R_5}{L_3} & -\dfrac{R_5R_6}{L_3} & -\dfrac{R_5R_6}{L_3} \\ \dfrac{R_5 + R_6}{L_4} & \dfrac{R_6}{L_4} & -\dfrac{R_5R_6}{L_4} & -\dfrac{R_5R_6}{L_4} \end{bmatrix}\begin{bmatrix} x_1 \\ x_2 \\ x_3 \\ x_4 \end{bmatrix} + \begin{bmatrix} 0 \\ \dfrac{R_5}{C_2(R_5 + R_6)} \\ \dfrac{-R_5R_6}{L_3(R_5 + R_6)} \\ \dfrac{-R_5R_6}{L_4(R_5 + R_6)} \end{bmatrix}u(t) \tag{5.2-17}$$

or, in a more compact form,

$$\dot{\mathbf{x}}(t) = \mathbf{A}\mathbf{x}(t) + \mathbf{b}u(t) \tag{5.2-18}$$

where $\mathbf{x}$ is a column vector whose elements are x_1, x_2, x_3, and x_4, $\mathbf{A}$ is the 4×4 matrix, and $\mathbf{b}$ is the column vector given in (5.2-17); this equation represents the dynamical equations of the network. Equation (5.2-18) henceforth will be called the *state equations* of linear time-invariant networks in the *normal form*. Solving (5.2-17) for $\mathbf{x}$, we can obtain any branch voltage or current by simple manipulations. For example, if the desired output of this network is v_6, the voltage across the terminating resistor, we can write KVL for loop l_2 to get

$$x_1 + L_4\dot{x}_4 + v_6 = 0$$

Then using the last equation in (5.2-17), we obtain

$$v_6 = -x_1 - x_1 - \frac{R_6}{R_5 + R_6}x_2 + \frac{R_5R_6}{R_5 + R_6}x_3 + \frac{R_5R_6}{R_5 + R_6}x_4 + \frac{R_5R_6}{R_5 + R_6}u$$

or $\quad v_6 = \mathbf{c}^T\mathbf{x} + du$

where

$$\mathbf{c}^T = \begin{bmatrix} -2 & \dfrac{-R_6}{R_5 + R_6} & \dfrac{R_5R_6}{R_5 + R_6} & \dfrac{R_5R_6}{R_5 + R_6} \end{bmatrix}$$

and

$$d = \frac{R_5R_6}{R_5 + R_6}$$

In general, we can consider any number of branch voltages and currents as the "output" and denote the corresponding output vector by $\mathbf{y}$; then this output vector can be written as a linear combination of the state vector and the "input" vector $\mathbf{u}(t)$:

$$\mathbf{y} = \mathbf{C}\mathbf{x} + \mathbf{D}\mathbf{u}$$

where $\mathbf{C}$ and $\mathbf{D}$ are matrices with appropriate dimensions.

The major difficulty in setting up the state equations of linear time-invariant networks will be in step 4, that is, the elimination of the nonstate variables. This difficulty will be resolved when we employ topological partitioning to achieve the elimination process. This method is particularly suitable for computer analysis of large-scale networks.

It was mentioned earlier that the state-variable method is particularly attractive in the analysis of linear time-varying and nonlinear networks. The steps required for writing the state equations of a linear time-varying network are identical to those required for writing the state equations of linear time-invariant networks mentioned previously, except that, in this case, the state variables must be chosen to be the capacitor charges and inductor fluxes. Let us illustrate this by means of an example.

Example 5.2-2. Consider the linear time-varying network shown in Fig. 5.2-2; it is desired to write the state equations of this network in the normal form.

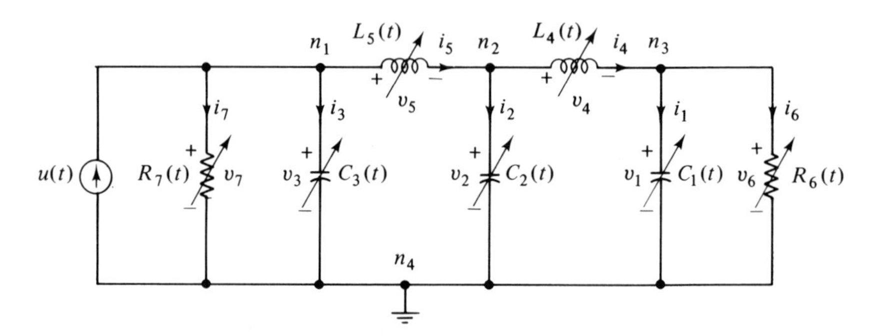

Figure 5.2-2 Linear time-varying low-pass filter whose state variables are capacitor charges and inductor fluxes.

Solution

There are no capacitor-only loops and no inductor-only cutsets in this network; the state variables, therefore, are capacitor charges and inductor fluxes.

Step 1. Choose the state vector $\mathbf{x}(t)$ to be

$$\mathbf{x}(t) = [x_1(t), x_2(t), x_3(t), x_4(t), x_5(t)]^T \tag{5.2-19}$$

where

$$\begin{aligned} x_1(t) &= q_1(t) \\ x_2(t) &= q_2(t) \\ x_3(t) &= q_3(t) \\ x_4(t) &= \phi_4(t) \\ x_5(t) &= \phi_5(t) \end{aligned} \tag{5.2-20}$$

Step 2. Independent KCL and KVL equations are

$$\begin{aligned} \text{KCL} \quad u &= i_7 + i_3 + i_5 \\ i_5 &= i_2 + i_4 \\ i_4 &= i_1 + i_6 \\ \text{KVL} \quad v_7 &= v_3 \\ v_3 &= v_5 + v_2 \\ v_2 &= v_4 + v_1 \\ v_1 &= v_6 \end{aligned}$$

Step 3. In this case, the branch relations of the energy-storing elements should be written in terms of the charges and fluxes; the current through each capacitor is the time derivative of the charge across it, and the voltage across each inductor is the time derivative of the flux through it. We then have

$$\begin{aligned}
i_1 &= \dot{x}_1, & v_1 &= \frac{1}{C_1}x_1 \\
i_2 &= \dot{x}_2, & v_2 &= \frac{1}{C_2}x_2 \\
i_3 &= \dot{x}_3, & v_3 &= \frac{1}{C_3}x_3 \\
v_4 &= \dot{x}_4, & i_4 &= \frac{1}{L_4}x_4 \\
v_5 &= \dot{x}_5, & i_5 &= \frac{1}{L_5}x_5 \\
v_6 &= R_6 i_6, & v_7 &= R_7 i_7
\end{aligned}$$

Step 4. Eliminating all the nonstate variables from the equations obtained in steps 2 and 3 and putting the results in matrix form, we obtain

$$\begin{bmatrix} \dot{x}_1 \\ \dot{x}_2 \\ \dot{x}_3 \\ \dot{x}_4 \\ \dot{x}_5 \end{bmatrix} = \begin{bmatrix} -\frac{1}{C_1(t)R_6(t)} & 0 & 0 & \frac{1}{L_4(t)} & 0 \\ 0 & 0 & 0 & -\frac{1}{L_4(t)} & \frac{1}{L_5(t)} \\ 0 & 0 & -\frac{1}{C_3(t)R_7(t)} & 0 & -\frac{1}{L_5(t)} \\ -\frac{1}{C_1(t)} & \frac{1}{C_2(t)} & 0 & 0 & 0 \\ 0 & -\frac{1}{C_2(t)} & \frac{1}{C_3(t)} & 0 & 0 \end{bmatrix} \begin{bmatrix} x_1 \\ x_2 \\ x_3 \\ x_4 \\ x_5 \end{bmatrix} + \begin{bmatrix} 0 \\ 0 \\ 1 \\ 0 \\ 0 \end{bmatrix} u(t) \tag{5.2-21}$$

or, equivalently,

$$\dot{\mathbf{x}} = \mathbf{A}(t)\mathbf{x} + \mathbf{b}u(t) \tag{5.2-22}$$

where $\mathbf{x}$ is the state vector defined in (5.2-19), $\mathbf{A}(t)$ is the 5×5 time-varying matrix, and $\mathbf{b}$ is the time-invariant column vector given in (5.2-21). Equation (5.2-21) is the state equation of the time-varying network in the normal form. Having solved this first-order differential equation for $\mathbf{x}(t)$, the branch voltages and currents can then be easily obtained using the equations derived in step 3. Solving (5.2-21), however, is not an easy task; a closed-form solution of (5.2-22) is yet to be found. Approximate solutions of such equations are usually obtained by numerical methods; some of these methods will be explored in Chapter 7.

If the network under consideration contains nonlinear elements, under certain conditions, the state equations can be represented in the *normal form*

$$\dot{\mathbf{x}} = \mathbf{f}(\mathbf{x}, \mathbf{u}) \tag{5.2-23}$$

where $\mathbf{x}$ represents the state vector, $\mathbf{u}$ represents the sources, and $\mathbf{f}(\cdot, \cdot)$ is a nonlinear vector function. It should be mentioned, howevei, that the normal form (5.2-23) may not exist for some nonlinear networks; the existence of the normal form of the state equations of a nonlinear network primarily depends on the nature of the network elements. Later in this section we give an example for which the normal form does not exist. Using a simple example, let us now show that the state equations of a class of nonlinear networks can be written by inspection.

Example 5.2-3 Nonlinear Network. Consider the network shown in Fig. 5.2-3. Let the resistors be voltage controlled, the capacitors be charge controlled, and the inductor be flux controlled with characteristics $v_1 = f_1(q_1)$, $v_2 = f_2(q_2)$, $i_3 = f_3(\phi_3)$, $i_4 = f_4(v_4)$, and $i_5 = f_5(v_5)$. The objective is to write the state equations of the network in the normal form.

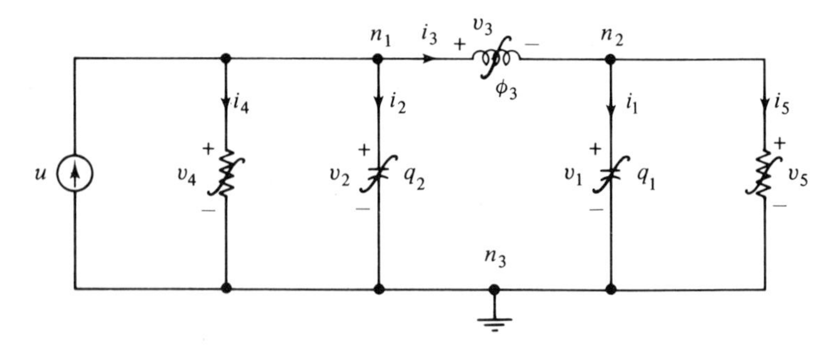

Figure 5.2-3 Nonlinear network whose state equation is derived in Example 5.2-3.

Solution

Since there are no capacitor-only loops and no inductor-only cutsets, the state variables can be chosen to be the capacitor charges and the inductor fluxes; more precisely,

Step 1. Choose the state vector $\mathbf{x}$ to be

$$\mathbf{x}(t) = [x_1(t), x_2(t), x_3(t)]^T \tag{5.2-24}$$

where

$$x_1(t) = q_1(t), \qquad x_2(t) = q_2(t), \qquad x_3(t) = \phi_3(t) \tag{5.2-25}$$

Step 2. Independent KCL and KVL equations are

$$u = i_2 + i_3 + i_4, \qquad v_4 = v_2$$
$$i_3 = i_1 + i_5, \qquad v_1 = v_5$$
$$v_2 = v_3 + v_1$$

Step 3. The branch relations are

$$\begin{aligned} i_1 &= \dot{x}_1, & v_2 &= f_2(x_2) \\ i_2 &= \dot{x}_2, & i_3 &= f_3(x_3) \\ v_3 &= \dot{x}_3, & i_4 &= f_4(v_4) \\ v_1 &= f_1(x_1), & i_5 &= f_5(v_5) \end{aligned} \tag{5.2-26}$$

Step 4. Eliminating nonstate variables among the equations derived in steps 2 and 3 yields

$$\begin{aligned} \dot{x}_1 &= f_3(x_3) - f_5[f_1(x_1)] \\ \dot{x}_2 &= -f_3(x_3) - f_4[f_2(x_2)] + u \\ \dot{x}_3 &= -f_1(x_1) + f_2(x_2) \end{aligned} \tag{5.2-27}$$

Equations (5.2-27) are the state equations of the network under study in the normal form. A unique solution of such nonlinear equations may or may not exist. In Chapter 6 we give sufficient conditions under which (5.2-23) possesses a unique solution. This solution, if it exists, can generally be obtained by numerical methods.

It was mentioned earlier that the existence of the state equations in the normal form depends on the nature of the network elements. To illustrate this fact, let us consider the same network considered in Example 5.2-3, but this time assume that resistor R_4 is nonlinear current controlled; that is, let

$$v_4 = f_4(i_4) \tag{5.2-28}$$

where $f_4(\cdot)$ is such that i_4 cannot be expressed as a single-valued function of v_4. Let all the other elements be the same as in Example 5.2-3. If we now follow steps 1 through 3, we obtain the same equations with the exception of equation (5.2-26), which must be replaced by (5.2-28). Going through the elimination process (step 4), we obtain

$$\dot{x}_1 = f_3(x_3) - f_5[f_1(x_1)]$$

$$\dot{x}_2 = -f_3(x_3) - i_4 + u$$

$$\dot{x}_3 = f_1(x_1) + f_2(x_2)$$

In these equations, however, there is no way of expressing i_4 in terms of the chosen state variables; hence, the state equations cannot be written in the normal form. Some general results concerning the sufficient conditions for existence of the normal form of the state equations will be discussed in Section 5.5.

The procedure outlined is not limited to *RLC* networks alone. The state equations of networks containing gyrators, operational amplifiers, or dependent sources can also be written using steps 1 through 4. In the following example we write the state equations of a second-order gyrator network to illustrate the procedure.

Example 5.2-4 Gyrator Network. Consider the network shown in Fig. 5.2-4. Let us write its state equations. Following step 1, the state variables are

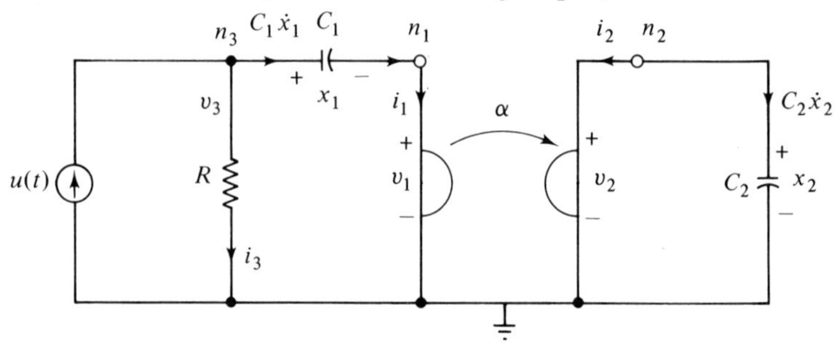

Figure 5.2-4 Gyrator network whose state equations are derived in Example 5.2-4.

chosen as the capacitor voltages x_1 and x_2. The currents through the capacitors are therefore $C_1\dot{x}_1$ and $C_2\dot{x}_2$, which are denoted in the figure for convenience. Writing node and loop equations, we get

$$\textit{KCL} \qquad u(t) = i_3 + C_1\dot{x}_1$$

$$i_1 = C_1\dot{x}_1$$

$$i_2 = -C_2\dot{x}_2$$

$$\textit{KVL} \qquad v_3 = x_1 + v_1$$

$$v_2 = x_2$$

The remaining voltage–current relations are

$$v_3 = Ri_3, \qquad v_1 = -\alpha i_2, \qquad v_2 = \alpha i_1$$

Eliminating i_1, i_2, i_3, v_1, v_2, and v_3 from these equations yields

$$\begin{bmatrix} \dot{x}_1 \\ \dot{x}_2 \end{bmatrix} = \begin{bmatrix} 0 & \dfrac{1}{\alpha C_1} \\ \dfrac{-1}{\alpha C_2} & \dfrac{R}{\alpha^2 C_2} \end{bmatrix} \begin{bmatrix} x_1 \\ x_2 \end{bmatrix} + \begin{bmatrix} 0 \\ \dfrac{R}{\alpha C_2} \end{bmatrix} u$$

which is the desired state equation.

Going through the detailed derivation of these examples, the reader undoubtedly agrees that there must be a more systematic method of deriving the state equations. Particularly if the number of the capacitors and inductors is large, elimination of the nonstate variables in step 4 becomes a formidable task. Such a method is considered in the next section. It should be emphasized at the outset that a formal derivation of the state equations should only be used if the number of the energy-storing elements in the network is large or if the informal technique discussed does not seem efficient.

5.3 STATE-VARIABLE FORMULATION OF PROPER NETWORKS

In this section we implement the topological results of Chapter 3 to obtain a systematic formulation of *proper networks*. Let us first define a proper network.

Definition 5.3-1 (Proper Network). A network comprised of lumped two-terminal elements is said to be *proper* if it contains no capacitor-only loops and no inductor-only cutsets.

Recall that according to Definitions 5.2-1 and 5.2-2, by a capacitor-only loop we mean a loop that contains only (*two* or more) capacitors and possibly some voltage sources. By an inductor-only cutset we mean a cutset that contains only (*two* or more) inductors and possibly some current sources.

To simplify the notations, we alter our previous convention and assume that the branches of the network under study are simple; that is, we assume that a branch is either an independent voltage or current source or a non-source element (resistor, capacitor, or inductor), but not a combination of these. With this new convention, however, we must assume that the initial conditions are present, since they cannot be included in the source elements.

Next, we proceed to obtain the general formulation of the state equations of a proper network. To do so, we must first define a proper tree for the network under consideration.

Definition 5.3-2 (Proper Tree). A *proper tree* of a lumped connected network comprised of two-terminal resistors, capacitors, inductors, and independent sources is a tree that contains *all* the voltage sources, *all* the capacitors, and possibly some resistors, but no inductors and no current sources.

If the network under study is a proper network, it has at least one proper tree (why?). We are now in a position to introduce a step-by-step method of state-variable formulation of proper networks; first, we consider

Linear Time-Varying Proper Networks

Step 1. Choose a proper tree for the network and number the branches of the network in the following order:

1. All the tree branch voltage sources.
2. All the tree branch capacitors.
3. All the tree branch resistors.
4. All the chord resistors.
5. All the chord inductors.
6. All the chord current sources.

Then partition the branch voltage vector $\mathbf{v}_b$ and the branch current vector $\mathbf{i}_b$ as follows:

$$\mathbf{v}_b = [\mathbf{v}_v \quad \mathbf{v}_c \quad \mathbf{v}_g \quad \mathbf{v}_r \quad \mathbf{v}_l \quad \mathbf{v}_i]^T \tag{5.3-1}$$

and

$$\mathbf{i}_b = [\mathbf{i}_v \quad \mathbf{i}_c \quad \mathbf{i}_g \quad \mathbf{i}_r \quad \mathbf{i}_l \quad \mathbf{i}_i]^T \tag{5.3-2}$$

where $\mathbf{v}_v$ and $\mathbf{i}_v$ are the voltage and current vectors of the voltage sources
$\mathbf{v}_c$ and $\mathbf{i}_c$ are the voltage and current vectors of the capacitor branches
$\mathbf{v}_g$ and $\mathbf{i}_g$ are the voltage and current vectors of the tree branch resistors
$\mathbf{v}_r$ and $\mathbf{i}_r$ are the voltage and current vectors of the chord resistors
$\mathbf{v}_l$ and $\mathbf{i}_l$ are the voltage and current vectors of the inductor chords
$\mathbf{v}_i$ and $\mathbf{i}_i$ are the voltage and current vectors of the current sources

For the sake of simplicity of notation, we assume that the number of the elements in each subvector is represented by its corresponding subscript. For example, $\mathbf{i}_v$ is a column vector with v elements and $\mathbf{i}_c$ is a column vector with c elements, and so on. In other words, we assume that the number of the voltage sources in the network is v, the number of the capacitors in the network is c, and so on. Now recall that the linearly independent KVL and KCL equations are given by

$$\mathbf{B}_f \mathbf{v}_b = \mathbf{0} \tag{5.3-3}$$

and

$$\mathbf{Q}_f \mathbf{i}_b = \mathbf{0} \tag{5.3-4}$$

where $\mathbf{B}_f$ and $\mathbf{Q}_f$ are the fundamental loop and cutset matrices corresponding to the proper tree chosen previously. In Chapter 3 we saw that, with the particular numbering of the branches adopted, $\mathbf{B}_f$ and $\mathbf{Q}_f$ can be partitioned as

$$\mathbf{B}_f = [-\mathbf{F}^T \mid \mathbf{1}] \tag{5.3-5}$$

and

$$\mathbf{Q}_f = [\mathbf{1} \mid \mathbf{F}] \tag{5.3-6}$$

where $\mathbf{1}$ denotes the identity matrix and $\mathbf{F}$ is the fundamental submatrix. If we now partition $\mathbf{1}$ and $\mathbf{F}$ with respect to the partitionings of $\mathbf{v}_b$ and $\mathbf{i}_b$, the KCL and KVL equations can be written as

$$\begin{bmatrix} \mathbf{1}_{vv} & \mathbf{0} & \mathbf{0} & \mathbf{F}_{vr} & \mathbf{F}_{vl} & \mathbf{F}_{vi} \\ \mathbf{0} & \mathbf{1}_{cc} & \mathbf{0} & \mathbf{F}_{cr} & \mathbf{F}_{cl} & \mathbf{F}_{ci} \\ \mathbf{0} & \mathbf{0} & \mathbf{1}_{gg} & \mathbf{F}_{gr} & \mathbf{F}_{gl} & \mathbf{F}_{gi} \end{bmatrix} \begin{bmatrix} \mathbf{i}_v \\ \mathbf{i}_c \\ \mathbf{i}_g \\ \mathbf{i}_r \\ \mathbf{i}_l \\ \mathbf{i}_i \end{bmatrix} = \mathbf{0} \tag{5.3-7}$$

and

$$\begin{bmatrix} -\mathbf{F}_{vr}^T & -\mathbf{F}_{cr}^T & -\mathbf{F}_{gr}^T & \mathbf{1}_{rr} & \mathbf{0} & \mathbf{0} \\ -\mathbf{F}_{vl}^T & -\mathbf{F}_{cl}^T & -\mathbf{F}_{gl}^T & \mathbf{0} & \mathbf{1}_{ll} & \mathbf{0} \\ -\mathbf{F}_{vi}^T & -\mathbf{F}_{ci}^T & -\mathbf{F}_{gi}^T & \mathbf{0} & \mathbf{0} & \mathbf{1}_{ii} \end{bmatrix} \begin{bmatrix} \mathbf{v}_v \\ \mathbf{v}_c \\ \mathbf{v}_g \\ \mathbf{v}_r \\ \mathbf{v}_l \\ \mathbf{v}_i \end{bmatrix} = \mathbf{0} \tag{5.3-8}$$

The dimensions of submatrices of $\mathbf{F}$ are determined by the dimensions of their associated subvectors in $\mathbf{i}_b$ and $\mathbf{v}_b$. For example, $\mathbf{F}_{vr}$ has v rows and r columns, $\mathbf{F}_{vl}$ has v rows and l columns, and so on.

Step 2. Choose the state variables to be the capacitor charges and the inductor fluxes; that is, let

$$\mathbf{x}(t) = \begin{bmatrix} \mathbf{q}_c \\ \boldsymbol{\phi}_l \end{bmatrix} \tag{5.3-9}$$

where $\mathbf{q}_c$ is a column vector whose elements are the tree branch capacitor charges and $\boldsymbol{\phi}_l$ is a column vector whose elements are the chord inductor fluxes. The branch VCR are therefore

$$\mathbf{i}_c = \dot{\mathbf{q}}_c \tag{5.3-10}$$

$$\mathbf{v}_l = \dot{\boldsymbol{\phi}}_l \tag{5.3-11}$$

$$\mathbf{q}_c = \mathbf{C}_c(t)\mathbf{v}_c \tag{5.3-12}$$

$$\boldsymbol{\phi}_l = \mathbf{L}_l(t)\mathbf{i}_l \tag{5.3-13}$$

$$\mathbf{v}_r = \mathbf{R}_r(t)\mathbf{i}_r \tag{5.3-14}$$

$$\mathbf{i}_g = \mathbf{G}_g(t)\mathbf{v}_g \tag{5.3-15}$$

where $\mathbf{C}_c(t)$ is a diagonal matrix whose diagonal elements are the capacitances of the network

$\mathbf{L}_l(t)$ is a square matrix whose diagonal elements are the self-inductances and the off-diagonal elements are the mutual inductances

$\mathbf{R}_r(t)$ and $\mathbf{G}_g(t)$ are diagonal matrices whose diagonal elements are chord resistances and tree branch conductances, respectively

Throughout the rest of this section we assume that these matrices are nonsingular for each t. This condition is satisfied for almost all practical networks.

Step 3. We must now use the equations obtained in steps 1 and 2 to express all the nonstate variables in terms of the state variables $\mathbf{q}_c$ and $\boldsymbol{\phi}_l$ and independent voltage and current sources $\mathbf{v}_v$ and $\mathbf{i}_i$. We start by writing (5.3-7) and (5.3-8) in a more explicit form:

$$\mathbf{i}_v + \mathbf{F}_{vr}\mathbf{i}_r + \mathbf{F}_{vl}\mathbf{i}_l + \mathbf{F}_{vi}\mathbf{i}_i = \mathbf{0} \tag{5.3-16}$$

$$\mathbf{i}_c + \mathbf{F}_{cr}\mathbf{i}_r + \mathbf{F}_{cl}\mathbf{i}_l + \mathbf{F}_{ci}\mathbf{i}_i = \mathbf{0} \tag{5.3-17}$$

$$\mathbf{i}_g + \mathbf{F}_{gr}\mathbf{i}_r + \mathbf{F}_{gl}\mathbf{i}_l + \mathbf{F}_{gi}\mathbf{i}_i = \mathbf{0} \tag{5.3-18}$$

and

$$\mathbf{v}_r - \mathbf{F}_{vr}^T\mathbf{v}_v - \mathbf{F}_{cr}^T\mathbf{v}_c - \mathbf{F}_{gr}^T\mathbf{v}_g = \mathbf{0} \tag{5.3-19}$$

$$\mathbf{v}_l - \mathbf{F}_{vl}^T\mathbf{v}_v - \mathbf{F}_{cl}^T\mathbf{v}_c - \mathbf{F}_{gl}^T\mathbf{v}_g = \mathbf{0} \tag{5.3-20}$$

$$\mathbf{v}_i - \mathbf{F}_{vi}^T\mathbf{v}_v - \mathbf{F}_{ci}^T\mathbf{v}_c - \mathbf{F}_{gi}^T\mathbf{v}_g = \mathbf{0} \tag{5.3-21}$$

Replacing $\mathbf{i}_c$ and $\mathbf{v}_L$ from (5.3-10) and (5.3-11) in (5.3-17) and (5.3-20), respectively, we obtain

$$\dot{\mathbf{q}}_c = -\mathbf{F}_{cr}\mathbf{i}_r - \mathbf{F}_{cl}\mathbf{i}_l - \mathbf{F}_{ci}\mathbf{i}_i \tag{5.3-22}$$

$$\dot{\boldsymbol{\phi}}_l = \mathbf{F}_{vl}^T\mathbf{v}_v + \mathbf{F}_{cl}^T\mathbf{v}_c + \mathbf{F}_{gl}^T\mathbf{v}_g \tag{5.3-23}$$

Since by assumption $\mathbf{L}_l(t)$ and $\mathbf{C}_c(t)$ matrices are nonsingular, (5.3-12) and (5.3-13) can be written as

$$\mathbf{v}_c = \mathbf{C}_c^{-1}\mathbf{q}_c \tag{5.3-24}$$

$$\mathbf{i}_l = \mathbf{L}_l^{-1}\boldsymbol{\phi}_l \tag{5.3-25}$$

(Since it is understood that the element matrices are time varying, we delete the argument t to simplify the notations.) Substituting $\mathbf{i}_l$ and $\mathbf{v}_c$ from (5.3-24) and (5.3-25) in (5.3-22) and (5.3-23) yields

$$\dot{\mathbf{q}}_c = -\mathbf{F}_{cl}\mathbf{L}_l^{-1}\boldsymbol{\phi}_l - \mathbf{F}_{cr}\mathbf{i}_r - \mathbf{F}_{ci}\mathbf{i}_i \tag{5.3-26}$$

$$\dot{\boldsymbol{\phi}}_l = \mathbf{F}_{cl}^T\mathbf{C}_c^{-1}\mathbf{q}_c + \mathbf{F}_{gl}^T\mathbf{v}_g + \mathbf{F}_{vl}^T\mathbf{v}_v \tag{5.3-27}$$

or, equivalently,

$$\begin{bmatrix} \dot{\mathbf{q}}_c \\ \dot{\boldsymbol{\phi}}_l \end{bmatrix} = \begin{bmatrix} \mathbf{0} & -\mathbf{F}_{cl}\mathbf{L}_l^{-1} \\ \mathbf{F}_{cl}^T\mathbf{C}_c^{-1} & \mathbf{0} \end{bmatrix} \begin{bmatrix} \mathbf{q}_c \\ \boldsymbol{\phi}_l \end{bmatrix} + \begin{bmatrix} \mathbf{0} & -\mathbf{F}_{ci} \\ \mathbf{F}_{vl}^T & \mathbf{0} \end{bmatrix} \begin{bmatrix} \mathbf{v}_v \\ \mathbf{i}_i \end{bmatrix} + \begin{bmatrix} \mathbf{0} & -\mathbf{F}_{cr} \\ \mathbf{F}_{gl}^T & \mathbf{0} \end{bmatrix} \begin{bmatrix} \mathbf{v}_g \\ \mathbf{i}_r \end{bmatrix} \tag{5.3-28}$$

In these equations the only nonstate variables are $\mathbf{i}_r$ and $\mathbf{v}_g$; $\mathbf{i}_i$ and $\mathbf{v}_v$ are the current and voltage source vectors. To eliminate the nonstate variables, we can replace $\mathbf{i}_g$ and $\mathbf{v}_r$ in (5.3-18) and (5.3-19) from (5.3-14) and (5.3-15) to get

$$\mathbf{G}_g\mathbf{v}_g + \mathbf{F}_{gr}\mathbf{i}_r = -\mathbf{F}_{gl}\mathbf{L}_l^{-1}\boldsymbol{\phi}_l - \mathbf{F}_{gi}\mathbf{i}_i \tag{5.3-29}$$

$$-\mathbf{F}_{gr}^T\mathbf{v}_g + \mathbf{R}_r\mathbf{i}_r = \mathbf{F}_{cr}^T\mathbf{C}_c^{-1}\mathbf{q}_c + \mathbf{F}_{vr}^T\mathbf{v}_v \tag{5.3-30}$$

Solving (5.3-29) and (5.3-30) for $\mathbf{v}_g$ and $\mathbf{i}_r$, we get

$$\mathbf{v}_g = -\mathbf{G}^{-1}[\mathbf{F}_{gr}\mathbf{R}_r^{-1}\mathbf{F}_{cr}^T\mathbf{C}_c^{-1}\mathbf{q}_c + \mathbf{F}_{gr}\mathbf{R}_r^{-1}\mathbf{F}_{vr}^T\mathbf{v}_v + \mathbf{F}_{gl}\mathbf{L}_l^{-1}\boldsymbol{\phi}_l + \mathbf{F}_{gi}\mathbf{i}_i] \tag{5.3-31}$$

$$\mathbf{i}_r = \mathbf{R}^{-1}[\mathbf{F}_{cr}^T\mathbf{C}_c^{-1}\mathbf{q}_c + \mathbf{F}_{vr}^T\mathbf{v}_v - \mathbf{F}_{gr}^T\mathbf{G}_g^{-1}\mathbf{F}_{gl}\mathbf{L}_l^{-1}\boldsymbol{\phi}_l - \mathbf{F}_{gr}^T\mathbf{G}_g^{-1}\mathbf{F}_{gi}\mathbf{i}_i] \tag{5.3-32}$$

where we have used the following notations:

$$\mathbf{G} \triangleq \mathbf{G}_g + \mathbf{F}_{gr}\mathbf{R}_r^{-1}\mathbf{F}_{gr}^T \tag{5.3-33}$$

$$\mathbf{R} \triangleq \mathbf{R}_r + \mathbf{F}_{gr}^T\mathbf{G}_g^{-1}\mathbf{F}_{gr} \tag{5.3-34}$$

Replacing $\mathbf{v}_g$ and $\mathbf{i}_r$ from (5.3-31) and (5.3-32) in (5.3-28), after some straightforward manipulations, we obtain

$$\begin{bmatrix} \dot{\mathbf{q}}_c \\ \dot{\boldsymbol{\phi}}_l \end{bmatrix} = \begin{bmatrix} \mathbf{H}_{cc}\mathbf{C}_c^{-1} & \mathbf{H}_{cl}\mathbf{L}_l^{-1} \\ \mathbf{H}_{lc}\mathbf{C}_c^{-1} & \mathbf{H}_{ll}\mathbf{L}_l^{-1} \end{bmatrix} \begin{bmatrix} \mathbf{q}_c \\ \boldsymbol{\phi}_l \end{bmatrix} + \begin{bmatrix} \mathbf{H}_{cv} & \mathbf{H}_{ci} \\ \mathbf{H}_{lv} & \mathbf{H}_{li} \end{bmatrix} \begin{bmatrix} \mathbf{v}_v \\ \mathbf{i}_i \end{bmatrix} \tag{5.3-35}$$

where the corresponding submatrices are defined by

$$\mathbf{H}_{cc} \triangleq -\mathbf{F}_{cr}\mathbf{R}^{-1}\mathbf{F}_{cr}^T \tag{5.3-36}$$

$$\mathbf{H}_{cl} \triangleq -\mathbf{F}_{cl} + \mathbf{F}_{cr}\mathbf{R}^{-1}\mathbf{F}_{gr}^T\mathbf{G}_g^{-1}\mathbf{F}_{gl} \tag{5.3-37}$$

$$\mathbf{H}_{ll} \triangleq -\mathbf{F}_{gl}^T\mathbf{G}^{-1}\mathbf{F}_{gl} \tag{5.3-38}$$

$$\mathbf{H}_{cv} \triangleq -\mathbf{F}_{cr}\mathbf{R}^{-1}\mathbf{F}_{vr}^T \tag{5.3-39}$$

$$\mathbf{H}_{lv} \triangleq \mathbf{F}_{vl}^T - \mathbf{F}_{gl}^T\mathbf{G}^{-1}\mathbf{F}_{gr}\mathbf{R}_r^{-1}\mathbf{F}_{vr}^T \tag{5.3-40}$$

$$\mathbf{H}_{ci} \triangleq -\mathbf{F}_{ci} + \mathbf{F}_{cr}\mathbf{R}^{-1}\mathbf{F}_{gr}^T\mathbf{G}_g^{-1}\mathbf{F}_{gi} \tag{5.3-41}$$

$$\mathbf{H}_{li} \triangleq -\mathbf{F}_{gl}^T\mathbf{G}^{-1}\mathbf{F}_{gi} \tag{5.3-42}$$

$$\mathbf{H}_{lc} \triangleq \mathbf{F}_{cl}^T - \mathbf{F}_{gl}^T\mathbf{G}^{-1}\mathbf{F}_{gr}\mathbf{R}_r^{-1}\mathbf{F}_{cr}^T \tag{5.3-43}$$

Equation (5.3-35) is the desired state equation in the normal form. Note that since the state variables are the capacitor charges and inductor fluxes (5.3-35) represents the state equations of a linear time-varying network. Writing (5.3-35) in a more compact form, we get

$$\dot{\mathbf{x}}(t) = \mathbf{A}(t)\mathbf{x}(t) + \mathbf{B}(t)\mathbf{u}(t) \tag{5.3-44}$$

where $\mathbf{x}(t)$ is the state vector defined in (5.3-9); the number of the elements in $\mathbf{x}(t)$ is equal to the total number of capacitors and inductors of the network. Elements of $\mathbf{u}(t)$ are the independent voltage and current sources; $\mathbf{A}(t)$ and $\mathbf{B}(t)$ are *time-varying* matrices with appropriate dimensions and are defined by

$$\mathbf{A}(t) \triangleq \left[\begin{array}{c|c} \mathbf{H}_{cc}\mathbf{C}_c^{-1} & \mathbf{H}_{cl}\mathbf{L}_l^{-1} \\ \hline \mathbf{H}_{lc}\mathbf{C}_c^{-1} & \mathbf{H}_{ll}\mathbf{L}_l^{-1} \end{array}\right] \tag{5.3-45}$$

and

$$\mathbf{B}(t) \triangleq \left[\begin{array}{c|c} \mathbf{H}_{cv} & \mathbf{H}_{ci} \\ \hline \mathbf{H}_{lv} & \mathbf{H}_{li} \end{array}\right] \tag{5.3-46}$$

Having solved (5.3-44) for $\mathbf{x}(t)$, all the branch voltages and currents can then be found by purely algebraic manipulations of (5.3-9) through (5.3-21).

Exercise 5.3-1. Show that if the element matrices $\mathbf{C}_c$, $\mathbf{L}_l$, $\mathbf{R}_r$, and $\breve{\mathbf{G}}_g$ of the network are symmetric then

$$\mathbf{H}_{lc} = -\mathbf{H}_{cl}^T \tag{5.3-47}$$

Linear Time-Invariant Proper Networks. If the network under study is comprised of linear time-invariant two-terminal elements and independent sources, the state variables can be chosen to be either the capacitor charges and inductor fluxes or capacitor voltages and inductor currents. If the first set of variables is chosen, the state equations will be identical to those of linear time-varying networks derived previously, except that the matrices $\mathbf{A}(t)$ and $\mathbf{B}(t)$ will be time-invariant. In many practical cases, however, it is more appropriate to choose the latter set of variables as the state, that is, the voltages across the capacitors and currents through the inductors. In this case the derivation of the state equations is also similar to the case of linear time-varying networks. From (5.3-12) and (5.3-13) we can write

$$\left[\begin{array}{c} \mathbf{q}_c \\ \hline \boldsymbol{\phi}_l \end{array}\right] = \left[\begin{array}{c|c} \mathbf{C}_c & \mathbf{0} \\ \hline \mathbf{0} & \mathbf{L}_l \end{array}\right] \left[\begin{array}{c} \mathbf{v}_c \\ \hline \mathbf{i}_l \end{array}\right]$$

Replacing this in (5.3-35), premultiplying both sides of the resulting equation by

$$\left[\begin{array}{c|c} \mathbf{C}_c & \mathbf{0} \\ \hline \mathbf{0} & \mathbf{L}_l \end{array}\right]^{-1}$$

and letting

$$\mathbf{x}(t) = \left[\begin{array}{c} \mathbf{v}_c \\ \mathbf{i}_l \end{array}\right]$$

we obtain

$$\dot{\mathbf{x}}(t) = \mathbf{A}\mathbf{x}(t) + \mathbf{B}\mathbf{u}(t) \tag{5.3-48}$$

where $\mathbf{A}$ is a *time-invariant* square matrix given by

$$\mathbf{A} = \left[\begin{array}{c|c} \mathbf{C}_c^{-1}\mathbf{H}_{cc} & \mathbf{C}_c^{-1}\mathbf{H}_{cl} \\ \hline \mathbf{L}_l^{-1}\mathbf{H}_{lc} & \mathbf{L}_l^{-1}\mathbf{H}_{ll} \end{array}\right] \tag{5.3-49}$$

and $\mathbf{B}$ is a time invariant matrix given by

$$\mathbf{B} = \left[\begin{array}{c|c} \mathbf{C}_c^{-1}\mathbf{H}_{cv} & \mathbf{C}_c^{-1}\mathbf{H}_{ci} \\ \hline \mathbf{L}_l^{-1}\mathbf{H}_{lv} & \mathbf{L}_l^{-1}\mathbf{H}_{li} \end{array}\right] \tag{5.3-50}$$

Equations (5.3-35) and (5.3-48) can now be used for systematic formulation of the state equations of linear time-varying and linear time-invariant networks, respectively. The usefulness and significance of these equations becomes manifest when one has to analyze large-scale networks. The formulation given here is also suitable for computer application. Next we work out a simple example to illustrate the outlined procedure, the results of this example can then be checked by the inspection method discussed in the previous section.

Example 5.3-1. Consider the network shown in Fig. 5.3-1. Let all the elements be linear time varying and nonzero for all $t \geq 0$. The problem is to use the outlined procedure to derive the state equations of this network.

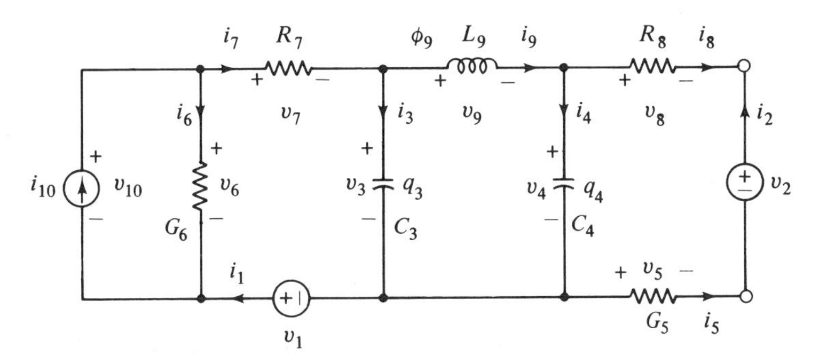

Figure 5.3-1 Linear time-varying proper network considered in Example 5.3-1.

Solution

Since there are no capacitor-only loops and no inductor-only cutsets and since the network is time varying, the state variables can be taken to be capacitor charges and the flux through the inductor; that is,

$$\mathbf{x}(t) = [q_3(t) \quad q_4(t) \quad \phi_9(t)]^T$$

The branches of this network are labeled according to the procedure given in step 1; a proper tree can then be drawn as in Fig. 5.3-2 (the solid line).

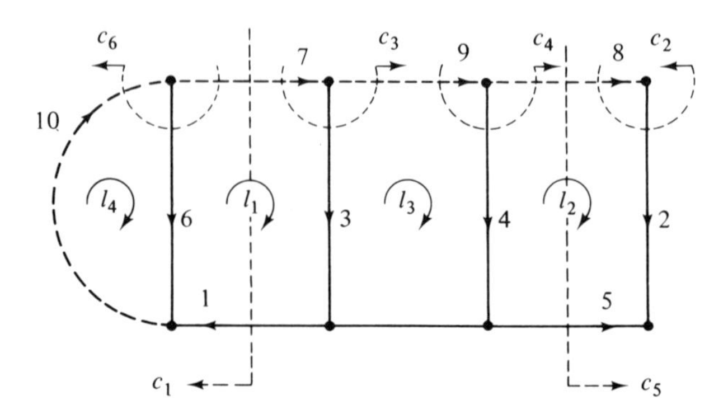

Figure 5.3-2 Proper tree for the network of Fig. 5.3-1.

The fundamental cutsets are denoted by $c_1, c_2, \ldots, c_6$ and fundamental loops by l_1 through l_4. The branch voltage vector and branch current vector corresponding to the network under study, partitioned as in (5.3-1) and (5.3-2), are

$$\mathbf{v}_b = [v_1 \quad v_2 \mid v_3 \quad v_4 \mid v_5 \quad v_6 \mid v_7 \quad v_8 \mid v_9 \mid v_{10}]^T$$

and
$$\mathbf{i}_b = [i_1 \quad i_2 \mid i_3 \quad i_4 \mid i_5 \quad i_6 \mid i_7 \quad i_8 \mid i_9 \mid i_{10}]^T$$

Hence, $v = 2$, $c = 2$, $g = 2$, $r = 2$, $l = 1$, and $i = 1$; then the fundamental cutset matrix $\mathbf{Q}_f$ is clearly seen to be

$$\mathbf{Q}_f = \begin{array}{c} \\ c_1 \\ c_2 \\ c_3 \\ c_4 \\ c_5 \\ c_6 \end{array} \begin{array}{c} \begin{array}{cccccc|cccc} b_1 & b_2 & b_3 & b_4 & b_5 & b_6 & b_7 & b_8 & b_9 & b_{10} \end{array} \\ \left[\begin{array}{cccccc|cccc} 1 & 0 & 0 & 0 & 0 & 0 & -1 & 0 & 0 & 0 \\ 0 & 1 & 0 & 0 & 0 & 0 & 0 & 1 & 0 & 0 \\ 0 & 0 & 1 & 0 & 0 & 0 & -1 & 0 & 1 & 0 \\ 0 & 0 & 0 & 1 & 0 & 0 & 0 & 1 & -1 & 0 \\ 0 & 0 & 0 & 0 & 1 & 0 & 0 & 1 & 0 & 0 \\ 0 & 0 & 0 & 0 & 0 & 1 & 1 & 0 & 0 & -1 \end{array}\right] \end{array}$$

identity matrix F matrix

The partitioned **F** matrix will then be

$$\mathbf{F} = \left[\begin{array}{c|c|c} \mathbf{F}_{vr} & \mathbf{F}_{vl} & \mathbf{F}_{vi} \\ \hline \mathbf{F}_{cr} & \mathbf{F}_{cl} & \mathbf{F}_{ci} \\ \hline \mathbf{F}_{gr} & \mathbf{F}_{gl} & \mathbf{F}_{gi} \end{array}\right] = \left[\begin{array}{cc|c|c} -1 & 1 & 0 & 0 \\ 0 & 1 & 0 & 0 \\ \hline -1 & 0 & 1 & 0 \\ 0 & 1 & -1 & 0 \\ \hline 0 & 1 & 0 & 0 \\ 1 & 0 & 0 & -1 \end{array}\right]$$

From Fig. 5.3-1 we immediately get

$$\mathbf{R}_r = \begin{bmatrix} R_7 & 0 \\ 0 & R_8 \end{bmatrix}, \qquad \mathbf{G}_g = \begin{bmatrix} G_5 & 0 \\ 0 & G_6 \end{bmatrix}$$

and

$$\mathbf{C}_c = \begin{bmatrix} C_3 & 0 \\ 0 & C_4 \end{bmatrix}, \qquad \mathbf{L}_l = L_9$$

where the time-varying nature of these matrices is understood. Then,

$$R_r^{-1} = \begin{bmatrix} R_7^{-1} & 0 \\ 0 & R_8^{-1} \end{bmatrix} \quad \text{and} \quad \mathbf{G}_g^{-1} = \begin{bmatrix} G_5^{-1} & 0 \\ 0 & G_6^{-1} \end{bmatrix}$$

Using (5.3-33) and (5.3-34), we obtain

$$\mathbf{G} = \begin{bmatrix} G_5 & 0 \\ 0 & G_6 \end{bmatrix} + \begin{bmatrix} 0 & 1 \\ 1 & 0 \end{bmatrix} \begin{bmatrix} R_7^{-1} & 0 \\ 0 & R_8^{-1} \end{bmatrix} \begin{bmatrix} 0 & 1 \\ 1 & 0 \end{bmatrix} = \begin{bmatrix} G_5 + R_8^{-1} & 0 \\ 0 & G_6 + R_7^{-1} \end{bmatrix}$$

Then

$$\mathbf{G}^{-1} = \begin{bmatrix} \dfrac{R_8}{1 + R_8 G_5} & 0 \\ 0 & \dfrac{R_7}{1 + G_6 R_7} \end{bmatrix}$$

and, similarly,

$$\mathbf{R}^{-1} = \begin{bmatrix} \dfrac{G_6}{1 + G_6 R_7} & 0 \\ 0 & \dfrac{G_5}{1 + G_5 R_8} \end{bmatrix}$$

Using these relations together with partitioned **F** in equations (5.3-36) to (5.3-43), we get

$$\mathbf{H}_{cc} = -\begin{bmatrix} -1 & 0 \\ 0 & 1 \end{bmatrix} \begin{bmatrix} \dfrac{G_6}{1 + G_6 R_7} & 0 \\ 0 & \dfrac{G_5}{1 + G_5 R_8} \end{bmatrix} \begin{bmatrix} -1 & 0 \\ 0 & 1 \end{bmatrix}$$

or

$$\mathbf{H}_{cc} = \begin{bmatrix} \dfrac{-G_6}{1 + G_6 R_7} & 0 \\ 0 & \dfrac{-G_5}{1 + G_5 R_8} \end{bmatrix}, \qquad \mathbf{H}_{cl} = \begin{bmatrix} -1 \\ 1 \end{bmatrix}, \qquad \mathbf{H}_{lc} = [1 \quad -1]$$

$$\mathbf{H}_{ll} = 0, \qquad \mathbf{H}_{cv} = \begin{bmatrix} \dfrac{-G_6}{1 + G_6 R_7} & 0 \\ 0 & \dfrac{-G_5}{1 + G_5 R_8} \end{bmatrix}$$

$$\mathbf{H}_{lv} = [0 \quad 0], \qquad \mathbf{H}_{ci} = \begin{bmatrix} \dfrac{1}{1 + G_6 R_7} \\ 0 \end{bmatrix}$$

$$\mathbf{H}_{li} = 0$$

Observe that, according to Exercise 5.3-1, since the resistive network is symmetric $\mathbf{H}_{cl} = -\mathbf{H}_{lc}^T$. We can then write

$$\mathbf{H}_{cc}\mathbf{C}_c^{-1} = \begin{bmatrix} \dfrac{-G_6}{C_3(1+G_6R_7)} & 0 \\ 0 & \dfrac{-G_5}{C_4(1+G_6R_7)} \end{bmatrix}$$

$$\mathbf{H}_{lc}\mathbf{C}_c^{-1} = \begin{bmatrix} \dfrac{1}{C_3} & -\dfrac{1}{C_4} \end{bmatrix}, \qquad \mathbf{H}_{cl}\mathbf{L}_l^{-1} = \begin{bmatrix} -\dfrac{1}{L_9} \\ \dfrac{1}{L_9} \end{bmatrix}$$

and $\qquad \mathbf{H}_{ll}\mathbf{L}_l^{-1} = 0$

Finally, using (5.3-35), the state equations of the network under study can be written as

$$\begin{bmatrix} \dot{q}_3 \\ \dot{q}_4 \\ \dot{\phi}_9 \end{bmatrix} = \begin{bmatrix} \dfrac{-G_6}{C_3(1+G_6R_7)} & 0 & -\dfrac{1}{L_9} \\ 0 & \dfrac{-G_5}{C_4(1+G_5R_8)} & \dfrac{1}{L_9} \\ \dfrac{1}{C_3} & -\dfrac{1}{C_4} & 0 \end{bmatrix} \begin{bmatrix} q_3 \\ q_4 \\ \phi_9 \end{bmatrix}$$
$$+ \begin{bmatrix} \dfrac{-G_6}{1+G_6R_7} & 0 & \dfrac{1}{1+G_6R_7} \\ 0 & \dfrac{-G_5}{1+G_5R_8} & 0 \\ 0 & 0 & 0 \end{bmatrix} \begin{bmatrix} v_1 \\ v_2 \\ i_{10} \end{bmatrix}$$

In this equation all the resistors, capacitors, and inductors are time varying. The equation is, clearly, in the normal form.

Let us now assume that all the elements of the network under study are linear and *time invariant;* then we may choose an alternative set of variables; capacitor voltages and inductor currents. Choosing this latter set of variables as the state variables, we can write

$$\begin{bmatrix} q_3 \\ q_4 \\ \phi_9 \end{bmatrix} = \begin{bmatrix} C_3 & 0 & 0 \\ 0 & C_4 & 0 \\ 0 & 0 & L_9 \end{bmatrix} \begin{bmatrix} v_3 \\ v_4 \\ i_9 \end{bmatrix}$$

Replacing q_3, q_4, and ϕ_9 in the state equations just obtained from the previous equations and simplifying the result, we obtain

$$\begin{bmatrix} \dot{v}_3 \\ \dot{v}_4 \\ \dot{i}_9 \end{bmatrix} = \begin{bmatrix} \dfrac{-G_6}{C_3(1 + G_6R_7)} & 0 & -\dfrac{1}{C_3} \\ 0 & \dfrac{-G_5}{C_4(1 + G_5R_8)} & \dfrac{1}{C_4} \\ \dfrac{1}{L_9} & -\dfrac{1}{L_9} & 0 \end{bmatrix} \begin{bmatrix} v_3 \\ v_4 \\ i_9 \end{bmatrix} + \begin{bmatrix} \dfrac{-G_6}{C_3(1 + G_6R_7)} & 0 & \dfrac{1}{C_3(1 + G_6R_7)} \\ 0 & \dfrac{-G_5}{C_4(1 + G_5R_8)} & 0 \\ 0 & 0 & 0 \end{bmatrix} \begin{bmatrix} v_1 \\ v_2 \\ i_{10} \end{bmatrix}$$

First, by solving for v_3, v_4, and i_9, and then by a simple algebraic manipulation, we can obtain the voltage and the current of any branch. For instance, if we are interested in the voltage across R_7, the KVL and VCR for loop l_1 yield

$$-v_1 + v_7 + v_3 - G_6^{-1}i_6 = 0$$

But

$$i_6 = i_{10} - R_7^{-1}v_7$$

Then

$$v_7 = \frac{G_6R_7}{1 + G_6R_7}(-v_3 + v_1 + i_{10})$$

where v_1 and i_{10} are given voltage and current sources and v_3 is the state variable. It was mentioned earlier that one of the features of state-variable representation is that nonlinear and time-varying networks can also be represented by this method. We have already shown how linear time-varying networks can be handled; next we introduce a similar method for writing the normal form of the state equations of a special class of nonlinear networks. As we shall see, only a certain class of nonlinear networks can be represented in the normal form.

Nonlinear Time-Invariant Proper Networks. The desired normal form of the state equations of a nonlinear network is

$$\dot{\mathbf{x}} = \mathbf{f}(\mathbf{x}, \mathbf{u}) \tag{5.3-51}$$

where $\mathbf{x}$ is an n vector representing the state, $\mathbf{u}$ is an m vector representing the independent sources, and $\mathbf{f}(\cdot, \cdot)$ is a nonlinear vector function. In contrast to the linear case, it is rather difficult and sometimes impossible to represent the state equations of a general nonlinear network in the normal form. Only in special cases can one express the state equations of such networks in the form (5.3-51). In this section we make several simplifying assumptions, on both the topology of the network and element characteristics, and give a general method for obtaining the state equations of proper networks that satisfy these conditions.

Let us now consider a network comprised of nonlinear time-invariant resistors, capacitors, inductors, and independent voltage and current sources. Assume that the network under study satisfies the following conditions:

1. There are no capacitor-only loops and no inductor-only cutsets.
2. All the capacitors are *charge controlled* and all the inductors are *flux-controlled.*
3. There exists a proper tree such that all its tree branch resistors are *current controlled* and all its chord resistors are *voltage controlled.*

Under these conditions the voltage vector of the capacitor branches, $\mathbf{v}_c$, can be written as

$$\mathbf{v}_c = \mathbf{f}_c(\mathbf{q}_c) \tag{5.3-52}$$

where $\mathbf{q}_c$ is the charge vector of the capacitive branches and $\mathbf{f}_c(\cdot)$ is a vector-valued nonlinear function. Also, the current vector of the inductor chords $\mathbf{i}_l$ can be written as

$$\mathbf{i}_l = \mathbf{f}_l(\boldsymbol{\phi}_l) \tag{5.3-53}$$

where $\boldsymbol{\phi}_l$ is the flux vector of the inductive chords and $\mathbf{f}_l(\cdot)$ is a vector-valued function. Similarly, the voltage vector of the tree branch resistors, $\mathbf{v}_g$, and the current vector of the chord resistors, $\mathbf{i}_r$, can be written as

$$\mathbf{v}_g = \mathbf{f}_g(\mathbf{i}_g) \tag{5.3-54}$$

and

$$\mathbf{i}_r = \mathbf{f}_r(\mathbf{v}_r) \tag{5.3-55}$$

The steps involved in writing the state equations of this class of networks are similar to those taken in writing the state equations of linear time-varying proper networks. Equations (5.3-17) through (5.3-20) hold regardless of the nature of the network elements; rewriting (5.3-17) and (5.3-20) here, we get

$$\mathbf{i}_c = -\mathbf{F}_{cl}\mathbf{i}_l - \mathbf{F}_{cr}\mathbf{i}_r - \mathbf{F}_{ci}\mathbf{i}_i \tag{5.3-56}$$

$$\mathbf{v}_l = \mathbf{F}_{cl}^T\mathbf{v}_c + \mathbf{F}_{gl}^T\mathbf{v}_g + \mathbf{F}_{vl}^T\mathbf{v}_v \tag{5.3-57}$$

But $\mathbf{i}_c$ and $\mathbf{v}_l$ are related to $\mathbf{q}_c$ and $\boldsymbol{\phi}_l$ by

$$\mathbf{i}_c = \dot{\mathbf{q}}_c \tag{5.3-58}$$

and

$$\mathbf{v}_l = \dot{\boldsymbol{\phi}}_l \tag{5.3-59}$$

Then using (5.3-52) through (5.3-59), we get

$$\dot{\mathbf{q}}_c = -\mathbf{F}_{cl}\mathbf{f}_l(\boldsymbol{\phi}_l) - \mathbf{F}_{cr}\mathbf{f}_r(\mathbf{v}_r) - \mathbf{F}_{ci}\mathbf{i}_i \tag{5.3-60}$$

$$\dot{\boldsymbol{\phi}}_l = \mathbf{F}_{cl}^T\mathbf{f}_c(\mathbf{q}_c) + \mathbf{F}_{gl}^T\mathbf{f}_g(\mathbf{i}_g) + \mathbf{F}_{vl}^T\mathbf{v}_v \tag{5.3-61}$$

If we now replace $\mathbf{v}_r$ and $\mathbf{i}_g$ in terms of $\mathbf{q}_c$, $\boldsymbol{\phi}_l$, $\mathbf{i}_i$, and $\mathbf{v}_v$, the desired state equations will be obtained. To do so, let us use (5.3-54) and (5.3-55) to write (5.3-19) and (5.3-18), respectively, as

$$\mathbf{v}_r - \mathbf{F}_{gr}^T\mathbf{f}_g(\mathbf{i}_g) = \mathbf{F}_{cr}^T\mathbf{f}_c(\mathbf{q}_c) + \mathbf{F}_{vr}^T\mathbf{v}_v \tag{5.3-62}$$

and

$$\mathbf{i}_g + \mathbf{F}_{gr}^T\mathbf{f}_r(\mathbf{v}_r) = -\mathbf{F}_{gl}\mathbf{f}_l(\boldsymbol{\phi}_l) - \mathbf{F}_{gi}\mathbf{i}_i \tag{5.3-63}$$

The next step is to solve (5.3-62) and (5.3-63) for $\mathbf{v}_r$ and $\mathbf{i}_g$; however, in many nonlinear networks this is the main stumbling block. These equations represent a set of coupled nonlinear algebraic equations for which a unique solution may or may not exist. Let us now digress for a moment to examine (5.3-62) and (5.3-63) more closely. The unknowns of these two coupled vector equations are $\mathbf{i}_g$ and $\mathbf{v}_r$, and the equations are in the general form of

$$\mathbf{z} + \mathbf{P}\mathbf{f}(\mathbf{y}) = \boldsymbol{\alpha} \tag{5.3-64}$$

$$\mathbf{y} - \mathbf{P}^T\mathbf{h}(\mathbf{z}) = \boldsymbol{\beta} \tag{5.3-65}$$

where $\mathbf{y}$ and $\mathbf{z}$ are the unknown vectors, $\mathbf{P}$ is a rectangular constant matrix, and $\boldsymbol{\alpha}$ and $\boldsymbol{\beta}$ are given vectors. The main question will then be whether these equations possess a unique solution. If the answer to this question is negative, the state equations of the network cannot be represented in the normal form. Before stating a theorem concerning the uniqueness of the solutions of (5.3-64) and (5.3-65), we need to define the following useful terms.

Definition 5.3-3 (Jacobian Matrix). A square matrix function $\mathbf{F}(\mathbf{y})$ is said to be the *Jacobian* of a vector values function $\mathbf{f}(\mathbf{y})$ if

$$\mathbf{F}(y) = \frac{d}{d\mathbf{y}}f(\mathbf{y}) \tag{5.3-66}$$

More explicitly, let

$$\mathbf{y} = [y_1, y_2, \ldots, y_n] \tag{5.3-67}$$

$$f(\mathbf{y}) = [f_1(\mathbf{y}), f_2(\mathbf{y}), \ldots, f_n(\mathbf{y})]^T \tag{5.3-68}$$

Then the ijth element of the Jacobian matrix F_{ij} $(\mathbf{y})$ is

$$F_{ij}(\mathbf{y}) = \left.\frac{\partial f_i}{\partial y_j}\right|_{\mathbf{y}} \tag{5.3-69}$$

Definition 5.3-4 (Positive Definite Matrix). An $n \times n$ matrix $\mathbf{M}$ is said to be *positive definite* if

$$\mathbf{x}^T\mathbf{M}\mathbf{x} > 0 \quad \text{for all } \mathbf{x} \neq \mathbf{0}$$

Similarly, an $n \times n$ matrix $\mathbf{M}$ is said to be *positive semidefinite* if

$$\mathbf{x}^T\mathbf{M}\mathbf{x} \geq 0 \quad \text{for all } \mathbf{x}$$

We can now state two sufficient conditions that guarantee a unique solution for equations (5.3-64) and (5.3-65).

Theorem 5.3-1. Consider the coupled nonlinear equations given in (5.3-64) and (5.3-65); these equations possess a unique solution if the vector-valued functions $\mathbf{f}(\cdot)$ and $\mathbf{h}(\cdot)$ satisfy *either* 1 and 2 *or* 1 and 3 of the following conditions:

1. The derivatives of $\mathbf{f}(\cdot)$ and $\mathbf{h}(\cdot)$ exist and are continuous, and the Jacobian matrices $\mathbf{F}(\mathbf{y})$ and $\mathbf{H}(\mathbf{z})$, corresponding to $\mathbf{f}(\mathbf{y})$ and $\mathbf{h}(\mathbf{z})$, respectively, are positive semidefinite for all $\mathbf{y}$ and $\mathbf{z}$.
2. Either $\mathbf{F}(\mathbf{y})$ is symmetric positive definite for all $\mathbf{y}$ or $\mathbf{H}(\mathbf{z})$ is symmetric positive definite for all $\mathbf{z}$.
3. Either $\mathbf{F}(\mathbf{y})$ is diagonal for all $\mathbf{y}$ or $\mathbf{H}(\mathbf{z})$ is diagonal for all $\mathbf{z}$.

A rigorous proof of this theorem is beyond the scope of this book; it can be found in [10].

Let us now return to the formulation of the state equations. If the nonlinear functions $\mathbf{f}_g(\mathbf{i}_g)$ and $\mathbf{f}_r(\mathbf{v}_r)$ of equations (5.3-64) and (5.3-65) satisfy the conditions of Theorem 5.3-1, $\mathbf{v}_r$ and $\mathbf{i}_g$ can be solved in terms of $\mathbf{q}_c$, $\boldsymbol{\phi}_l$, $\mathbf{v}_v$, and $\mathbf{i}_i$; that is, we can write

$$\mathbf{v}_r = \hat{\mathbf{f}}_r(\mathbf{q}_c, \boldsymbol{\phi}_l; \mathbf{v}_v, \mathbf{i}_i) \tag{5.3-70}$$

and

$$\mathbf{i}_g = \hat{\mathbf{f}}_g(\mathbf{q}_c, \boldsymbol{\phi}_l; \mathbf{v}_v, \mathbf{i}_i) \tag{5.3-71}$$

Replacing $\mathbf{v}_r$ and $\mathbf{i}_g$ from these equations into (5.3-60) and (5.3-61), we obtain the state equations of the network under consideration in the normal form:

$$\dot{\mathbf{q}}_c = -\mathbf{F}_{cl}(\boldsymbol{\phi}_l) - \mathbf{F}_{cr}\hat{\mathbf{f}}_r[\mathbf{f}_r(\mathbf{q}_c, \boldsymbol{\phi}_l; \mathbf{v}_v, \mathbf{i}_i)] - \mathbf{F}_{ci}\mathbf{i}_i \tag{5.3-72}$$

$$\dot{\boldsymbol{\phi}}_l = \mathbf{F}_{cl}^T\mathbf{f}_c(\mathbf{q}_c) + \mathbf{F}_{gl}^T\hat{\mathbf{f}}_g[\mathbf{f}_g(\mathbf{q}_c, \boldsymbol{\phi}_l; \mathbf{v}_v, \mathbf{i}_i)] + \mathbf{F}_{vl}^T\mathbf{v}_v \tag{5.3-73}$$

The solution of such nonlinear state equations (if the solution exists and is unique) can usually be found by iterative methods; some of these methods will be discussed in Chapter 7.

Let us now consider another simplifying assumption, which greatly reduces the effort required in eliminating the nonstate variables $\mathbf{v}_r$ and $\mathbf{i}_g$. A quick look at (5.3-62) and (5.3-63) shows that if

$$\mathbf{F}_{gr} = \mathbf{0} \tag{5.3-74}$$

then

$$\mathbf{v}_r = \mathbf{F}_{cr}^T\mathbf{f}_c(\mathbf{q}_c) + \mathbf{F}_{vr}^T\mathbf{v}_v \tag{5.3-75}$$

and

$$\mathbf{i}_g = -\mathbf{F}_{gl}\mathbf{f}_l(\boldsymbol{\phi}_l) - \mathbf{F}_{gi}\mathbf{i}_i \tag{5.3-76}$$

Upon replacement of these in (5.3-60) and (5.3-61), we obtain

$$\dot{\mathbf{q}}_c = -\mathbf{F}_{cr}\mathbf{f}_r[\mathbf{F}_{cr}^T\mathbf{f}_c(\mathbf{q}_c) + \mathbf{F}_{vr}^T\mathbf{v}_v] - \mathbf{F}_{cl}\mathbf{f}_l(\boldsymbol{\phi}_l) - \mathbf{F}_{ci}\mathbf{i}_i \tag{5.3-77}$$

$$\dot{\boldsymbol{\phi}}_l = \mathbf{F}_{cl}^T\mathbf{f}_c(\mathbf{q}_c) + \mathbf{F}_{gl}^T\mathbf{f}_g[-\mathbf{F}_{gl}\mathbf{f}_l(\boldsymbol{\phi}_l) - \mathbf{F}_{gi}\mathbf{i}_i] + \mathbf{F}_{vl}^T\mathbf{v}_v \tag{5.3-78}$$

which is the desired state equation in the normal form. Now, to fully appreciate the physical significance of the assumption made in (5.3-74), consider the matrix relation (5.3-7); $\mathbf{F}_{gr}$ *is zero if, for the chosen proper tree, none of the fundamental cutsets defined by a tree branch resistor contains any chord resistance.* For example, in Fig. 5.3-2 the fundamental cutset corresponding to tree branch G_5 contains the chord resistor R_8 and the cutset corresponding to G_6 contains R_7; for this reason, the corresponding $\mathbf{F}_{gr}$ is not zero. *A particular case in which this condition is always satisfied is when every resistor in the network under consideration is either in series with an inductor or in parallel with a capacitor.* Let us now illustrate the outlined procedure by an example.

Example 5.3-2. Consider the network shown in Fig. 5.3-3. Let all the elements be nonlinear time invariant with characteristics

$$v_3 = f_3(q_3), \qquad v_4 = f_4(q_4), \qquad v_5 = f_5(i_5)$$
$$v_6 = f_6(i_6), \qquad i_7 = f_7(v_7), \qquad i_8 = f_8(\phi_8), \qquad i_9 = f_9(\phi_9)$$

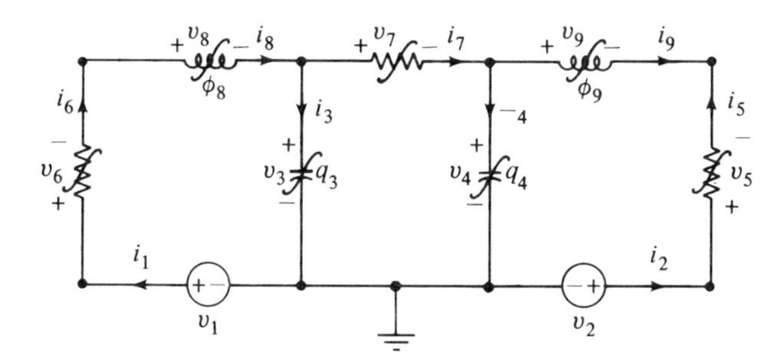

Figure 5.3-3 Nonlinear network considered in Example 5.3-2.

The problem is to write the state equations of this network in normal form using the general procedure outlined previously.

Solution

A proper tree of this network is shown in Fig. 5.3-4. The branch voltage vector $\mathbf{v}_b$ and the branch current vector $\mathbf{i}_b$ are

$$\mathbf{v}_b = [v_1 \quad v_2 \mid v_3 \quad v_4 \mid v_5 \quad v_6 \mid v_7 \mid v_8 \quad v_9]^T$$

and

$$\mathbf{i}_b = [i_1 \quad i_2 \mid i_3 \quad i_4 \mid i_5 \quad i_6 \mid i_7 \mid i_8 \quad i_9]^T$$

Notice that there are no independent current sources in the network; hence, the corresponding $\mathbf{v}_i$ and $\mathbf{i}_i$ vectors are zero and therefore are deleted from $\mathbf{v}_b$ and $\mathbf{i}_b$ just given. There are six fundamental cutsets and nine branches;

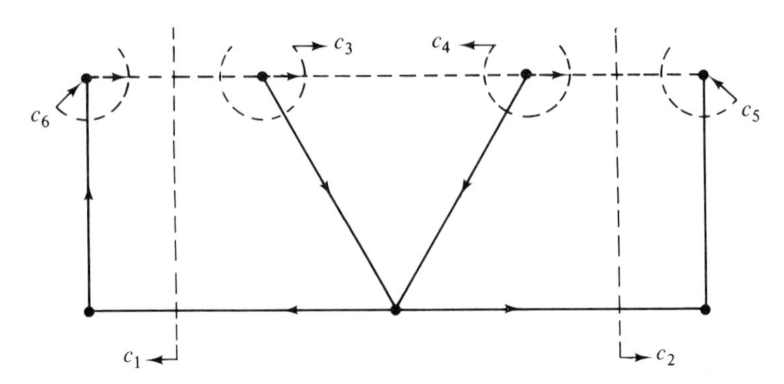

Figure 5.3-4 Proper tree of the network of Fig. 5.3-3 and its corresponding cutsets.

hence, the fundamental cutset matrix is a 6×9 matrix given by

$$\mathbf{Q}_f = \left[\begin{array}{cccccc|ccc} 1 & 0 & 0 & 0 & 0 & 0 & 0 & -1 & 0 \\ 0 & 1 & 0 & 0 & 0 & 0 & 0 & 0 & 1 \\ 0 & 0 & 1 & 0 & 0 & 0 & 1 & -1 & 0 \\ 0 & 0 & 0 & 1 & 0 & 0 & -1 & 0 & 1 \\ 0 & 0 & 0 & 0 & 1 & 0 & 0 & 0 & 1 \\ 0 & 0 & 0 & 0 & 0 & 1 & 0 & -1 & 0 \end{array}\right]$$
$$\underbrace{\hphantom{1\ 0\ 0\ 0\ 0\ 0}}_{\text{identity matrix}} \qquad \underbrace{\hphantom{0\ -1\ 0}}_{\mathbf{F}\text{ matrix}}$$

Again, since there are no current sources present, all the submatrices of **F** with indices i are identically zero; that is,

$$\mathbf{F}_{vi} = \mathbf{0}, \qquad \mathbf{F}_{ci} = \mathbf{0}, \qquad \mathbf{F}_{gi} = \mathbf{0}$$

We therefore have

$$\mathbf{F} = \left[\begin{array}{c|c} \mathbf{F}_{vr} & \mathbf{F}_{vl} \\ \hline \mathbf{F}_{cr} & \mathbf{F}_{cl} \\ \hline \mathbf{F}_{gr} & \mathbf{F}_{gl} \end{array}\right] = \left[\begin{array}{c|cc} 0 & -1 & 0 \\ 0 & 0 & 1 \\ \hline 1 & -1 & 0 \\ -1 & 0 & 1 \\ \hline 0 & 0 & 1 \\ 0 & -1 & 0 \end{array}\right]$$

Note that, since none of the cutsets defined by a tree branch resistor (i.e., cutsets C_5 and C_6) contains a chord resistor, the submatrix $\mathbf{F}_{gr}$ is zero; hence, the desired state equations are in the form of (5.3-77) and (5.3-78). To derive the explicit form of these equations for the network under study, we observe that the state variables are the capacitor charges and the inductor fluxes;

that is,

$$\mathbf{q}_c = [q_3 \quad q_4]^T \quad \text{and} \quad \boldsymbol{\phi}_l = [\phi_8 \quad \phi_9]^T$$

We also have

$$\mathbf{v}_c = \mathbf{f}_c(\mathbf{q}_c) = \begin{bmatrix} f_3(q_3) \\ f_4(q_4) \end{bmatrix}, \qquad \mathbf{i}_l = \mathbf{f}_l(\boldsymbol{\phi}_l) = \begin{bmatrix} f_8(\phi_8) \\ f_9(\phi_9) \end{bmatrix}$$

$$\mathbf{v}_v = \begin{bmatrix} v_1 \\ v_2 \end{bmatrix}, \qquad \mathbf{v}_g = \mathbf{f}_g(\mathbf{i}_g) = \begin{bmatrix} f_5(i_5) \\ f_6(i_6) \end{bmatrix}, \qquad \mathbf{i}_r = \mathbf{f}_r(\mathbf{v}_r) = f_7(v_7)$$

Using these relations in (5.3-75) and (5.3-76), we get

$$\mathbf{v}_r = [1 \quad -1]\begin{bmatrix} f_3(q_3) \\ f_4(q_4) \end{bmatrix} + [0 \quad 0]\begin{bmatrix} v_1 \\ v_2 \end{bmatrix} = f_3(q_3) - f_4(q_4)$$

and

$$\mathbf{i}_g = \begin{bmatrix} i_5 \\ i_6 \end{bmatrix} = -\begin{bmatrix} 0 & 1 \\ -1 & 0 \end{bmatrix}\begin{bmatrix} f_8(\phi_8) \\ f_9(\phi_9) \end{bmatrix} = \begin{bmatrix} -f_9(\phi_9) \\ f_8(\phi_8) \end{bmatrix}$$

Plugging $\mathbf{v}_r$ and $\mathbf{i}_g$ just obtained in (5.3-60) and (5.3-61), we obtain

$$\dot{\mathbf{q}}_c = \begin{bmatrix} 1 & 0 \\ 0 & -1 \end{bmatrix}\begin{bmatrix} f_8(\phi_8) \\ f_9(\phi_9) \end{bmatrix} + \begin{bmatrix} -1 \\ 1 \end{bmatrix} f_7[f_3(q_3) - f_4(q_4)]$$

$$\dot{\boldsymbol{\phi}}_l = \begin{bmatrix} -1 & 0 \\ 0 & 1 \end{bmatrix}\begin{bmatrix} f_3(q_3) \\ f_4(q_4) \end{bmatrix} + \begin{bmatrix} 0 & -1 \\ 1 & 0 \end{bmatrix}\begin{bmatrix} f_5[-f_9(\phi_9)] \\ f_6[f_8(\phi_8)] \end{bmatrix} + \begin{bmatrix} -v_1 \\ v_2 \end{bmatrix}$$

Finally, simplifying these relations, the desired state equations can be written in the following form:

$$\begin{aligned} \dot{q}_3 &= f_8(\phi_8) - f_7[f_3(q_3) - f_4(q_4)] \\ \dot{q}_4 &= f_9(\phi_9) + f_7[f_3(q_3) - f_4(q_4)] \\ \dot{\phi}_8 &= -f_3(q_3) - f_6[f_8(\phi_8)] - v_1 \\ \dot{\phi}_9 &= f_4(q_4) + f_5[-f_9(\phi_9)] + v_2 \end{aligned}$$

Remark 5.3-1. In the preceding equations, because of nonlinear resistors, some of the terms are composite nonlinear functions. If we assume that all the resistive elements of the network under study are linear, equations (5.3-62) and (5.3-63) reduce to a set of *linear* equations in $\mathbf{v}_r$ and $\mathbf{i}_g$ and can always be solved. Hence, we can conclude that

Assertion 5.3-1. The state equations of a proper network comprised of charge-controlled capacitors, flux-controlled inductors, linear resistors, and independent sources can always be represented in the normal form.

For the example in hand, if we assume that the resistors are linear, that is, if $\mathbf{i}_r = g_7\mathbf{v}_r$ and $\mathbf{v}_g = [R_5 i_5,\ R_6 i_6]^T$, then the state equations can be written as

$$\dot{q}_3 = -G_7 f_3(q_3) + G_7 f_4(q_4) + f_8(\phi_8)$$
$$\dot{q}_4 = G_7 f_3(q_3) - G_7 f_4(q_4) - f_9(\phi_9)$$
$$\dot{\phi}_8 = -f_3(q_3) - R_6 f_8(\phi_8) - v_1$$
$$\dot{\phi}_9 = f_4(q_4) - R_5 f_9(\phi_9) + v_2$$

Throughout Sections 5.2 and 5.3 we have assumed that the network under study has no capacitor-only loops and no inductor-only cutsets. Under these conditions, either the voltage or charge across *each* capacitor and either the current or flux through *each* inductor were chosen as a state variable. The resulting state vector **x** then forms a *basis;* that is, all the chosen state variables are linearly independent in the sense that none of the elements of **x**, say x_i, can be obtained from a linear combination of the others. Furthermore, the resulting state equations completely describe the behavior of the network; that is, any of the network variables can be obtained from a linear combination of the elements of **x** (such a set of variables is sometimes called a *complete* set of state variables [8]). We can, therefore, assert that *the number of the independent state variables in a proper network is equal to the number of its energy-storing elements.*

If there are some capacitor-only loops or inductor-only cutsets in a network, the number of the linearly independent state variables that can be chosen is equal to the order of complexity of the network. The definition of the order of complexity of a network and related topics are discussed in the next section.

5.4 CONCEPT OF STATE AND ORDER OF COMPLEXITY OF A NETWORK

The fundamental problem of network analysis is to obtain the branch voltages and currents of a given network for given sources. One direct method of solving this problem is to use KVL, KCL, and VCR to obtain a set of $2B$ linearly independent integro-differential equations in $2B$ unknowns that can be solved for all the branch voltages and branch currents of a network with $N + 1$ nodes and B branches. Alternatively, one can choose the node voltages or loop currents as auxiliary variables and obtain a set of integro-differential equations in terms of these variables. The branch voltages and currents can then be obtained through algebraic manipulations or simple differentiation or integration of these auxiliary variables. It was mentioned in the beginning of this chapter that, for a number of reasons, it is more advantageous to choose these auxiliary variables in such a way that the dynamical equations of the network will be in the form of a set of first-order differential equations. It was shown that the capacitor charges (or voltages) and inductor fluxes (or currents) are appropriate auxiliary variables for achieving this objective. Such auxiliary variables are called the *state* variables of the network under consideration and are denoted by $\mathbf{x}$. The state equations of linear networks, written in the normal form, can therefore be represented by

$$\dot{\mathbf{x}} = \mathbf{A}\mathbf{x} + \mathbf{B}\mathbf{u} \tag{5.4-1}$$

and those of nonlinear networks by

$$\dot{\mathbf{x}} = \mathbf{f}(\mathbf{x}, \mathbf{u}) \tag{5.4-2}$$

where in both cases $\mathbf{x}$ represents the n-dimensional state vector, and $\mathbf{u}$ is the m-dimensional input vector, whose elements are independent voltage and current sources of the network. Given the state at time t_0, $\mathbf{x}(t_0)$, the state at time $t \geq t_0$ can then be obtained by solving the preceding differential equations. Furthermore, having solved (5.4-1) and (5.4-2) for $\mathbf{x}(t)$, every branch voltage or current can be obtained from $\mathbf{x}(t)$ through purely algebraic equations. More precisely, let $\mathbf{y}$ denote the desired branch voltages and/or currents; then for linear networks we have

$$\mathbf{y} = \mathbf{C}\mathbf{x} + \mathbf{D}\mathbf{u} \tag{5.4-3}$$

and for nonlinear networks $\mathbf{y}$ is given by

$$\mathbf{y} = \mathbf{g}(\mathbf{x}, \mathbf{u}) \tag{5.4-4}$$

Consequently, equations (5.4-1) or (5.4-2) completely describe the dynamical behavior of the network—these equations together with (5.4-3) and (5.4-4) are called the *input–output state equations*.

From the theory of differential equations we know that to solve (5.4-1) or (5.4-2) we need n *independent* initial conditions $\mathbf{x}(t_0) = [x_1(t_0), x_2(t_0), \ldots, x_n(t_0)]^T$; given $\mathbf{x}(t_0)$ and $\mathbf{u}(t)$ for $t \geq t_0$, the state $\mathbf{x}(t)$ and consequently the output $\mathbf{y}(t)$ can be obtained for all $t \geq t_0$. For this reason, $\mathbf{x}(t_0)$ is called the *state of the network* at time t_0. More precisely;

Definition 5.4-1 (State). The *state* of a network at time t^* is any amount of information that together with the input function is sufficient to specify the output for all $t \geq t^*$.

Order of Complexity of a Network. In the preceding sections we observed that the number of the first-order differential equations that completely describe the dynamical behavior of the network is equal to the number of the energy-storing elements of the network. If the total number of capacitors and inductors in a network is n, and if there are no capacitor-only loops or inductor-only cutsets in the network, the number of the linearly independent first-order differential equations describing the network is exactly n. To solve these equations, we need n independent initial conditions. If the auxiliary variables are chosen to be, say, the capacitor voltages and inductor currents, we need to specify the initial voltages on the capacitors and initial currents through the inductors. These initial conditions can also be taken to be the state of the network at time t_0; hence, the state variables chosen in this manner are linearly independent and completely describe the dynamical behavior of the network. For this reason, we say that the *order of complexity* of such a network is equal to the number of its energy-storing elements, or, equivalently, it is equal to the number of the independent initial conditions of the network.

On the other hand, if there are some capacitor-only loops or inductor-only cutsets in the network, the order of complexity of the network is less than n, since some of the initial conditions can be obtained as a linear combination of the others. Hence, the state equations obtained by assigning all the capacitor voltages or charges and all the inductor currents or fluxes as the state variables are not linearly independent. To see this, consider the network in Fig. 5.4-1. There is one capacitor-only loop and one inductor-only

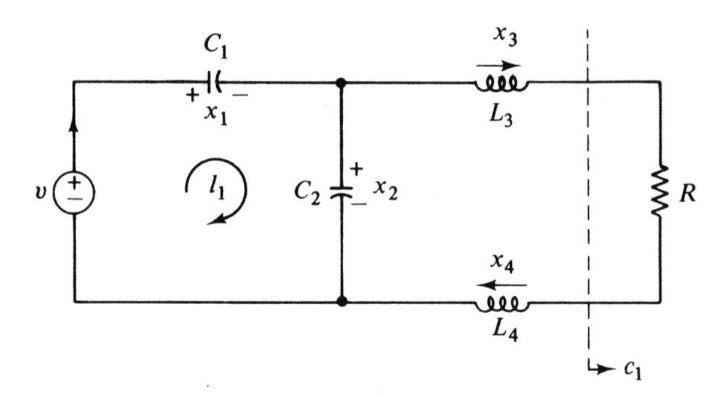

Figure 5.4-1 Network with one capacitor-only loop and one inductor-only cutset.

cutset in this network. Let us assume, for simplicity, that all the elements of this network are linear and time invariant. Then if we choose the capacitor voltages and inductor currents as the state variables, the resulting state equations will not be linearly independent. More precisely, writing KVL for l_1 and KCL for c_1, we get

$$x_2 = -x_1 + v$$

and

$$x_3 = x_4$$

That is, we can have only two independent state variables for this network. Also, among the four initial conditions $x_1(t_0)$, $x_2(t_0)$, $x_3(t_0)$, and $x_4(t_0)$ only two are independent, say $x_1(t_0)$ and $x_3(t_0)$.

In general, we can state that the number of the independent initial conditions necessary for the unique solution of a network is equal to the number of the independent state variables of the network. We can now give a precise definition of order of complexity of a network.

Definition 5.4-2 (Order of Complexity). The *order of complexity* of a network is equal to the number of *independent* initial conditions that can be specified in the network; or, equivalently,

The order of complexity of a network is the number of the *independent* state variables that completely specify the dynamical behavior of the network; or, equivalently,

The order of complexity of a network is the number of the *independent* first-order differential equations that completely specify the behavior of the network.

From the foregoing discussions we can assert the following about the order of complexity of a network.

Assertion 5.4-1. The order of complexity of a network is equal to the total number of its energy-storing elements minus the total number of *independent* capacitor-only loops and inductor-only cutsets; equivalently,

$$n = n_{lc} - (n_c + n_l) \tag{5.4-5}$$

where n = order of complexity of the network
n_{lc} = total number of inductors and capacitors
n_c = number of independent capacitor-only loops
n_l = number of independent inductor-only cutsets

The proof of this assertion is obvious, since the state variables assigned to each capacitor of a capacitor-only loop are related through a KVL equation and the state variables assigned to each inductor in an inductor-only cutset are related through a KCL equation. Hence, if these capacitor-only loops and inductor-only cutsets are independent, there are $n_c + n_l$ independent equations relating n_{lc} state variables. Then n, given by (5.4-5), represents the number of linearly independent state variables and consequently the order of complexity of the network.

Notice that since the order of complexity of a network is the same as the number of independent state variables that completely describes the network, we now have a method for determining the number of independent state variables of a general network. For example, in Fig. 5.4-1 the order of complexity (and hence the number of independent state variables) is 2, since there are four energy-storing elements, one capacitor-only loop, and one inductor-only cutset. In this simple example it was easy to detect the *independent* capacitor-only loops and inductor-only cutsets. In more complicated networks this is not usually an easy task—one generally has to resort to topological partitioning of the network. Perhaps the best way to see this is through an example.

Example 5.4-1. Consider the network shown in Fig. 5.4-2. The capacitors, inductors, and resistors can be linear, nonlinear, time varying, or time invariant. The objective is to determine the order of complexity and hence the number of the independent state variables that completely specify the dynamical behavior of the network.

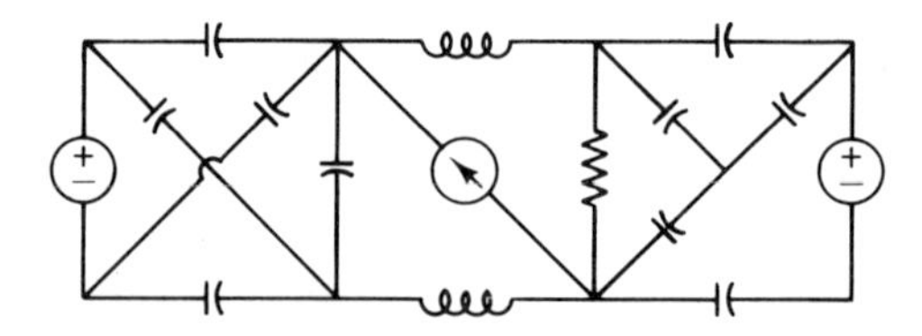

Figure 5.4-2 Network considered in Example 5.4-1; this network has 12 energy-storing elements but only 6 independent state variables.

Solution

There are 2 inductors and 10 capacitors in this network. If we choose capacitor charges and inductor fluxes as the state variables, we get a total of 12 state variables, which, obviously, are not independent of one another. According to Assertion 5.4-1, the number of the independent state variables, n, is

$$n = 12 - (n_c + n_l) \tag{5.4-6}$$

From Fig. 5.4-2 it is easy to see that there is one inductor-only cutset; hence, $n_l = 1$. However, it is not easy to discern, at first glance, the number of *independent* capacitor-only loops—there are a total of 10 capacitor-only loops in this network! We then must ask how many of these loops are independent. Before answering this question, let us outline a systematic procedure for obtaining n_c and n_l.

Consider a network $\mathfrak{N}$ and let $\mathfrak{N}_c$ and $\mathfrak{N}_l$ be subnetworks of $\mathfrak{N}$ defined as follows:

$\mathfrak{N}_c$ is obtained from $\mathfrak{N}$ by open circuiting all the inductors, independent current sources, and resistors in $\mathfrak{N}$.

$\mathfrak{N}_l$ is obtained from $\mathfrak{N}$ by short circuiting all the capacitors, independent voltage sources, and resistors in $\mathfrak{N}$.

The subgraphs $\mathfrak{N}_c$ and $\mathfrak{N}_l$, therefore, contain only capacitors and voltage sources and inductors and current sources, respectively. Note that each of these subgraphs may be a collection of several disjoint subgraphs. For instance, in our example $\mathfrak{N}_c$ is made up of two disjoint subgraphs; these subgraphs are given in Fig. 5.4-3. It is now quite easy to see that n_c and n_l are given by

$$n_c = \text{number of fundamental loops in } \mathfrak{N}_c$$

$$n_l = \text{number of fundamental cutsets in } \mathfrak{N}_l$$

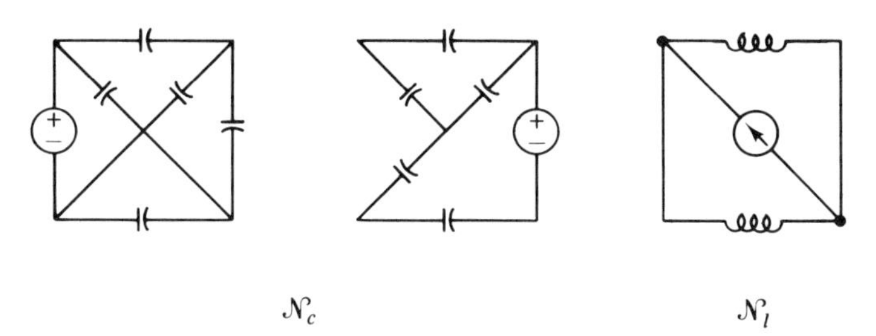

$\mathcal{N}_c$ $\mathcal{N}_l$

Figure 5.4-3 Capacitive and inductive subgraphs of the network of Fig. 5.4-2.

For the example in hand, we get $n_c = 5$ and $n_l = 1$. The order of complexity (and, hence, the number of independent state variables), therefore, is

$$n = 12 - (5 + 1) = 6$$

Next we introduce a method for writing the state equations of a class of networks that may contain capacitor-only loops and inductor-only cutsets.

5.5 STATE-VARIABLE FORMULATION OF GENERAL NETWORKS

In Section 5.3 we introduced a method for obtaining the normal form of the state equations of a class of networks that have no capacitor-only loops and no inductor-only cutsets. In this section we relax this condition and give a systematic procedure for obtaining the state equation of any network comprised of two-terminal capacitors, inductors, resistors, and independent voltage and current sources.

Recall that the first step in deriving the state equations of a proper network was to choose a proper tree. If the network under study has some capacitor-only loops or inductor-only cutsets, it is not possible to obtain such a tree, and the procedure outlined in Section 5.3 fails. To remove this obstacle, we define a normal tree, which was first introduced by Bryant [2].

Definition 5.5-1 (Normal Tree). A *normal tree* of a lumped connected network comprised of two-terminal resistors, capacitors, inductors, and independent voltage and current sources is a tree that contains all the voltage sources, none of the current sources, as many as possible capacitors, and as few as possible inductors.

Notice that in this definition we excluded the possibility of having any loops of voltage sources only or any cutsets of current sources only. Before discussing the general derivation of state equation for such networks, let us present an informal description of the steps involved in writing the state equations. The crucial step in writing the state equation of networks containing capacitor-only loops and inductor-only cutsets is to choose a set of appropriate linearly independent state variables. Once this is done, the step-by-step method of Section 5.2 can be employed to obtain the desired state equations. To choose the desired state variables, it is best to use the normal tree of the network just defined. More specifically, for a particular normal tree, we *choose all the tree branch capacitor voltages and all the chord inductor currents as the state variables.* With this choice of state variables we then proceed to write linearly independent KCL, KVL, and VCR equations and eliminate all the nonstate variables among them. Let us illustrate this procedure by means of an example.

Example 5.5-1. Consider the linear time-invariant network shown in Fig. 5.5-1. There is one capacitor only loop and one inductor-only cutset in this network. Consequently, we can specify only four linearly independent state

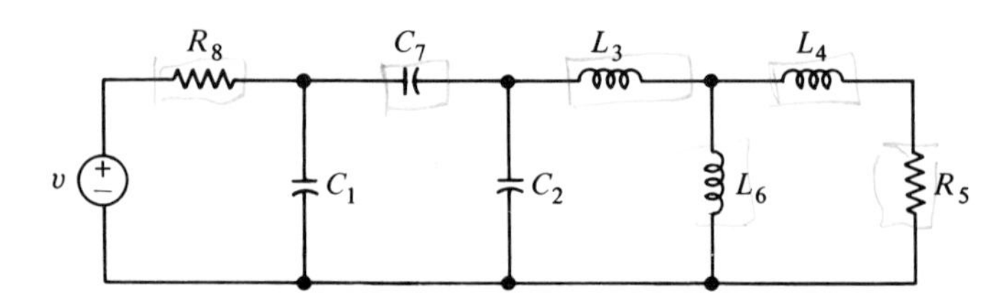

Figure 5.5-1 Network considered in Example 5.5-1.

variables. In this simple case there is no need for using the concept of the normal tree; the voltages across any two of the three capacitors and the currents through any two of the three inductors can be chosen as the state variables. However, to illustrate the application of a normal tree, let us choose the normal tree shown in Fig. 5.5-2. Note that there are several possibilities in choosing a normal tree for the network under study. In each case the tree will have the voltage source, two capacitors, and one inductor. Since the net-

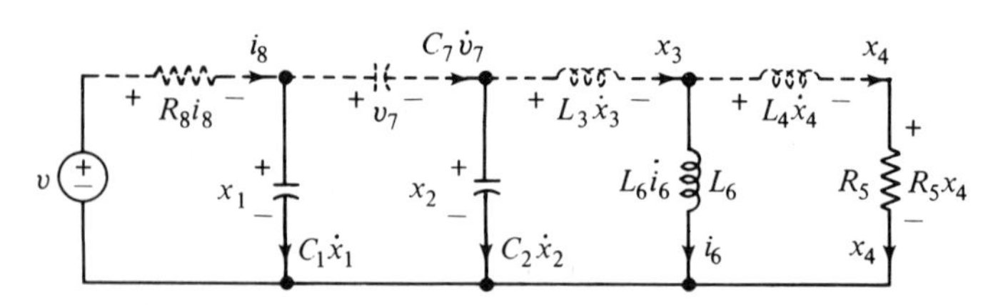

Figure 5.5-2 Normal tree (drawn in solid line) of the network of Fig. 5.5-1.

work is linear and time invariant, we choose the capacitor voltages and inductor currents as the state variables. These state variables (tree branch capacitor voltages and chord inductor currents) are denoted by x_1 through x_4 and are shown in Fig. 5.5-2. As in Section 5.2, the next step is to write the linearly independent KVL and KCL (the VCR have already been used in the figure). Thus, we have

$$\text{KCL:} \qquad i_8 = C_1\dot{x}_1 + C_7\dot{v}_7$$

$$C_7\dot{v}_7 = C_2\dot{x}_2 + x_3$$

$$x_3 = i_6 + x_4$$

$$\text{KVL:} \qquad v = R_8 i_8 + x_1$$

$$x_1 = v_7 + x_2$$

$$x_2 = L_3\dot{x}_3 + L_6\dot{i}_6$$

$$L_6\dot{i}_6 = L_4\dot{x}_4 + R_5 x_4$$

Next we eliminate the nonstate variables v_7, i_8, and i_6 from these equations to get

$$C_7\dot{x}_1 - (C_7 + C_2)\dot{x}_2 = x_3$$

$$(C_7 + C_1)\dot{x}_1 - C_7\dot{x}_2 = -\frac{1}{R_8}x_1 + \frac{1}{R_8}v$$

$$(L_6 + L_3)\dot{x}_3 - L_6\dot{x}_4 = x_2$$

$$L_6\dot{x}_3 - (L_4 + L_6)\dot{x}_4 = R_5 x_4$$

To put these equations in the normal form, let us assign specific values to the network elements; take $C_1 = C_2 = 1$ F, $C_7 = 2$ F, $L_3 = L_4 = \frac{1}{2}$ H, $L_6 = 1$ H, and $R_5 = R_8 = 1\ \Omega$. The equations can then be written as

$$\begin{bmatrix} \dot{x}_1 \\ \dot{x}_2 \\ \dot{x}_3 \\ \dot{x}_4 \end{bmatrix} = \begin{bmatrix} -\frac{3}{5} & 0 & -\frac{2}{5} & 0 \\ -\frac{2}{5} & 0 & -\frac{3}{5} & 0 \\ 0 & \frac{6}{5} & 0 & -\frac{4}{5} \\ 0 & \frac{4}{5} & 0 & -\frac{6}{5} \end{bmatrix} \begin{bmatrix} x_1 \\ x_2 \\ x_3 \\ x_4 \end{bmatrix} + \begin{bmatrix} \frac{3}{5} \\ \frac{2}{5} \\ 0 \\ 0 \end{bmatrix} v$$

which is the desired state equation in the normal form.

Let us now use a topological method similar to that of proper networks to obtain a general formulation of the state equations in this case. Once again it should be emphasized that these systematic approaches are most suitable for large-scale networks or automated computer analysis programs. Hence, whenever we can, we should check the feasibility of writing the state equations by inspection in order to avoid the general formulation (which can be quite lengthy).

Linear Time-Varying Networks

Step 1. Choose a normal tree and number the branches of the network in the following order:

1. All the voltage sources with voltage and current vectors $\mathbf{v}_v$ and $\mathbf{i}_v$.
2. Tree branch capacitors with voltage and current vectors $\mathbf{v}_c$ and $\mathbf{i}_c$.
3. Tree branch resistors with voltage and current vectors $\mathbf{v}_g$ and $\mathbf{i}_g$.
4. Tree branch inductors with voltage and current vectors $\mathbf{v}_\gamma$ and $\mathbf{i}_\gamma$.
5. Chord capacitors with voltage and current vectors $\mathbf{v}_s$ and $\mathbf{i}_s$
6. Chord resistors with voltage and current vectors $\mathbf{v}_r$ and $\mathbf{i}_r$.
7. Chord inductors with voltage and current vectors $\mathbf{v}_l$ and $\mathbf{i}_l$.
8. All the current sources with voltage and current vectors $\mathbf{v}_i$ and $\mathbf{i}_i$.

The branch voltage and curient vectors can then be partitioned in the form

$$\mathbf{v}_b = [\mathbf{v}_v \mid \mathbf{v}_c \mid \mathbf{v}_g \mid \mathbf{v}_\gamma \mid \mathbf{v}_s \mid \mathbf{v}_r \mid \mathbf{v}_l \mid \mathbf{v}_i]^T \tag{5.5-1}$$

and

$$\mathbf{i}_b = [\mathbf{i}_v \mid \mathbf{i}_c \mid \mathbf{i}_g \mid \mathbf{i}_\gamma \mid \mathbf{i}_s \mid \mathbf{i}_r \mid \mathbf{i}_l \mid \mathbf{i}_i]^T \tag{5.5-2}$$

Let v, c, g, γ, s, r, l, and i denote the number of elements in $\mathbf{v}_v$, $\mathbf{v}_c$, $\mathbf{v}_g$, $\mathbf{v}_\gamma$, $\mathbf{v}_s$, $\mathbf{v}_r$, $\mathbf{v}_l$, and $\mathbf{v}_i$, respectively. As in the case of proper networks, we can write the independent KCL and KVL equations in the form

$$\mathbf{Q}_f \mathbf{i}_b = \mathbf{0}$$

and

$$\mathbf{B}_f \mathbf{v}_b = \mathbf{0}$$

where $\mathbf{Q}_f$ and $\mathbf{B}_f$ are

$$\mathbf{Q}_f = [\mathbf{1} \mid \mathbf{F}]$$

and

$$\mathbf{B}_f = [-\mathbf{F}^T \mid \mathbf{1}]$$

The particular partitioning of $\mathbf{v}_b$ and $\mathbf{i}_b$ in (5.5-1) and (5.5-2) guarantees that KCL and KVL can be written as

$$\begin{bmatrix} \mathbf{1}_{vv} & \mathbf{0} & \mathbf{0} & \mathbf{0} & \mathbf{F}_{vs} & \mathbf{F}_{vr} & \mathbf{F}_{vl} & \mathbf{F}_{vi} \\ 0 & \mathbf{1}_{cc} & \mathbf{0} & \mathbf{0} & \mathbf{F}_{cs} & \mathbf{F}_{cr} & \mathbf{F}_{cl} & \mathbf{F}_{ci} \\ \mathbf{0} & \mathbf{0} & \mathbf{1}_{gg} & \mathbf{0} & \mathbf{F}_{gs} & \mathbf{F}_{gr} & \mathbf{F}_{gl} & \mathbf{F}_{gi} \\ \mathbf{0} & \mathbf{0} & \mathbf{0} & \mathbf{1}_{\gamma\gamma} & \mathbf{F}_{\gamma s} & \mathbf{F}_{\gamma r} & \mathbf{F}_{\gamma l} & \mathbf{F}_{\gamma i} \end{bmatrix} \begin{bmatrix} \mathbf{i}_v \\ \mathbf{i}_c \\ \mathbf{i}_g \\ \mathbf{i}_\gamma \\ \mathbf{i}_s \\ \mathbf{i}_r \\ \mathbf{i}_l \\ \mathbf{i}_i \end{bmatrix} = \mathbf{0} \tag{5.5-3}$$

and

$$\begin{bmatrix} -\mathbf{F}_{vs}^T & -\mathbf{F}_{cs}^T & -\mathbf{F}_{gs}^T & -\mathbf{F}_{\gamma s}^T & \mathbf{1}_{ss} & \mathbf{0} & \mathbf{0} & \mathbf{0} \\ -\mathbf{F}_{vr}^T & -\mathbf{F}_{cr}^T & -\mathbf{F}_{gr}^T & -\mathbf{F}_{\gamma r}^T & \mathbf{0} & \mathbf{1}_{rr} & \mathbf{0} & \mathbf{0} \\ -\mathbf{F}_{vl}^T & -\mathbf{F}_{cl}^T & -\mathbf{F}_{gl}^T & -\mathbf{F}_{\gamma l}^T & \mathbf{0} & \mathbf{0} & \mathbf{1}_{ll} & \mathbf{0} \\ -\mathbf{F}_{vi}^T & -\mathbf{F}_{ci}^T & -\mathbf{F}_{gi}^T & -\mathbf{F}_{\gamma i}^T & \mathbf{0} & \mathbf{0} & \mathbf{0} & \mathbf{1}_{ii} \end{bmatrix} \begin{bmatrix} \mathbf{v}_v \\ \mathbf{v}_c \\ \mathbf{v}_g \\ \mathbf{v}_\gamma \\ \mathbf{v}_s \\ \mathbf{v}_r \\ \mathbf{v}_l \\ \mathbf{v}_i \end{bmatrix} = \mathbf{0} \quad (5.5\text{-}4)$$

where the dimensions of submatrices of $\mathbf{F}$ are determined by the order of their associated subvectors in $\mathbf{i}_b$ and $\mathbf{v}_b$. More precisely, the first subscript of the submatrices of $\mathbf{F}$ denotes the number of their rows and the second subscript denotes the number of their columns. For example, $\mathbf{F}_{vl}$ has v rows and l columns.

Exercise 5.5-1. Use the definition of a normal tree to show that

$$\mathbf{F}_{gs} = \mathbf{0}, \qquad \mathbf{F}_{\gamma s} = \mathbf{0}, \qquad \mathbf{F}_{\gamma r} = \mathbf{0}$$

Step 2. Choose all the *tree branch capacitor charges* and all the *chord inductor fluxes* as the state variables. That is, choose the state vector $\mathbf{x}$ to be

$$\mathbf{x} = [\mathbf{q}_c, \boldsymbol{\phi}_l]^T \quad (5.5\text{-}5)$$

The branch VCR can then be written as

$$\mathbf{i}_c = \dot{\mathbf{q}}_c \quad (5.5\text{-}6)$$

$$\mathbf{v}_l = \dot{\boldsymbol{\phi}}_l \quad (5.5\text{-}7)$$

$$\mathbf{q}_c = \mathbf{C}_c\mathbf{v}_c \quad (5.5\text{-}8)$$

$$\boldsymbol{\phi}_l = \mathbf{L}_l\mathbf{i}_l \quad (5.5\text{-}9)$$

$$\mathbf{v}_r = \mathbf{R}_r\mathbf{i}_r \quad (5.5\text{-}10)$$

$$\mathbf{i}_g = \mathbf{G}_g\mathbf{v}_g \quad (5.5\text{-}11)$$

$$\mathbf{v}_\gamma = \dot{\boldsymbol{\phi}}_\gamma \quad (5.5\text{-}12)$$

$$\mathbf{i}_s = \dot{\mathbf{q}}_s \quad (5.5\text{-}13)$$

where $\mathbf{C}_c$, $\mathbf{L}_l$, $\mathbf{R}_r$, and $\mathbf{G}_g$ are, respectively, tree branch capacitance matrix, chord inductance matrix, chord resistance matrix, and tree branch conductance matrix. In the last two equations, $\boldsymbol{\phi}_\gamma$ and $\mathbf{q}_s$ denote the tree branch inductor fluxes and chord capacitor charges.

Step 3. As in the previous case this is the most crucial step in the formulation of the state equations. We must eliminate all the nonstate variables among equations (5.5-3) to (5.5-13). For this purpose, let us rewrite the KCL and KVL equations derived previously; these equations together with the result of Exercise 5.5-1 yield

$$\mathbf{i}_v + \mathbf{F}_{vs}\mathbf{i}_s + \mathbf{F}_{vr}\mathbf{i}_r + \mathbf{F}_{vl}\mathbf{i}_l + \mathbf{F}_{vi}\mathbf{i}_i = \mathbf{0} \tag{5.5-14}$$

$$\mathbf{i}_c + \mathbf{F}_{cs}\mathbf{i}_s + \mathbf{F}_{cr}\mathbf{i}_r + \mathbf{F}_{cl}\mathbf{i}_l + \mathbf{F}_{ci}\mathbf{i}_i = \mathbf{0} \tag{5.5-15}$$

$$\mathbf{i}_g + \mathbf{F}_{gr}\mathbf{i}_r + \mathbf{F}_{gl}\mathbf{i}_l + \mathbf{F}_{gi}\mathbf{i}_i = \mathbf{0} \tag{5.5-16}$$

$$\mathbf{i}_y + \mathbf{F}_{yl}\mathbf{i}_l + \mathbf{F}_{yi}\mathbf{i}_i = \mathbf{0} \tag{5.5-17}$$

and

$$-\mathbf{F}_{vs}^T\mathbf{v}_v - \mathbf{F}_{cs}^T\mathbf{v}_c + \mathbf{v}_s = \mathbf{0} \tag{5.5-18}$$

$$-\mathbf{F}_{vr}^T\mathbf{v}_v - \mathbf{F}_{cr}^T\mathbf{v}_c - \mathbf{F}_{gr}^T\mathbf{v}_g + \mathbf{v}_r = \mathbf{0} \tag{5.5-19}$$

$$-\mathbf{F}_{vl}^T\mathbf{v}_v - \mathbf{F}_{cl}^T\mathbf{v}_c - \mathbf{F}_{gl}^T\mathbf{v}_g - \mathbf{F}_{yl}^T\mathbf{v}_y + \mathbf{v}_l = \mathbf{0} \tag{5.5-20}$$

$$-\mathbf{F}_{vi}^T\mathbf{v}_v - \mathbf{F}_{ci}^T\mathbf{v}_c - \mathbf{F}_{gi}^T\mathbf{v}_g - \mathbf{F}_{yi}^T\mathbf{v}_y + \mathbf{v}_i = \mathbf{0} \tag{5.5-21}$$

As a first step in the elimination process, let us replace $\mathbf{i}_c$ and $\mathbf{v}_l$ from (5.5-6) and (5.5-7) into (5.5-15) and (5.5-20) to get

$$\dot{\mathbf{q}}_c = -\mathbf{F}_{cs}\mathbf{i}_s - \mathbf{F}_{cr}\mathbf{i}_r - \mathbf{F}_{cl}\mathbf{i}_l - \mathbf{F}_{ci}\mathbf{i}_i \tag{5.5-22}$$

$$\dot{\boldsymbol{\phi}}_l = \mathbf{F}_{yl}^T\mathbf{v}_y + \mathbf{F}_{gl}^T\mathbf{v}_g + \mathbf{F}_{cl}^T\mathbf{v}_c + \mathbf{F}_{vl}^T\mathbf{v}_v \tag{5.5-23}$$

Also, rewrite (5.5-17) and (5.5-18) as

$$\mathbf{i}_y = -\mathbf{F}_{yl}\mathbf{i}_l - \mathbf{F}_{yi}\mathbf{i}_i$$

$$\mathbf{v}_s = \mathbf{F}_{cs}^T\mathbf{v}_c + \mathbf{F}_{vs}^T\mathbf{v}_v$$

Upon premultiplication by $\mathbf{L}_y$ and $\mathbf{C}_s$ and noting that

$$\boldsymbol{\phi}_y = \mathbf{L}_y\mathbf{i}_y \quad \text{and} \quad \mathbf{q}_s = \mathbf{C}_s\mathbf{v}_s$$

these equations yield

$$\mathbf{i}_s = \dot{\mathbf{q}}_s = \frac{d}{dt}[\mathbf{C}_s\mathbf{F}_{cs}^T\mathbf{C}_c^{-1}\mathbf{q}_c + \mathbf{C}_s\mathbf{F}_{vs}^T\mathbf{v}_v] \tag{5.5-24}$$

and

$$\mathbf{v}_y = \dot{\boldsymbol{\phi}}_y = -\frac{d}{dt}[\mathbf{L}_y\mathbf{F}_{yl}\mathbf{L}_l^{-1}\boldsymbol{\phi}_l + \mathbf{L}_y\mathbf{F}_{yi}\mathbf{i}_i] \tag{5.5-25}$$

Replacing $\mathbf{i}_s$ and $\mathbf{v}_y$ from these equations into (5.5-22) and (5.5-23), we get

$$\frac{d}{dt}[\mathbf{q}_c + \mathbf{F}_{cs}\mathbf{C}_s\mathbf{F}_{cs}^T\mathbf{C}_c^{-1}\mathbf{q}_c] = -\mathbf{F}_{cr}\mathbf{i}_r - \mathbf{F}_{cl}\mathbf{i}_l - \mathbf{F}_{ci}\mathbf{i}_i - \frac{d}{dt}[\mathbf{F}_{cs}\mathbf{C}_s\mathbf{F}_{vs}^T\mathbf{v}_v]$$

$$\frac{d}{dt}[\boldsymbol{\phi}_l + \mathbf{F}_{yl}^T\mathbf{L}_y\mathbf{F}_{yl}\mathbf{L}_l^{-1}\boldsymbol{\phi}_l] = \mathbf{F}_{gl}^T\mathbf{v}_g + \mathbf{F}_{cl}^T\mathbf{v}_c + \mathbf{F}_{vl}^T\mathbf{v}_v - \frac{d}{dt}[\mathbf{F}_{yl}^T\mathbf{L}_y\mathbf{F}_{yi}\mathbf{i}_i]$$

Let us now adopt the following convenient notations:

$$\mathbf{q} \triangleq [\mathbf{1} + \mathbf{F}_{cs}\mathbf{C}_s\mathbf{F}_{cs}^T\mathbf{C}_c^{-1}]q_c \tag{5.5-26}$$

and
$$\boldsymbol{\phi} \triangleq [\mathbf{1} + \mathbf{F}_{yl}^T\mathbf{L}_y\mathbf{F}_{yl}\mathbf{L}_l^{-1}]\boldsymbol{\phi}_l \tag{5.5-27}$$

Thus, the previous set of differential equations can be written as

$$\dot{\mathbf{q}} = -\mathbf{F}_{cr}\mathbf{i}_r - \mathbf{F}_{cl}\mathbf{L}_l^{-1}\boldsymbol{\phi}_l - \frac{d}{dt}[\mathbf{F}_{cs}\mathbf{C}_s\mathbf{F}_{vs}^T\mathbf{v}_v] - \mathbf{F}_{ci}\mathbf{i}_i \tag{5.5-28}$$

$$\dot{\boldsymbol{\phi}} = \mathbf{F}_{gl}^T\mathbf{v}_g + \mathbf{F}_{cl}^T\mathbf{C}_c^{-1}\mathbf{q}_c - \frac{d}{dt}[\mathbf{F}_{yl}^T\mathbf{L}_y\mathbf{F}_{yi}\mathbf{i}_i] + \mathbf{F}_{vl}^T\mathbf{v}_l \tag{5.5-29}$$

Solving (5.5-16) and (5.5-19) for $\mathbf{v}_g$ and $\mathbf{v}_r$, we obtain

$$\mathbf{v}_g = -\mathbf{G}^{-1}[\mathbf{F}_{gr}\mathbf{R}_r^{-1}\mathbf{F}_{cr}^T\mathbf{C}_c^{-1}\mathbf{q}_c + \mathbf{F}_{gr}\mathbf{R}_r^{-1}\mathbf{F}_{vr}^T\mathbf{v}_v + \mathbf{F}_{gl}\mathbf{L}_l^{-1}\boldsymbol{\phi}_l + \mathbf{F}_{gi}\mathbf{i}_i] \tag{5.5-30}$$
$$\mathbf{i}_r = \mathbf{R}^{-1}[\mathbf{F}_{cr}^T\mathbf{C}_c^{-1}\mathbf{q}_c + \mathbf{F}_{vr}^T\mathbf{v}_v - \mathbf{F}_{gr}^T\mathbf{G}_g^{-1}\mathbf{F}_{gl}\mathbf{L}_l^{-1}\boldsymbol{\phi}_l - \mathbf{F}_{gr}^T\mathbf{G}_g^{-1}\mathbf{F}_{gi}\mathbf{i}_i] \tag{5.5-31}$$

where $\mathbf{G}$ and $\mathbf{R}$ are defined in (5.3-33) and (5.3-34), respectively. Replacing $\mathbf{v}_g$ and $\mathbf{i}_r$ from these equations into (5.5-28) and (5.5-29), we get

$$\dot{\mathbf{q}} = \mathbf{H}_{cc}\mathbf{C}_c^{-1}\mathbf{q}_c + \mathbf{H}_{cl}\mathbf{L}_l^{-1}\boldsymbol{\phi}_l + \mathbf{H}_{cv}\mathbf{v}_v + \mathbf{H}_{ci}\mathbf{i}_i - \dot{\bar{\mathbf{v}}}_v \tag{5.5-32}$$
$$\dot{\boldsymbol{\phi}} = \mathbf{H}_{lc}\mathbf{C}_c^{-1}\mathbf{q}_c + \mathbf{H}_{ll}\mathbf{L}_l^{-1}\boldsymbol{\phi}_l + \mathbf{H}_{lv}\mathbf{v}_v + \mathbf{H}_{li}\mathbf{i}_i - \dot{\bar{\mathbf{i}}}_i \tag{5.5-33}$$

where $\mathbf{H}$ matrices with appropriate subscripts are defined in equations (5.3-36) through (5.3-43) and

$$\dot{\bar{\mathbf{v}}}_v \triangleq \frac{d}{dt}[\mathbf{F}_{cs}\mathbf{C}_s\mathbf{F}_{vs}^T\mathbf{v}_v] \tag{5.5-34}$$

$$\dot{\bar{\mathbf{i}}}_l \triangleq \frac{d}{dt}[\mathbf{F}_{yl}^T\mathbf{L}_y\mathbf{F}_{yi}\mathbf{i}_i] \tag{5.5-35}$$

To write the left-hand side of (5.5-32) and (5.5-33) in terms of the state variables $\mathbf{q}_c$ and $\boldsymbol{\phi}_l$, we rewrite (5.5-26) and (5.5-27) in the form

$$\mathbf{q} = \mathbf{M}\mathbf{q}_c \tag{5.5-36}$$
$$\boldsymbol{\phi} = \mathbf{P}\boldsymbol{\phi}_l \tag{5.5-37}$$

where
$$\mathbf{M} \triangleq \mathbf{1} + \mathbf{F}_{cs}\mathbf{C}_s\mathbf{F}_{cs}^T\mathbf{C}_c^{-1} \tag{5.5-38}$$

and
$$\mathbf{P} \triangleq \mathbf{1} + \mathbf{F}_{yl}^T\mathbf{L}_y\mathbf{F}_{yl}\mathbf{L}_l^{-1} \tag{5.5-39}$$

Hence,
$$\dot{\mathbf{q}} = \dot{\mathbf{M}}\mathbf{q}_c + \mathbf{M}\dot{\mathbf{q}}_c \tag{5.5-40}$$

and
$$\dot{\boldsymbol{\phi}} = \dot{\mathbf{p}}\boldsymbol{\phi}_l + \mathbf{P}\dot{\boldsymbol{\phi}}_l \tag{5.5-41}$$

Now, if $\mathbf{M}$ and $\mathbf{P}$ are nonsingular, we can replace $\mathbf{q}$ and $\boldsymbol{\phi}$ in terms of $\mathbf{q}_c$ and $\boldsymbol{\phi}_l$ in (5.5-32) and (5.5-33) and write the results in the form

$$\begin{aligned}\dot{\mathbf{q}}_c = {} & \mathbf{M}^{-1}(-\dot{\mathbf{M}} + \mathbf{H}_{cc}\mathbf{C}_c^{-1})\mathbf{q}_c + \mathbf{M}^{-1}\mathbf{H}_{cl}\mathbf{L}_l^{-1}\boldsymbol{\phi}_l + \mathbf{M}^{-1}\mathbf{H}_{cv}\mathbf{v}_v \\ & + \mathbf{M}^{-1}\mathbf{H}_{ci}\mathbf{i}_i - \mathbf{M}^{-1}\dot{\bar{\mathbf{v}}}_v\end{aligned} \tag{5.5-42}$$

$$\begin{aligned}\dot{\boldsymbol{\phi}}_l = {} & \mathbf{P}^{-1}\mathbf{H}_{lc}\mathbf{C}_c^{-1}\mathbf{q}_c - \mathbf{P}^{-1}(\dot{\mathbf{P}} - \mathbf{H}_{ll}\mathbf{L}_l^{-1})\boldsymbol{\phi}_l + \mathbf{P}^{-1}\mathbf{H}_{lv}\mathbf{v}_v \\ & + \mathbf{P}^{-1}\mathbf{H}_{li}\mathbf{i}_i - \mathbf{P}^{-1}\dot{\bar{\mathbf{i}}}_i\end{aligned} \tag{5.5-43}$$

Writing these equations in matrix form yields

$$\begin{bmatrix}\dot{\mathbf{q}}_c\\ \dot{\boldsymbol{\phi}}_l\end{bmatrix}=\begin{bmatrix}\mathbf{M}^{-1}(-\dot{\mathbf{M}}+\mathbf{H}_{cc}\mathbf{C}_c^{-1}) & \mathbf{M}^{-1}\mathbf{H}_{cl}\mathbf{L}_l^{-1}\\ \mathbf{P}^{-1}\mathbf{H}_{lc}\mathbf{C}_c^{-1} & -\mathbf{P}^{-1}(\dot{\mathbf{P}}-\mathbf{H}_{ll}\mathbf{L}_l^{-1})\end{bmatrix}\begin{bmatrix}\mathbf{q}_c\\ \boldsymbol{\phi}_l\end{bmatrix}$$
$$+\begin{bmatrix}\mathbf{M}^{-1}\mathbf{H}_{cv} & \mathbf{M}^{-1}\mathbf{H}_{ci} & -\mathbf{M}^{-1} & \mathbf{0}\\ \mathbf{P}^{-1}\mathbf{H}_{lv} & \mathbf{P}^{-1}\mathbf{H}_{li} & \mathbf{0} & -\mathbf{P}^{-1}\end{bmatrix}\begin{bmatrix}\mathbf{v}_v\\ \mathbf{i}_i\\ \dot{\tilde{\mathbf{v}}}_v\\ \dot{\tilde{\mathbf{i}}}_i\end{bmatrix} \tag{5.5-44}$$

which is in the form

$$\dot{\mathbf{x}}(t)=\mathbf{A}(t)\mathbf{x}(t)+\mathbf{B}(t)\mathbf{u}(t) \tag{5.5-45}$$

where

$$\mathbf{x}(t)\triangleq[\mathbf{q}_c \quad \boldsymbol{\phi}_l]^T \tag{5.5-46}$$

$$\mathbf{u}(t)\triangleq[\mathbf{v}_v \quad \mathbf{i}_i \quad \dot{\tilde{\mathbf{v}}}_v \quad \dot{\tilde{\mathbf{i}}}_i]^T \tag{5.5-47}$$

$$\mathbf{A}(t)\triangleq\begin{bmatrix}\mathbf{M}^{-1}(-\dot{\mathbf{M}}+\mathbf{H}_{cc}\mathbf{C}_c^{-1}) & \mathbf{M}^{-1}\mathbf{H}_{cl}\mathbf{L}_l^{-1}\\ \mathbf{P}^{-1}\mathbf{H}_{lc}\mathbf{C}_c^{-1} & -\mathbf{P}^{-1}(\dot{\mathbf{P}}-\mathbf{H}_{ll}\mathbf{L}_l^{-1})\end{bmatrix} \tag{5.5-48}$$

and

$$\mathbf{B}(t)\triangleq\begin{bmatrix}\mathbf{M}^{-1}\mathbf{H}_{cv} & \mathbf{M}^{-1}\mathbf{H}_{ci} & -\mathbf{M}^{-1} & \mathbf{0}\\ \mathbf{P}^{-1}\mathbf{H}_{lv} & \mathbf{P}^{-1}\mathbf{H}_{li} & \mathbf{0} & -\mathbf{P}^{-1}\end{bmatrix} \tag{5.5-49}$$

Note that if the network under study is proper in the sense of Definition 5.3-1, then $\mathbf{F}_{cs}=\mathbf{0}$ and $\mathbf{F}_{\gamma l}=\mathbf{0}$; hence, by (5.5-34), (5.5-35), (5.5-38), and (5.5-39), we can write

$$\mathbf{M}=\mathbf{1},\quad \mathbf{P}=\mathbf{1},\quad \dot{\tilde{\mathbf{v}}}_v=\mathbf{0},\quad \dot{\tilde{\mathbf{i}}}_i=\mathbf{0},\quad \dot{\mathbf{M}}=\mathbf{0},\quad \text{and}\quad \dot{\mathbf{P}}=\mathbf{0}$$

Consequently, $\mathbf{A}(t)$ and $\mathbf{B}(t)$ just defined will become identical to those defined in (5.3-45) and (5.3-46). Furthermore, $\mathbf{u}(t)$ will be reduced to $\mathbf{u}(t)=[\mathbf{v}_v \quad \mathbf{i}_i]^T$, as in the case of proper networks. It is interesting to see that in $\mathbf{u}(t)$ defined by (5.5-47) all the independent sources *and* their derivatives are present.

Linear Time-Invariant Networks. If the network under consideration is linear time invariant, it is more convenient to choose the state variables as follows:

$$\mathbf{x}(t)=[\mathbf{v}_c \quad \mathbf{i}_l]^T \tag{5.5-50}$$

where $\mathbf{v}_c$ is the tree branch capacitor voltages and $\mathbf{i}_l$ is the chord inductor currents. We can then write

$$\mathbf{q}_c=\mathbf{C}_c\mathbf{v}_c \tag{5.5-51}$$

and

$$\boldsymbol{\phi}_l=\mathbf{L}_l\mathbf{i}_l \tag{5.5-52}$$

Since $\mathbf{C}_c$ and $\mathbf{L}_l$ are assumed to be time-invariant,

$$\dot{\mathbf{M}}=\mathbf{0} \tag{5.5-53}$$

$$\dot{\mathbf{P}}=\mathbf{0} \tag{5.5-54}$$

$$\dot{\tilde{\mathbf{v}}}_v=\mathbf{F}_{cs}\mathbf{C}_s\mathbf{F}_{vs}^T\dot{\mathbf{v}}_v \tag{5.5-55}$$

$$\dot{\tilde{\mathbf{i}}}_i=\mathbf{F}_{\gamma l}^T\mathbf{L}_\gamma\mathbf{F}_{\gamma i}\dot{\mathbf{i}}_i \tag{5.5-56}$$

Furthermore, since $\mathbf{C}_c$ and $\mathbf{L}_l$ are assumed to be nonsingular, (5.5-42) and (5.5-43) can be written as

$$\dot{\mathbf{v}}_c = \mathbf{C}_c^{-1}\mathbf{M}^{-1}\mathbf{H}_{cc}\mathbf{v}_c + \mathbf{C}_c^{-1}\mathbf{M}^{-1}\mathbf{H}_{cl}\mathbf{i}_l + \mathbf{C}_c^{-1}\mathbf{M}^{-1}\mathbf{H}_{cv}\mathbf{v}_v + \mathbf{C}_c^{-1}\mathbf{M}^{-1}\mathbf{H}_{ci}\mathbf{i}_i - \mathbf{C}_c^{-1}\mathbf{M}^{-1}\mathbf{F}_{cs}\mathbf{C}_s\mathbf{F}_{vs}^T\dot{\mathbf{v}}_v$$

$$\dot{\mathbf{i}}_l = \mathbf{L}_l^{-1}\mathbf{P}^{-1}\mathbf{H}_{lc}\mathbf{v}_c + \mathbf{L}_l^{-1}\mathbf{P}^{-1}\mathbf{H}_{ll}\mathbf{i}_l + \mathbf{L}_l^{-1}\mathbf{P}^{-1}\mathbf{H}_{lv}\mathbf{v}_v + \mathbf{L}_l^{-1}\mathbf{P}^{-1}\mathbf{H}_{li}\mathbf{i}_i - \mathbf{L}_l^{-1}\mathbf{P}^{-1}\mathbf{F}_{yl}^T\mathbf{L}_y\mathbf{F}_{yi}\dot{\mathbf{i}}_i$$

Putting these equations in matrix form, we get

$$\mathbf{x}(t) = \mathbf{A}\mathbf{x}(t) + \mathbf{B}\mathbf{u}(t) \tag{5.5-57}$$

where

$$\mathbf{A} = \begin{bmatrix} \mathbf{C}_c^{-1}\mathbf{M}^{-1}\mathbf{H}_{cc} & \mathbf{C}_c^{-1}\mathbf{M}^{-1}\mathbf{H}_{cl} \\ \mathbf{L}_l^{-1}\mathbf{P}^{-1}\mathbf{H}_{lc} & \mathbf{L}_l^{-1}\mathbf{P}^{-1}\mathbf{H}_{ll} \end{bmatrix} \tag{5.5-58}$$

$$\mathbf{B} = \begin{bmatrix} \mathbf{C}_c^{-1}\mathbf{M}^{-1}\mathbf{H}_{cv} & \mathbf{C}_c^{-1}\mathbf{M}^{-1}\mathbf{H}_{ci} & -\mathbf{C}_c^{-1}\mathbf{M}^{-1}\mathbf{F}_{cs}\mathbf{C}_s\mathbf{F}_{vs}^T & \mathbf{0} \\ \mathbf{L}_l^{-1}\mathbf{P}^{-1}\mathbf{H}_{lv} & \mathbf{L}_l^{-1}\mathbf{P}^{-1}\mathbf{H}_{li} & \mathbf{0} & -\mathbf{L}_l^{-1}\mathbf{P}^{-1}\mathbf{F}_{yl}^T\mathbf{L}_y\mathbf{F}_{yi} \end{bmatrix} \tag{5.5-59}$$

and

$$\mathbf{u}(t) = [\mathbf{v}_v \quad \mathbf{i}_i \quad \dot{\mathbf{v}}_v \quad \dot{\mathbf{i}}_i]^T \tag{5.5-60}$$

Let us now work out an example to illustrate the procedure.

Example 5.5-2. Consider the network shown in Fig. 5.5-3. For simplicity, let us first assume that all the elements are linear and time invariant. It is desired to write the state equations of this network in the normal form.

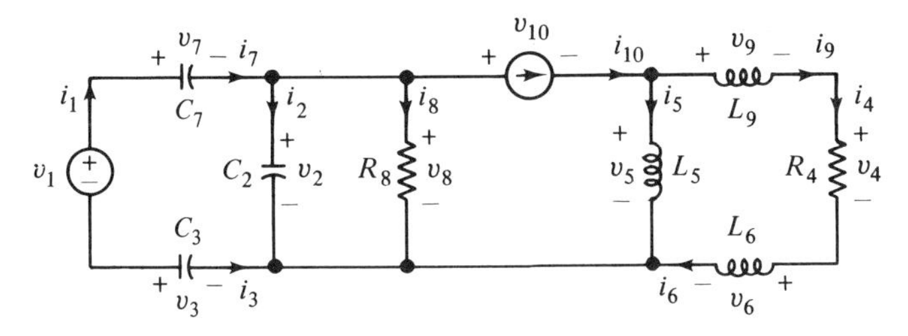

Figure 5.5-3 Network considered in Example 5.5-2.

Solution:

Notice that there is a capacitor-only loop and two inductor-only cutsets; hence, this network is not proper and we must use the procedure outlined in the present section to write the state equations in the normal form. There are six energy-storing elements in the network; that is, $n_{lc} = 6$. From the figure it is obvious that $n_c = 1$ and $n_l = 2$; hence, using (5.4-1), the order of complexity of the network is

$$n = 6 - (2 + 1) = 3$$

Therefore, the state equations will be in the form of three linearly independent first-order differential equations.

Choose a normal tree as in Fig. 5.5-4. The state variables are then the tree branch capacitor voltages and the chord inductor currents; that is,

$$\mathbf{x}^T(t) = [v_2 \quad v_3 \quad i_9]$$

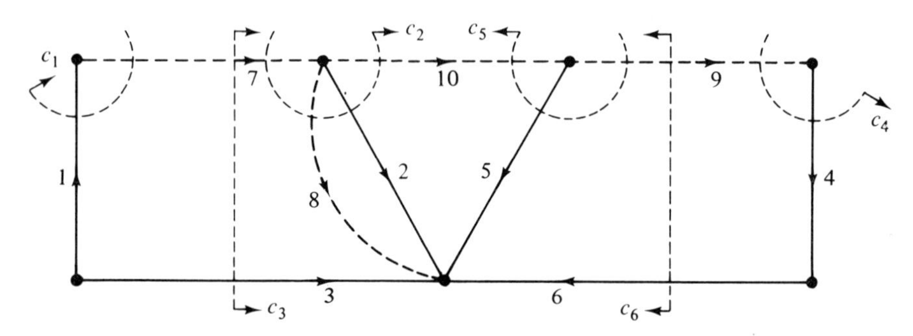

Figure 5.5-4 Normal tree corresponding to the network of Fig. 5.5-3.

The branch voltage and current vectors can then be partitioned as

$$\mathbf{v}_b = [v_1 \mid v_2 \quad v_3 \mid v_4 \mid v_5 \quad v_6 \mid v_7 \mid v_8 \mid v_9 \mid v_{10}]^T$$

and
$$\mathbf{i}_b = [i_1 \mid i_2 \quad i_3 \mid i_4 \mid i_5 \quad i_6 \mid i_7 \mid i_8 \mid i_9 \mid i_{10}]^T$$

The fundamental cutset matrix $\mathbf{Q}_f$ of the network corresponding to the normal tree chosen in Fig. 5.5-4 will be

$$\mathbf{Q}_f = \underbrace{\left[\begin{array}{cccccc|cccc} 1 & 0 & 0 & 0 & 0 & 0 & -1 & 0 & 0 & 0 \\ 0 & 1 & 0 & 0 & 0 & 0 & -1 & 1 & 0 & 1 \\ 0 & 0 & 1 & 0 & 0 & 0 & 1 & 0 & 0 & 0 \\ 0 & 0 & 0 & 1 & 0 & 0 & 0 & 0 & -1 & 0 \\ 0 & 0 & 0 & 0 & 1 & 0 & 0 & 0 & 1 & -1 \\ 0 & 0 & 0 & 0 & 0 & 1 & 0 & 0 & -1 & 0 \end{array}\right]}_{\text{identity matrix} \quad | \quad \mathbf{F} \text{ matrix}}$$

Partitioning $\mathbf{F}$ according to (5.5-3) yields

$$\left[\begin{array}{c|c|c|c} \mathbf{F}_{vs} & \mathbf{F}_{vr} & \mathbf{F}_{vl} & \mathbf{F}_{vi} \\ \hline \mathbf{F}_{cs} & \mathbf{F}_{cr} & \mathbf{F}_{cl} & \mathbf{F}_{ci} \\ \hline \mathbf{F}_{gs} & \mathbf{F}_{gr} & \mathbf{F}_{gl} & \mathbf{F}_{gi} \\ \hline \mathbf{F}_{\gamma s} & \mathbf{F}_{\gamma r} & \mathbf{F}_{\gamma l} & \mathbf{F}_{\gamma i} \end{array}\right] = \left[\begin{array}{c|c|c|c} -1 & 0 & 0 & 0 \\ \hline -1 & 1 & 0 & 1 \\ 1 & 0 & 0 & 0 \\ \hline 0 & 0 & -1 & 0 \\ \hline 0 & 0 & 1 & -1 \\ 0 & 0 & -1 & 0 \end{array}\right]$$

As we mentioned previously, $\mathbf{F}_{gs} = \mathbf{0}$, $\mathbf{F}_{\gamma s} = \mathbf{0}$, and $\mathbf{F}_{\gamma r} = \mathbf{0}$. From Fig. 5.5-3 we have

$$\mathbf{G}_g = \frac{1}{R_4} \quad \text{and} \quad \mathbf{R}_r = R_8$$

Also, since $\mathbf{F}_{gr} = \mathbf{0}$, from (5.3-33) and (5.3-34) we obtain

$$\mathbf{G} = \frac{1}{R_4} \quad \text{and} \quad \mathbf{R} = R_8$$

Furthermore, all the $\mathbf{H}$ matrices defined in (5.3-36) through (5.3-43) can be seen to be

$$\mathbf{H}_{cc} = \begin{bmatrix} -\frac{1}{R_8} & 0 \\ 0 & 0 \end{bmatrix}, \quad \mathbf{H}_{cl} = \begin{bmatrix} 0 \\ 0 \end{bmatrix}, \quad \mathbf{H}_{ci} = \begin{bmatrix} -1 \\ 0 \end{bmatrix}, \quad \mathbf{H}_{cv} = \begin{bmatrix} 0 \\ 0 \end{bmatrix}$$

$$\mathbf{H}_{ll} = -R_4, \qquad \mathbf{H}_{lv} = 0, \qquad \mathbf{H}_{li} = 0, \quad \text{and} \quad \mathbf{H}_{lc} = [0 \quad 0]$$

From the network we can write

$$\mathbf{C}_c = \begin{bmatrix} C_2 & 0 \\ 0 & C_3 \end{bmatrix} \quad \text{and} \quad \mathbf{L}_l = L_9, \quad \mathbf{C}_s = C_7, \quad \mathbf{L}_\gamma = \begin{bmatrix} L_5 & 0 \\ 0 & L_6 \end{bmatrix}$$

From (5.5-38) and (5.5-39) and the equations just derived, we have

$$\mathbf{M} = \begin{bmatrix} 1 + \frac{C_7}{C_2} & -\frac{C_7}{C_3} \\ -\frac{C_7}{C_2} & 1 + \frac{C_7}{C_3} \end{bmatrix}$$

and

$$\mathbf{M}^{-1} = \frac{C_2C_3}{C_2C_3 + C_2C_7 + C_3C_7} \begin{bmatrix} 1 + \frac{C_7}{C_3} & \frac{C_7}{C_3} \\ \frac{C_7}{C_2} & 1 + \frac{C_7}{C_2} \end{bmatrix}$$

Also

$$\mathbf{P} = \frac{L_5 + L_6 + L_9}{L_9} \quad \text{and} \quad \mathbf{P}^{-1} = \frac{L_9}{L_5 + L_6 + L_9}$$

Then using (5.5-58), after some simple manipulations we get

$$\mathbf{A} = \begin{bmatrix} \frac{-(C_7 + C_3)}{R_8(C_2C_3 + C_2C_7 + C_3C_7)} & 0 & 0 \\ \frac{-C_7}{R_8(C_2C_3 + C_2C_7 + C_3C_7)} & 0 & 0 \\ 0 & 0 & \frac{-R_4}{L_5 + L_6 + L_9} \end{bmatrix}$$

Also, using (5.5-59) and the relations just derived, we get

$$\mathbf{B} = \begin{bmatrix} 0 & \dfrac{-(C_3 + C_7)}{C_2C_3 + C_2C_7 + C_3C_7} & -\dfrac{C_3C_7}{C_2C_3 + C_2C_7 + C_3C_7} & 0 \\ 0 & \dfrac{-C_7}{C_2C_3 + C_2C_7 + C_3C_7} & \dfrac{C_2C_7}{C_2C_3 + C_2C_7 + C_3C_7} & 0 \\ 0 & 0 & 0 & \dfrac{L_5}{L_5 + L_6 + L_9} \end{bmatrix}$$

and

$$\mathbf{u}(t) = [v_1 \quad i_{10} \quad \dot{v}_1 \quad \dot{i}_{10}]^T$$

This result can easily be checked by writing the state equations of the network under study by inspection.

Let us now assign some specific values to the network elements and get the numerical values of **A** and **B**. Let $C_2 = 1$ F, $C_3 = C_7 = \frac{1}{2}$ F, $L_5 = L_6 = L_9 = \frac{1}{3}$ H, $R_4 = R_8 = 1\ \Omega$, $v_1(t) = \sin t$, and $i_{10}(t) = \cos t$. Note that since the first column of the **B** matrix is zero it can be eliminated together with the first element of $\mathbf{u}(t)$; we then obtain

$$\begin{bmatrix} \dot{x}_1 \\ \dot{x}_2 \\ \dot{x}_3 \end{bmatrix} = \begin{bmatrix} -\frac{4}{5} & 0 & 0 \\ -\frac{2}{5} & 0 & 0 \\ 0 & 0 & -1 \end{bmatrix} \begin{bmatrix} x_1 \\ x_2 \\ x_3 \end{bmatrix} + \begin{bmatrix} -\frac{5}{4} & -\frac{1}{5} & 0 \\ -\frac{2}{5} & \frac{2}{5} & 0 \\ 0 & 0 & \frac{1}{3} \end{bmatrix} \begin{bmatrix} \cos t \\ \cos t \\ -\sin t \end{bmatrix}$$

Along the same lines, a similar (but more elaborate) formulation can be obtained for networks containing nonlinear elements and dependent sources. The formal derivation of such formulation is given in a number of references [5, 11, 12]. In this chapter we work out some examples to show how the state equation of such networks can be written by inspection.

Example 5.5-3. Consider the active filter shown in Fig. 5.5-5. The voltage-controlled voltage source is the equivalent representation of an operational amplifier, indicating that the voltage at node 2 is kv_1. The nonlinear resistor is assumed to be voltage controlled with the *i-v* relation $i = f(v)$. We wish to write the state representation of this network.

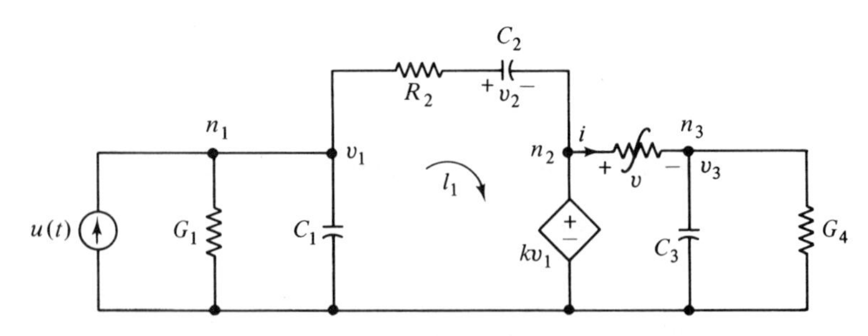

Figure 5.5-5 Nonlinear active network considered in Example 5.5-3.

Since there are no capacitor-only loops and no inductor-only cutsets, we choose the state variables to be the capacitor voltages; that is, $x_1 = v_1$, $x_2 = v_2$, and $x_3 = v_3$. Writing the node equations at nodes n_1 and n_3 and the loop equation for loop l_1, we get

$$u(t) = C_1\dot{x}_1 + G_1x_1 + C_2\dot{x}_2$$
$$f(kx_1 - x_3) = C_3\dot{x}_3 + G_4x_3$$
$$x_1 = R_2C_2\dot{x}_2 + x_2 + kx_1$$

Rearranging these equations, we obtain the state equations in the standard form:

$$\dot{x}_1 = -\left(\frac{G_1}{C_1} + \frac{1-k}{C_1R_2}\right)x_1 + \frac{1}{R_2C_1}x_2 + \frac{1}{C_1}u$$
$$\dot{x}_2 = \frac{1-k}{R_2C_2}x_1 - \frac{1}{R_2C_2}x_2$$
$$\dot{x}_3 = \frac{1}{C_3}f(kx_1 - x_3) - \frac{G_4}{C_3}x_3$$

Example 5.5-4. Consider the same network discussed in the last example but this time assume that $R_2 = 0$. With this assumption loop l_1 will become a capacitor-only loop, and we get

$$x_1 = x_2 + kx_1$$

That is, x_2 can be obtained from x_1. Consequently, we choose x_1 and x_3 as the linear independent state variables. Then writing the node equations at n_1 and n_3 and using the previous equation, we obtain

$$\dot{x}_1 = -\frac{G_1}{C_1 + C_2 - C_2k}x_1 + \frac{1}{C_1 + C_2 - C_2k}u(t)$$
$$\dot{x}_3 = \frac{1}{C_3}f(kx_1 - x_3) - \frac{G_4}{C_3}x_3$$

which are the desired state equations in the normal form.

REFERENCES

[1] Bashkow, T. R., "The **A** Matrix, a New Network Description," *IRE Trans. Circuit Theory*, **CT-4**, No. 3 (1957), 117–120.

[2] Bryant, P. R., "The Order of Complexity of Electrical Networks," *Proc. Inst. Elec. Engrs.* (*London*), **106C** (1959), 174–188.

[3] Bryant, P. R., "The Explicit Form of Bashkow's **A** Matrix," *IRE Trans. Circuit Theory*, **CT-9**, No. 3 (1962), 303–306.

[4] Kuh, E. S., and R. A. Rohrer, "The State Variable Approach to Network Analysis," *Proc. IEEE*, **53**, No. 7 (1965), 672–686.

[5] DERVISHOGLU, A., "Bashkow's **A**-matrix for Active *RLC* Networks," *IEEE Trans. Circuit Theory*, **CT-11** (Sept. 1964), 404–406.

[6] DESOER, C. A., and J. KATZENELSON, "Nonlinear *RLC* Networks," *Bell System Tech. J.*, **44** (Jan. 1965), 161–198.

[7] CHUA, L. O., and R. A. ROHRER, "On the Dynamic Equations of a Class of Nonlinear *RLC* Networks," *IEEE Trans. Circuit Theory*, **CT-12**, No. 4 (1965), 475–489.

[8] STERN, T. E., "On the Equations of Nonlinear Networks," *IEEE Trans. on Circuit Theory*, **CT-13**, No. 1, (1966), 74–81.

[9] BRAYTON, R. K., and J. K. MOSER, "A Theory of Nonlinear Networks, (I and II)," *Quart. Appl. Math*, **22** (April 1964) 1–33; (July 1964), 81–104.

[10] VARAIYA, P. P., and R. LIU, "Normal Form and Stability of a Class of Coupled Nonlinear Networks," *IEEE Trans. Circuit Theory*, **CT-13**, No. 14 (1966), 413–418.

[11] PURSLOW, E. J., "Solvability and Analysis of Linear Active Networks by Use of the State Equations," *IEEE Trans. Circuit Theory*, **CT-17**, No. 4 (1970), 469–475.

[12] DESOER, C. A., and E. S. KUH, *Basic Circuit Theory*, McGraw-Hill Book Company, New York, 1969.

[13] STERN, T. E., *Theory of Nonlinear Networks and Systems*, Addison-Wesley Publishing Company, Inc., Reading, Mass., 1965.

[14] CHUA, L. O., *Introduction to Nonlinear Network Theory*, McGraw-Hill Book Company, New York, 1969.

[15] BALABANIAN, N., T. A. BICKART, and S. SESHU, *Electrical Network Theory*, John Wiley & Sons, Inc., New York, 1968.

[16] ROHRER, R. A., *Circuit Theory: An Introduction to the State Variable Approach*, McGraw-Hill Book Company, New York, 1970.

[17] CHEN, C. T., *Introduction to Linear System Theory*, Holt, Rinehart and Winston, Inc., New York, 1970.

[18] CALAHAN, D. A., *Computer-Aided Network Design*, McGraw-Hill Book Company, New York, 1968.

PROBLEMS

5.1 Obtain the state representation of the single-input single-output networks whose transfer functions are

(a) $H(s) = \dfrac{s^2 + 2s + 3}{s^3 + 3s^2 + 2s + 1}$

(b) $H(s) = \dfrac{s^3 + 2}{s^3 + 3s + 2}$

(c) $H(s) = \dfrac{1}{s^4 + 2s^3 + 5s^2 + s + 1}$

5.2 Use a set of auxiliary variables to put the following integro-differential equations in the form of a set of first-order differential equations.

(a) $\dfrac{d^2y}{dt^2} + a_1\dfrac{dy}{dt} + a_0 y = b_1\dfrac{du}{dt} + b_0 u$

(b) $\dfrac{d^3y}{dt^3} + a_2\dfrac{d^2y}{dt^2} + a_0 y + \displaystyle\int_0^t y(\tau)\,d\tau = b_1\dfrac{du}{dt}$

5.3 Consider the linear time-invariant network shown in Fig. P5.3. Take the charges and the fluxes as the state variables and obtain the normal form of the state equations by inspection.

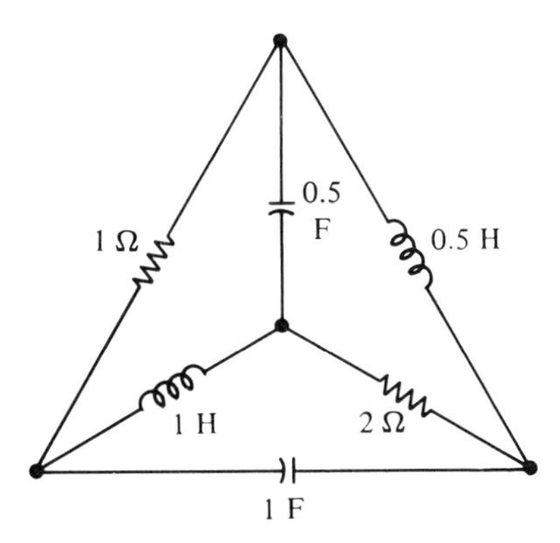

Figure P5.3

5.4 Repeat Problem 5.3, but this time take capacitor voltages and inductor currents as the state variables.

5.5 Consider the network shown in Fig. P5.5. Let $C_1 = 1$ F, $C_2 = 2$ F, $C_3 = 0.5$ F, $L_1 = 1$ H, $L_2 = 0.5$ H, $v_1 = \sin t$, and $i_1 = i_2 = 2\cos t$. Choose the capacitor charges and the inductor fluxes to write the normal form of the state equations of the network by inspection.

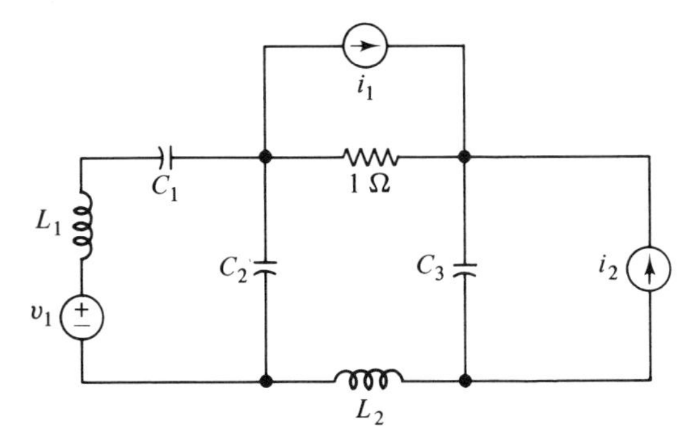

Figure P5.5

5.6 Repeat Problem 5.5, but this time assume that the capacitors are time varying with capacitances

$$C_1(t) = C_2(t) = C_3(t) = 1 + 0.5 \sin t$$

5.7 Repeat Problem 5.5, but this time assume that the inductors are nonlinear with characteristics

$$i_{L_1} = \phi_{L_1}^3 \quad \text{and} \quad i_{L_2} = \phi_{L_2} + \tanh \phi_{L_2}$$

5.8 Consider the network shown in Fig. P5.8. Let $R_1 = R_2 = R_3 = R$, $C_1 = C_2 = C$, and $L_1 = L_2 = L$.

(a) Write the normal form of the state equations of this network by inspection.

(b) Let v_1 be the input, and the voltage across the terminating resistor R_3 be the output. Write the input–output state equations of the network.

(c) Let $R = 1\ \Omega$, $C = 1$ F, $L = 1$ H, and $v_1 = \sin t$; find the steady-state response of this network.

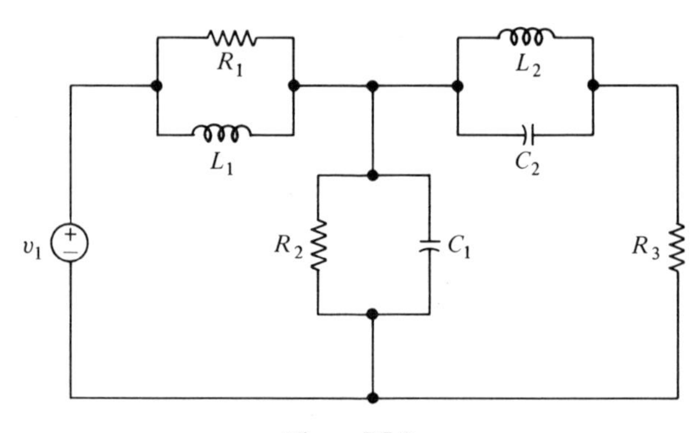

Figure P5.8

5.9 Consider the network shown in Fig. P5.8. Let all the elements be nonlinear time invariant with monotonically increasing characteristics. Choose the capacitor charges and inductor fluxes as the state variables and write the normal form of the state equations by inspection.

5.10 Consider the network shown in Fig. P5.10. Let the mutual inductance M be 0.5 H; choose the capacitor voltages and the inductor currents as the state variable to write the normal form of the state equations by inspection.

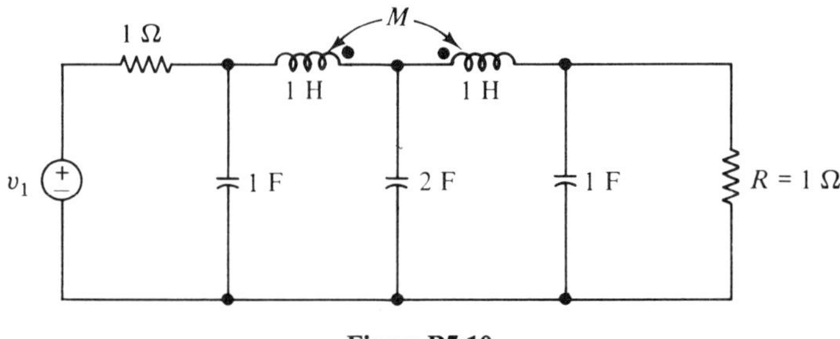

Figure P5.10

5.11 Consider the network shown in Fig. P5.10; let the input be the voltage source v_1 and the output be the voltage across the terminating resistor R. Use the results of Problem 5.10 to write the input–output state equations of this network.

5.12 Consider the network shown in Fig. P5.12. Let the nonlinear capacitor be charge controlled with the characteristics $v_2 = f_2(q_2)$. Write the normal form of the state equations of this nonlinear time-varying network by inspection.

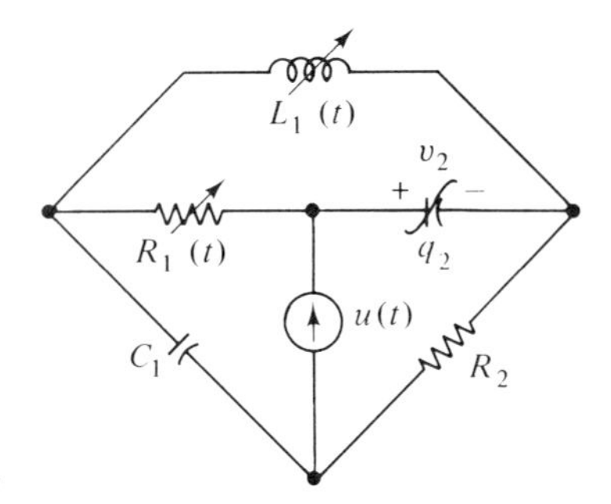

Figure P5.12

5.13 Consider the network shown in Fig. P5.12. Let the input be the current source $i(t)$ and the output be the current through R_2. Use the results of Problem 5.12 to obtain the input–output state equations of this network.

5.14 Consider the network shown in Fig. P5.14. Let the capacitors be charge controlled with characteristics

$$v_1 = f_1(q_1) \quad \text{and} \quad v_2 = f_2(q_2)$$

Write the normal form of the state equations of this network by inspection.

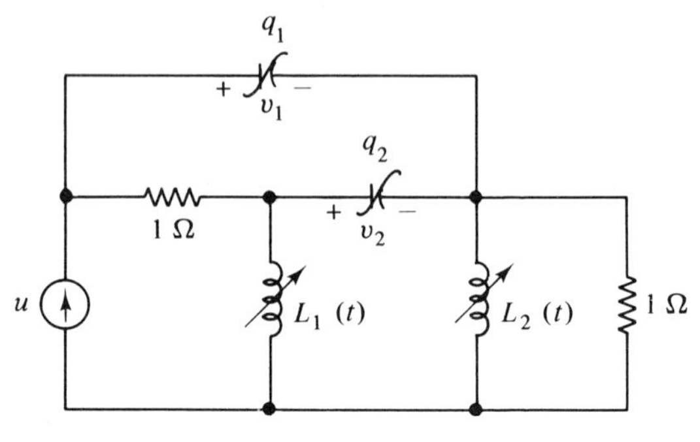

Figure P5.14

5.15 Consider the network shown in Fig. P5.15. Write the normal form of the state equations of this network by inspection.

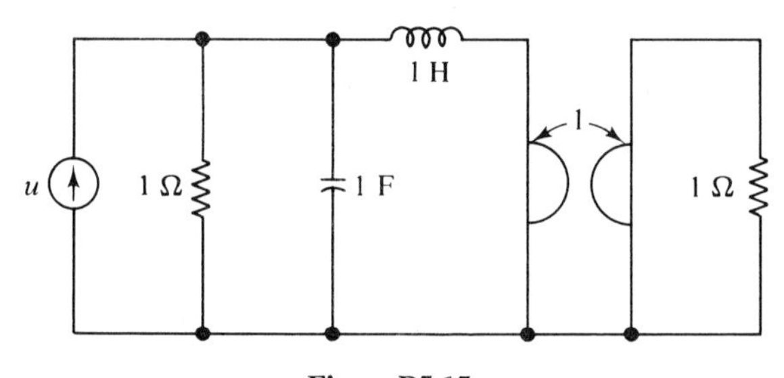

Figure P5.15

5.16 Draw a proper tree for each of the networks given in Figs. P5.3, P5.5, P5.8, P5.10, P5.12, and P5.14.

5.17 Use equations (5.3-29) and (5.3-30) to derive (5.3-31) and (5.3-32).

5.18 Write down the detailed derivation of equation (5.3-35).

5.19 Repeat Problem 5.3 using (5.3-35).

5.20 Write the normal form of the state equations of the network shown in Fig. P 5.3 by choosing the capacitor voltages and inductor currents as state variables and using (5.3-48).

5.21 Repeat Problem 5.20 for the network shown in Fig. P5.5.

5.22 Use capacitor charges and inductor fluxes and (5.3-35) to write the state equations of the network shown in Fig. P5.5.

5.23 Use (5.3-35) to write the state equations of the network shown in Fig. P5.8.

5.24 Consider the network shown in Fig. P5.24. Choose the capacitor voltages and inductor currents to write the state equations in the normal form.

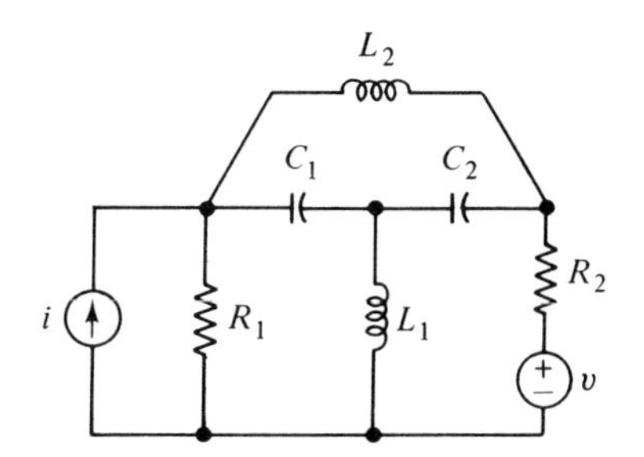

Figure P5.24

5.25 Determine the order of complexity of the network shown in Fig. P5.25.

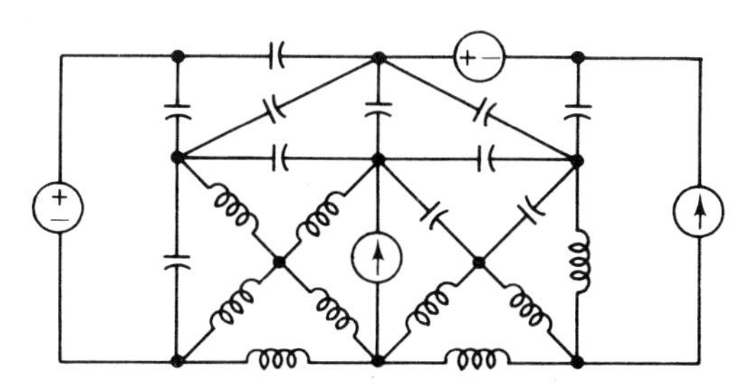

Figure P5.25

5.26 Consider the network shown in Fig. P5.26; choose the capacitor charges and the inductor fluxes as the state variables to write a complete set of linearly independent state equations by inspection.

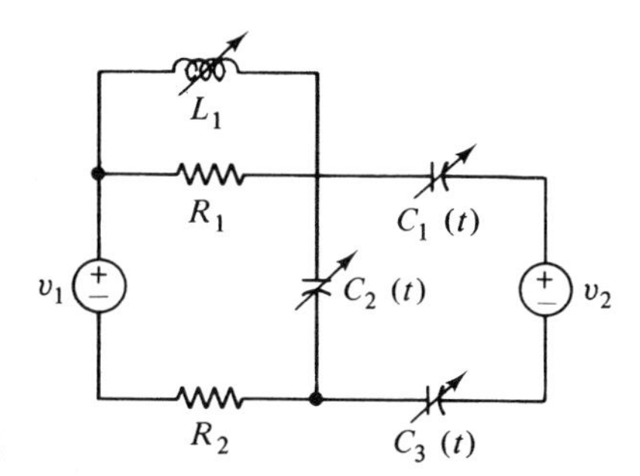

Figure P5.26

5.27 Consider the network shown in Fig. P5.27.

(a) Determine the order of complexity of this network.

(b) Choose appropriate charges and fluxes to write the state equations in the normal form.

(c) Choose appropriate voltages and currents to write the state equations in the normal form.

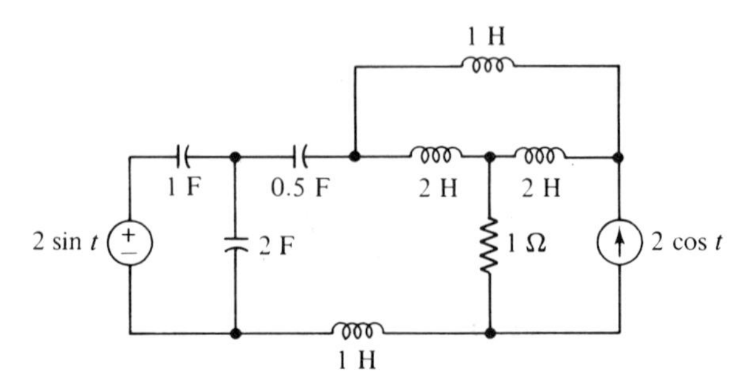

Figure P5.27

5.28 Consider the linear time-invariant network shown in Fig. P5.28.

(a) Determine the order of complexity of this network.

(b) Choose appropriate voltages and currents to write the state equations in the normal form.

(c) Use the Laplace-transform method to solve the state equations. Assume that both sources are step functions.

(d) Determine the voltage across the 1 **Ω** resistor for all $t \geq 0$.

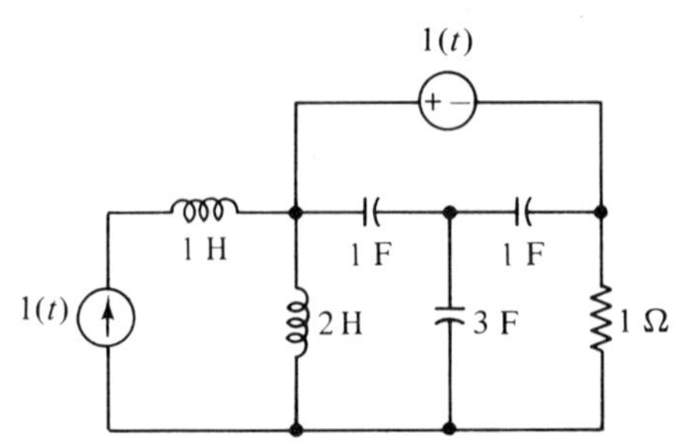

Figure P5.28

5.29 Repeat Problem 5.28 for the network shown in Fig. P5.29. But this time determine the voltage across the terminating 2 **Ω** resistor for $t \geq 0$. The independent voltage source is a unit step function.

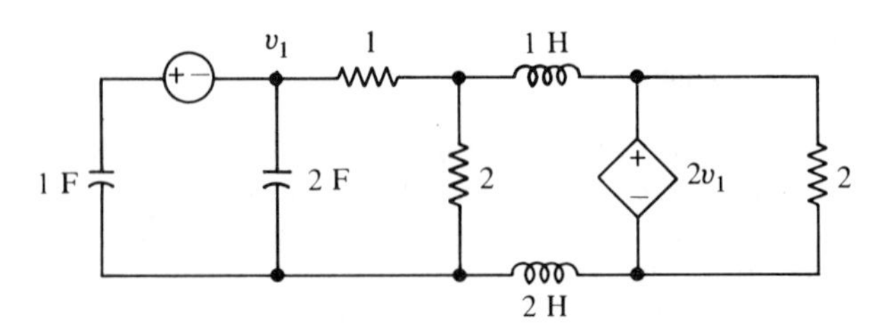

Figure P5.29

5.30 Determine the order of complexity and write the state equation of the network shown in Fig. P5.30.

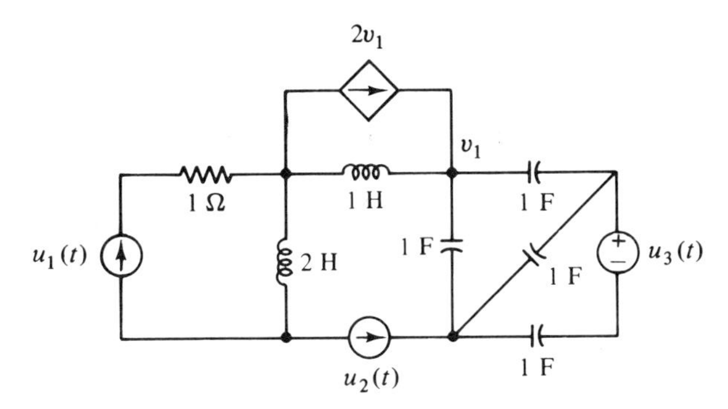

Figure P5.30

Time-Domain Solution of State Equations 6

6.1 INTRODUCTION

In the previous chapter we saw that a network comprised of lumped elements can be represented by a set of first-order differential equations called the state equations. If the network under study contains only linear time-invariant elements, its state equations can be represented in the form

$$\dot{\mathbf{x}}(t) = \mathbf{A}\mathbf{x}(t) + \mathbf{B}\mathbf{u}(t), \qquad \mathbf{x}(t_0) = \mathbf{x}_0 \tag{6.1-1}$$

$$\mathbf{y}(t) = \mathbf{C}\mathbf{x}(t) + \mathbf{D}\mathbf{u}(t) \tag{6.1-2}$$

where $\mathbf{x}(t)$ is an n-column vector denoting the state of the network, $\mathbf{u}(t)$ is an m-column vector representing the sources and possibly their derivatives, $\mathbf{y}(t)$ is a p-column vector denoting the outputs of the network, and $\mathbf{A}$, $\mathbf{B}$, $\mathbf{C}$, and $\mathbf{D}$ are time-invariant matrices with appropriate dimensions. The initial state $\mathbf{x}_0$ represents the initial conditions on the capacitors and inductors.

If the linear network under consideration contains one or more time-varying elements, the coefficient matrices will generally be time dependent and the state equations can be written as

$$\dot{\mathbf{x}}(t) = \mathbf{A}(t)\mathbf{x}(t) + \mathbf{B}(t)\mathbf{u}(t), \qquad \mathbf{x}(t_0) = \mathbf{x}_0 \tag{6.1-3}$$

$$\mathbf{y}(t) = \mathbf{C}(t)\mathbf{x}(t) + \mathbf{D}(t)\mathbf{u}(t) \tag{6.1-4}$$

where the dimensions of the state vector and the coefficient matrices, as in the previous case, depend on the order of complexity of the network, the number of independent sources, and the number of desired outputs. If, on the other hand, the network contains one or more nonlinear elements, its state equations can be written as

$$\dot{\mathbf{x}}(t) = \mathbf{f}(\mathbf{x}(t), \mathbf{u}(t)), \qquad \mathbf{x}(t_0) = \mathbf{x}_0 \tag{6.1-5}$$

$$\mathbf{y}(t) = \mathbf{g}(\mathbf{x}(t), \mathbf{u}(t)) \tag{6.1-6}$$

where $\mathbf{f}(\cdot, \cdot)$ and $\mathbf{g}(\cdot, \cdot)$ are vector-valued functions.

In all three cases, the equations are called the *input–output state equations.* Given the initial conditions and the inputs, we can generally solve the input-state differential equations for the state $\mathbf{x}(t)$. The output, $\mathbf{y}(t)$, can then be obtained by solving a set of purely algebraic equations. The major task, therefore, is the solution of the differential state equations. In this chapter we present a systematic method of solving such differential equations. It will be shown that an analytic solution to the linear time-invariant state equations of the form (6.1-1) can be easily obtained. To obtain the solution of linear time-varying and nonlinear state equations, in general, we have to resort to numerical methods. The existence and uniqueness of the solution of nonlinear differential equations will be considered in the latter part of this chapter.

Before giving a detailed description of the solution of state equations, let us see what we mean by a solution of a differential equation—this is made precise in the following. Consider the general differential equation given in (6.1-5) with the corresponding initial conditions; then

Definition 6.1-1. A unique continuous vector-valued function $\mathbf{x}^*(t)$ defined on the interval $[t_0, t_f]$ is said to be the solution of (6.1-5) if

(a) $\mathbf{x}^*(t)$ satisfies the initial conditions; that is, $\mathbf{x}^*(t_0) = \mathbf{x}_0$, and
(b) $\mathbf{x}^*(t)$ satisfies the differential equation; that is, $\dot{\mathbf{x}}^* = f(\mathbf{x}^*, \mathbf{u})$.

Later in this chapter we shall show that for such a solution to exist and be unique, the function $f(\cdot, \cdot)$ must satisfy certain "smoothness" conditions.

6.2 SOLUTION OF LINEAR TIME-INVARIANT STATE EQUATIONS

Consider the linear time-invariant state equations

$$\dot{\mathbf{x}}(t) = \mathbf{A}\mathbf{x}(t) + \mathbf{B}\mathbf{u}(t), \qquad \mathbf{x}(t_0) = \mathbf{x}_0 \tag{6.2-1}$$

where $\mathbf{x}(t)$ is an n vector, $\mathbf{u}(t)$ is an m vector, and $\mathbf{A}$ and $\mathbf{B}$ are $n \times n$ and $n \times m$ constant matrices, respectively. The solution of (6.2-1) is then given in the following:

Theorem 6.2-1. The solution of the linear time-invariant differential equation (6.2-1) is given by

$$\mathbf{x}(t) = e^{\mathbf{A}(t-t_0)}\mathbf{x}_0 + \int_{t_0}^{t} e^{\mathbf{A}(t-\tau)}\mathbf{B}\mathbf{u}(\tau)\,d\tau \tag{6.2-2}$$

where $e^{\mathbf{A}t}$ is an $n \times n$ matrix called the *state transition matrix* and is defined by

$$e^{\mathbf{A}t} \triangleq \mathbf{1} + \mathbf{A}t + \frac{1}{2!}\mathbf{A}^2t^2 + \frac{1}{3!}\mathbf{A}^3t^3 + \cdots$$

or, in a more compact form,

$$e^{\mathbf{A}t} \triangleq \sum_{k=0}^{\infty} \frac{1}{k!}(\mathbf{A}t)^k \tag{6.2-3}$$

where the infinite series can be shown to be convergent for each finite t [1].

Before proving this theorem, let us investigate some of the useful properties of the state transition matrix just defined. From (6.2-3) it is clear that

$$e^{\mathbf{0}} = \mathbf{1} \tag{6.2-4}$$

where $\mathbf{0}$ denotes the $n \times n$ zero matrix. Now let us take the time derivative of both sides of (6.2-3) to get

$$\begin{aligned}\frac{d}{dt}e^{\mathbf{A}t} &= \mathbf{0} + \mathbf{A} + \mathbf{A}^2t + \frac{1}{2!}\mathbf{A}^3t^2 + \cdots \\ &= \mathbf{A}\left(\mathbf{1} + \mathbf{A}t + \frac{1}{2!}\mathbf{A}^2t^2 + \cdots\right)\end{aligned}$$

or, equivalently,

$$\frac{d}{dt}e^{\mathbf{A}t} = \mathbf{A}e^{\mathbf{A}t} = e^{\mathbf{A}t}\mathbf{A} \tag{6.2-5}$$

Finally, for each finite t_1 and t_2 it can be shown that

$$e^{\mathbf{A}(t_1+t_2)} = e^{\mathbf{A}t_1}\cdot e^{\mathbf{A}t_2} \tag{6.2-6}$$

Exercise 6.2-1. Prove equation (6.2-6) by using an infinite series expansion of the form (6.2-3).

We can now proceed with the proof of the theorem.

Proof

To prove that $\mathbf{x}(t)$ given in (6.2-2) is the solution of (6.2-1), we must show that (a) it satisfies the initial conditions, and (b) it satisfies the differential equation. The first condition is obviously satisfied since

$$\mathbf{x}(t_0) = e^{\mathbf{A}(t_0-t_0)}x_0 + \int_{t_0}^{t_0} e^{\mathbf{A}(t_0-\tau)}\mathbf{B}\mathbf{u}(\tau)\,d\tau = \mathbf{x}_0$$

by (6.2-4) and the fact that the result of the integral is zero, since its lower and upper limits are the same. To check the second condition, we differentiate both sides of (6.2-2) and use (6.2-5) to get

$$\dot{\mathbf{x}}(t) = \mathbf{A}e^{\mathbf{A}(t-t_0)}\mathbf{x}_0 + e^{\mathbf{A}(t-\tau)}\mathbf{B}\mathbf{u}(\tau)|_{\tau=t} + \int_{t_0}^{t} \mathbf{A}e^{\mathbf{A}(t-\tau)}\mathbf{B}\mathbf{u}(\tau)\,d\tau$$

or

$$\dot{\mathbf{x}}(t) = \mathbf{A}[e^{\mathbf{A}(t-t_0)}\mathbf{x}_0 + \int_{t_0}^{t} e^{\mathbf{A}(t-\tau)}\mathbf{B}\mathbf{u}(\tau)\,d\tau] + \mathbf{B}\mathbf{u}(t)$$

The expression inside the bracket is identical to the right-hand side of (6.2-2); hence, we obtain

$$\dot{\mathbf{x}}(t) = \mathbf{A}\mathbf{x}(t) + \mathbf{B}\mathbf{u}(t)$$

This proves the theorem.

The response or the output of the linear time-invariant network under study can then be obtained using (6.1-2) and (6.2-2). In fact, $\mathbf{y}(t)$ can be written as

$$\mathbf{y}(t) = \mathbf{C}e^{\mathbf{A}(t-t_0)}\mathbf{x}_0 + \mathbf{C}\int_{t_0}^{t} e^{\mathbf{A}(t-\tau)}\mathbf{B}\mathbf{u}(\tau)\,d\tau + \mathbf{D}\mathbf{u}(t) \qquad (6.2\text{-}7)$$

Hence, given the coefficient matrices **A**, **B**, **C**, and **D**, the initial conditions $\mathbf{x}_0$, and the input vector $\mathbf{u}(t)$, the output vector $\mathbf{y}(t)$ can be completely determined for all $t \geq t_0$ by (6.2-7).

Notice that the response of the network consists of two separate parts; first, the *zero input response*, $\mathbf{y}_{0_i}(t)$, given by

$$\mathbf{y}_{0_i}(t) \triangleq \mathbf{C}e^{\mathbf{A}(t-t_0)}\mathbf{x}_0 \qquad (6.2\text{-}8)$$

and, second, the *zero state response*, $\mathbf{y}_{0_s}(t)$, given by

$$\mathbf{y}_{0_s}(t) \triangleq \mathbf{C}\int_{t_0}^{t} e^{\mathbf{A}(t-\tau)}\mathbf{B}\mathbf{u}(\tau)\,d\tau + \mathbf{D}\mathbf{u}(t) \qquad (6.2\text{-}9)$$

From (6.2-7) it is clear that computing the response of a linear time-invariant network involves computing the state transition matrix $e^{\mathbf{A}t}$ and simple integration and matrix multiplications. Typically, computing $e^{\mathbf{A}t}$ is the major task in obtaining the response of linear time-invariant networks. For this reason, several different methods for computing the state transition matrix have been devised, some of which will be discussed next.

Methods of Computing $e^{\mathbf{A}t}$

Power-Series Method

Since the infinite series (6.2-3) is convergent for each finite t, it can be approximated by the finite series

$$e^{\mathbf{A}t} \simeq \mathbf{1} + \mathbf{A}t + \frac{1}{2!}\mathbf{A}^2t^2 + \cdots + \frac{1}{n!}\mathbf{A}^nt^n \qquad (6.2\text{-}10)$$

where n is chosen large enough so that the right-hand side of the equation is approximately equal to the left-hand side. This relation can be used to compute $e^{\mathbf{A}t}$ for small t. However, this method is only useful if a digital computer is employed (hand computation of such a series is a hopeless task). The major advantage of this method is the simplicity of programming, and its main disadvantages are the lack of accuracy for large t and the large computation time required for evaluating (6.2-10). This method is seldom used in the actual computation of $e^{\mathbf{A}t}$.

Laplace-Transform Method

Let us consider the homogeneous part of the linear time-invariant differential equation given in (6.2-1); for convenience, assume that the initial time t_0 is equal to zero; that is,

$$\dot{\mathbf{x}}(t) = \mathbf{A}\mathbf{x}(t), \qquad \mathbf{x}(0) = \mathbf{x}_0 \tag{6.2-11}$$

Taking the Laplace transform of both sides of (6.2-11) yields

$$s\mathbf{x}(s) - \mathbf{x}_0 = \mathbf{A}\mathbf{x}(s)$$

or

$$(s\mathbf{1} - \mathbf{A})\mathbf{x}(s) = \mathbf{x}_0$$

Hence,

$$\mathbf{x}(s) = (s\mathbf{1} - \mathbf{A})^{-1}\mathbf{x}_0$$

Taking the inverse Laplace transform, we obtain

$$\mathbf{x}(t) = [\mathcal{L}^{-1}(s\mathbf{1} - \mathbf{A})^{-1}]\mathbf{x}_0$$

Comparing this equation with the zero input solution of (6.2-1) yields

$$e^{\mathbf{A}t} = \mathcal{L}^{-1}(s\mathbf{1} - \mathbf{A})^{-1} \tag{6.2-12}$$

Hence, to obtain $e^{\mathbf{A}t}$, we can first invert the $n \times n$ complex matrix $(s\mathbf{1} - \mathbf{A})$ and then take its inverse Laplace transform. An example will illustrate a simple application of this method.

Example 6.2-1. Compute $e^{\mathbf{A}t}$ for $\mathbf{A}$ given by

$$\mathbf{A} = \begin{bmatrix} 1 & 0 & 0 \\ 0 & 1 & 0 \\ 0 & 1 & 2 \end{bmatrix}$$

using the Laplace-transform method.

Solution

We have

$$(s\mathbf{1} - \mathbf{A}) = \begin{bmatrix} s-1 & 0 & 0 \\ 0 & s-1 & 0 \\ 0 & -1 & s-2 \end{bmatrix}$$

Then

$$(s\mathbf{1} - \mathbf{A})^{-1} = \frac{1}{(s-1)^2(s-2)} \begin{bmatrix} (s-1)(s-2) & 0 & 0 \\ 0 & (s-1)(s-2) & 0 \\ 0 & (s-1) & (s-1)^2 \end{bmatrix}$$

or

$$(s\mathbf{1} - \mathbf{A})^{-1} = \begin{bmatrix} \dfrac{1}{s-1} & 0 & 0 \\ 0 & \dfrac{1}{s-1} & 0 \\ 0 & \dfrac{1}{(s-1)(s-2)} & \dfrac{1}{s-2} \end{bmatrix}$$

Taking the inverse Laplace transform yields

$$e^{\mathbf{A}t} = \mathcal{L}^{-1}(s\mathbf{1} - \mathbf{A})^{-1} = \begin{bmatrix} e^t & 0 & 0 \\ 0 & e^t & 0 \\ 0 & e^{2t} - e^t & e^{2t} \end{bmatrix}$$

Remark 6.2-1. The Laplace-transform method of computing $e^{\mathbf{A}t}$ is practical only if the dimension of **A** is small. The main difficulty in computing $e^{\mathbf{A}t}$ for matrices with large dimensions arises in inverting $(s\mathbf{1} - \mathbf{A})$. More precisely, for a matrix **A** of dimension n we must compute one determinant of order n and n^2 determinants of order $n - 1$. We must also compute the roots of a polynomial of degree n in order to be able to find the inverse Laplace transform. For large matrices, therefore, this amounts to a formidable job. There are several shortcuts for obtaining $(s\mathbf{1} - \mathbf{A})^{-1}$ for a given **A**. In this chapter, however, we omit these in favor of more general methods that are useful in machine computations and refer the interested reader to the literature (see, for example, [2], pp. 324–327).

Exercise 6.2-2. Show that for a diagonal matrix

$$\mathbf{A} = \begin{bmatrix} \lambda_1 & 0 & 0 & \cdots & 0 \\ 0 & \lambda_2 & 0 & \cdots & 0 \\ 0 & 0 & & \cdots & 0 \\ \vdots & & & & \vdots \\ 0 & 0 & 0 & \cdots & \lambda_n \end{bmatrix}$$

the matrix $e^{\mathbf{A}t}$ is given by

$$e^{\mathbf{A}t} = \begin{bmatrix} e^{\lambda_1 t} & 0 & \cdots & 0 \\ 0 & e^{\lambda_2 t} & \cdots & 0 \\ \vdots & & & \vdots \\ 0 & 0 & \cdots & e^{\lambda_n t} \end{bmatrix}$$

To proceed with introducing other methods of computing $e^{\mathbf{A}t}$, we first discuss some methods of computing a function of a matrix; computing $e^{\mathbf{A}t}$ will then be considered as a special case of this general method.

Function of a Matrix and Related Topics

Definition 6.2-1 (Eigenvalue). The *eigenvalues* of an $n \times n$ matrix $\mathbf{A}$ are defined to be the roots of its characteristic polynomial given by

$$q(\lambda) \triangleq \det[\mathbf{A} - \lambda\mathbf{1}] \tag{6.2-13}$$

The polynomial $q(\lambda)$ is of degree n; hence, it has n roots. These roots may be real or complex, distinct or identical. In any case $q(\lambda)$ can be written as

$$q(\lambda) = \sum_{i=0}^{n} a_i \lambda^i \tag{6.2-14}$$

where the a_i are scalar constants. The following important theorem is needed for obtaining an efficient method of computing $e^{\mathbf{A}t}$.

Theorem 6.2-2 (Cayley–Hamilton). Any square matrix $\mathbf{A}$ satisfies its own characteristic polynomial; that is,

$$q(\mathbf{A}) = \mathbf{0} \tag{6.2-15}$$

where the right-hand side of (6.2-15) is an $n \times n$ zero matrix.

Proof

By definition of the inverse of a matrix, we can write

$$(\mathbf{A} - \lambda\mathbf{1})^{-1} = \frac{\operatorname{adj}(\mathbf{A} - \lambda\mathbf{1})}{\det(\mathbf{A} - \lambda\mathbf{1})}$$

Using (6.2-13), we get

$$q(\lambda)(\mathbf{A} - \lambda\mathbf{1})^{-1} = \operatorname{adj}(\mathbf{A} - \lambda\mathbf{1})$$

or simply

$$q(\lambda)\mathbf{1} = (\mathbf{A} - \lambda\mathbf{1}) \operatorname{adj}(\mathbf{A} - \lambda\mathbf{1}) \tag{6.2-16}$$

Notice that $\operatorname{adj}(\mathbf{A} - \lambda\mathbf{1})$ is an $n \times n$ matrix whose elements are polynomials of degree $n - 1$ in λ. We can rewrite this matrix as the sum of n matrices as

$$\operatorname{adj}(\mathbf{A} - \lambda\mathbf{1}) = \mathbf{B}_0 + \lambda\mathbf{B}_1 + \lambda^2\mathbf{B}_2 + \cdots + \lambda^{n-1}\mathbf{B}_{n-1} \tag{6.2-17}$$

where the $\mathbf{B}_i$ are constant $n \times n$ matrices. From (6.2-14), (6.2-16), and (6.2-17) we can write

$$\begin{aligned}\sum_{i=0}^{n} a_i \lambda^i \mathbf{1} = \mathbf{A}\mathbf{B}_0 &+ (\mathbf{A}\mathbf{B}_1 - \mathbf{B}_0)\lambda + (\mathbf{A}\mathbf{B}_2 - \mathbf{B}_1)\lambda^2 + \cdots \\ &+ (\mathbf{A}\mathbf{B}_{n-1} - \mathbf{B}_{n-2})\lambda^{n-1} - \mathbf{B}_{n-1}\lambda^n\end{aligned}$$

Since this equality must hold for all values of λ except the eigenvalues of $\mathbf{A}$

we can equate the coefficients of λ^i in both sides of this equation and obtain

$$\begin{aligned} a_0\mathbf{1} &= \mathbf{AB}_0 \\ a_1\mathbf{1} &= \mathbf{AB}_1 - \mathbf{B}_0 \\ a_2\mathbf{1} &= \mathbf{AB}_2 - \mathbf{B}_1 \\ \cdots &= \cdots \\ a_{n-1}\mathbf{1} &= \mathbf{AB}_{n-1} - \mathbf{B}_{n-2} \\ a_n\mathbf{1} &= -\mathbf{B}_{n-1}\lambda^n \end{aligned}$$

If we premultiply the second of these equations by **A**, the third by $\mathbf{A}^2$, and so on, and add the resulting equations together, we get

$$a_0\mathbf{1} + a_1\mathbf{A} + a_2\mathbf{A}^2 + \cdots + a_{n-1}\mathbf{A}^{n-1} + a_n\mathbf{A}^n = \mathbf{0} \qquad (6.2\text{-}18)$$

or simply

$$q(\mathbf{A}) = \mathbf{0}$$

This proves the theorem. Note that the proof of the theorem does not depend on whether the eigenvalues of **A** are distinct or not.

There are numerous consequences of the Cayley–Hamilton theorem. An application of this theorem is a simple method for inverting a nonsingular matrix **A**. To see this, let us multiply both sides of (6.2-18) by $\mathbf{A}^{-1}$ and rearrange the equation to get

$$\mathbf{A}^{-1} = -\frac{1}{a_0}(a_1\mathbf{1} + a_2\mathbf{A} + a_3\mathbf{A}^2 + \cdots + a_n\mathbf{A}^{n-1}) \qquad (6.2\text{-}19)$$

Consequently, if the coefficients of the characteristic equation of **A** are known (i.e., if the a_i are known), the inverse of **A** can easily be found using (6.2-19). Notice that this method of inverting a matrix requires only simple matrix multiplications.

Before defining a function of a matrix, let us introduce some notations which will be useful in the analysis that follows. Consider an $n \times n$ matrix **A**; denote the *distinct* eigenvalues of **A** by $\lambda_1, \lambda_2, \ldots, \lambda_p$. Let the multiplicity of the jth eigenvalue, λ_j, be m_j; then all the eigenvalues of **A** can be arranged in the form

$$\overbrace{\lambda_1, \lambda_1, \ldots, \lambda_1}^{m_1}, \quad \overbrace{\lambda_2, \lambda_2, \ldots, \lambda_2}^{m_2}, \ldots, \overbrace{\lambda_j, \lambda_j, \ldots, \lambda_j}^{m_j}, \ldots, \overbrace{\lambda_p, \lambda_p, \ldots, \lambda_p}^{m_p}$$

Since the total number of the eigenvalues of any $n \times n$ matrix is n, we must have

$$n = \sum_{j=1}^{p} m_j$$

Let us now give a precise definition of a function of a matrix. The exponential matrix $e^{\mathbf{A}t}$ can then be considered as a special case of this definition.

Definition 6.2-2 (Function of a Matrix). Consider a scalar function $f(x)$. If $f(x)$ can be expanded as a *convergent* power series

$$f(x) = \sum_{k=0}^{\infty} a_k x^k \tag{6.2-20}$$

then $f(\mathbf{A})$ is called a function of the matrix $\mathbf{A}$ and is written as

$$f(\mathbf{A}) = \sum_{k=0}^{\infty} a_k \mathbf{A}^k \tag{6.2-21}$$

where $\mathbf{A}^0$ is the identity matrix.

As an example, consider an $n \times n$ constant matrix $\mathbf{A}$ and the exponential function e^x. Since

$$e^x = \sum_{k=0}^{\infty} \frac{1}{k!} x^k$$

is a convergent power-series expansion of e^x, then the function of matrix $e^{\mathbf{A}}$ is well defined and is given by

$$e^{\mathbf{A}} = \sum_{k=0}^{\infty} \frac{1}{k!} \mathbf{A}^k \tag{6.2-22}$$

Let us now state and prove an important theorem, which can be used to compute a function of a matrix.

Theorem 6.2-3. Let $f(\mathbf{A})$ be a function of an $n \times n$ matrix $\mathbf{A}$. Then $f(\mathbf{A})$ can be written as

$$f(\mathbf{A}) = \alpha_0 \mathbf{1} + \alpha_1 \mathbf{A} + \alpha_2 \mathbf{A}^2 + \cdots + \alpha_{n-1} \mathbf{A}^{n-1} \tag{6.2-23}$$

where α_j, $j = 0, \ldots, n - 1$, are scalars.

Proof

Let us start from (6.2-20). Consider a finite series $f_i(x)$ obtained from $f(x)$ by truncating the infinite series at x^i, where i is an integer larger than n. Then $f_i(x)$ can be written as

$$f_i(x) \triangleq \sum_{k=0}^{i} a_k x^k = g_i(x) q(x) + r_i(x) \tag{6.2-24}$$

where $q(x)$ is the characteristic polynomial of $\mathbf{A}$, $g_i(x)$ is a polynomial of degree $i - n$, and $r_i(x)$ is the "remainder," which is a polynomial of degree $n - 1$. Hence, (6.2-24) can be written as

$$f_i(x) = g_i(x) q(x) + \sum_{k=0}^{n-1} \alpha_{ik} x^k$$

Since the series (6.2-20) is convergent, we can take the limit of both sides of this equation as i goes to ∞:

$$\lim_{i \to \infty} f_i(x) = (\lim_{i \to \infty} g_i(x)) q(x) + \sum_{k=0}^{n-1} (\lim_{i \to \infty} \alpha_{ik}) x^k \tag{6.2-25}$$

Consequently, if we choose the notations

$$g(x) \triangleq \lim_{i \to \infty} g_i(x)$$

and

$$\alpha_k = \lim_{i \to \infty} \alpha_{ik}, \qquad k = 0, 1, \ldots, n-1$$

equation (6.2-25) can be written as

$$f(x) = g(x)q(x) + \sum_{k=0}^{n-1} \alpha_k x^k \tag{6.2-26}$$

If we now replace x by $\mathbf{A}$ in (6.2-26), it yields

$$f(\mathbf{A}) = g(\mathbf{A})q(\mathbf{A}) + \sum_{k=0}^{n-1} \alpha_k \mathbf{A}^k \tag{6.2-27}$$

But since $q(x)$ is the characteristic polynomial of $\mathbf{A}$, by the Cayley–Hamilton theorem we have

$$q(\mathbf{A}) = \mathbf{0}$$

Consequently, (6.2-27) becomes

$$f(\mathbf{A}) = \sum_{k=0}^{n-1} \alpha_k \mathbf{A}^k$$

This proves the theorem.

Note that a special case of a function of a matrix is the exponential function $e^{\mathbf{A}t}$. According to the theorem just proved, $e^{\mathbf{A}t}$ can be written as

$$e^{\mathbf{A}t} = \alpha_0(t)\mathbf{1} + \alpha_1(t)\mathbf{A} + \alpha_2(t)\mathbf{A}^2 + \cdots + \alpha_{n-1}(t)\mathbf{A}^{n-1} \tag{6.2-28}$$

where the scalars α_k, $k = 1, 2, \ldots, n-1$, are, in this case, functions of time.

The next crucial step in computing a function of a matrix is to evaluate the coefficients α_k. To do this, we consider two separate cases. First, we assume that the eigenvalues of $\mathbf{A}$ are all distinct and give a method for computing $f(\mathbf{A})$; we then generalize this method to the case where some of the eigenvalues of $\mathbf{A}$ are of multiplicity two or more.

Case 1: All the Eigenvalues of A Are Distinct. Denote these eigenvalues by $\lambda_1, \lambda_2, \ldots, \lambda_n$. Then the characteristic polynomial of $\mathbf{A}$ can be written as

$$q(x) = (x - \lambda_1)(x - \lambda_2)\cdots(x - \lambda_n)$$

Equation (6.2-26) can then be written as

$$f(x) = g(x)(x - \lambda_1)(x - \lambda_2)\cdots(x - \lambda_n) + \sum_{k=0}^{n-1} \alpha_k x^k \tag{6.2-29}$$

From (6.2-29) we can obtain n equations in n unknowns $\alpha_0, \alpha_1, \ldots, \alpha_{n-1}$ by replacing x with $\lambda_1, \lambda_2, \ldots, \lambda_n$. Since the first term in the right-hand side of

(6.2-29) drops out, we have

$$
\begin{aligned}
\alpha_0 + \alpha_1\lambda_1 + \alpha_2\lambda_1^2 + \cdots + \alpha_{n-1}\lambda_1^{n-1} &= f(\lambda_1) \\
\alpha_0 + \alpha_1\lambda_2 + \alpha_2\lambda_2^2 + \cdots + \alpha_{n-1}\lambda_2^{n-1} &= f(\lambda_2) \\
\alpha_0 + \alpha_1\lambda_3 + \alpha_2\lambda_3^2 + \cdots + \alpha_{n-1}\lambda_3^{n-1} &= f(\lambda_3) \\
\vdots \qquad\qquad\qquad\qquad &\quad \vdots \\
\alpha_0 + \alpha_1\lambda_n + \alpha_2\lambda_n^2 + \cdots + \alpha_{n-1}\lambda_n^{n-1} &= f(\lambda_n)
\end{aligned}
\tag{6.2-30}
$$

These n equations can then be solved for $\alpha_1, \alpha_2, \ldots, \alpha_n$. To obtain a unique solution, we must show that the equations given in (6.2-30) are linearly independent. For this reason we write (6.2-30) in matrix form:

$$
\begin{bmatrix}
1 & \lambda_1 & \lambda_1^2 & \cdots & \lambda_1^{n-1} \\
1 & \lambda_2 & \lambda_2^2 & \cdots & \lambda_2^{n-1} \\
1 & \lambda_3 & \lambda_3^2 & \cdots & \lambda_3^{n-1} \\
\vdots & & & & \vdots \\
1 & \lambda_n & \lambda_n^2 & \cdots & \lambda_n^{n-1}
\end{bmatrix}
\begin{bmatrix}
\alpha_0 \\ \alpha_1 \\ \alpha_2 \\ \vdots \\ \alpha_{n-1}
\end{bmatrix}
=
\begin{bmatrix}
f(\lambda_1) \\ f(\lambda_2) \\ f(\lambda_3) \\ \vdots \\ f(\lambda_n)
\end{bmatrix}
\tag{6.2-31}
$$

Since the λ_i are distinct, the determinant of the $n \times n$ matrix in (6.2-31) is the well-known Vandermonde determinant, which can be shown to be nonzero (see Problem 6.14). Consequently, (6.2-31) has a unique solution.

Example 6.2-2. Compute $e^{\mathbf{A}t}$ for the 2×2 matrix given by

$$
\mathbf{A} = \begin{bmatrix} 1 & 2 \\ 0 & -3 \end{bmatrix}
$$

using the method just discussed.

Solution

Let us first obtain the eigenvalues of **A**. The characteristic equation of **A** is

$$
q(\lambda) = \det[\mathbf{A} - \lambda\mathbf{1}] = \det\begin{bmatrix} 1-\lambda & 2 \\ 0 & -3-\lambda \end{bmatrix} = (\lambda - 1)(\lambda + 3) = 0
$$

The eigenvalues then are

$$
\lambda_1 = 1, \qquad \lambda_2 = -3
$$

Since the dimension of **A** is 2, according to Theorem 6.2-3, $e^{\mathbf{A}t}$ can be written as

$$
e^{\mathbf{A}t} = \alpha_0(t)\mathbf{1} + \alpha_1(t)\mathbf{A} \tag{6.2-32}
$$

To compute the coefficients α_0 and α_1, we observe that the eigenvalues of **A** are distinct and thus (6.2-30) can be employed to get

$$
\begin{aligned}
\alpha_0 + 1\cdot\alpha_1 &= e^t \\
\alpha_0 - 3\cdot\alpha_1 &= e^{-3t}
\end{aligned}
$$

Solving these equations for α_0 and α_1 yields

$$\alpha_0(t) = \tfrac{1}{4}(3e^t + e^{-3t}) \quad \text{and} \quad \alpha_1(t) = \tfrac{1}{4}(e^t - e^{-3t})$$

From (6.2-32) we can then obtain

$$e^{\mathbf{A}t} = \frac{1}{4}\begin{bmatrix} 3e^t + e^{-3t} & 0 \\ 0 & 3e^t + e^{-3t} \end{bmatrix} + \frac{1}{4}\begin{bmatrix} e^t - e^{-3t} & 2e^t - 2e^{-3t} \\ 0 & -3e^t - 3e^{-3t} \end{bmatrix}$$

$$= \begin{bmatrix} e^t & \tfrac{1}{2}(e^t - e^{-3t}) \\ 0 & e^{-3t} \end{bmatrix}$$

Case 2: A Has Multiple Eigenvalues. In this case $f(x)$ given in (6.2-26) can be written as

$$f(x) = g(x)(x - \lambda_1)^{m_1}(x - \lambda_2)^{m_2} \cdots (x - \lambda_p)^{m_p} + \sum_{k=0}^{n-1} \alpha_k x^k \tag{6.2-33}$$

where m_j denotes the multiplicity of the jth eigenvalue and p is the number of the distinct eigenvalues. By replacing x in both sides of (6.2-33) with λ_1, $\lambda_2, \ldots, \lambda_p$, we obtain p linearly independent equations in n unknowns α_0, $\alpha_1, \ldots, \alpha_{n-1}$. To obtain $n - p$ other linearly independent equations needed for solving all the coefficients, we can first take $m_1 - 1$ consecutive derivatives of both sides of (6.2-33) and then replace x by λ_1. This will result in $m_1 - 1$ more linearly independent equations. In general, if we take $m_j - 1$ consecutive derivatives of (6.2-33) and then replace x by λ_j, we obtain $m_j - 1$ new linearly independent equations. All together we can then obtain n linearly independent equations. Of course, after replacing x by λ_j, the contribution of the first term in the right-hand side of (6.2-33) will be zero (why?). The desired n linearly independent equations in $\alpha_0, \alpha_1, \ldots, \alpha_{n-1}$ then are

$$\begin{aligned}
\alpha_0 + \alpha_1\lambda_1 + \alpha_2\lambda_1^2 + \cdots + \alpha_{n-1}\lambda_1^{n-1} &= f(\lambda_1) \\
\frac{d}{d\lambda_1}[\alpha_0 + \alpha_1\lambda_1 + \alpha_2\lambda_1^2 + \cdots + \alpha_{n-1}\lambda_1^{n-1}] &= \frac{d}{d\lambda_1}f(\lambda_1) \\
&\vdots \\
\frac{d^{m_1-1}}{d\lambda_1^{m_1-1}}[\alpha_0 + \alpha_1\lambda_1 + \alpha_2\lambda_1^2 + \cdots + \alpha_{n-1}\lambda_1^{n-1}] &= \frac{d^{m_1-1}}{d\lambda_1^{m_1-1}}f(\lambda_1) \\
&\vdots \\
\alpha_0 + \alpha_1\lambda_p + \alpha_2\lambda_p^2 + \cdots + \alpha_{n-1}\lambda_p^{n-1} &= f(\lambda_p) \\
\frac{d}{d\lambda_p}[\alpha_0 + \alpha_1\lambda_p + \alpha_2\lambda_p^2 + \cdots + \alpha_{n-1}\lambda_p^{n-1}] &= \frac{d}{d\lambda_p}f(\lambda_p) \\
&\vdots \\
\frac{d^{m_p-1}}{d\lambda_p^{m_p-1}}[\alpha_0 + \alpha_1\lambda_p + \alpha_2\lambda_p^2 + \cdots + \alpha_{n-1}\lambda_p^{n-1}] &= \frac{d^{m_p-1}}{d\lambda_p^{m_p-1}}f(\lambda_p)
\end{aligned} \tag{6.2-34}$$

Example 6.2-3. Find the exponential matrix $e^{\mathbf{A}t}$, where $\mathbf{A}$ is given by

$$\mathbf{A} = \begin{bmatrix} 1 & 0 & 0 \\ 0 & 1 & 0 \\ 0 & 1 & 2 \end{bmatrix}$$

Solution

The characteristic equation of $\mathbf{A}$ is

$$\det[\mathbf{A} - \lambda\mathbf{1}] = (1 - \lambda)(1 - \lambda)(2 - \lambda) = 0$$

Hence, the eigenvalues are

$$\lambda_1 = 2 \quad \text{with multiplicity } 1;\ m_1 = 1$$

$$\lambda_2 = 1 \quad \text{with multiplicity } 2;\ m_2 = 2$$

Since $n = 3$, according to Theorem 6.2-3 we can write

$$e^{\mathbf{A}t} = \alpha_0(t)\mathbf{1} + \alpha_1(t)\mathbf{A} + \alpha_2(t)\mathbf{A}^2 \tag{6.2-35}$$

Also, since the eigenvalues of $\mathbf{A}$ are not distinct, to obtain the coefficients α_0, α_1, and α_2, we must use (6.2-34):

$$\alpha_0 + \alpha_1\lambda_1 + \alpha_2\lambda_1^2 = e^{\lambda_1 t}$$

$$\alpha_0 + \alpha_1\lambda_2 + \alpha_2\lambda_2^2 = e^{\lambda_2 t}$$

$$\frac{d}{d\lambda_2}[\alpha_0 + \alpha_1\lambda_2 + \alpha_2\lambda_2^2] = \frac{d}{d\lambda_2}e^{\lambda_2 t}$$

or, equivalently,

$$\alpha_0 + 2\alpha_1 + 4\alpha_2 = e^{2t}$$

$$\alpha_0 + \alpha_1 + \alpha_2 = e^{t}$$

$$\alpha_1 + 2\alpha_2 = te^{t}$$

Solving this set of three linearly independent equations yields

$$\alpha_0(t) = e^{2t} - 2te^{t}$$

$$\alpha_1(t) = 2e^{t} + 3te^{t} - 2e^{2t}$$

$$\alpha_2(t) = -e^{t} - te^{t} + e^{2t}$$

The exponential function $e^{\mathbf{A}t}$ can then easily be computed using (6.2-35); the result is

$$e^{\mathbf{A}t} = \begin{bmatrix} e^{t} & 0 & 0 \\ 0 & e^{t} & 0 \\ 0 & e^{2t} - e^{t} & e^{2t} \end{bmatrix}$$

Note that this answer agrees with the result of Example 6.2-1, as it should.

Computation of f(A) Using the Minimal Polynomial. Recall that according to the Cayley–Hamilton theorem, every $n \times n$ matrix $\mathbf{A}$ satisfies its characteristic polynomial. This polynomial is generally of degree n. In some cases $\mathbf{A}$ might satisfy polynomials of lesser degree than n. The polynomial of the least degree that $\mathbf{A}$ satisfies is called the *minimal polynomial* of $\mathbf{A}$. The precise definition of a minimal polynomial is

Definition 6.2-3 (Minimal Polynomial). The *minimal polynomial* of an $n \times n$ matrix $\mathbf{A}$ is the polynomial $\Phi(\mathbf{A})$ of least degree such that

$$\Phi(\mathbf{A}) \triangleq a_0\mathbf{1} + a_1\mathbf{A} + a_2\mathbf{A}^2 + \cdots + a_r\mathbf{A}^r = \mathbf{0} \qquad (6.2\text{-}36)$$

where $r \leq n$ and the a_i are scalar coefficients.

Example 6.2-4. The minimal polynomial of matrix $\mathbf{A}$ given by

$$\mathbf{A} = \begin{bmatrix} 1 & 0 & 0 \\ 0 & 1 & 0 \\ 0 & 1 & 2 \end{bmatrix}$$

is

$$\Phi(\mathbf{A}) = 2\mathbf{1} - 3\mathbf{A} + \mathbf{A}^2 = \mathbf{0}$$

This polynomial is of order 2 in contrast to its characteristic polynomial

$$q(\mathbf{A}) = 2\mathbf{1} - 5\mathbf{A} + 4\mathbf{A}^2 - \mathbf{A}^3 = \mathbf{0}$$

which is of order 3.

Let us now consider the scalar polynomial $\Phi(\lambda)$ defined by

$$\Phi(\lambda) = a_0 + a_1\lambda + a_2\lambda^2 + \cdots + a_r\lambda^r = 0$$

where the a_i are the same as those in (6.2-36). Denote the roots of $\Phi(\lambda)$ by $\lambda_1, \lambda_2, \ldots, \lambda_r$. These roots may or may not be distinct. The important point is that their total number is less than n, and this fact can be useful in computing $f(\mathbf{A})$ in a more efficient way.

We can now state a theorem concerning the function of a matrix that is similar to Theorem 6.2-3, except that it is stated in terms of the minimal polynomial.

Theorem 6.2-4. Let $f(\mathbf{A})$ be a function of an $n \times n$ matrix $\mathbf{A}$ whose minimal polynomial is of degree $r \leq n$; then

$$f(\mathbf{A}) = \beta_0\mathbf{1} + \beta_1\mathbf{A} + \beta_2\mathbf{A}^2 + \cdots + \beta_{r-1}\mathbf{A}^{r-1} \qquad (6.2\text{-}37)$$

where the β_i are scalar coefficients.

The proof of this theorem is similar to that of Theorem 6.2-3 and hence will be omitted. To compute the coefficients $\beta_0, \beta_1, \ldots, \beta_{r-1}$, we consider two cases. If the roots of the minimal polynomial are distinct, the β_i can be computed from the following r linearly independent equations:

$$\begin{aligned}
\beta_0 + \beta_1\lambda_1 + \beta_2\lambda_1^2 + \cdots + \beta_{r-1}\lambda_1^{r-1} &= f(\lambda_1) \\
\beta_0 + \beta_1\lambda_2 + \beta_2\lambda_2^2 + \cdots + \beta_{r-1}\lambda_2^{r-1} &= f(\lambda_2) \\
\vdots \qquad\qquad\qquad\qquad & \quad \vdots \\
\beta_0 + \beta_1\lambda_r + \beta_2\lambda_r^2 + \cdots + \beta_{r-1}\lambda_r^{r-1} &= f(\lambda_r)
\end{aligned} \tag{6.2-38}$$

where $\lambda_1, \lambda_2, \ldots, \lambda_r$ are the roots of the minimal polynomials of **A**. Notice that if the degree of the minimal polynomial of **A** is *less* than the degree of its characteristic polynomial ($r < n$), the amount of effort involved in computing $f(\mathbf{A})$ is considerably reduced by using (6.2-38) instead of (6.2-30). This can be illustrated by an example.

Example 6.2-5. Once again consider the matrix **A** given by

$$\mathbf{A} = \begin{bmatrix} 1 & 0 & 0 \\ 0 & 1 & 0 \\ 0 & 1 & 2 \end{bmatrix}$$

Compute $e^{\mathbf{A}t}$ using the minimal-polynomial approach.

Solution

The minimal polynomial of **A** was just found to be

$$\Phi(\lambda) = 2 - 3\lambda + \lambda^2 = 0$$

The roots of this equation are $\lambda_1 = 1$ and $\lambda_2 = 2$. According to Theorem 6.2-4, the exponential function $e^{\mathbf{A}t}$ can be written as

$$e^{\mathbf{A}t} = \beta_0(t)\mathbf{1} + \beta_1(t)\mathbf{A}$$

To obtain the coefficients β_0 and β_1, since the roots of the minimal polynomial are distinct, we use (6.2-38):

$$\beta_0 + \beta_1\lambda_1 = e^{\lambda_1 t}$$

$$\beta_0 + \beta_1\lambda_2 = e^{\lambda_2 t}$$

Since $\lambda_1 = 1$ and $\lambda_2 = 2$, we get

$$\beta_0 + \beta_1 = e^t$$

$$\beta_0 + 2\beta_1 = e^{2t}$$

Consequently,

$$\beta_0 = 2e^t - e^{2t} \quad \text{and} \quad \beta_1 = e^{2t} - e^t$$

Hence,

$$e^{\mathbf{A}t} = \beta_0\mathbf{1} + \beta_1\mathbf{A} = \begin{bmatrix} e^t & 0 & 0 \\ 0 & e^t & 0 \\ 0 & e^{2t} - e^t & e^{2t} \end{bmatrix}$$

which is the same as the result of the previous example.

If the roots of the minimal polynomial of a matrix are not distinct, the set of linearly independent equations that can be solved for the coefficients β_i are similar to those given in (6.2-34). To be more specific, let the multiplicity of the jth root of the minimal polynomial of $\mathbf{A}$ be l_j; then we can write

$$\begin{aligned}
\beta_0 + \beta_1\lambda_1 + \beta_2\lambda_1^2 + \cdots + \beta_{r-1}\lambda_1^{r-1} &= f(\lambda_1) \\
\frac{d}{d\lambda_1}[\beta_0 + \beta_1\lambda_1 + \beta_2\lambda_1^2 + \cdots + \beta_{r-1}\lambda_1^{r-1}] &= \frac{d}{d\lambda_1}f(\lambda_1) \\
&\vdots \\
\frac{d^{l_1-1}}{d\lambda_1^{l_1-1}}[\beta_0 + \beta_1\lambda_1 + \beta_2\lambda_1^2 + \cdots + \beta_{r-1}\lambda_1^{r-1}] &= \frac{d^{l_1-1}}{d\lambda_1^{l_1-1}}f(\lambda_1) \\
&\vdots \\
\beta_0 + \beta_1\lambda_r + \beta_2\lambda_r^2 + \cdots + \beta_{r-1}\lambda_r^{r-1} &= f(\lambda_r) \\
&\vdots \\
\frac{d^{l_r-1}}{d\lambda_r^{l_r-1}}[\beta_0 + \beta_1\lambda_r + \beta_2\lambda_r^2 + \cdots + \beta_{r-1}\lambda_r^{r-1}] &= \frac{d^{l_r-1}}{d\lambda_r^{l_r-1}}f(\lambda_r)
\end{aligned} \tag{6.2-39}$$

where the λ are the roots of the minimal polynomial. Let us now work out an example to illustrate the application of (6.2-39).

Example 6.2-6. Use the minimal polynomial approach to obtain the exponential matrix $e^{\mathbf{A}t}$, where $\mathbf{A}$ is given by

$$\mathbf{A} = \begin{bmatrix} 1 & 1 & 0 & 0 \\ 0 & 1 & 0 & 0 \\ 0 & 0 & 1 & 1 \\ 0 & 0 & 0 & 1 \end{bmatrix}$$

Solution

The characteristic polynomial of this matrix is of degree 4 and is given by

$$q(\lambda) \triangleq \det(\mathbf{A} - \lambda\mathbf{1}) = (1 - \lambda)^4$$

whereas the minimal polynomial of $\mathbf{A}$ is a polynomial of degree 2 given by

$$\Phi(\lambda) = (1 - \lambda)^2$$

because $(\mathbf{1} - \mathbf{A})^2 = \mathbf{0}$ is the polynomial of the least degree that $\mathbf{A}$ satisfies. From (6.2-37) we can write

$$e^{\mathbf{A}} = \beta_0(t)\mathbf{1} + \beta_1(t)\mathbf{A}$$

Since the roots of the minimal polynomial are not unique, we must use (6.2-39).

Hence,
$$\beta_0 + \beta_1\lambda_1 = e^{\lambda_1 t}$$
$$\frac{d}{d\lambda_1}[\beta_0 + \beta_1\lambda_1] = \frac{d}{d\lambda_1}e^{\lambda_1 t}$$

Since $\lambda_1 = 1$, we get

$$\beta_0 + \beta_1 = e^t$$
$$\beta_1 = te^t$$

Consequently,

$$\beta_0 = e^t(1 - t) \quad \text{and} \quad \beta_1 = te^t$$

and, finally,

$$e^{\mathbf{A}t} = \beta_0 \mathbf{1} + \beta_1 \mathbf{A} = \begin{bmatrix} e^t & te^t & 0 & 0 \\ 0 & e^t & 0 & 0 \\ 0 & 0 & e^t & te^t \\ 0 & 0 & 0 & e^t \end{bmatrix}$$

In summary, to compute the state transition matrix $e^{\mathbf{A}t}$ or any other function of a matrix $\mathbf{A}$, we can use (6.2-23) where the coefficients α_i can be computed using (6.2-30) or (6.2-34), depending on whether the eigenvalues of $\mathbf{A}$ are distinct or not. If the minimal polynomial of $\mathbf{A}$ is known, the computation of $f(\mathbf{A})$ can be greatly reduced if (6.2-37) is used.

As was mentioned earlier, the bulk of the computation of the solution of a linear time-invariant matrix differential equation

$$\dot{\mathbf{x}} = \mathbf{A}\mathbf{x} + \mathbf{B}\mathbf{u}, \qquad \mathbf{x}(t_0) = \mathbf{x}_0$$

is in computing $e^{\mathbf{A}t}$. Having done this, the total solution can be obtained using (6.2-2). As an illustration, let us work out a simple example.

Example 6.2-7. Find the total solution of the differential equation

$$\begin{bmatrix} \dot{x}_1 \\ \dot{x}_2 \\ \dot{x}_3 \end{bmatrix} = \begin{bmatrix} 1 & 0 & 0 \\ 0 & 1 & 0 \\ 0 & 1 & 2 \end{bmatrix} \begin{bmatrix} x_1 \\ x_2 \\ x_3 \end{bmatrix} + \begin{bmatrix} 1 \\ 0 \\ 0 \end{bmatrix} 1(t) \qquad \begin{bmatrix} x_1(0) \\ x_2(0) \\ x_3(0) \end{bmatrix} = \begin{bmatrix} 0 \\ 0 \\ 1 \end{bmatrix}$$

where $1(t)$ denotes the unit step function.

Solution

The state transition matrix for this equation was found in Example 6.2-3 to be

$$e^{\mathbf{A}t} = \begin{bmatrix} e^t & 0 & 0 \\ 0 & e^t & 0 \\ 0 & e^{2t} - e^t & e^{2t} \end{bmatrix}$$

Using (6.2-2), we get

$$\mathbf{x}(t) = \begin{bmatrix} e^t & 0 & 0 \\ 0 & e^t & 0 \\ 0 & e^{2t} - e^t & e^{2t} \end{bmatrix} \begin{bmatrix} 0 \\ 0 \\ 1 \end{bmatrix} + \int_0^t \begin{bmatrix} e^{t-\tau} & 0 & 0 \\ 0 & e^{t-\tau} & 0 \\ 0 & e^{2(t-\tau)} - e^{t-\tau} & e^{2(t-\tau)} \end{bmatrix} \begin{bmatrix} 1 \\ 0 \\ 0 \end{bmatrix} 1(\tau)\, d\tau$$

or, after multiplying the matrices through,

$$\mathbf{x}(t) = \begin{bmatrix} 0 \\ 0 \\ e^{2t} \end{bmatrix} + \int_0^t \begin{bmatrix} e^{t-\tau} \\ 0 \\ 0 \end{bmatrix} d\tau$$

Integrating and adding yields

$$\mathbf{x}(t) = \begin{bmatrix} e^t - 1 \\ 0 \\ e^{2t} \end{bmatrix} \quad \text{for all } t \geq 0$$

Next we consider the solution of linear time-varying state equations.

6.3 SOLUTION OF LINEAR TIME-VARYING STATE EQUATIONS

In contrast to the case of linear time-invariant state equations, the analytic solution of general linear time-varying first-order differential equations is not easy to find. Except in some specific cases, the solution of such state equations must be obtained by numerical methods. In this section we give the *form* of the solution of linear time-varying state equations and obtain the actual solution in some specific cases. Let us start by defining the state transition matrix of a linear time-varying state equation.

Consider a linear time-varying network whose state equations are given by

$$\dot{\mathbf{x}}(t) = \mathbf{A}(t)\mathbf{x}(t) + \mathbf{B}(t)\mathbf{u}(t), \qquad \mathbf{x}(t_0) = \mathbf{x}_0 \tag{6.3-1}$$

where $\mathbf{x}(t)$ and $\mathbf{u}(t)$ are n- and m-column vectors representing the state and the inputs, respectively. $\mathbf{A}(t)$ and $\mathbf{B}(t)$ are time-varying $n \times n$ and $n \times m$ matrices. The homogeneous part of (6.3-1) is written as

$$\dot{\mathbf{x}} = \mathbf{A}(t)\mathbf{x} \tag{6.3-2}$$

In can be shown that (6.3-2) has n linearly independent solutions $\mathbf{x}^1(t)$, $\mathbf{x}^2(t), \ldots, \mathbf{x}^n(t)$ so that any other solution of (6.3-2) can be written in terms of these solutions [2].

Definition 6.3-1 (Fundamental Matrix). The $n \times n$ time-varying matrix $\mathbf{W}(t)$ defined by

$$\mathbf{W}(t) \triangleq [\mathbf{x}^1(t), \mathbf{x}^2(t), \ldots, \mathbf{x}^n(t)] \tag{6.3-3}$$

is called the *fundamental matrix* of (6.3-2).

Since $\mathbf{x}^1(t), \mathbf{x}^2(t), \ldots, \mathbf{x}^n(t)$ are linearly independent for each t, then

$$\det \mathbf{W}(t) \neq 0 \quad \text{for all } t$$

Also, since each $\mathbf{x}^j(t)$ is a solution of (6.3-2), $\mathbf{W}(t)$ satisfies the matrix differential equation

$$\frac{d}{dt}\mathbf{W}(t) = \mathbf{A}(t)\mathbf{W}(t) \tag{6.3-4}$$

We can now define the state transition matrix of (6.3-1) as follows:

Definition 6.3-2 (State Transition Matrix). The *state transition matrix* of (6.3-1) is an $n \times n$ matrix defined by

$$\mathbf{\Phi}(t, \tau) \triangleq \mathbf{W}(t)\mathbf{W}^{-1}(\tau) \quad \text{for all } t \text{ and } \tau \tag{6.3-5}$$

where $\mathbf{W}(t)$ is the fundamental matrix of (6.3-1).

As we shall soon see, the state transition matrix $\mathbf{\Phi}(t, \tau)$ plays an important role in the formulation of the solution of (6.3-1). Let us first examine some of the properties of $\mathbf{\Phi}(t, \tau)$.

Properties of the State Transition Matrix

1. For each τ the state transition matrix $\mathbf{\Phi}(t, \tau)$ satisfies the differential equation

$$\frac{\partial}{\partial t}\mathbf{\Phi}(t, \tau) = \mathbf{A}(t)\mathbf{\Phi}(t, \tau) \tag{6.3-6}$$

This result can easily be obtained by postmultiplying (6.3-4) by $\mathbf{W}^{-1}(\tau)$ and using (6.3-5).

2. Putting $\tau = t$ in (6.3-5), we get

$$\mathbf{\Phi}(t, t) = \mathbf{1} \quad \text{for all } t \tag{6.3-7}$$

3. Inverting both sides of (6.3-5) yields

$$\mathbf{\Phi}^{-1}(t, \tau) = \mathbf{W}(\tau)\mathbf{W}^{-1}(t)$$

Then, comparing the right-hand side of this equation with that of (6.3-5), we get

$$\mathbf{\Phi}^{-1}(t, \tau) = \mathbf{\Phi}(\tau, t) \quad \text{for all } t \text{ and } \tau \tag{6.3-8}$$

4. For any t, ξ, and τ the following relation holds:

$$\mathbf{\Phi}(t, \xi)\mathbf{\Phi}(\xi, \tau) = \mathbf{\Phi}(t, \tau) \tag{6.3-9}$$

To prove this, notice that the left-hand side of this equation can be written as

$$\mathbf{\Phi}(t, \xi)\mathbf{\Phi}(\xi, \tau) = \mathbf{W}(t)\mathbf{W}^{-1}(\xi)\mathbf{W}(\xi)\mathbf{W}^{-1}(\tau) = \mathbf{W}(t)\mathbf{W}^{-1}(\tau) = \mathbf{\Phi}(t, \tau)$$

5. For any t and τ we have

$$\frac{\partial\mathbf{\Phi}(t, \tau)}{\partial\tau} = -\mathbf{\Phi}(t, \tau)\mathbf{A}(\tau) \tag{6.3-10}$$

To show this, differentiate both sides of (6.3-5) with respect to τ to get

$$\frac{\partial}{\partial\tau}\mathbf{\Phi}(t, \tau) = \mathbf{W}(t)\frac{d}{d\tau}\mathbf{W}^{-1}(\tau)$$

Now observe that

$$\frac{\partial}{\partial\tau}[\mathbf{W}(\tau)\mathbf{W}^{-1}(\tau)] = \left[\frac{d}{d\tau}\mathbf{W}(\tau)\right]\mathbf{W}^{-1}(\tau) + \mathbf{W}(\tau)\frac{d}{d\tau}\mathbf{W}^{-1}(\tau) = \mathbf{0}$$

Then

$$\begin{aligned}\frac{d}{d\tau}\mathbf{W}^{-1}(\tau) &= -\mathbf{W}^{-1}(\tau)\left[\frac{d}{d\tau}\mathbf{W}(\tau)\right]\mathbf{W}^{-1}(\tau) = -\mathbf{W}^{-1}(\tau)[\mathbf{A}(\tau)\mathbf{W}(\tau)]\mathbf{W}^{-1}(\tau) \\ &= -\mathbf{W}^{-1}(\tau)\mathbf{A}(\tau)\end{aligned}$$

Consequently,

$$\frac{\partial}{\partial\tau}\mathbf{\Phi}(t, \tau) = -\mathbf{W}(t)\mathbf{W}^{-1}(\tau)\mathbf{A}(\tau) = -\mathbf{\Phi}(t, \tau)\mathbf{A}(\tau)$$

An important property of the state transition matrix is that the solution of (6.3-1) can be written in terms of $\mathbf{\Phi}(t, \tau)$. This is shown in the following:

Theorem 6.3-1. The solution of (6.3-1) is given by

$$\mathbf{x}(t) = \mathbf{\Phi}(t, t_0)\mathbf{x}_0 + \int_{t_0}^{t} \mathbf{\Phi}(t, \tau)\mathbf{B}(\tau)\mathbf{u}(\tau)\, d\tau \quad \text{for } t \geq t_0 \qquad (6.3\text{-}11)$$

Proof

$\mathbf{x}(t)$ given in (6.3-11) certainly satisfies the initial conditions, since

$$\mathbf{x}(t_0) = \mathbf{\Phi}(t_0, t_0)\mathbf{x}_0 + \int_{t_0}^{t_0} \mathbf{\Phi}(t_0, \tau)\mathbf{B}(\tau)\mathbf{u}(\tau)\, d\tau = \mathbf{x}_0$$

To show that $\mathbf{x}(t)$ of (6.3-11) satisfies (6.3-1), we differentiate both sides of (6.3-11) with respect to t to get

$$\dot{\mathbf{x}}(t) = \dot{\mathbf{\Phi}}(t, t_0)\mathbf{x}_0 + \mathbf{\Phi}(t, \tau)\mathbf{B}(\tau)\mathbf{u}(\tau)\big|_{\tau=t} + \int_{t_0}^{t} \dot{\mathbf{\Phi}}(t, \tau)\mathbf{B}(\tau)\mathbf{u}(\tau)\, d\tau$$

This equation together with (6.3-6) and (6.3-7) yields

$$\dot{\mathbf{x}}(t) = \mathbf{A}(t)\mathbf{\Phi}(t, t_0)\mathbf{x}_0 + \mathbf{B}(t)\mathbf{u}(t) + \mathbf{A}(t)\int_{t_0}^{t} \mathbf{\Phi}(t, \tau)\mathbf{B}(\tau)\mathbf{u}(\tau)\, d\tau$$

Rearranging and using (6.3-11), we get

$$\dot{\mathbf{x}}(t) = \mathbf{A}(t)\mathbf{x}(t) + \mathbf{B}(t)\mathbf{u}(t)$$

This proves the theorem.

Now if the output state equations of the network under consideration are given by

$$\mathbf{y}(t) = \mathbf{C}(t)\mathbf{x}(t) + \mathbf{D}(t)\mathbf{u}(t)$$

the response of the network can be written as

$$\mathbf{y}(t) = \mathbf{C}(t)\mathbf{\Phi}(t, t_0)\mathbf{x}_0 + \mathbf{C}(t)\int_{t_0}^{t} \mathbf{\Phi}(t, \tau)\mathbf{B}(\tau)\mathbf{u}(\tau)\, d\tau + \mathbf{D}(t)\mathbf{u}(t) \tag{6.3-12}$$

Since the network under study is linear, the response can be separated into two parts, the zero input response $\mathbf{y}_{0i}$ and the zero state response $\mathbf{y}_{0s}$:

$$\mathbf{y}(t) = \mathbf{y}_{0i} + \mathbf{y}_{0s}$$

where

$$\mathbf{y}_{0i} \triangleq \mathbf{C}(t)\mathbf{\Phi}(t, t_0)\mathbf{x}_0 \tag{6.3-13}$$

and

$$\mathbf{y}_{0s} \triangleq \mathbf{C}(t)\int_{t_0}^{t} \mathbf{\Phi}(t, \tau)\mathbf{B}(\tau)\mathbf{u}(\tau)\, d\tau + \mathbf{D}(t)\mathbf{u}(t) \tag{6.3-14}$$

Remark 6.3-1. Although (6.3-11) represents the *form* of the solution of the linear time-varying state equation under study, unless $\mathbf{\Phi}(t, \tau)$ is known it is not of much use. In contrast to the state transition matrix of linear time-invariant state equations, as yet the problem of finding $\mathbf{\Phi}(t, \tau)$ in closed form is unsolved.

We shall now consider some specific forms of $\mathbf{A}(t)$ for which $\mathbf{\Phi}(t, \tau)$ can be computed.

Theorem 6.3-2. Consider the linear time-varying homogeneous differential equation

$$\dot{\mathbf{x}} = \mathbf{A}(t)\mathbf{x}(t) \tag{6.3-15}$$

If $\mathbf{A}(t)$ and $\int_{\tau}^{t} \mathbf{A}(\sigma)\, d\sigma$ commute, the state transition matrix of (6.3-15) is

$$\mathbf{\Phi}(t,\tau) = \exp\left[\int_{\tau}^{t} \mathbf{A}(\sigma) d\sigma\right] \tag{6.3-16}$$

Proof

To show that (6.3-16) is the state transition matrix of (6.3-15), we must show that

$$\mathbf{\Phi}(t, t) = \mathbf{1} \quad \text{and} \quad \frac{\partial}{\partial t}\mathbf{\Phi}(t, \tau) = \mathbf{A}(t)\mathbf{\Phi}(t, \tau)$$

The first condition is trivially satisfied since

$$\mathbf{\Phi}(t, t) = \exp\left[\int_{t}^{t} \mathbf{A}(\sigma)\, d\sigma\right] = e^0 = \mathbf{1}$$

To show the second condition, let

$$\mathbf{B}(t, \tau) \triangleq \int_{\tau}^{t} \mathbf{A}(\sigma)\, d\sigma$$

Then (6.3-16) can be written as

$$\mathbf{\Phi}(t, \tau) = e^{\mathbf{B}(t, \tau)}$$

Expanding the right-hand side of this equation in an infinite power series, we obtain

$$\mathbf{\Phi}(t, \tau) = \mathbf{1} + \mathbf{B}(t, \tau) + \frac{1}{2!}\mathbf{B}^2(t, \tau) + \frac{1}{3!}\mathbf{B}^3(t, \tau) + \cdots$$

Differentiating both sides of this equation with respect to t and noting that

$$\frac{\partial}{\partial t}\mathbf{B}(t, \tau) = \mathbf{A}(t)$$

we obtain

$$\frac{\partial}{\partial t}\mathbf{\Phi}(t, \tau) = \mathbf{A}(t) + \frac{1}{2!}[\mathbf{A}(t)\mathbf{B}(t, \tau) + \mathbf{B}(t, \tau)\mathbf{A}(t)] + \frac{1}{3!}[\mathbf{A}(t)\mathbf{B}^2(t, \tau) + \mathbf{B}(t)\mathbf{A}(t)\mathbf{B}(t, \tau) + \mathbf{B}^2(t, \tau)\mathbf{A}(t)] + \cdots$$

By assumption of the theorem, we have

$$\mathbf{B}(t, \tau)\mathbf{A}(t) = \mathbf{A}(t)\mathbf{B}(t, \tau)$$

Then $$\dot{\mathbf{\Phi}}(t, \tau) = \mathbf{A}(t) + \mathbf{A}(t)\mathbf{B}(t, \tau) + \frac{1}{2!}\mathbf{A}(t)\mathbf{B}^2(t, \tau) + \cdots$$

Factoring $\mathbf{A}(t)$, we get

$$\dot{\mathbf{\Phi}}(t, \tau) = \mathbf{A}(t)\left[\mathbf{1} + \mathbf{B}(t, \tau) + \frac{1}{2!}\mathbf{B}^2(t, \tau) + \cdots\right]$$

Hence, $$\dot{\mathbf{\Phi}}(t, \tau) = \mathbf{A}(t)\mathbf{\Phi}(t, \tau)$$

This proves the theorem.

Example 6.3-1. Obtain the solution of the following linear time-varying homogeneous differential equation:

$$\begin{bmatrix} \dot{x}_1 \\ \dot{x}_2 \\ \dot{x}_3 \end{bmatrix} = \begin{bmatrix} t & 0 & 0 \\ 0 & \cos t & 0 \\ 0 & 0 & 1 \end{bmatrix} \begin{bmatrix} x_1 \\ x_2 \\ x_3 \end{bmatrix}, \qquad \begin{bmatrix} x_1(0) \\ x_2(0) \\ x_3(0) \end{bmatrix} = \begin{bmatrix} 1 \\ 1 \\ 2 \end{bmatrix}$$

Solution

In this case, $\mathbf{A}(t)$ and $\int_\tau^t \mathbf{A}(\sigma)\, d\sigma$ obviously commute. Hence,

$$\mathbf{\Phi}(t, \tau) = \exp\left[\int_\tau^t \mathbf{A}(\sigma)\, d\sigma\right] = \begin{bmatrix} e^{(1/2)(t^2-\tau^2)} & 0 & 0 \\ 0 & e^{\sin t - \sin \tau} & 0 \\ 0 & 0 & e^{t-\tau} \end{bmatrix}$$

Consequently, the solution of the equation is given by

$$\mathbf{x}(t) = \mathbf{\Phi}(t, 0)\mathbf{x}(0) = \begin{bmatrix} e^{(1/2)t^2} \\ e^{\sin t} \\ 2e^t \end{bmatrix} \quad \text{for } t \geq 0$$

Example 6.3-2. Obtain the state transition matrix for the differential equation

$$\begin{bmatrix} \dot{x}_1 \\ \dot{x}_2 \\ \dot{x}_3 \end{bmatrix} = \begin{bmatrix} t & 0 & 0 \\ 1 & t & 0 \\ 0 & 0 & 1 \end{bmatrix} \begin{bmatrix} x_1 \\ x_2 \\ x_3 \end{bmatrix}$$

Solution

Let us check (6.3-16):

$$\int_\tau^t \mathbf{A}(\sigma)\,d\sigma = \begin{bmatrix} \frac{1}{2}(t^2 - \tau^2) & 0 & 0 \\ t - \tau & \frac{1}{2}(t^2 - \tau^2) & 0 \\ 0 & 0 & t - \tau \end{bmatrix}$$

A simple manipulation shows that

$$\mathbf{A}(t)\int_\tau^t \mathbf{A}(\sigma)\,d\sigma = \begin{bmatrix} \frac{1}{2}t(t^2 - \tau^2) & 0 & 0 \\ \frac{3}{2}t^2 - \frac{1}{2}\tau^2 - t\tau & \frac{1}{2}t(t^2 - \tau^2) & 0 \\ 0 & 0 & t - \tau \end{bmatrix} = \left(\int_\tau^t \mathbf{A}(\sigma)\,d\sigma\right)\mathbf{A}(t)$$

Consequently,

$$\mathbf{\Phi}(t, \tau) = \exp\left[\int_\tau^t \mathbf{A}(\sigma)\,d\sigma\right]$$

which can be considered as a function of a matrix and may be computed using the methods discussed in the last section.

Weighting Function of Linear Time-Varying Networks. To introduce the basic role of the weighting function in the analysis of linear networks, it is best to consider a single-input single-output network. For this reason, let Fig. 6.3-1

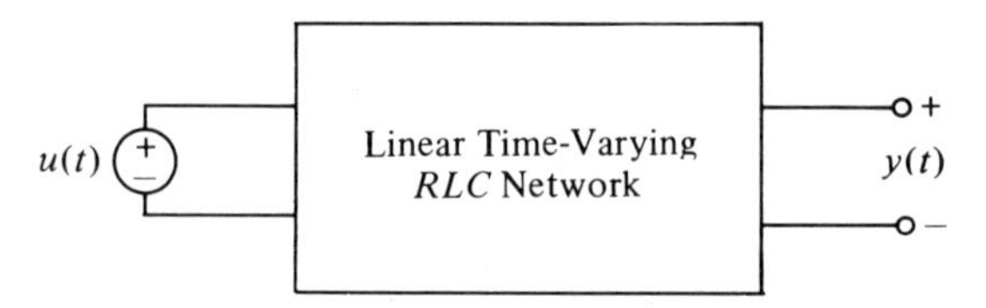

Figure 6.3-1 Single-input single-output linear time-varying network.

be the schematic representation of the network under consideration. The input–output state equations of this network can be written as

$$\dot{\mathbf{x}}(t) = \mathbf{A}(t)\mathbf{x}(t) + \mathbf{b}(t)u(t), \qquad \mathbf{x}(t_0) = \mathbf{x}_0 \tag{6.3-17}$$

$$y(t) = \mathbf{c}^T(t)\mathbf{x}(t) + d(t)u(t) \tag{6.3-18}$$

where $\mathbf{A}(t)$ is an $n \times n$ matrix, $\mathbf{b}(t)$ is an n-column vector, $\mathbf{c}^T(t)$ is an n-row vector, and $d(t)$ is a scalar function. According to (6.3-11), the solution of

(6.3-17) can be written as

$$\mathbf{x}(t) = \mathbf{\Phi}(t, t_0)\mathbf{x}_0 + \int_0^t \mathbf{\Phi}(t, \tau)\mathbf{b}(\tau)u(\tau)\, d\tau$$

where $\mathbf{\Phi}(\cdot, \cdot)$ represents the state transition matrix of (6.3-17). For simplicity, assume that prior to applying the input $u(t)$ the network is *relaxed* in the sense that all the capacitor voltages and inductor currents are zero; hence, $\mathbf{x}_0 = \mathbf{0}$. With this assumption we can extend the lower limit of the integral to $-\infty$, and $\mathbf{x}(t)$ can be written as

$$\mathbf{x}(t) = \int_{-\infty}^t \mathbf{\Phi}(t, \tau)\mathbf{b}(\tau)u(\tau)\, d\tau$$

If we now let the input be a *unit impulse*, $\delta(t - t')$, defined by

$$\delta(t - t') = 0 \quad \text{for all } t \neq t' \tag{6.3-19}$$

and

$$\int_{-\infty}^{\infty} \delta(t - t')\, dt = 1 \tag{6.3-20}$$

the state $\mathbf{x}(t)$ can be written as

$$\mathbf{x}(t, t') = \int_{-\infty}^t \mathbf{\Phi}(t, \tau)b(\tau)\delta(\tau - t')\, d\tau$$

where the first argument in $\mathbf{x}(t, t')$ denotes the time of the observation and the second argument, t' denotes the time of the application of the impulse function. Using the basic definition of the impulse function, $\mathbf{x}(t, t')$ becomes

$$\mathbf{x}(t, t') = \mathbf{\Phi}(t, t')\mathbf{b}(t')$$

To obtain the output or the *response* of the network, we can replace this $\mathbf{x}(t, t')$ in (6.3-18) to get

$$y(t, t') = \mathbf{c}^T(t)\mathbf{\Phi}(t, t')\mathbf{b}(t') + d(t)\,\delta(t - t')$$

This particular response of the network is given a specific name:

Definition 6.3-3 (Weighting Function). The *weighting function* of a linear time-varying network whose state equations are given in (6.3-17) and (6.3-18) is defined by

$$\psi(t, t') \triangleq \mathbf{c}^T(t)\mathbf{\Phi}(t, t')\mathbf{b}(t') + d(t)\,\delta(t - t') \tag{6.3-21}$$

where all the initial conditions are assumed to be zero.

Quite clearly, given the weighting function of a network, $\psi(t, t')$, for all $t \geq t'$ and for all $t' \geq t_0$, the zero state response of the network to an input $u(t)$ can be obtained from

$$y(t) = \int_{-\infty}^t \psi(t, \tau)u(\tau)\, d\tau$$

That is, the input is being "weighted" by $\psi(t, \tau)$. If the input $u(\tau)$ is zero prior to some initial time t_0 and the initial conditions at t_0 are $\mathbf{x}_0$, the total response of the network can be written as

$$y(t; t_0, \mathbf{x}_0, u) = \mathbf{c}^T\mathbf{\Phi}(t, t_0)\mathbf{x}_0 + \int_{t_0}^{t} \psi(t, \tau)u(\tau)\, d\tau \qquad (6.3.22)$$

where the first argument in $y(t; t_0, \mathbf{x}_0, u)$ denotes the observation time, the second and third represent the initial time and initial conditions, respectively, and the fourth represents the input. Notice that if the network under consideration is linear time invariant, the weighting function can be written as

$$\psi(t - t') = \mathbf{c}^T\mathbf{\Phi}(t - t')\mathbf{b} + d\delta(t - t') \qquad (6.3\text{-}23)$$

Comparing (6.3-21) and (6.3-23) shows that the *shape* of the weighting function of a linear time-invariant network, $\psi(t - t')$, is independent of the time of the application of the impulse t', whereas the shape of the weighting function of a linear time-varying network, $\psi(t, t')$, depends on t'. This fact is illustrated in Figs. 6.3-2 and 6.3-3. Notice that in Fig. 6.3-2 the weighting functions for $t' = 1, 2, \ldots$ are just duplicates of the first weighting function shifted by 1, 2, . . . seconds. In Fig. 6.3-3, however, the shape of $\psi(t, t')$ depends on t'.

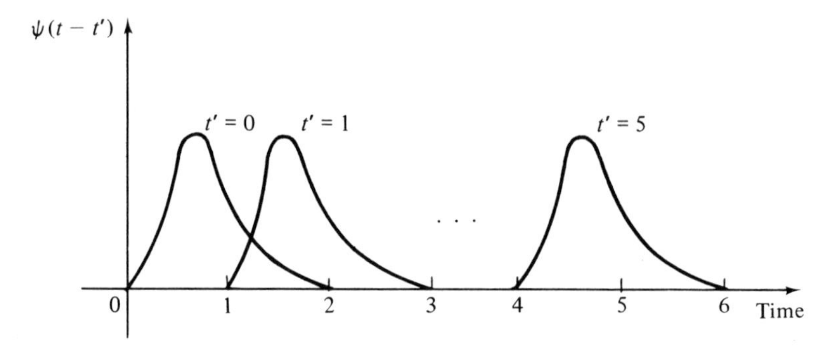

Figure 6.3-2 Weighting function of a linear time-invariant network.

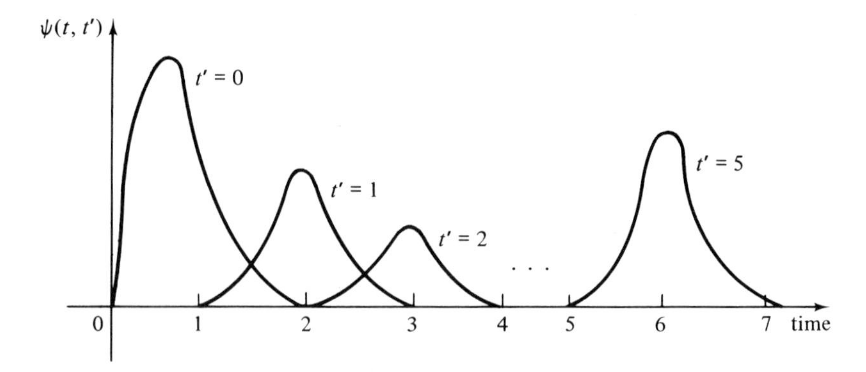

Figure 6.3-3 Weighting function of a linear time-varying network.

For a nonanticipative network the weighting function is zero prior to the application of the input; hence, the integral of (6.3-22) should be evaluated in the region $t \geq \tau$ in the (t, τ) plane. This region is shown as the shaded area in Fig. 6.3-4.

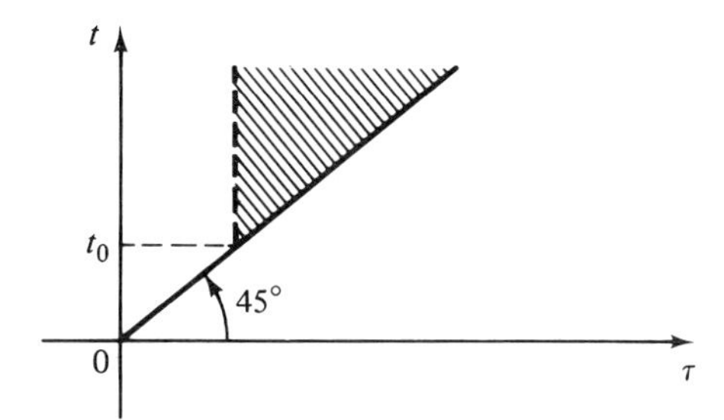

Figure 6.3-4 Region over which $\psi(t, \tau)$ of a nonanticipative network is defined is the shaded area.

As a final remark on the solution of linear time-varying state equations, we once again emphasize that closed-form solution of such equations can only be obtained for special cases. One such special case is discussed in Theorem 6.3-2. A general method for solving these equations in closed form is yet to be found. On the other hand, numerical solution of such equations can easily be found using a digital computer. We shall discuss some of the efficient methods of solving linear time-varying differential equations in Chapter 7. Next we turn our attention to the solution of nonlinear state equations.

6.4 SOLUTION OF NONLINEAR STATE EQUATIONS

The input–output state equations of a general nonlinear network can be written as

$$\dot{\mathbf{x}}(t) = \mathbf{f}(\mathbf{x}(t), \mathbf{u}(t)), \qquad \mathbf{x}(t_0) = \mathbf{x}_0 \tag{6.4-1}$$

$$\mathbf{y}(t) = \mathbf{g}(\mathbf{x}(t), \mathbf{u}(t)) \tag{6.4-2}$$

where $\mathbf{x}(t)$ is a vector representing the state, $\mathbf{u}(t)$ is an m vector representing the sources and possibly their derivatives, $\mathbf{y}(t)$ is the output vector, and $\mathbf{f}(\cdot, \cdot)$ and $\mathbf{g}(\cdot, \cdot)$ are two nonlinear vector functions. If (6.4-1) can be solved for $\mathbf{x}(t)$, then obtaining the output $\mathbf{y}(t)$ from (6.4-2) requires little effort. For this reason, in this section we concern ourselves primarily with the solution of (6.4-1).

It should be mentioned at this point that to search for an analytical solution of the general nonlinear differential equation (6.4-1) is a hopeless task! However, if we can show that the nonlinear differential equation under consideration has indeed a unique solution, this solution can then be found approximately using available numerical techniques. The question of the existence and uniqueness of the solution of nonlinear differential equations is of considerable importance. To convince ourselves of this fact, let us consider some simple examples.

Example 6.4-1. Consider the simple circuit shown in Fig. 6.4-1. Let the v-q characteristics of the nonlinear capacitor be given by

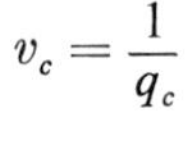

$$v_c = \frac{1}{q_c}$$

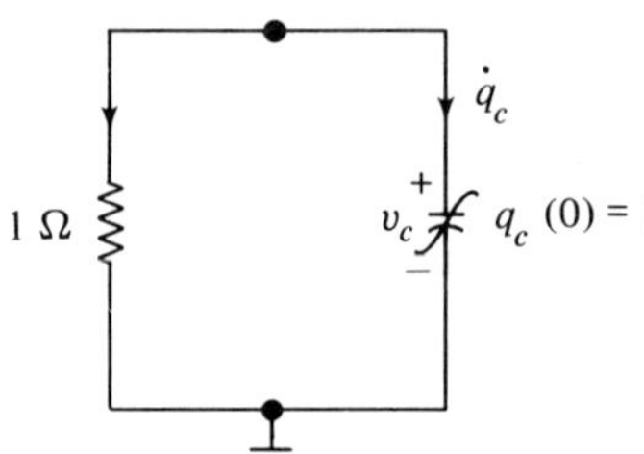

Figure 6.4-1 Simple nonlinear circuit for which a solution does not exist.

Then the differential equation describing this network is

$$\dot{q}_c = -\frac{1}{q_c}, \qquad q_c(0) = 1$$

Solving this equation for $q_c(t)$, we get

$$\tfrac{1}{2}q_c^2(t) = -t + k$$

Considering the initial conditions, this equation reduces to

$$q_c(t) = \pm\sqrt{1 - 2t}$$

Obviously, for $t > 0.5$, the charge on the capacitor will be imaginary, which physically has no meaning. Consequently, the state equation of this network has no real solution for all $t > 0.5$ second.

Example 6.4-2. Consider a first-order nonlinear network whose state equation is given by

$$\dot{x} = x^{1/3}, \qquad x(0) = 0$$

The following two functions are both solutions of this differential equation:

$$x_1(t) = 0, \qquad t \geq 0$$

$$x_2(t) = (\tfrac{2}{3}t)^{3/2}, \qquad t \geq 0$$

Consequently, there are two distinct solutions to the state equation under consideration; which one to choose?

Example 6.4-3. Consider the nonlinear differential equation

$$\dot{x} = x^2, \qquad x(0) = 1$$

The solution of this equation is

$$x(t) = \frac{1}{1-t}$$

This solution is unique and well defined for the interval [0, 1], but for $t = 1$ it is not defined; $x(t)$ in this case is said to have a *finite escape* time.

From the outcome of the previous examples it is evident that before trying to solve a nonlinear differential equation, either numerically or analytically, we should check to see whether a unique solution exists. For this purpose we shall next introduce a sufficient condition for the existence and uniqueness of the solution of nonlinear differential equations. To do so, let us rewrite (6.4-1) in a slightly different form:

$$\dot{\mathbf{x}} = \mathbf{f}(\mathbf{x}, t), \qquad \mathbf{x}(t_0) = \mathbf{x}_0 \tag{6.4-3}$$

In this representation the effect of the input $\mathbf{u}(t)$ is introduced by the second argument, t. The result of the following theorem, then, gives a sufficient (but not necessary) condition for the existence and uniqueness of the solution of the differential equation given in (6.4-3).

Theorem 6.4-1 (Existence and Uniqueness Theorem). Consider the differential equation (6.4-3). Assume that

(a) $\mathbf{f}(\mathbf{x}, \cdot)$ is a continuous function for each $\mathbf{x}$.

(b) There exists a scalar constant k such that

$$|\mathbf{f}(\mathbf{x}_p, t) - \mathbf{f}(\mathbf{x}_q, t)|_\infty \leq k|\mathbf{x}_p - \mathbf{x}_q|_\infty \tag{6.4-4}$$

for any vector functions $\mathbf{x}_p$ and $\mathbf{x}_q$ and for all $t \geq 0$ where the l_∞ norm, $|\cdot|_\infty$, is the sup norm defined by (2.3-5). Under these conditions, for any initial condition $\mathbf{x}_0$ and for any initial time t_0, (6.4-3) has a unique solution.

Condition (b) is called the *Lipschitz* condition and must be satisfied for all $\mathbf{x}_p$ and $\mathbf{x}_q$ and all t. To prove this theorem, we must first show that there exists a solution to (6.4-3) and then show that this solution is unique. Before proving this theorem, recall that a *continuous* function $\mathbf{x}(t)$ qualifies as a solution of (6.4-3) if it satisfies the initial condition *and* also the differential equation; that is, we must have

$$\mathbf{x}(t_0) = \mathbf{x}_0$$

and

$$\dot{\mathbf{x}}(t) = \mathbf{f}(\mathbf{x}(t), t)$$

Proof (Existence)

To prove that conditions (a) and (b) are indeed sufficient for the existence of a solution, we construct such a solution using an iterative technique called Picard's iteration. It goes as follows:

Consider the finite time interval $[t_0, t_1]$ and form the infinite sequence of vector functions $\mathbf{x}_1(t), \mathbf{x}_2(t), \ldots$ defined by

$$\begin{aligned}
\mathbf{x}_1(t) &= \mathbf{x}_0 + \int_{t_0}^{t} \mathbf{f}(\mathbf{x}_0, \sigma)\, d\sigma \\
\mathbf{x}_2(t) &= \mathbf{x}_0 + \int_{t_0}^{t} \mathbf{f}(\mathbf{x}_1(\sigma), \sigma)\, d\sigma \\
&\vdots \\
\mathbf{x}_{n+1}(t) &= \mathbf{x}_0 + \int_{t_0}^{t} \mathbf{f}(\mathbf{x}_n(\sigma), \sigma)\, d\sigma \\
&\vdots
\end{aligned} \tag{6.4-5}$$

where $\mathbf{x}_j(t)$ is an n-column vector whose elements are continuous functions defined over the interval $[t_0, t_1]$. We next show that this sequence is a Cauchy sequence which converges to a continuous vector function $\mathbf{x}(t)$, satisfying conditions (a) and (b). To do so, we must show that

$$\|\mathbf{x}_m - \mathbf{x}_n\|_\infty \longrightarrow 0 \quad \text{as } m \text{ and } n \longrightarrow \infty \tag{6.4-6}$$

where the L_∞ norm, $\|\cdot\|_\infty$, is defined in (2.3-11). For this purpose, subtract the nth term from the $(n+1)$st term of (6.4-5) to get

$$\mathbf{x}_{n+1}(t) - \mathbf{x}_n(t) = \int_{t_0}^{t} [\mathbf{f}(\mathbf{x}_n(\sigma), \sigma) - \mathbf{f}(\mathbf{x}_{n-1}(\sigma), \sigma)]\, d\sigma$$

Taking the l_∞ norm of both sides of this equation and using the Schwartz inequality, we get

$$|\mathbf{x}_{n+1}(t) - \mathbf{x}_n(t)|_\infty \leq \int_{t_0}^{t} |\mathbf{f}(\mathbf{x}_n(\sigma), \sigma) - \mathbf{f}(\mathbf{x}_{n-1}(\sigma), \sigma)|_\infty\, d\sigma$$

Now by Lipschitz condition there exists a finite number k such that the integrand in the integral becomes less than $k|\mathbf{x}_n(t) - \mathbf{x}_{n-1}(t)|_\infty$. Hence, the last inequality can be written as

$$|\mathbf{x}_{n+1}(t) - \mathbf{x}_n(t)|_\infty \leq k \int_{t_0}^{t} |\mathbf{x}_n(\sigma) - \mathbf{x}_{n-1}(\sigma)|_\infty\, d\sigma \tag{6.4-7}$$

But from (6.4-5) we have

$$|\mathbf{x}_1(t) - \mathbf{x}_0|_\infty \leq \int_{t_0}^{t} |\mathbf{f}(\mathbf{x}_0, \sigma)|_\infty\, d\sigma \leq \int_{t_0}^{t_1} |\mathbf{f}(\mathbf{x}_0, \sigma)|_\infty\, d\sigma \leq M \tag{6.4-8}$$

where M is a finite constant, since the interval $[t_0, t_1]$ is finite and $|\mathbf{f}(\mathbf{x}_0, \sigma)|_\infty$ is assumed to be finite. From (6.4-7) and (6.4-8), for $n = 1$, we get

$$|\mathbf{x}_2(t) - \mathbf{x}_1(t)|_\infty \leq k \int_{t_0}^{t} M\, d\sigma$$

Integrating the right-hand side of this inequality and repeating this process for $n = 2, 3, \ldots$ yields

$$
\begin{aligned}
|\mathbf{x}_2(t) - \mathbf{x}_1(t)|_\infty &\le kM(t - t_0) \\
|\mathbf{x}_3(t) - \mathbf{x}_2(t)|_\infty &\le M\frac{[k(t - t_0)]^2}{2!} \\
&\vdots \\
|\mathbf{x}_{n+1}(t) - \mathbf{x}_n(t)|_\infty &\le M\frac{[k(t - t_0)]^n}{n!}
\end{aligned}
\tag{6.4-9}
$$

Taking the L_∞ norm of both sides of the last inequality in (6.4-9) over the interval $T \triangleq [t_0, t_1]$ results in

$$\|\mathbf{x}_{n+1} - \mathbf{x}_n\|_\infty \le M\frac{(kT)^n}{n!} \tag{6.4-10}$$

To prove (6.4-6), let $m = n + j$; then (6.4-10) can be written as

$$\|\mathbf{x}_m - \mathbf{x}_n\|_\infty = \|\mathbf{x}_{n+j} - \mathbf{x}_n\|_\infty = \|(\mathbf{x}_{n+j} - \mathbf{x}_{n+j-1}) + \cdots + (\mathbf{x}_{n+1} - \mathbf{x}_n)\|_\infty$$

Using the triangle inequality, this equation can be written as

$$\|\mathbf{x}_m - \mathbf{x}_n\|_\infty \le \sum_{i=1}^{j} \|\mathbf{x}_{n+i} - \mathbf{x}_{n+i-1}\|_\infty \tag{6.4-11}$$

This equation together with (6.4-10) yields

$$\|\mathbf{x}_m - \mathbf{x}_n\|_\infty \le M \sum_{i=1}^{j} \frac{(kT)^{n+i-1}}{(n + i - 1)!} \tag{6.4-12}$$

Since

$$\frac{1}{(n + i - 1)!} \le \frac{1}{n!} \cdot \frac{1}{(i - 1)!}$$

we can write (6.4-12) as

$$\|\mathbf{x}_m - \mathbf{x}_n\|_\infty \le M\frac{(kT)^n}{n!} \sum_{i=1}^{j} \frac{(kT)^{i-1}}{(i - 1)!}$$

If we let $j \longrightarrow \infty$ in the right-hand side, we get

$$\|\mathbf{x}_m - \mathbf{x}_n\|_\infty \le M\frac{(kT)^n}{n!} \sum_{i=1}^{\infty} \frac{(kT)^{i-1}}{(i - 1)!}$$

But

$$\sum_{i=1}^{\infty} \frac{(kT)^{i-1}}{(i - 1)!} \triangleq e^{kT}$$

and hence

$$\|\mathbf{x}_m - \mathbf{x}_n\|_\infty \le Me^{kT}\frac{(kT)^n}{n!} \tag{6.4-13}$$

Since the right-hand side of this inequality tends to zero as $n \longrightarrow \infty$ (why?), the left-hand side will also go to zero. Hence, (6.4-6) is satisfied, and therefore the sequence of continuous functions $\mathbf{x}_1(t), \mathbf{x}_2(t), \ldots, \mathbf{x}_n(t)$ given in (6.4-5) form a Cauchy sequence. Consequently, this sequence will converge to a continuous function $\mathbf{x}(t)$ defined on $[t_0, t_1]$ (see [3]). More precisely, as $n \longrightarrow \infty$, $\mathbf{x}_n(t)$ given in (6.4-5) tends to $\mathbf{x}(t)$ given by

$$\mathbf{x}(t) = \mathbf{x}_n + \int_{t_0}^{t} \mathbf{f}(\mathbf{x}(\sigma), \sigma)\, d\sigma \tag{6.4-14}$$

The function $\mathbf{x}(t)$ is indeed *a* solution to the differential equation (6.4-1), since differentiating both sides of (6.4-14) yields

$$\dot{\mathbf{x}}(t) = \mathbf{f}(\mathbf{x}(t), t)$$

and, furthermore,

$$\mathbf{x}(t_0) = \mathbf{x}_0$$

This proves the first part of the theorem.

To get a feeling for the sequence of continuous functions given in (6.4-5) and the Picard iteration, let us consider a simple example.

Example 6.4-4. Use the Picard iteration discussed in the proof of Theorem 6.4-1 to find a solution to the scalar linear differential equation

$$\dot{x} = x(t), \qquad x(t_0) = x_0$$

Solution

From (6.4-5) we get

$$x_1(t) = x_0 + \int_{t_0}^{t} x_0\, d\sigma = x_0[1 + (t - t_0)]$$

$$x_2(t) = x_0 + \int_{t_0}^{t} x_0[1 + (\sigma - t_0)]\, d\sigma = x_0\left[1 + (t - t_0) + \frac{1}{2!}(t - t_0)^2\right]$$

$$\vdots \qquad\qquad \vdots$$

$$x_n(t) = x_0\left[1 + (t - t_0) + \frac{1}{2!}(t - t_0)^2 + \cdots + \frac{1}{n!}(t - t_0)^n\right]$$

The last equation can be written in a more compact form:

$$x_n(t) = x_0 \sum_{k=0}^{n} \frac{1}{k!}(t - t_0)^k$$

As $n \longrightarrow \infty$, $x_n(t)$ tends to $x(t)$ given by

$$x(t) = x_0 \sum_{k=0}^{\infty} \frac{1}{k!}(t - t_0)^k = x_0 e^{(t - t_0)}$$

which we already know is the solution of the differential equation under consideration.

To prove the uniqueness of the solution, we must use an important lemma, known as the Bellman–Gronwall lemma. Let us first state and prove this lemma, and then continue with the proof of the second part of the theorem.

Bellman–Gronwall Lemma. Consider two continuous scalar functions $\alpha(t)$ and $\beta(t)$ defined on the interval $[t_0, t_1]$; if

$$\text{(a)} \qquad \alpha(t) \geq 0 \quad \text{and} \quad \beta(t) \geq 0 \quad \text{for } t_0 \leq t \leq t_1 \tag{6.4-15}$$

and

$$\text{(b)} \qquad \alpha(t) \leq \epsilon + \int_{t_0}^{t} \alpha(\sigma)\beta(\sigma)\, d\sigma \quad \text{for } t_0 \leq t \leq t_1 \tag{6.4-16}$$

where ϵ is a positive constant, then

$$\alpha(t) \leq \epsilon \exp\left[\int_{t_0}^{t} \beta(\sigma)\, d\sigma\right] \quad \text{for } t_0 \leq t \leq t_1 \tag{6.4-17}$$

To prove this lemma, let us define a function $p(t)$ as

$$p(t) \triangleq \epsilon + \int_{t_0}^{t} \alpha(\sigma)\beta(\sigma)\, d\sigma \tag{6.4-18}$$

Then (6.4-16) becomes

$$\alpha(t) \leq p(t) \tag{6.4-19}$$

Differentiating both sides of (6.4-18) and considering the last inequality, we get

$$\dot{p}(t) = \alpha(t)\beta(t) \leq p(t)\beta(t)$$

and, since $p(t) \geq \epsilon > 0$, we can divide both sides of the inequality by $p(t)$; this results in

$$\frac{\dot{p}(t)}{p(t)} \leq \beta(t)$$

Integrating the last inequality and noting that $p(t_0) = \epsilon$, we get

$$p(t) \leq \epsilon \exp\left[\int_{t_0}^{t} \beta(d\sigma)\, d\sigma\right]$$

Then (6.4-19) implies that

$$\alpha(t) \leq \epsilon \exp\left[\int_{t_0}^{t} \beta(\sigma)\, d\sigma\right]$$

This proves the lemma. We can now prove the second part of the theorem.

Proof (Uniqueness)

To show the uniqueness of the solution of (6.4-3) under conditions (a) and (b) of the theorem, assume that there are two distinct solutions $\mathbf{x}(t)$ and $\hat{\mathbf{x}}(t)$ satisfying (6.4-3). From (6.4-14) we have

$$\mathbf{x}(t) = \mathbf{x}_0 + \int_{t_0}^{t} \mathbf{f}(\mathbf{x}(\sigma), \sigma)\, d\sigma$$

and

$$\hat{\mathbf{x}}(t) = \mathbf{x}_0 + \int_{t_0}^{t} \mathbf{f}(\hat{\mathbf{x}}(\sigma), \sigma)\, d\sigma$$

Next we prove that $\mathbf{x}(t)$ and $\hat{\mathbf{x}}(t)$ are indeed the same. Subtracting $\mathbf{x}(t)$ from

$\hat{\mathbf{x}}(t)$ and taking the l_∞ norm of both sides of the resulting equation, we get

$$|\hat{\mathbf{x}}(t) - \mathbf{x}(t)|_\infty \leq \int_{t_0}^{t} |\mathbf{f}(\hat{\mathbf{x}}(\sigma), \sigma) - \mathbf{f}(\mathbf{x}(\sigma), \sigma)|_\infty \, d\sigma$$

By the Lipschitz condition, we have

$$|\hat{\mathbf{x}}(t) - \mathbf{x}(t)|_\infty \leq \int_{t_0}^{t} k\,|\hat{\mathbf{x}}(\sigma) - \mathbf{x}(\sigma)|_\infty \, d\sigma$$

or

$$|\hat{\mathbf{x}}(t) - \mathbf{x}(t)|_\infty \leq \epsilon + \int_{t_0}^{t} k\,|\hat{\mathbf{x}}(\sigma) - \mathbf{x}(\sigma)|_\infty \, d\sigma$$

where ϵ is an arbitrary positive number. Comparing the last equation to (6.4-16) and using the conclusion of the Bellman–Gronwall lemma, we get

$$|\hat{\mathbf{x}}(t) - \mathbf{x}(t)|_\infty \leq \epsilon \exp\left[\int_{t_0}^{t} k\,|\hat{\mathbf{x}}(\sigma) - \mathbf{x}(\sigma)|_\infty \, d\sigma\right.$$

Letting $\epsilon \longrightarrow 0$, the right-hand side of the last inequality will go to zero, and hence

$$|\hat{\mathbf{x}}(t) - \mathbf{x}(t)|_\infty = 0 \quad \text{for all } t_0 < t < t_1$$

or, equivalently,

$$\hat{\mathbf{x}}(t) = \mathbf{x}(t) \quad \text{for all } t_0 < t < t_1$$

This proves the uniqueness of the solution.

Remark 6.4-1. In Theorem 6.4-1 only *sufficient* conditions for existence and uniqueness were stated. If the state equations of a network satisfy these conditions, the solutions exist and are unique. If, however, the conditions are not satisfied, the result is inconclusive—we should use other conditions for determining the existence and uniqueness of the solution of the differential equation under study.

Remark 6.4-2. The Lipschitz condition stated in Theorem 6.4-1 implies that the function $\mathbf{f}(\cdot, t)$ is continuous for each t (show this). If a function $\mathbf{f}(\mathbf{x}, t)$ is not continuous in $\mathbf{x}$, there is no need to proceed further; the Lipschitz condition will not be satisfied. But this does not imply that a solution does not exist.

Example 6.4-5. Consider the nonlinear differential equation

$$\dot{x}(t) = \sin(x(t)), \qquad x(0) = 1$$

Since this nonlinear differential equation does not explicitly depend on time, condition (a) of Theorem 6.4-1 is automatically satisfied. To check condition (b), note that the right-hand side of the following inequality is finite for all x_1 and x_2:

$$\frac{|\sin(x_1) - \sin(x_2)|}{|x_1 - x_2|} \leq k$$

Consequently, the differential equation under study has a unique solution for all $t \geq 0$.

Exercise 6.4-1. Show that for $x_1 = x_2$ the left-hand side of the preceding inequality is finite.

Example 6.4-6. Consider the state equation of the nonlinear network discussed in Example 6.4-1:

$$\dot{x}(t) = \frac{1}{x(t)} \qquad x(0) = 1$$

To check the Lipschitz conditions, take two arbitrary functions $x_1(t), x_2(t)$ and form the ratio

$$\frac{\left| \dfrac{1}{x_1(t)} - \dfrac{1}{x_2(t)} \right|}{|x_1(t) - x_2(t)|} = \frac{1}{|x_1(t)x_2(t)|}$$

Now if either $x_1(t)$ or $x_2(t)$ is zero, the ratio will be infinite; therefore, the Lipschitz condition is not satisfied for any region that contains the origin, $x(t) = 0$. The conclusion of Theorem 6.4-1 can also be used to determine the existence and uniqueness of the solution of linear state equations. More explicitly,

Corollary 6.4-1. Consider the linear time-varying differential equation

$$\dot{\mathbf{x}}(t) = \mathbf{A}(t)\mathbf{x}(t) + \mathbf{B}(t)\mathbf{u}(t), \qquad \mathbf{x}(t_0) = \mathbf{x}_0 \tag{6.4-20}$$

If $\mathbf{A}(t)$ and $\mathbf{B}(t)\mathbf{u}(t)$ are continuous for $t \geq t_0$, then (6.4-20) has a unique solution on $t_0 \leq t < \infty$.

The proof of this corollary is similar to the proof of Theorem 6.4-1 and hence will be omitted.

Having established the existence and uniqueness of the solution of nonlinear differential equations, we can then try to obtain this solution either analytically or numerically. Several numerical methods for computing these solutions will be presented in Chapter 7.

REFERENCES

[1] ZADEH, L. A., and C. A. DESOER, *Linear Systems Theory*, McGraw-Hill Book Company, New York, 1963.

[2] OGATA, K., *State-Space Analysis of Control Systems*, Prentice-Hall, Inc., Englewood Cliffs, N.J., 1967.

[3] BARTEL, R. G., *The Elements of Real Analysis*, John Wiley & Sons, Inc., New York, 1964.

[4] CODDINGTON, E. A., and N. LEVINSON, *Theory of Ordinary Differential Equations*, McGraw-Hill Book Company, New York, 1955.

PROBLEMS

6.1 Consider the matrix differential equation

$$\dot{\mathbf{X}}(t) = \mathbf{A}_1\mathbf{X}(t) + \mathbf{X}(t)\mathbf{A}_2 + \mathbf{F}(t), \qquad \mathbf{X}(0) = \mathbf{X}_0$$

where $\mathbf{X}(t)$, $\mathbf{A}_1$, $\mathbf{A}_2$, and $\mathbf{F}(t)$ are $n \times n$ matrices. Show that the solution of this differential equation is

$$\mathbf{X}(t) = e^{\mathbf{A}_1 t}\mathbf{X}_0 e^{\mathbf{A}_2 t} + \int_0^t e^{\mathbf{A}_1(t-\tau)}\mathbf{F}(\tau)e^{\mathbf{A}_2(t-\tau)}\, d\tau$$

6.2 Consider the algebraic equation

$$\mathbf{A}_1\mathbf{X} + \mathbf{X}\mathbf{A}_2 = -\mathbf{C}$$

where $\mathbf{A}_1$, $\mathbf{A}_2$ and $\mathbf{C}$ are given $n \times n$ matrices and $\mathbf{X}$ is an $n \times n$ *constant* matrix. Assume that $e^{\mathbf{A}_1 t} \longrightarrow \mathbf{0}$ and $e^{\mathbf{A}_2 t} \longrightarrow \mathbf{0}$ as $t \longrightarrow \infty$. Show that the solution of this equation is

$$\mathbf{X} = \int_0^\infty e^{\mathbf{A}_1 t}\mathbf{C}e^{\mathbf{A}_2 t}\, dt$$

6.3 Consider the linear time-invariant n-vector differential equation

$$\dot{\mathbf{x}} = \mathbf{A}\mathbf{x}$$

Show that if $\boldsymbol{\psi}_1(t), \boldsymbol{\psi}_2(t), \ldots, \boldsymbol{\psi}_n(t)$ are solutions of this differential equation then

$$\boldsymbol{\psi} = \sum_{i=1}^{n} \alpha_i \boldsymbol{\psi}_i(t)$$

where the α_i are scalar constants, is also a solution.

6.4 Show that if $\mathbf{AB} = \mathbf{BA}$ then

$$e^{\mathbf{A}+\mathbf{B}} = e^{\mathbf{A}} \cdot e^{\mathbf{B}}$$

for all $n \times n$ matrices $\mathbf{A}$ and $\mathbf{B}$.

6.5 Find two square matrices $\mathbf{A}$ and $\mathbf{B}$ such that

$$e^{\mathbf{A}+\mathbf{B}} \neq e^{\mathbf{A}} \cdot e^{\mathbf{B}}$$

6.6 Use the infinite-series method to show that

(a) For $\mathbf{A} = \begin{bmatrix} a & 0 \\ 0 & b \end{bmatrix}, \quad e^{\mathbf{A}t} = \begin{bmatrix} e^{at} & 0 \\ 0 & e^{bt} \end{bmatrix}$

(b) For $\mathbf{A} = \begin{bmatrix} 0 & \omega \\ -\omega & 0 \end{bmatrix}, \quad e^{\mathbf{A}t} = \begin{bmatrix} \cos \omega t & \sin \omega t \\ -\sin \omega t & \cos \omega t \end{bmatrix}$

(c) For $\mathbf{A} = \begin{bmatrix} 1 & 0 & 0 & 0 \\ 0 & -1 & 0 & 0 \\ 0 & 0 & 0 & 2 \\ 0 & 0 & -2 & 0 \end{bmatrix}, \quad e^{\mathbf{A}t} = \begin{bmatrix} e^{t} & 0 & 0 & 0 \\ 0 & e^{-t} & 0 & 0 \\ 0 & 0 & \cos 2t & \sin 2t \\ 0 & 0 & -\sin 2t & \cos 2t \end{bmatrix}$

6.7 Given a linear time-invariant network whose input–output state equations are

$$\dot{\mathbf{x}} = \begin{bmatrix} 0 & 0 \\ -1 & 0 \end{bmatrix}\mathbf{x} + \begin{bmatrix} 1 \\ 1 \end{bmatrix}u(t), \qquad \mathbf{x}(0) = \begin{bmatrix} 1 \\ 1 \end{bmatrix}$$

$$y = [1 \quad 0]\mathbf{x}$$

where $u(t) = \sin t$,

(a) Determine the zero-input response.

(b) Determine the complete response.

6.8 Given a linear time-invariant network whose state equations are

$$\dot{\mathbf{x}} = \begin{bmatrix} -2 & -3 \\ 0 & -2 \end{bmatrix}\mathbf{x} + \begin{bmatrix} 1 \\ 1 \end{bmatrix}e^{-2t}1(t), \qquad \mathbf{x}(0) = \begin{bmatrix} 1 \\ 1 \end{bmatrix}$$

$$y = [1 \quad 0]\mathbf{x}$$

where $1(t)$ is the unit step function,

(a) Determine the zero-input response.

(b) Determine the total response.

[Ans:. (a) $y_{oi}(t) = (1 - 3t)e^{-2t}$, (b) $y(t) = (1 - 2t - \frac{3}{2}t^2)e^{-2t}$]

6.9 Use the Laplace-transform method to show that

(a) For $\mathbf{A} = \begin{bmatrix} a & 1 \\ 0 & a \end{bmatrix}, \qquad e^{\mathbf{A}t} = \begin{bmatrix} 1 & t \\ 0 & 1 \end{bmatrix}e^{at}$

(b) For $\mathbf{A} = \begin{bmatrix} a & 1 & 0 \\ 0 & a & 1 \\ 0 & 0 & a \end{bmatrix}, \qquad e^{\mathbf{A}t} = \begin{bmatrix} 1 & t & \frac{1}{2}t^2 \\ 0 & 1 & t \\ 0 & 0 & 1 \end{bmatrix}e^{at}$

6.10 Use the results of the Cayley–Hamilton theorem to show that for any positive integer k the $n \times n$ matrix $\mathbf{A}^k$ can be written as

$$\mathbf{A}^k = a_1\mathbf{A}^{n-1} + a_2\mathbf{A}^{n-2} + \cdots + a_k\mathbf{A}^0, \qquad k \geq n$$

6.11 Use the conclusions of Theorem 6.2-3 to compute

(a) $\sin \mathbf{A}$

(b) $\cos \mathbf{A}$

(c) $\mathbf{A}^{79}$

where

$$\mathbf{A} = \begin{bmatrix} 1 & 3 \\ 0 & 2 \end{bmatrix}$$

6.12 Use the results of the Cayley–Hamilton theorem to compute $\mathbf{A}^{-1}$ where

$$\mathbf{A} = \begin{bmatrix} 1 & 3 & 2 & 1 \\ 0 & 2 & 1 & 2 \\ 0 & 0 & 3 & 0 \\ 0 & 0 & 0 & 4 \end{bmatrix}$$

6.13 Use the Cayley–Hamilton theorem to compute $e^{\mathbf{A}t}$ for

(a) $\mathbf{A} = \begin{bmatrix} 1 & 2 \\ 0 & 3 \end{bmatrix}$

(b) $\mathbf{A} = \begin{bmatrix} 1 & 0 & 2 \\ 0 & 1 & 1 \\ 0 & 0 & 2 \end{bmatrix}$

(c) $\mathbf{A} = \begin{bmatrix} 0 & 1 & 0 \\ 0 & 0 & 1 \\ 6 & 11 & 6 \end{bmatrix}$

(d) $\mathbf{A} = \begin{bmatrix} 0 & 1 & 0 \\ 0 & 0 & 1 \\ 0 & -2 & -3 \end{bmatrix}$

6.14 Show that the Vandermonde matrix

$$\mathbf{V} \triangleq \begin{bmatrix} 1 & \lambda_1 & \lambda_1^2 & \cdots & \lambda_1^{n-1} \\ 1 & \lambda_2 & \lambda_2^2 & \cdots & \lambda_2^{n-1} \\ 1 & \lambda_3 & \lambda_3^2 & \cdots & \lambda_3^{n-1} \\ \cdot & & & & \\ \cdot & & & & \\ \cdot & & & & \\ 1 & \lambda_n & \lambda_n^2 & \cdots & \lambda_n^{n-1} \end{bmatrix}$$

is nonsingular provided that all the λ_i are distinct. [Hint: Show that det $\mathbf{V} = \prod_{i=1}^{n} \prod_{j=i+1}^{n} (\lambda_j - \lambda_i)$].

6.15 Let λ_1 and λ_2 be the eigenvalues of

$$\mathbf{A} = \begin{bmatrix} 1 & 2 \\ 2 & -1 \end{bmatrix}$$

Show that the eigenvalues of $\mathbf{A}^2$ are λ_1^2 and λ_2^2, and the eigenvalues of $\mathbf{A}^3$ are λ_1^3 and λ_2^3.

6.16 Let $\lambda_1, \lambda_2, \ldots, \lambda_n$ be eigenvalues of a matrix $\mathbf{A}$, and $f(\mathbf{A})$ be any polynomial of degree r in $\mathbf{A}$. Show that the eigenvalues of $f(\mathbf{A})$ are $f(\lambda_1), f(\lambda_2), \ldots, f(\lambda_n)$.

6.17 Find the 2×2 matrix $\mathbf{A}$ if $e^{\mathbf{A}}$ is given as

$$e^{\mathbf{A}} = \begin{bmatrix} 1 & 0 \\ 1 & 2 \end{bmatrix}$$

[Hint: Use the conclusions of the Cayley–Hamilton theorem and the results of Problem 6.16.]

6.18 Show that the minimal polynomial of

$$\mathbf{A} = \begin{bmatrix} 2 & 0 & 0 \\ 0 & 2 & 0 \\ 0 & 3 & 1 \end{bmatrix}$$

is $\Phi(\lambda) = \lambda^2 - 3\lambda + 2$.

6.19 Find the minimal polynomial of

$$\mathbf{A} = \begin{bmatrix} 1 & 0 & 1 & 0 \\ 0 & 1 & 0 & 1 \\ 0 & 0 & 1 & 0 \\ 0 & 0 & 0 & 1 \end{bmatrix}$$

6.20 Show that the minimal polynomial of

$$\mathbf{A} = \begin{bmatrix} 1 & 1 & 0 & 0 \\ 0 & 1 & 0 & 0 \\ 0 & 0 & 1 & 1 \\ 0 & 0 & 0 & 1 \end{bmatrix}$$

is $\Phi(\lambda) = (\lambda - 1)^2$.

6.21 Consider the linear time-invariant network shown in Fig. P6.21.

(a) Write the input–output state equation of this network.

(b) Let $u_1(t) = \delta(t)$ and $u_2(t) = 0$. Solve for the output vector $\mathbf{y} = [y_1 \quad y_2]^T$. Assume all the initial conditions are zero.

(c) Let $u_1(t) = 1(t)$ (unit step function) and $u_2(t) = 0$. Solve for the output $\mathbf{y}(t) = [y_1 \quad y_2]^T$. Assume zero initial conditions.

(d) Assume that the initial voltage on the capacitor is 1 V. Repeat part (c).

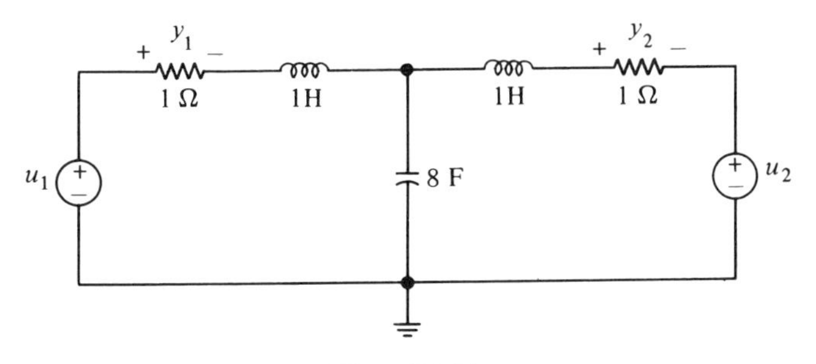

Figure P6.21

6.22 Show that

$$\mathbf{W}(t) = \begin{bmatrix} \cos t + 3 \sin t & 2 \cos t + 4 \sin t \\ -\sin t + 3 \cos t & -2 \sin t + 4 \cos t \end{bmatrix}$$

is the fundamental matrix for the linear time-invariant vector differential equation

$$\dot{\mathbf{x}} = \begin{bmatrix} 0 & 1 \\ -1 & 0 \end{bmatrix} \mathbf{x}$$

6.23 Show that

$$\mathbf{W}(t) = \begin{bmatrix} t^2 + 1 & t + 1 \\ t - 1 & 1 \end{bmatrix}$$

is the fundamental matrix for

$$\dot{\mathbf{x}} = \frac{1}{2}\begin{bmatrix} t+1 & -t^2 - 2t + 1 \\ 1 & -t - 1 \end{bmatrix}\mathbf{x}$$

6.24 Determine the state transition matrix of the linear time-varying homogeneous differential equation

$$\dot{\mathbf{x}} = \begin{bmatrix} 0 & 1 & 0 \\ -1 & 0 & 0 \\ 0 & 0 & t \end{bmatrix}\mathbf{x}$$

6.25 Show that the fundamental matrix of

$$\dot{\mathbf{x}} = \begin{bmatrix} 0 & 1 \\ 6t^{-2} & 0 \end{bmatrix}\mathbf{x}$$

is

$$\mathbf{W}(t) = \begin{bmatrix} t^3 & t^{-2} \\ 3t^2 & -2t^{-3} \end{bmatrix}$$

6.26 Consider the linear time-varying differential equation

$$\dot{\mathbf{x}} = \begin{bmatrix} -t & t \\ 0 & -2t \end{bmatrix}\mathbf{x}, \qquad \mathbf{x}(0) = \begin{bmatrix} 1 \\ 1 \end{bmatrix}$$

$$y(t) = [t \quad 1]\mathbf{x}$$

[Ans: $y(t) = 2te^{-t^2} + (1 - t)e^{-t^2}$]

6.27 Consider a linear time-varying network whose state equation is

$$\begin{bmatrix} \dot{x}_1 \\ \dot{x}_2 \end{bmatrix} = \begin{bmatrix} t & 1 \\ 1 & t \end{bmatrix}\begin{bmatrix} x_1 \\ x_2 \end{bmatrix}$$

Show that the state transition matrix of this network is

$$\mathbf{\Phi}(t, \tau) = \exp\begin{bmatrix} \frac{1}{2}(t^2 - \tau^2) & t - \tau \\ t - \tau & \frac{1}{2}(t^2 - \tau^2) \end{bmatrix}$$

6.28 Find the state transition matrix of

$$\begin{bmatrix} \dot{x}_1 \\ \dot{x}_2 \end{bmatrix} = \begin{bmatrix} 0 & \omega(t) \\ -\omega(t) & 0 \end{bmatrix}\begin{bmatrix} x_1 \\ x_2 \end{bmatrix}$$

6.29 Find the weighting function of the network whose input–output state equations are given in Problem 6.26.

6.30 Solve the linear time-varying differential equation

$$\begin{bmatrix} \dot{x}_1 \\ \dot{x}_2 \\ \dot{x}_3 \end{bmatrix} = \begin{bmatrix} -1 & 0 & 0 \\ 2 & -3 & 0 \\ 0 & 0 & -t \end{bmatrix}\begin{bmatrix} x_1 \\ x_2 \\ x_3 \end{bmatrix}, \qquad x_1(0) = x_2(0) = x_3(0) = 1$$

6.31 Show that the nonlinear state equation

$$\dot{x} = x^{1/n};\ n \geq 2 \text{ and } x(0) = 0$$

does not have a unique solution.

6.32 Determine whether the nonlinear state equation

$$\dot{x} = -x + x^3$$

satisfies the Lipschitz condition. What is your conclusion regarding the existance and uniqueness of its solution?

Computer-Aided Network Analysis 7

7.1 INTRODUCTION

Since the first implementation of digital computers in the analysis and design of electrical networks over a decade ago, the techniques of computer-aided analysis and design have advanced drastically. Today, seldom is any relatively complicated network designed without some sort of computer implementation. In network analysis, computers are used to simulate a given network and determine its response to a set of admissible inputs. By simulating a network on a digital computer the design engineer can adjust the parameters of the network so that a desired response is obtained. This method avoids the costly and inefficient process of constructing and testing a prototype network during the preliminary stages of design.

There are numerous computer programs, such as ECAP, NASAP, AEDNET, NET-I, MECCA, and the like, available for the analysis of linear and nonlinear networks. Each program is developed for the analysis of a specific class of networks; each has its own particular features and drawbacks. However, since the field of computer-aided analysis and design is constantly under evolution due to the development of highly efficient numerical methods, these packaged programs become less attractive as new programs emerge. For this reason, in this chapter we concern ourselves only with the basic ideas underlying the main computer analysis techniques and refer the potential user of these "canned" programs to their sources. In each section, however, sample programs will be presented to illustrate the application of specific numerical techniques. These programs are presented in simple FORTRAN language to provide the reader with a general understanding of the basic idea.

Before discussing any numerical technique in detail, let us briefly mention the various errors that one might encounter in computer-aided network analysis.

Types of Errors. In virtually all the numerical methods, a given continuous operator is first approximated by a discrete operator, and then a solution to the corresponding problem is sought. This operator can be a differential operator, integral operator, and so on. For example, to evaluate the integral of a function $f(t)$ over an interval (α, β), the interval can first be divided into n equal segments, each of length $h = (\beta - \alpha)/n$, and then the following relation can be employed:

$$\int_{\alpha}^{\beta} f(t)\,dt = \sum_{k=0}^{n-1} hf(\alpha + kh) + e_t \tag{7.1-1}$$

where e_t represents the error involved in the approximation. This error is commonly referred to as truncation error. As another example, consider the evaluation of $e^{0.1}$ using an infinite Taylor series:

$$e^{0.1} = 1 + 0.1 + \frac{(0.1)^2}{2!} + \frac{(0.1)^3}{3!} + \cdots$$

If only the first two terms in this infinite series are taken as the approximate value of $e^{0.1}$, the truncation error will then be

$$e_t = \frac{(0.1)^2}{2!} + \frac{(0.1)^3}{3!} + \cdots$$

Consequently, the *truncation error* is the error caused by truncating an infinite process to get a finite process. One major objective in computer-aided network analysis should, therefore, be to reduce or, if possible, eliminate the truncation error. This objective is usually achieved by adopting more refined methods of approximating an infinite process by a finite process. For example, the truncation error in (7.1-1) can be substantially reduced by using a trapezoidal rule of integration. The study of different methods of reducing the truncation error is the subject of the numerical analysis discussed in several texts [1, 2].

Using a digital computer to perform arithmetic operations, only a certain number of digits are retained and the rest are discarded. For example, if a computer is programmed to retain three decimals from each multiplication, then multiplying, say, 2.316 and 1.207 will be approximated by 2.795; that is,

$$2.316 \times 1.207 = 2.795412 = 2.795 + e_r$$

Here 2.795412 is rounded to 2.795 and the round-off error e_r is 0.000412.

On this basis we can state that the *round-off* error in an arithmetic operation is the error caused by rounding the result of that operation. Sometimes, particularly when the true value of the result of an arithmetic operation is either very small or very large, it is more meaningful to consider the relative error of the operation rather than the absolute error. The *relative error* of an arithmetic operation is the ratio of the absolute error to the true answer. For our example the relative error is

$$\text{relative error} = \frac{0.000412}{2.795412}$$

Let us now proceed with the main objective of this chapter: computer-aided analysis of networks. This is done in the next four sections. Section 7.2 considers the analysis of linear resistive networks, the result of which is then generalized in Section 7.3 to analyze linear time-invariant networks under steady-state conditions. The solution of nonlinear resistive networks and the solution of general state equations are considered in Sections 7.4 and 7.5, respectively.

7.2 COMPUTER-AIDED ANALYSIS OF LINEAR RESISTIVE NETWORKS

Analytical formulations of general linear time-invariant networks were discussed in Chapter 4. In this section we consider only the analysis of linear resistive networks. Among the three methods of analysis discussed in Chapter 4, nodal analysis is the most common, since it can be applied to planar and nonplanar networks alike. Furthermore, as we shall soon see, the formulation of nodal equations can easily be obtained by inspection. Let us now, for convenience, rederive the nodal equations for a linear resistive network.

Nodal Analysis Formulation. Consider a resistive network with $N + 1$ nodes and B branches. Assume that a typical branch of this network contains an independent voltage source and an independent current source (see Fig. 7.2-1). Referring to the figure, we conclude that

$$\mathbf{v}_b = \mathbf{v} - \mathbf{v}_s \tag{7.2-1}$$

$$\mathbf{i}_b = \mathbf{i} + \mathbf{i}_s \tag{7.2-2}$$

and

$$\mathbf{i} = \mathbf{G}\mathbf{v} \tag{7.2-3}$$

where $\mathbf{v}$ and $\mathbf{i}$ are element voltage and current vectors, $\mathbf{v}_s$ and $\mathbf{i}_s$ are independent voltage and current source vectors, and $\mathbf{v}_b$ and $\mathbf{i}_b$ are branch voltage and current vectors. The $n \times n$ matrix $\mathbf{G}$ is a constant matrix whose diagonal element g_{kk} is the conductance of the kth branch and whose ijth off-diagonal

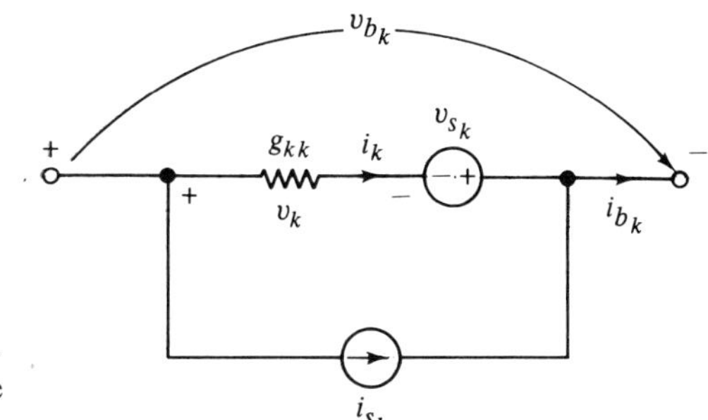

Figure 7.2-1 General branch considered in the analysis of resistive networks.

element g_{ij} represents the coupling between the ith and jth branch. (In a resistive network these couplings exist if the network contains some dependent sources, such as gyrators, operational amplifiers, etc.) Let us denote the node voltage vector by $\mathbf{v}_n$; then, from (3.5-8), $\mathbf{v}_b$ is given in terms of $\mathbf{v}_n$ by

$$\mathbf{v}_b = \mathbf{A}^T\mathbf{v}_n \tag{7.2-4}$$

where $\mathbf{A}$ represents the incidence matrix of the network under study. To obtain $\mathbf{v}_n$ in terms of the known parameters of the network, let us replace $\mathbf{v}_b$ in (7.2-4) from (7.2-1) and premultiply the resulting equation by $\mathbf{AG}$;

$$\mathbf{AGA}^T\mathbf{v}_n = \mathbf{AGv} - \mathbf{AGv}_s \tag{7.2-5}$$

From KCL we have $\mathbf{Ai}_b = \mathbf{0}$. In view of (7.2-2) and (7.2-3) this can be written as $\mathbf{AGv} = -\mathbf{Ai}_s$. Comparing the last equation with (7.2-5), we get $\mathbf{AGA}^T\mathbf{v}_n = -\mathbf{A}(\mathbf{i}_s + \mathbf{Gv}_s)$. Hence, given $\mathbf{A}$, $\mathbf{G}$, $\mathbf{i}_s$, and $\mathbf{v}_s$, provided that $\mathbf{AGA}^T$ is nonsingular, this equation can be solved for $\mathbf{v}_n$.

In relatively simple networks it is more convenient to obtain the nodal equations by inspection. This is shown in the following:

Example 7.2-1. Consider the linear time-invariant resistive network shown in Fig. 7.2-2; the resistances are all in ohms. Find a set of linearly independent equations that can be solved for the node voltages.

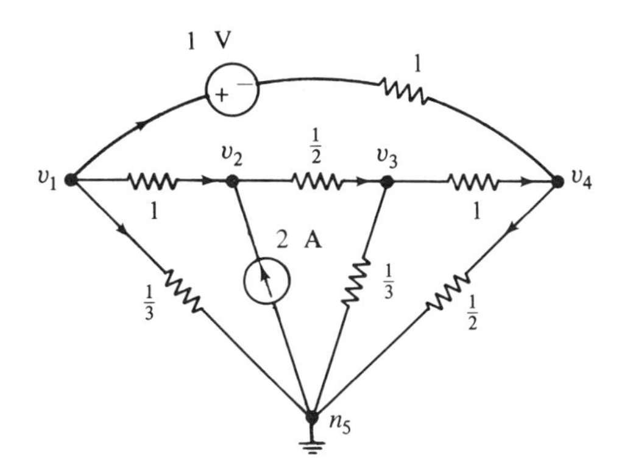

Figure 7.2-2 Linear resistive network considered in Example 7.2-1

Solution

Take the node n_5 as the datum and use KCL and VCR at each node to get

$$\begin{aligned}
&\text{node 1:} && (v_1 + 1 - v_4) + (v_1 - v_2) + 3v_1 = 0 \\
&\text{node 2:} && -(v_1 - v_2) + 2(v_2 - v_3) - 2 = 0 \\
&\text{node 3:} && -2(v_2 - v_3) + 3v_3 + (v_3 - v_4) = 0 \\
&\text{node 4:} && -(v_1 + 1 - v_4) - (v_3 - v_4) + 2v_4 = 0
\end{aligned}$$

Rearranging the equations and putting the result in matrix form, we get

$$\begin{bmatrix} 5 & -1 & 0 & -1 \\ -1 & 3 & -2 & 0 \\ 0 & -2 & 6 & -1 \\ -1 & 0 & -1 & 4 \end{bmatrix} \begin{bmatrix} v_1 \\ v_2 \\ v_3 \\ v_4 \end{bmatrix} = \begin{bmatrix} -1 \\ 2 \\ 0 \\ 1 \end{bmatrix}$$

These are the desired nodal equations. Hence, in such cases it is much easier to obtain the nodal equations by inspection.

Remark 7.2-1. If the network under consideration has no coupling elements and if the source elements are restricted to current sources only, a simple algorithm can be outlined for obtaining the nodal admittance matrix. Consider a linear resistive network containing no voltage sources and no coupled elements whose nodal equations can be written as

$$\mathbf{Y}_n \mathbf{v}_n(t) = \mathbf{i}_s(t)$$

where $\mathbf{v}_n(t)$ is a column vector whose elements are node voltages, $\mathbf{i}_s(t)$ is the current source vector whose jth element is i_{sj}, and $\mathbf{Y}_n$ is the nodal conductance matrix whose ijth element is y_{ij}. We can then state the following algorithm for obtaining y_{ij}:

1. y_{ii} is the sum of the conductances of all the branches incident to node i.
2. $-y_{jk}$ is the sum of the conductances of all branches incident to nodes j *and* k.
3. $-i_{sj}$ is the algebraic sum of all the current sources entering or leaving node j. (Recall that if a current leaves a node we assign a positive sign to it; otherwise, we assign a negative sign.)

If the network under consideration contains some independent voltage sources in series with some resistors, we can use the source transformation discussed in Chapter 4 to replace these voltage sources by their equivalent current sources. For instance, in Example 7.2-1 the branch containing the voltage source can be replaced by a 1 Ω resistor *in parallel* with a 1 A current source.

In any case, the analysis of linear resistive networks reduces to solving a set of linear algebraic equations in the form

$$\mathbf{Px} = \mathbf{b} \tag{7.2-6}$$

where $\mathbf{P}$ is an $n \times n$ matrix whose elements are given; for instance, in the nodal analysis $\mathbf{P} = \mathbf{AGA}^T$ as in (7.2-5). The n-column vector $\mathbf{x}$ represents the unknowns of the network, which are typically either node voltages or loop currents; $\mathbf{b}$ is a column vector whose elements can be obtained from the given voltage and current sources. To solve (7.2-6) for $\mathbf{x}$, if $\mathbf{P}$ is nonsingular, we can write

$$\mathbf{x} = \mathbf{P}^{-1}\mathbf{b} \tag{7.2-7}$$

That is, to obtain $\mathbf{x}$, it is sufficient to obtain $\mathbf{P}^{-1}$ from $\mathbf{P}$ and then postmultiply it by $\mathbf{b}$. This method, although quite simple in theory, becomes very time consuming if the dimension of $\mathbf{P}$ is relatively large. To get an estimate of the number of multiplication operations needed for inverting an $n \times n$ matrix, note that computing just the determinant by expanding along a row or column requires

$$n! \sum_{k=1}^{n-1} \frac{1}{(n-k)!}$$

or, for large n, approximately $(e - 1)(n!)$ multiplications. For a 10×10 matrix this number amounts to over 9 million multiplications, which using a relatively fast computer (around 10^{-6} second per multiplication) requires about 9 seconds (not counting the computation time required for additions and subtractions). If, on the other hand, the determinant of a 20×20 matrix is to be computed by expanding along a row or column, the computation time will be well over 10,000 years! Thus it is practically impossible to invert large matrices using the classical rules of matrix inversion. However, the total number of multiplications in inverting $\mathbf{P}$, with dimension 4 or higher, can be reduced to n^3 by cleverly avoiding the computation of determinants (see [2], Chapter 2). Consequently, the total number of multiplications required for solving (7.2-6) by inverting $\mathbf{P}$ is $n^3 + n^2$. This number can be further reduced by a factor of 3 by using a procedure called the Gaussian elimination method. We shall next give a detailed discussion of this method.

Gaussian Elimination Method. The basic idea behind the Gaussian elimination is to use a series of elementary row and column operations on (7.2-6) to transfer $\mathbf{P}$ to an upper triangular matrix. An *upper triangular matrix* is a matrix all of whose elements below the diagonal are zero. By doing so, the last equation will have only one unknown x_n, which can easily be solved. The $(n - 1)$st equation contains x_n and x_{n-1} only; substituting x_n from the last equation, it can then be solved for x_{n-1}, and so on. Before discussing the general method, let us work out a simple example to illustrate the procedure.

Example 7.2-2. Solve the following simultaneous equations using the Gaussian elimination method.

$$\begin{aligned} x_1 + 2x_2 + x_3 &= 2 \\ -x_1 - 3x_2 + x_3 &= -1 \\ x_1 - x_2 + 2x_3 &= 1 \end{aligned}$$

Solution

The first step is to eliminate x_1 from the second and third equations. This can be achieved by adding the first equation to the second and subtracting the first equation from the third; this yields

$$\begin{aligned} x_1 + 2x_2 + x_3 &= 2 \\ -x_2 + 2x_3 &= 1 \\ -3x_2 + x_3 &= -1 \end{aligned}$$

Next we eliminate x_2 from the third equation by subtracting three times the second equation from the third one; the result is

$$\begin{aligned} x_1 + 2x_2 - x_3 &= 2 \\ -x_2 + 2x_3 &= 1 \\ -5x_3 &= -4 \end{aligned}$$

These equations are now in the upper triangular form. The next step is to solve for x_3, x_2, and x_1 by *back substitution*. The last equation yields

$$x_3 = \tfrac{4}{5}$$

Substituting this value of x_3 in the second equation and solving for x_2 yields

$$x_2 = \tfrac{3}{5}$$

Finally, substituting x_2 and x_3 in the first equation, we get

$$x_1 = 0$$

In general, let us rewrite (7.2-6) in the following form:

$$\begin{aligned} E_1: &\quad p_{11}x_1 + p_{12}x_2 + \cdots + p_{1n}x_n = b_1 \\ E_2: &\quad p_{21}x_1 + p_{22}x_2 + \cdots + p_{2n}x_n = b_2 \\ &\quad \vdots \\ E_n: &\quad p_{n1}x_1 + p_{n2}x_2 + \cdots + p_{nn}x_n = b_n \end{aligned} \tag{7.2-8}$$

Suppose that $p_{11} \neq 0$; to eliminate x_1 from equations E_2 through E_n we subtract p_{i1}/p_{11} times the first equation from the ith equation; this yields

$$
\begin{aligned}
E_1:&\quad p_{11}x_1 + p_{12}x_2 + \cdots + p_{1n}x_n = b_1 \\
E_2^1:&\quad p_{22}^1x_2 + \cdots + p_{2n}^1x_n = b_2^1 \\
&\quad\vdots \\
E_n^1:&\quad p_{n2}^1x_2 + \cdots + p_{nn}^1x_n = b_n^1
\end{aligned}
\tag{7.2-9}
$$

where the new coefficients in equations E_2^1 through E_n^1 are given by

$$p_{ij}^1 = p_{ij} - \frac{p_{i1}}{p_{11}}p_{1j}, \qquad i, j = 2, 3, \ldots, n$$

and

$$b_i^1 = b_i - \frac{p_{i1}}{p_{11}}b_1, \qquad i = 2, 3, \ldots, n$$

If $p_{11} = 0$, we interchange and relabel the equations so that the coefficient appearing in the upper left-hand corner of (7.2-8) is nonzero. This can always be done, since by assumption $\mathbf{P}$ is nonsingular and, hence, not all the coefficients of x_1 are zero.

The next step is to eliminate x_2 from equations E_3 through E_n. This can be done by subtracting p_{i2}^1/p_{22}^1 times E_2^1 from the ith equation in (7.2-9) for $i = 3, 4, \ldots, n$. The result is

$$
\begin{aligned}
E_1:&\quad p_{11}x_1 + p_{12}x_2 + p_{13}x_3 + \cdots + p_{1n}x_n = b_1 \\
E_2^1:&\quad p_{22}^1x_2 + p_{23}^1x_3 + \cdots + p_{2n}^1x_n = b_2^1 \\
E_3^2:&\quad p_{33}^2x_3 + \cdots + p_{3n}^2x_n = b_3^2 \\
&\quad\vdots \\
E_n^2:&\quad p_{n3}^2x_3 + \cdots + p_{nn}^2x_n = b_n^2
\end{aligned}
$$

where the new coefficients in equations E_3^2 through E_n^2 are

$$p_{ij}^2 = p_{ij}^1 - \frac{p_{i2}^1}{p_{22}^1}p_{2j}^1, \qquad i, j = 3, 4, \ldots, n$$

and

$$b_i^2 = b_i^1 - \frac{p_{i2}^1}{p_{22}^1}b_2^1, \qquad i = 3, 4, \ldots, n$$

Repeating this process to eliminate x_3 through x_{n-1}, we get

$$
\begin{aligned}
E_1:&\quad p_{11}x_1 + p_{12}x_2 + p_{13}x_3 + \cdots + p_{1n}x_n = b_1 \\
E_2^1:&\quad p_{22}^1x_2 + p_{23}^1x_3 + \cdots + p_{2n}^1x_n = b_2^1 \\
E_3^2:&\quad p_{33}^2x_3 + \cdots + p_{3n}^2x_n = b_3^2 \\
&\quad\vdots \\
E_{n-1}^{n-2}:&\quad p_{(n-1)(n-1)}^{n-2}x_{n-1} + p_{(n-1)n}^{n-2}x_n = b_{n-1}^{n-2} \\
E_n^{n-1}:&\quad p_{nn}^{n-1}x_n = b_n^{n-1}
\end{aligned}
\tag{7.2-10}
$$

where

$$p_{ij}^{k} = p_{ij}^{k-1} - \frac{p_{ik}^{k-1}}{p_{kk}^{k-1}} p_{kj}^{k-1}, \qquad \begin{array}{l} k = 1, 2, \ldots, n-1 \\ i, j = k+1, \ldots, n \end{array} \qquad (7.2\text{-}11)$$

and

$$b_{i}^{k} = b_{i}^{k-1} - \frac{p_{ik}^{k-1}}{p_{kk}^{k-1}} b_{k}^{k-1}, \qquad \begin{array}{l} k = 1, 2, \ldots, n \\ i = k+1, \ldots, n \end{array} \qquad (7.2\text{-}12)$$

This completes the first step in the Gaussian elimination method. The next step is to obtain x_n from the last equation in (7.2-10) and substitute it in the next to the last equation to get x_{n-1}, and so on. In general, x_i is given by

$$x_i = \frac{1}{p_{ii}^{i-1}}\left(b_i^{i-1} - \sum_{j=i+1}^{n} p_{ij}^{i-1} x_j\right), \qquad i = n, n-1, \ldots, 1 \qquad (7.2\text{-}13)$$

where p_{ij}^{i-1} and b_i^{i-1} are given in (7.2-11) and (7.2-12) and $x_{n+1} = 0$. This process can easily be programmed on a digital computer to obtain all the unknowns in (7.2-6). It can be shown [2] that the total number of multiplications and divisions in this process is $(n^3/3) + n^2 - (n/3)$, which for large n is much smaller than $n^3 + n^2$ multiplications that are required for inverting **P**.

The Gaussian elimination method is quite efficient and accurate if the divisors p_{ii}^{i-1} are not too small in comparison with p_{ik}^{i-1}; otherwise, the round-off errors occurring in the divisions may affect the accuracy of the results. This problem can be overcome by the *positioning-for-size* method. This method consists of interchanging rows and columns of **P** and successive matrices to get the element of greatest magnitude in the kkth position [2]. The severity of this drawback is illustrated in the following simple example.

Example 7.2-3. Consider the set of linear independent equations

$$0.01x_1 + x_2 = 1$$
$$x_1 + x_2 = 2$$

which has the solution (within two decimal places) $x_1 = 1.01$ and $x_2 = 0.99$. If a Gaussian elimination without positioning for size is used, we get

$$0.01x_1 + x_2 = 1$$
$$99x_2 = 98$$

which yields $x_2 = 0.98$ and $x_1 = 2$, that is, a relative error of almost 100 per cent in computation of x_1. On the other hand, if the equations are first positioned for size by interchanging the position of x_1 and x_2, we get

$$x_2 + 0.01x_1 = 1$$
$$x_2 + x_1 = 2$$

The Gaussian elimination of these equations then yields $x_1 = 1.01$ and $x_2 = 0.99$, which is the correct answer.

Remark 7.2-2. So far we have made no mention of the time dependence of the matrix **P** or vectors **x** and **b** in (7.2-6). If the resistive network under consideration has one or more time-varying elements or if some of the sources are time dependent, the resulting nodal or loop equation will be in the form

$$\mathbf{P}(t)\mathbf{x}(t) = \mathbf{b}(t) \tag{7.2-14}$$

In this case the procedure outlined is still valid. For each given t, say t_1, (7.2-14) represents a set of linear equations with constant coefficients that can be solved by the Gaussian elimination method. The process can then be repeated for $t_2, t_3, \ldots, t_n$. Hence, the unknown vector $\mathbf{x}(t)$ can be determined at discrete times $t_1, t_2, \ldots, t_n$. Unfortunately, this is the best one can do using a digital computer. No computer program has yet been developed that gives $\mathbf{x}(t)$ in a closed form. In most practical problems, however, computing $\mathbf{x}(t)$ at discrete times is sufficient. Let us now develop a computer program for solving (7.2-6) using the Gaussian elimination method.

Sample Computer Program. There are many ways of writing a computer program for solving (7.2-6) by the Gaussian elimination method. These programs depend on the digital computer available and the language in which they are written. In this section we develop a program in FORTRAN. This is best shown by means of an example.

Example 7.2-4. Use the Gaussian elimination method to obtain the node voltages of the network shown in Fig. 7.2-2.

Solution

The nodal equations for this network were found in Example 7.2-1 to be $\mathbf{Px} = \mathbf{b}$, where

$$\mathbf{P} = \begin{bmatrix} 5 & -1 & 0 & -1 \\ -1 & 3 & -2 & 0 \\ 0 & -2 & 6 & -1 \\ -1 & 0 & -1 & 4 \end{bmatrix}, \qquad \mathbf{b} = \begin{bmatrix} -1 \\ 2 \\ 0 \\ 1 \end{bmatrix}$$

and $\quad \mathbf{x} = [v_1 \quad v_2 \quad v_3 \quad v_4]^T$

The corresponding computer program is given as

```
C       THIS PROGRAM IS FOR EXAMPLE 7.2-4
C       THIS PROGRAM SOLVES A SET OF LINEARLY INDEPENDENT EQUATIONS WHERE
C       N IS LESS OR EQUAL TO 20, N DESIGNATES THE NO. OF EQUATIONS TO BE
C       SOLVED. P IS THE COEFFICIENT MATRIX, X IS THE UNKNOWN
C       VECTOR AND B IS THE KNOWN VECTOR. P IS AUGUMENTED BY B AS
C       ITS N+1 COLUMN
        DIMENSION P(20,21)
        READ 20,N
   20   FORMAT(I2)
        M=N+1
        DO 1 I=1,N
```

```
    1 READ 21,(P(I,J),J=1,M)
   21 FORMAT(8F10.4)
      PRINT 25
   25 FORMAT(5X,*DATA*,/)
      PRINT 26,N
   26 FORMAT(9X,*N=*,I2)
      PRINT 27
   27 FORMAT(3X,*MATRIX P AUGMENTED BY VECTOR B IS*)
      DO 6 I=1,N
    6 PRINT 24,(P(I,J),J=1,M)
   24 FORMAT(10X,8(F10.4,2X))
C     USE GAUSSIAN ELIMINATION TO PUT P IN AN UPPER TRIANGULAR FORM.
      DO 14 I=2,N
      DO 14 J=I,N
C     CHECK THE DIAGONAL ELEMENT,IF IT IS ZERO,CHANGE ROWS.
      IF(P(I-1,I-1))5,2,5
    2 K=I-1
      MC=0
      MB=N-I+1
      DO 11 L=I,N
      IF(P(L,K))4,7,4
    7 MC=MC+1
      GO TO 11
    4 DO 13 K1=K,M
      TEMP=P(L,K1)
      P(L,K1)=P(K,K1)
   13 P(K,K1)=TEMP
   11 CONTINUE
      IF(MB-MC)8,8,5
    5 Q=P(J,I-1)/P(I-1,I-1)
      II=I-1
      DO 14 K2=II,M
   14 P(J,K2)=P(J,K2)-Q*P(I-1,K2)
C     USE THE BACK SUBSTITUTION TO COMPUTE X
      DO 15 I=2,N
      K=N-I+2
      IF(P(K,K) .EQ. 0.0) GO TO 8
      Q=P(K,M)/P(K,K)
      DO 15 J=I,N
      M2=N-J+1
   15 P(M2,M)=P(M2,M)-Q*P(M2,K)
      PRINT 28
   28 FORMAT(//,* THE SOLUTION OF THE MATRIX EQUATION PX=B*,/)
      DO 16 I=1,N
      X=P(I,M)/P(I,I)
   16 PRINT 23,I,X
   23 FORMAT(10X,*X(*,I1,*) =*,F10.4)
      GO TO 9
    8 PRINT 22
   22 FORMAT(10X,* MATRIX P IS SINGULAR*)
    9 STOP
      END
```

```
DATA

      N= 4
MATRIX P AUGMENTED BY VECTOR B IS
        5.0000     -1.0000      0.0000     -1.0000     -1.0000
       -1.0000      3.0000     -2.0000      0.0000      2.0000
        0.0000     -2.0000      6.0000     -1.0000      0.0000
       -1.0000      0.0000     -1.0000      4.0000      1.0000

        THE SOLUTION OF THE MATRIX EQUATION PX=B

               X(1) =     .0580
               X(2) =     .9330
               X(3) =     .3705
               X(4) =     .3571
```

Remark 7.2-3. Although the Gaussian elimination method requires fewer multiplications than matrix inversion, in situations where $\mathbf{Px} = \mathbf{b}$ should be solved for several different $\mathbf{b}$ vectors, it might be cheaper to invert $\mathbf{P}$ first. This situation might arise in networks for which the node voltages should be computed for several different sources.

As we observed in the beginning of this section, computing the inverse of a large matrix by first finding its determinant requires a huge amount of computation time. We therefore need to devise a new technique for inverting large matrices that requires a reasonable number of multiplications and divisions. One such method is discussed next.

LU Transformation. Consider the set of linearly independent equations

$$\mathbf{Px} = \mathbf{b} \tag{7.2-15}$$

where $\mathbf{P}$ is assumed to be nonsingular. Suppose that $\mathbf{P}$ can be decomposed into the product of two matrices $\mathbf{L}$ and $\mathbf{U}$, where $\mathbf{L}$ is a *lower triangular* matrix and $\mathbf{U}$ is an *upper triangular* matrix whose diagonal elements are all 1; that is,

$$\mathbf{P} = \mathbf{LU} \tag{7.2-16}$$

Inverting both sides of this equation yields

$$\mathbf{P}^{-1} = \mathbf{U}^{-1}\mathbf{L}^{-1} \tag{7.2-17}$$

But since both $\mathbf{U}$ and $\mathbf{L}$ are triangular matrices, their inverses can be computed with little effort. The solution of (7.2-15) is then

$$\mathbf{x} = \mathbf{U}^{-1}\mathbf{L}^{-1}\mathbf{b} \tag{7.2-18}$$

To obtain the elements of $\mathbf{U}$ and $\mathbf{L}$, let us consider (7.2-16) in the expanded form

$$\begin{bmatrix} p_{11} & p_{12} & \cdots & p_{1n} \\ p_{21} & p_{22} & \cdots & p_{2n} \\ \cdot & \cdot & & \cdot \\ \cdot & \cdot & & \cdot \\ \cdot & \cdot & & \cdot \\ p_{n1} & p_{n2} & \cdots & p_{nn} \end{bmatrix} = \begin{bmatrix} l_{11} & 0 & \cdots & 0 \\ l_{21} & l_{22} & \cdots & 0 \\ \cdot & \cdot & & \cdot \\ \cdot & \cdot & & \cdot \\ \cdot & \cdot & & \cdot \\ l_{n1} & l_{n2} & \cdots & l_{nn} \end{bmatrix} \begin{bmatrix} 1 & u_{12} & \cdots & u_{1n} \\ 0 & 1 & \cdots & u_{2n} \\ \cdot & \cdot & & \cdot \\ \cdot & \cdot & & \cdot \\ \cdot & \cdot & & \cdot \\ 0 & 0 & \cdots & 1 \end{bmatrix}$$

Multiplying the rows of **L** by the first column of **U** and comparing the result with the first column of **P** yields

$$l_{i1} = p_{i1}, \qquad i = 1, 2, \ldots, n$$

To get the first row of **U**, multiply its columns by the first row of **L** and compare the result with the first row of **P**; this will give

$$u_{ij} = p_{ij}/l_{11} \qquad j = 2, 3, \ldots, n$$

Hence, the first column of **L** and the first row of **U** have been determined so far. Proceeding in the same fashion, the rest of the elements of **L** and **U** can be determined. In general, l_{ij} and u_{ij} can be found from

$$l_{ij} = p_{ij} - \sum_{k=1}^{j-1} l_{ik}u_{kj}, \qquad i = 1, 2, \ldots, n; j = 1, 2, \ldots, i \tag{7.2-19}$$

and $$u_{ij} = \frac{1}{l_{ii}}\left(p_{ij} - \sum_{k=1}^{i-1} l_{ik}u_{kj}\right), \qquad j = 1, 2, \ldots, n; i = 1, 2, \ldots, j \tag{7.2-20}$$

Note that if the network under consideration is made up of reciprocal two-terminal resistors the matrix **P** will always be symmetric. In this case the calculations can be greatly simplified by observing that **P** can be decomposed in the form

$$\mathbf{P} = \mathbf{L}\mathbf{L}^T \tag{7.2-21}$$

Exercise 7.2-1. Use a procedure similar to that just discussed to obtain the elements of **L** in (7.2-21) in terms of the elements of **P**.

The LU transformation has the added advantage of reducing the number of memory cores required for inverting a matrix on a digital computer. This can be seen by noting that there is no need for storing the zero elements in either **L** or **U**; furthermore, the p_{ij} are used only once in computing l_{ij} and u_{ij}. Hence, the memory storage used for storing **P** can be used to store **U** and **L**. The total number of multiplications and divisions necessary for solving $\mathbf{Px} = \mathbf{b}$ by LU transformation turns out to be the same as that of the Gaussian elimination method; $(n^3/3) + n^2 - (n/3)$. As was mentioned earlier, the advantage of LU transformation over the Gaussian elimination method becomes obvious if $\mathbf{Px} = \mathbf{b}$ is to be solved for several different source vectors. If there are m different source vectors, the total number of multiplications and divisions required in the Gaussian elimination method will be $m[(n^3/3) + n^2 - (n/3)]$, whereas in LU transformation this number is reduced to $(n^3/3) + mn^2 - (n/3)$.

Sparse-Matrix Methods. For some large-scale electrical networks the coefficient matrix $\mathbf{P}$ contains many zero elements. In such cases the matrix $\mathbf{P}$ is referred to as a *sparse matrix*. Such matrices are usually encountered in the nodal analysis of power systems and communication networks. The Gaussian elimination method will have two serious drawbacks if the matrix $\mathbf{P}$ is sparse and is of large dimension, say, 1000 or more. First, this method does not take advantage of the sparsity of the matrix. Indeed, in the elimination process many of the zero elements in $\mathbf{P}$ will be changed to nonzero elements. Second, in many practical problems, elements of $\mathbf{P}$ are determined from the topology of the network and need not be stored in the computer. However, if a Gaussian method is used, a large portion of these elements must be stored; for matrices of high dimension this can be a serious problem.

Several investigators [3, 4] have taken advantage of the sparsity of such matrices to obtain highly efficient methods of solving $\mathbf{Px} = \mathbf{b}$ for matrices of dimension 1000 or more. The basic idea behind various sparse-matrix techniques is to store and process only nonzero elements of $\mathbf{P}$. By doing so the memory storage requirements and the number of multiplications and divisions can be greatly reduced. One method of storing nonzero elements of $\mathbf{P}$ is to introduce a single array $A(j)$; the first section of A is reserved for the diagonal elements of $\mathbf{P}$:

$$A(1) = p_{11},\ A(2) = p_{22},\ \ldots,\ A(n) = p_{nn}$$

The second section of A is reserved for the *nonzero* elements of the upper triangular portion of $\mathbf{P}$, and the last section of A is reserved to store nonzero elements of the lower triangular portion of $\mathbf{P}$. A table is also introduced to record the exact location of each nonzero off-diagonal element. An LU transformation is then used to decompose $\mathbf{P}$ into two triangular matrices $\mathbf{L}$ and $\mathbf{U}$. In the sparse-matrix case, however, only operations that result in a change in the elements of the matrix $\mathbf{P}$ are carried out. A renumbering of the nodes of the network under consideration is usually performed, which facilitates this process. A detailed description of sparse-matrix techniques can be found in the literature [4, 5, 6].

Gauss–Seidel Iterative Method. In this method, in contrast to the Gaussian elimination method, the original matrix $\mathbf{P}$ remains unchanged. Instead, we start with a reasonable guess, $\mathbf{x}^0$, for the solution of $\mathbf{Px} = \mathbf{b}$ and try to improve this initial guess iteratively. More precisely, let us write the equation

$\mathbf{Px} = \mathbf{b}$ in the expanded form

$$\begin{aligned}
p_{11}x_1 + p_{12}x_2 + p_{13}x_3 + \cdots + p_{1n}x_n &= b_1 \\
p_{21}x_1 + p_{22}x_2 + p_{23}x_3 + \cdots + p_{2n}x_n &= b_2 \\
\vdots \qquad\qquad\qquad\qquad & \quad \vdots \\
p_{n1}x_1 + p_{n2}x_2 + p_{n3}x_3 + \cdots + p_{nn}x_n &= b_n
\end{aligned} \tag{7.2-22}$$

Next choose an arbitrary (but reasonable) first guess $\mathbf{x}^0 = [x_1^0 \quad x_2^0 \ldots x_n^0]^T$ for the solution of these equations. If the l_∞ norm of the error vector $\mathbf{e}^0 = \mathbf{Px}^0 - \mathbf{b}$ is sufficiently small, the initial choice $\mathbf{x}^0$ is good and we stop the iterations. On the other hand, if $|\mathbf{e}^0|_\infty$ is not small, we choose a new vector $\mathbf{x}^1$ whose elements are given by

$$\begin{aligned}
x_1^1 &= \frac{1}{p_{11}}(b_1 - p_{12}x_2^0 - p_{13}x_3^0 - p_{14}x_4^0 - \cdots - p_{1n}x_n^0) \\
x_2^1 &= \frac{1}{p_{22}}(b_2 - p_{21}x_1^1 - p_{23}x_3^0 - p_{24}x_4^0 - \cdots - p_{2n}x_n^0) \\
x_3^1 &= \frac{1}{p_{33}}(b_3 - p_{31}x_1^1 - p_{32}x_2^1 - p_{34}x_4^0 - \cdots - p_{3n}x_n^0) \\
&\vdots \\
x_n^1 &= \frac{1}{p_{nn}}(b_n - p_{n1}x_1^1 - p_{n2}x_2^1 - p_{n3}x_3^1 - \cdots - p_{n(n-1)}x_{n-1}^1)
\end{aligned} \tag{7.2-23}$$

Note that the result of the first equation is immediately used in the second equation, the result of the first two equations is used in the third equation, and so on. Again we compute the norm $|\mathbf{e}^1|_\infty = |\mathbf{Px}^1 - \mathbf{b}|_\infty$ and repeat the process until $|\mathbf{e}^k|_\infty$ is sufficiently small. In the kth iteration the equations are

$$\begin{aligned}
x_1^{k+1} &= \frac{1}{p_{11}}(b_1 - p_{12}x_2^k - p_{13}x_3^k - p_{14}x_4^k - \cdots - p_{1n}x_n^k) \\
x_2^{k+1} &= \frac{1}{p_{22}}(b_2 - p_{21}x_1^{k+1} - p_{23}x_3^k - p_{24}x_4^k - \cdots - p_{2n}x_n^k) \\
&\vdots \\
x_n^{k+1} &= \frac{1}{p_{nn}}(b_n - p_{n1}x_1^{k+1} - p_{n2}x_2^{k+1} - \cdots - p_{n(n-1)}x_{n-1}^{k+1})
\end{aligned} \tag{7.2-24}$$

It can be shown [1] that if the matrix $\mathbf{P}$ is positive definite the Gauss–Seidel iterations will converge to the true value of $\mathbf{x}$. Checking the positive definiteness of a matrix with large dimensions, however, is not very practical. A more practical sufficient condition for checking the convergence of the Gauss–Seidel method is the dominance property of the matrix $\mathbf{P}$. A matrix $\mathbf{P}$ is said to be *dominant* if for each row the diagonal element is larger or equal to the sum of the absolute values of all the other elements in that row. More specifically, **P** is dominant if

$$p_{ii} \geq \sum_{\substack{j=1 \\ j \neq i}}^{n} |p_{ij}| \quad \text{for } i = 1, 2, \ldots, n \tag{7.2-25}$$

and the strict inequality is satisfied for at least one i. A sufficient condition for the convergence of this iterative method is given in the following:

Theorem 7.2-1. If the matrix **P** in $\mathbf{Px} = \mathbf{b}$ is dominant, the Gauss–Seidel iterative procedure is convergent.

In contrast to the first condition, the dominance condition is easy to check. Indeed, it can be proved that the nodal or loop analysis of a linear time-invariant resistive network with positive resistors and no dependent sources results in a dominant matrix. Consequently, the Gauss–Seidel method can always be applied to analyze such networks.

Exercise 7.2-2. Show that for a linear resistive network comprised of positive resistors and independent sources the loop or nodal analysis results in the matrix equation $\mathbf{Px} = \mathbf{b}$ in which **P** is a dominant matrix.

Let us now work out an example of the application of the Gauss–Seidel method.

Example 7.2-5. Consider the resistive network shown in Fig. 7.2-3. All the element values are in mhos. Perform a nodal analysis to obtain all the node voltages.

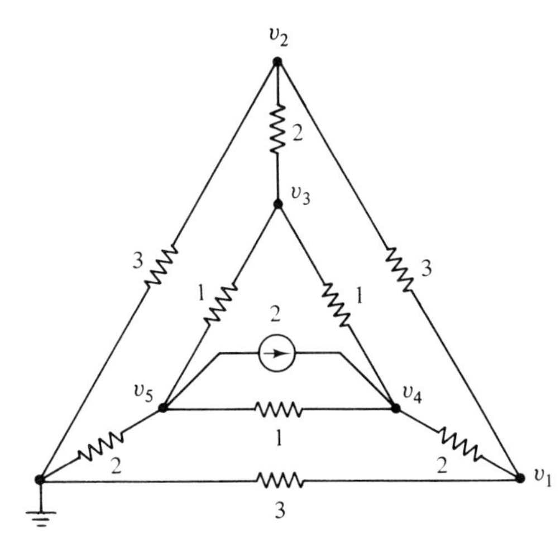

Figure 7.2-3 Linear resistive network considered in Example 7.2-5.

Solution

The corresponding node equation can easily be written as

$$\begin{bmatrix} 8 & -3 & 0 & -2 & 0 \\ -3 & 8 & -2 & 0 & 0 \\ 0 & -2 & 4 & -1 & -1 \\ -2 & 0 & -1 & 4 & -1 \\ 0 & 0 & -1 & -1 & 4 \end{bmatrix} \begin{bmatrix} v_1 \\ v_2 \\ v_3 \\ v_4 \\ v_5 \end{bmatrix} = \begin{bmatrix} 0 \\ 0 \\ 0 \\ 2 \\ -2 \end{bmatrix}$$

The coefficient matrix is clearly dominant in the sense of equation (7.2-25), and therefore the Gauss–Seidel iterative method must converge regardless of the initial guess $\mathbf{x}^0$. The following FORTRAN program can then be written to solve for the node voltages.

```
C     THIS PROGRAM IS FOR EXAMPLE 7.2-5. THE NODE VOLTAGES OF NETWORK
C     OF FIG. 7.2-3 ARE FOUND BY SOLVING PV=B USING THE GAUSS-SEIDEL
C     ITERATIVE METHOD
      DIMENSION P(20,20),B(20),X(20)
      READ 115,N,EPT
      DO 120 I=1,N
  120 READ 100,(P(I,J),J=1,N)
      READ 100,(B(I),I=1,N)
      READ 100,(X(I),I=1,N)
  115 FORMAT(I2,F10.5)
  100 FORMAT(8F10.5)
      PRINT 630
  630 FORMAT( * INITIAL GUESS IS*,/)
      PRINT 600,(X(I),I=1,N)
      CALL GASEDL(P,B,N,X,EPT,IER)
      PRINT 1200,IER
 1200 FORMAT(* NO. OF ITERATIONS IS*,I3,/)
      PRINT 610
  610 FORMAT(  * P MATRIX IS*,/)
      DO 20 I=1,N
   20 PRINT 600,(P(I,J),J=1,N)
  600 FORMAT(5X,8F12.6)
      PRINT 620
  620 FORMAT(* B VECTOR IS *,/)
      PRINT 600,(B(I),I=1,N)
      PRINT 110
  110 FORMAT(//,* THE SOLUTION OF EQUATIONS ARE*)
      DO 60 I=1,N
   60 PRINT 800,I,X(I)
  800 FORMAT(/,5X,*V*,I2,*=*,F12.6)
      END

      SUBROUTINE GASEDL(P,B,N,X,EPT,IER)
C     SOLVES LINEAR EQUATIONS PX=B BY GAUSSS SEIDL METHOD
C     METHOD CONVERGES IF P IS POSITIVE DEFINITE OR DOMINANT MATRIX
C              N=NO. OF UNKNOWNS
C         P(I,J)=P MATRIX
C           B(I)=B VECTOR
C           X(I)=INITIAL GUESS,CONTAINS FINAL X UPON RETURN
C            EPT= ERROR ACCURACY
C            IER=NO. OF ITERATIONS
      DIMENSION P(20,20),B(20),E(20),Z(20),X(20)
```

```
      IER=0
C     TERMINATE PROGRAM IF DIAGONAL TERM IS ZERO
      DO 10 I=1,N
      IF(P(I,I) .EQ. 0.0) GO TO 1000
   10 CONTINUE
C                CHECK SOLUTION SATISIFIES GIVEN EQUATIONS
    5 DO 30 I=1,N
      SUM=0.0
      DO 20 J=1,N
   20 SUM=SUM+P(I,J)*X(J)
      E(I)=ABS(SUM-B(I))
      IF(E(I) .GT. EPT) GO TO 40
   30 CONTINUE
      RETURN
C                NEW VALUE OF X CALCULATED
   40 DO 60 I=1,N
      SUM=0.0
      DO 50 J=1,N
      IF(I .EQ. J) GO TO 50
      SUM=SUM+P(I,J)*X(J)
   50 CONTINUE
      Z(I)=(B(I)-SUM)/P(I,I)
   60 CONTINUE
      DO 70 I=1,N
   70 X(I)=Z(I)
      IER=IER+1
      GO TO 5
 1000 PRINT 1010
 1010 FORMAT( * DIAGONAL TERM ZERO,METHOD DOES NOT WORK *)
      RETURN
      END
```

For this particular problem the initial guess was chosen to be $v_1^0 = 4.00$, $v_2^0 = 5.00$, $v_3^0 = 0.00$, $v_4^0 = -3.100$, and $v_5^0 = 14.00$. The l_∞ norm of the error vector was required to be less than 0.01; that is, $|\mathbf{e}^k|_\infty < 0.01$. After 23 iterations the result is

```
THE SOLUTION OF EQUATIONS ARE

     V 1=      .160542

     V 2=      .082174

     V 3=      .083960

     V 4=      .515407

     V 5=     -.349230
```

Note that although the number of iterations seems large, the actual computation time on a moderately fast computer (5-microsecond multiplication time) is less than 2 seconds.

Remark 7.2-4. If any of the diagonal elements of **P** are zero, the computer program just developed will not work. This situation is unlikely to occur in the analysis of actual resistive networks. Nevertheless, if some of the diagonal elements of **P** are zero one can relabel the components of **x** and change the rows of **P** to place nonzero elements on its diagonal.

To conclude this section, it should be mentioned that the analysis methods discussed here can be applied to the dc analysis of linear RLC networks or for obtaining the operating point of networks with nonlinear energy-storing elements and linear resistors. As we shall see in Chapter 10, dc analysis of *RLC* networks amounts to the analysis of the resistive subnetwork obtained by open circuiting all the capacitors and short circuiting all the inductors of the original network. Similarly, obtaining the operating point of a nonlinear time-invariant network with dc sources reduces to solving the resistive subnetwork obtained by open circuiting the capacitors and short circuiting the inductors of the original network.

Two more important forms of analysis of linear *RLC* networks remain to be studied. First, the steady-state analysis with sinusoidal excitation and, second, transient analysis with arbitrary excitation. In the next section we consider the sinusoidal steady-state analysis of linear time-invariant networks, and postpone the transient analysis until Section 7.5, where we take up the solution of state equations in general.

7.3 COMPUTER SOLUTION OF SINUSOIDAL STEADY-STATE RESPONSE OF LINEAR TIME-INVARIANT NETWORKS

Recall from Chapter 4 that, using a nodal analysis, the sinusoidal steady-state response of linear time-invariant RLC networks is given by

$$\mathbf{v}_n^{ss}(t) = \mathbf{r}(j\omega_0)e^{j\omega_0 t} \tag{7.3-1}$$

in which $\mathbf{v}_n^{ss}(t)$ represents the complex node voltage vector, ω_0 is the angular frequency of the sinusoidal inputs, and $\mathbf{r}(j\omega_0)$ is the residue vector, which is the solution of the following matrix equation:

$$\mathbf{Y}_n(j\omega_0)\mathbf{r}(j\omega_0) = -[\mathbf{A}\boldsymbol{I} + \mathbf{A}\mathbf{Y}(j\omega_0)\boldsymbol{V}] \tag{7.3-2}$$

where $\mathbf{Y}_n(j\omega_0)$ is the node admittance matrix, $\mathbf{Y}(j\omega_0)$ is the element admittance matrix, $\mathbf{A}$ is the network incidence matrix, and $\boldsymbol{I}$ and $\boldsymbol{V}$ are the phasor representation of the input current and voltage vectors. For a given network and for a specific frequency ω_0, the right-hand side of (7.3-2) can be obtained by simple matrix multiplication and addition. $\mathbf{Y}_n(j\omega_0)$ can also be determined from $\mathbf{Y}_n(j\omega_0) = \mathbf{A}^T\mathbf{Y}(j\omega_0)\mathbf{A}$. Consequently, using either nodal analysis or loop analysis the matrix equation describing the network can be written as

$$\mathbf{P}(j\omega_0)\mathbf{x}(j\omega_0) = \mathbf{b}(j\omega_0) \tag{7.3-3}$$

For example, for the nodal analysis mentioned previously the corresponding matrices are $\mathbf{P}(j\omega_0) = \mathbf{Y}_n(j\omega_0)$, $\mathbf{x}(j\omega_0) = \mathbf{r}(j\omega_0)$, and $\mathbf{b}(j\omega_0) = -[\mathbf{A}\boldsymbol{I} + \mathbf{A}\mathbf{Y}(j\omega_0)\boldsymbol{V}]$.

The only difference between solving (7.3-3) and the resistive networks discussed in the previous section is that, in the present case, the elements of the corresponding matrices are complex numbers. This, however, does not present any difficulty as far the computer solution of the problem is concerned. All one has to do is to program the computer in complex arithmetic and use any of the methods discussed in the previous section, such as the Gaussian elimination, LU transformation, sparse matrix, and the like. The computation time for complex arithmetics, however, is usually four times the computation time required for real arithmetic. Furthermore, the memory storage required for complex arithmetic is twice that for real arithmetic. For this reason, the savings in computation time and memory storage requirements become much more important in the present case.

Let us now work out an example to show the application of the Gaussian elimination method in the steady-state analysis of networks.

Example 7.3-1. Consider the fifth-order low-pass Butterworth filter shown in Fig. 7.3-1. The element values are $C_1 = C_5 = 0.618$ F, $C_3 = 2$ F, and $L_2 = L_4 = 1.618$ H. It is desired to obtain the voltage across the terminating resistor v_3. Assume that $\omega_0 = 1$ radian per second.

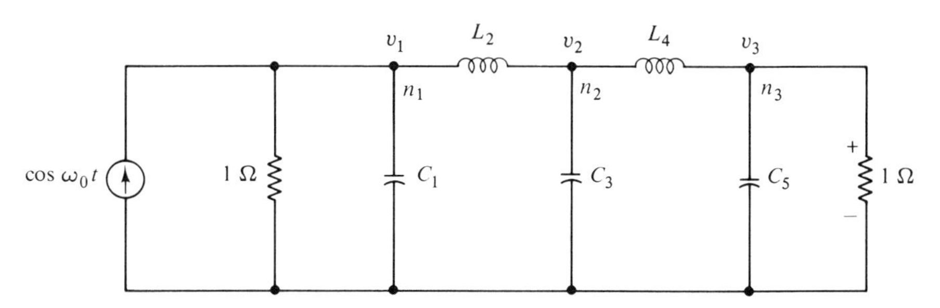

Figure 7.3-1 Low-pass Butterworth filter considered in Example 7.3-1.

Solution

Using phasor notation and performing nodal analysis at nodes n_1, n_2, and n_3 yields

$$\begin{bmatrix} 1 + j\omega_0 C_1 + \dfrac{1}{j\omega_0 L_2} & -\dfrac{1}{j\omega_0 L_2} & 0 \\ -\dfrac{1}{j\omega_0 L_2} & \dfrac{1}{j\omega_0 L_2} + j\omega_0 C_3 + \dfrac{1}{j\omega_0 L_4} & -\dfrac{1}{j\omega_0 L_4} \\ 0 & -\dfrac{1}{j\omega_0 L_4} & \dfrac{1}{j\omega_0 L_4} + j\omega_0 C_5 + 1 \end{bmatrix} \times \begin{bmatrix} V_1(j\omega_0) \\ V_2(j\omega_0) \\ V_3(j\omega_0) \end{bmatrix} = \begin{bmatrix} 1 \\ 0 \\ 0 \end{bmatrix}$$

For a given ω_0 this is a set of three equations in three unknowns that can be solved using any of the methods discussed in the last section. For instance, using a complex arithmetic, the following Gaussian elimination routine can be used to compute $v_3(j\omega_0)$ for the particular value given above.

```
C       THIS PROGRAM IS FOR EXAMPLE 7.3-1.IT USES GAUSSIAN ELIMINATION
C       WITH COMPLEX ARITHMETIC TO SOLVE A SET OF EQUATIONS.
        COMPLEX    P(20,21),TEMP,Q,X
        READ 20, N
     20 FORMAT(I2)
        M=N+1
        PRINT 25
     25 FORMAT(9X,*DATA*,/)
        PRINT 26,N
     26 FORMAT(9X,*N=*,I2)
        DO 1 I=1,N
      1 READ 21,(P(I,J),J=1,M)
     21 FORMAT(8F10.4)
        PRINT 27
     27 FORMAT(3X,*MATRIX P AUGMENTED BY VECTOR B IS*)
        DO 6 I=1,N
      6 PRINT 24,(P(I,J),J=1,M)
     24 FORMAT(2X,8F8.3)
C       USE GAUSSIAN ELIMINATION TO PUT P IN AN UPPER TRIANGLE FORM
        DO 14  I=2,N
        DO 14  J=I,N
C       CHECK THE DIAGONAL ELEMENT, IF IT IS ZERO , CHANGE ROWS
        IF (P(I-1,I-1)) 5,2,5
      2 K=I-1
        MC=0
        MB=N-I+1
        DO 11  L=I,N
        IF(P(L,K))4,7,4
      7 MC=MC+1
        GO TO 11
      4 DO 13  K1=K,M
        TEMP= P(L,K1)
        P(L,K1)= P(K,K1)
     13 P(K,K1)= TEMP
     11 CONTINUE
        IF(MB-MC)8,8,5
      5 Q=P(J,I-1)/P(I-1,I-1)
        II=I-1
        DO 14 K2=II,M
     14 P(J,K2)= P(J,K2)-Q*P(I-1,K2)
```

```
C     USE THE BACK SUBSTITUTION TO COMPUTE X
      DO 15  I=2,N
      K=N-I+2
      IF(P(K,K) .EQ. 0.0) GO TO 8
      Q= P(K,M)/P(K,K)
      DO 15  J=I,N
      M2=N-J+1
   15 P(M2,M)= P(M2,M)-Q*P(M2,K)
      PRINT 28
   28 FORMAT(//,* THE SOLUTION OF THE MATRIX EQUATION  PX=B*,/)
      DO 16  I=1,N
      X=P(I,M)/P(I,I)
   16 PRINT 23, I,X
   23 FORMAT(10X,*X(*,I1,*) =*,2F10.4)
      GO TO 9
    8 PRINT 22
   22 FORMAT(10X,* MATRIX P IS SINGULAR*)
    9 STOP
      END
```

```
THE SOLUTION OF THE MATRIX EQUATION  PX=B

          X(1) =      .7500      .2500
          X(2) =    -.4045     -.4045
          X(3) =    -.2500      .2500
```

Note: The first column of the result indicates the real part and the second column indicates the imaginary part of **x**.

7.4 COMPUTER SOLUTION OF NONLINEAR RESISTIVE NETWORKS

Analysis of nonlinear resistive networks is of considerable importance for several reasons. First, the majority of transistor and integrated circuits can be modeled as nonlinear resistive networks. Such networks are becoming more and more popular because they exclude expensive and bulky energy-storing elements. Second, as we shall see in Chapter 10, to obtain the equilibrium point or the operating point of a nonlinear network, we must first solve the corresponding resistive subnetwork, which is obtained by open circuiting all the capacitors and short circuiting all the inductors in the original network. Finally, many large-scale-system problems, such as transportation, scheduling, economic, and biomedical systems, can be modeled as nonlinear resistive networks that must be solved for their branch currents or node voltages.

In contrast to linear resistive networks, the analysis of general nonlinear resistive networks is quite difficult. The difficulties arising in the analysis of such networks usually stem from the fact that a nonlinear network may have no solution, a unique solution, several solutions, and possibly an infinite number of solutions. Many investigators [7–9] have considered the solution of such networks, and several computer programs are presently available [10, 11]. Almost all the existing methods use iterative techniques for solving a set of nonlinear equations that describe a given network. The basic difficulty in these methods is the convergence of the iterations. In this section we present some of these methods and discuss their advantages and disadvantages.

Let us first formulate the problem and give necessary and sufficient conditions for the uniqueness of its solution. Using either loop analysis or nodal analysis, the equations of a resistive network can be written as

$$\mathbf{f}(\mathbf{x}) = \mathbf{u} \tag{7.4-1}$$

where $\mathbf{x}$ is an n vector representing the chosen network variable to be determined (either loop currents, branch currents, or node voltages). The sources (either voltage source, current source, or both) are represented by the n vector $\mathbf{u}$. The function $\mathbf{f}(\cdot)$ is a mapping from the Euclidean n space into itself.

Note that for (7.4-1) to have a unique solution there must exist a function $\mathbf{g}(\cdot)$ such that

$$\mathbf{x} = \mathbf{g}(\mathbf{u}) \quad \text{for all admissible } \mathbf{u} \tag{7.4-2}$$

The existence of such a function is discussed in the following theorem due to Palais [12].

Theorem 7.4-1. The necessary and sufficient conditions for the function

$$\mathbf{f}(\mathbf{x}) = [f_1(\mathbf{x}), f_2(\mathbf{x}), \ldots, f_n(\mathbf{x})]^T \tag{7.4-3}$$

to have an inverse with continuous first derivative are

(a) Each $f_k(\cdot)$ has continuous first derivative; that is, $f'_k(\cdot)$ exists and is continuous.

(b) The Jacobian matrix of $\mathbf{f}(\mathbf{x})$ is nonsingular for all $\mathbf{x}$; that is,

$$\det\left[\frac{\partial \mathbf{f}}{\partial \mathbf{x}}\right] \neq 0 \quad \text{for all } \mathbf{x} \tag{7.4-4}$$

(c) The l_2 norm of $\mathbf{f}(\mathbf{x})$ approaches infinity as the l_2 norm of $\mathbf{x}$ approaches infinity; that is,

$$|\mathbf{f}(\mathbf{x})|_2 \longrightarrow \infty \quad \text{as} \quad |\mathbf{x}|_2 \longrightarrow \infty \tag{7.4-5}$$

For scalar nonlinear differential equations, conditions (a), (b), and (c) imply that the graph $\mathbf{f}(\mathbf{x})$ must be a continuously differentiable curve which is strictly monotonically increasing or decreasing, extending from $-\infty$ to $+\infty$. The proof of this theorem is given in [12]. There are several variations and generalizations of the theorem, some of which are mentioned in [9]. Next we introduce a set of sufficient conditions, in terms of network elements, for the existence and uniqueness of the solution of a given nonlinear resistive network.

Theorem 7.4-2. If (a) the resistive network under consideration contains no controlled sources, (b) the v-i characteristics of all its resistors are strictly monotonically increasing, and (c) $|i| \longrightarrow \infty$ as $|v| \longrightarrow \infty$ for each nonlinear resistor, then the network possesses a unique solution for all possible inputs.

The conditions of this theorem, although quite restricted, are very easy to check; most diode circuits, for instance, satisfy these conditions. Of course, these are only sufficient conditions; there is a large class of networks that do not satisfy these conditions and yet possess a unique solution.

Solution of Nonlinear Resistive Networks. Having settled the question of the existence and uniqueness of the solution of a given nonlinear resistive network, we must then seek a method of obtaining this solution. Virtually all practical resistive networks for which a solution exists must be solved numerically—only a very restricted class of networks can be solved by analytical means. Numerical methods used in solving such problems typically use one form or other of functional iteration methods. One of the most popular of these methods is the Newton–Raphson method, which we discuss next.

Newton–Raphson Method. To understand the basic idea behind this method, let us first consider the solution of a nonlinear first-order network whose equation is given by

$$f(x) = u \tag{7.4-6}$$

This equation may be expressed in the form

$$h(x) \triangleq f(x) - u = 0 \tag{7.4-7}$$

Denote the solution of (7.4-7) by x^*; then x^* is the point where the graph of $h(x)$ crosses the x axis (see Fig. 7.4-1).

The Newton–Raphson method can be summarized as follows:

Step 1. Guess a reasonable number x^0 for the solution of (7.4-7). If $h(x^0) = 0$, the solution is found; $x^* = x^0$. If $h(x^0) \neq 0$, go to step 2.

Step 2. Expand $h(x)$ around x^0 by a Taylor expansion; that is, write

$$h(x) = h(x^0) + (x - x^0)h'(x^0) + \frac{1}{2!}(x - x^0)^2 h''(x^0) + \cdots \tag{7.4-8}$$

Now choose x^1 that satisfies the linear part of (7.4-8):

$$h(x^0) + (x^1 - x^0)h'(x^0) = 0 \tag{7.4-9}$$

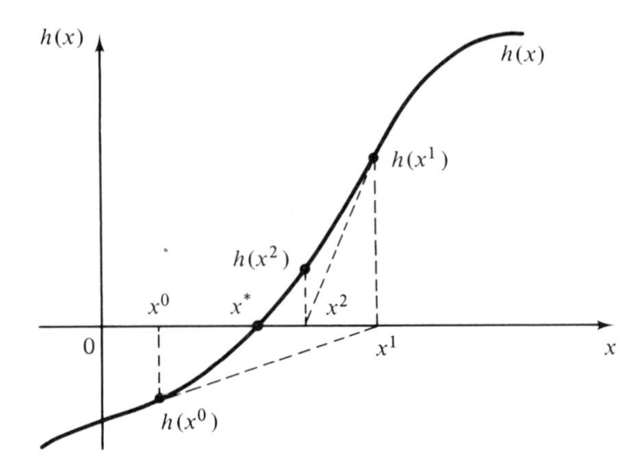

Figure 7.4-1 Graphical illustration of Newton-Raphson method.

or, equivalently,

$$x^1 = x^0 - \frac{h(x^0)}{h'(x^0)} \tag{7.4-10}$$

Since $f(x)$ and, consequently, $h(x)$ are strictly monotonical in the neighborhood of x^*, $h'(x) \neq 0$ and x^1 is well defined. This method of obtaining x^1 is shown in Fig. 7.4-1. Next we choose x^1 as the new initial guess and obtain x^2 as previously, and repeat this process. At the $(k + 1)st$ iteration, x^{k+1} is given by

$$x^{k+1} = x^k - \frac{h(x^k)}{h'(x^k)} \tag{7.4-11}$$

Step 3. If $h(x^k) = 0$, the desired solution is x^k and we stop the iterations. In practice, however, the iterations will be stopped if $h(x^k)$ is sufficiently small; that is, if

$$|h(x^k)| \leq \epsilon \tag{7.4-12}$$

where ϵ is a preassigned small positive number.

Remark 7.4-1. The Newton–Raphson method usually converges under two conditions. First, if the function $h(x)$ is convex over its domain of definition, no matter how far the initial guess x^0 is from the true solution x^*, the iterations will converge. Second, if the function is convex in the neighborhood of x^* and if x^0 is in this neighborhood, the iterations will converge. A situation for which these conditions are not met is shown in Fig. 7.4-2. In the situation shown, the initial guess x^0 will never converge to the true solution x^*.

Using the Newton–Raphson method for the analysis of resistive networks with nonconvex nonlinearities requires a prior knowledge of the approximate

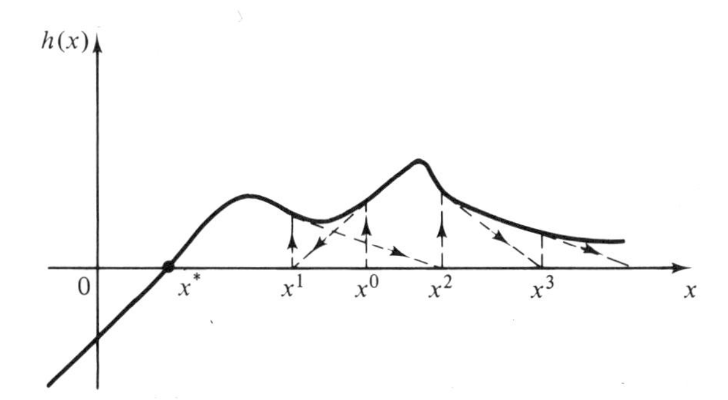

Figure 7.4-2 Situation for which the Newton-Raphson iteration method does not converge.

location of the solution. In practice, however, if the chosen initial guess x^0 leads to a divergent iteration, usually iteration is stopped and a new x^0 is chosen. This feature and several recent improvements [13, 14] have made the Newton–Raphson method a popular technique for analysis of nonlinear resistive networks. Before proceeding to the analysis of nth-order resistive networks by the Newton–Raphson method, let us work out an example to illustrate the procedure.

Example 7.4-1. Consider the simple diode circuit shown in Fig. 7.4-3, where the i-v relation of the nonlinear resistor is

$$i = 2(e^{0.1v} - 1)$$

Obtain the voltage v.

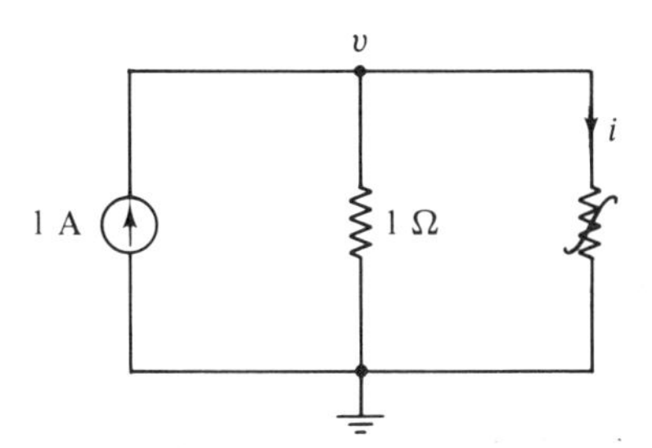

Figure 7.4-3 Diode circuit considered in Example 7.4-1.

Solution

The node equation of the circuit is

$$1 = v + 2(e^{0.1v} - 1)$$

or, equivalently,

$$v + 2e^{0.1v} - 3 = 0$$

This equation is in the form $h(v) = v + 2e^{0.1v} - 3 = 0$. To use the Newton–

Raphson method for solving this equation, observe that $h(v)$ is a strictly monotonically increasing function; hence, it has a unique solution. Furthermore, since $h(v)$ is convex, the iterations will converge even if the initial guess is not near the actual solution.

For this example let us take the initial guess v_0 to be

$$v^0 = 3 \text{ V}$$

The following computer program can then be used to compute the approximate solution. Note that in this program we stop the iteration if

$$|v + 2e^{0.1v} - 3| < 10^{-8}$$

or if the number of iterations is more than 20, whichever comes first. The second condition is imposed to prevent too many iterations if the equations are ill conditioned or if the initial guess is too far from the actual solution.

```
C     THIS PROGRAM IS FOR EXAMPLE 7.4-1. IT SOLVES THE NONLINEAR
C     EQUATION H(X)=0 BY NEWTON-RAPHSON METHOD.
      DIMENSION V(100),H(100),HP(100)
      K=1
      V(1)=3
    1 H(K)=V(K)+2.0*EXP(.1*V(K))-3.0
      HP(K)=1.0+.2*EXP(.1*V(K))
      V(K+1)= V(K)-H(K)/HP(K)
      IF(ABS(H(K)) .LT. 1.0E-08)GO TO 2
      IF(HP(K) .EQ. 0.0) GO TO 7
      K=K+1
      K1=20-K
      IF(K1) 1,3,1
    2 PRINT 4,V(K),K
    4 FORMAT(10X,///,10X,*V=*,E15.8,10X,*NO. OF ITERATION=*,I3)
      GO TO 99
    7 PRINT 8
    8 FORMAT(5X,////,10X,* GRADIENT IS ZERO*)
      GO TO 99
    3 PRINT 5
    5 FORMAT(10X,*THE PRESENT INITIAL GUESS GENERATES A DIVERGENT
     +ITERATION*)
      PRINT 4,V(K),K
   99 STOP
      END
```

The answer is

```
V=  .82746680E+00          NO. OF ITERATION=  4
```

Solution of n Simultaneous Nonlinear Equations. Let us now apply the Newton–Raphson method to solve a set of simultaneous nonlinear equations. Consider a nonlinear resistive network whose loop or nodal equations are

$$\mathbf{f}(\mathbf{x}) = \mathbf{u} \tag{7.4-13}$$

where $\mathbf{x}$ represents the unknown variables, such as loop currents or node voltages, $\mathbf{u}$ denotes the sources, and $\mathbf{f}(\cdot)$ is a vector-valued function. As in the previous case, let us rewrite (7.4-13) as

$$\mathbf{h}(\mathbf{x}) \triangleq \mathbf{f}(\mathbf{x}) - \mathbf{u} = \mathbf{0} \tag{7.4-14}$$

Let us now choose an initial guess $\mathbf{x}^0$; then using a similar procedure as in the first-order case, the $(k+1)$st iteration yields

$$\mathbf{x}^{k+1} = \mathbf{x}^k - \mathbf{J}^{-1}(\mathbf{x}^k)\cdot\mathbf{h}(\mathbf{x}^k) \tag{7.4-15}$$

where $\mathbf{J}(\mathbf{x}^k)$ denotes the Jacobian matrix of $\mathbf{h}$ evaluated at $\mathbf{x}^k$; that is,

$$\mathbf{J}(\mathbf{x}^k) = \begin{bmatrix} \frac{\partial h_1}{\partial x_1} & \frac{\partial h_1}{\partial x_2} & \cdots & \frac{\partial h_1}{\partial x_n} \\ \frac{\partial h_2}{\partial x_1} & \frac{\partial h_2}{\partial x_2} & \cdots & \frac{\partial h_2}{\partial x_n} \\ \cdot & & & \cdot \\ \cdot & & & \cdot \\ \cdot & & & \cdot \\ \frac{\partial h_n}{\partial x_1} & \frac{\partial h_n}{\partial x_2} & \cdots & \frac{\partial h_n}{\partial x_n} \end{bmatrix} \tag{7.4-16}$$

where h_k and x_k denote the kth components of $\mathbf{h}$ and $\mathbf{x}$, respectively.

Consequently, if the initial guess is sufficiently close to the actual solution $\mathbf{x}^*$, the iterative formula (7.4-15) will converge to $\mathbf{x}^*$. In the actual computer application of (7.4-15), the iterations will be stopped if the norm of the difference between $\mathbf{x}^{k+1}$ and $\mathbf{x}^k$ is sufficiently small or if the number of iterations exceeds a preassigned number. The latter case is to detect divergent iterations.

In contrast to the first-order case, it is quite difficult to obtain a reasonable initial guess $\mathbf{x}^0$ for an nth-order system of nonlinear equations. There are several modifications of this method that improve the chances of getting a convergent iteration. Such methods are discussed in great detail in the literature [13, 14]. Here we limit our discussion to the classical Newton–Raphson method.

To solve (7.4-15) at each iteration, one has to invert an $n \times n$ matrix $\mathbf{J}(\mathbf{x}^k)$. As was mentioned in the beginning of this chapter, inverting matrices can be costly and time consuming. An alternative method is to rewrite (7.4-15) in the form

$$\mathbf{J}(\mathbf{x}^k)[\mathbf{x}^k - \mathbf{x}^{k+1}] = \mathbf{h}(\mathbf{x}^k) \tag{7.4-17}$$

and solve it by the Gaussian elimination method, LU transformation, or the Gauss–Seidel technique discussed in the previous sections. Let us now work out a second-order example to illustrate the method.

Example 7.4-2. Consider the transistor circuit shown in Fig. 7.4-4. The Ebers–Moll model of this circuit can be given as in Fig. 7.4-5, where the *i-v*

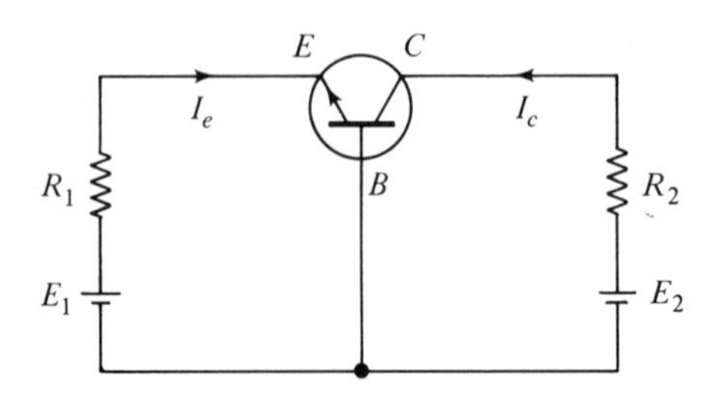

Figure 7.4-4 Simple transistor circuit considered in Example 7.4-2.

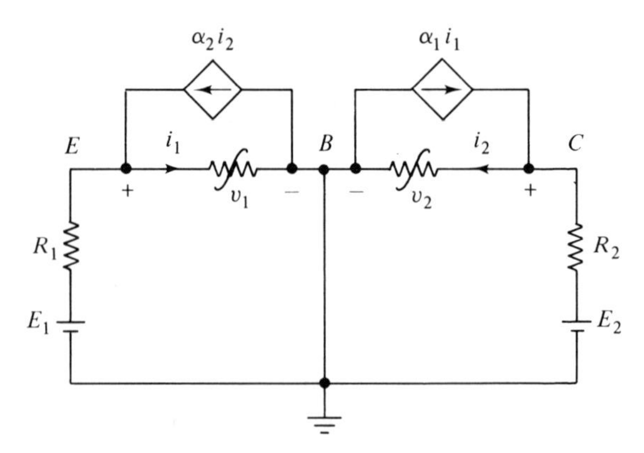

Figure 7.4-5 Ebers-Moll model of the transistor circuit of Figure 7.4-4.

relations of the nonlinear resistors are

$$i_1 = \beta_1(e^{\gamma_1 v_1} - 1)$$

and

$$i_2 = \beta_2(e^{\gamma_2 v_2} - 1)$$

Determine the voltages v_1 and v_2 if

$$E_1 = E_2 = 1 \text{ V}, \quad R_1 = R_2 = \tfrac{1}{2}\Omega, \quad \beta_1 = 2, \quad \beta_2 = 1, \quad \gamma_1 = \gamma_2 = 0.1,$$
$$\alpha_1 = 2, \quad \alpha_2 = 1$$

Solution

Using KCL and VCR at nodes E and C, we get

$$\frac{v_1 - E_1}{R_1} + \beta_1(e^{\gamma_1 v_1} - 1) - \alpha_2\beta_2(e^{\gamma_2 v_2} - 1) = 0$$

$$\frac{v_2 - E_2}{R_2} + \beta_2(e^{\gamma_2 v_2} - 1) - \alpha_1\beta_1(e^{\gamma_1 v_1} - 1) = 0$$

Using the given numerical values, these equations can be written as

$$\begin{aligned} 2v_1 + 2e^{0.1v_1} - e^{0.1v_2} - 3 &= 0 \\ 2v_2 - 4e^{0.1v_1} + e^{0.1v_2} + 1 &= 0 \end{aligned} \tag{7.4-18}$$

Equation (7.4-17), applied to the problem in hand, becomes

$$\begin{bmatrix} 2 + 0.2e^{0.1v_1{}^k} & -0.1e^{v_2{}^k} \\ -0.4e^{0.1v_1{}^k} & 2 + 0.1e^{0.1v_2{}^k} \end{bmatrix} \begin{bmatrix} v_1^k - v_1^{k+1} \\ v_2^k - v_2^{k+1} \end{bmatrix} = \begin{bmatrix} 2v_1^k + 2e^{0.1v_1{}^k} - e^{0.1v_2{}^k} - 3 \\ 2v_2^k - 4e^{0.1v_1{}^k} + e^{0.1v_2{}^k} + 1 \end{bmatrix}$$

Let us now choose as our initial guess

$$v_1^0 = 1 \quad \text{and} \quad v_2^0 = 1$$

These equations can be solved for v_1^1 and v_2^1 using a Gaussian elimination method; these values are then used for computing v_1^2 and v_2^2, and so on. The following program uses the Newton–Raphson method to obtain v_1 and v_2.

```
C     THIS PROGRAM IS FOR EXAMPLE 7.4-2.
C     THIS PROGRAM SOLVES A SET OF NONLINEAR EQUATIONS BY NEWTON RAPHSON
C     METHOD. A GAUSSIAN ELIMINATION ROUTINE IS USED TO SOLVE LINEARIZED
C     EQUATIONS.
      DIMENSION  P(20,21)
      DIMENSION V(10,100),X(10)
C     READ THE NO. OF EQUATIONS
      READ 20, N
   20 FORMAT(I2)
      M=N+1
C     READ THE INITIAL GUESSES.
      READ 50,(V(I,1),I=1,N)
   50 FORMAT(8F10.5)
      PRINT 49
   49 FORMAT(5X,////,10X,*INITIAL GUESS IS *)
      DO 44 I=1,N
   44 PRINT 51,I,V(I,1)
   51 FORMAT(10X,*V*,I2,5X,E15.8)
      KK=0
 1    KK=KK+1
      P(1,1)=2.0+.2*EXP(.1*V(1,KK))
      P(1,2)=-.1*EXP(.1*V(2,KK)).
      P(1,3)=-2.*V(1,KK)-2.0*EXP(.1*V(1,KK))+EXP(.1*V(2,KK))+3.0
      P(2,1)=-.4*EXP(.1*V(1,KK))
      P(2,2)=2.0+.1*EXP(.1*V(2,KK))
      P(2,3)=-(2.0*V(2,KK)-4.0*EXP(.1*V(1,KK))+EXP(.1*V(2,KK))+1.0)
      DO 52 I=1,N
      IF(ABS(P(I,M)) .GE. 1.0E-04) GO TO 53
   52 CONTINUE
      GO TO 34
   53 DO 14 I=2,N
      DO 14  J=I,N
C     CHECK THE DIAGONAL ELEMENT, IF IT IS ZERO , CHANGE ROWS
      IF (P(I-1,I-1)) 5,2,5
    2 K=I-1
      MC=0
      MB=N-I+1
      DO 11  L=I,N
      IF(P(L,K))4,7,4
    7 MC=MC+1
      GO TO 11
    4 DO 13  K1=K,M
      TEMP= P(L,K1)
      P(L,K1)= P(K,K1)
   13 P(K,K1)= TEMP
   11  CONTINUE
      IF(MB-MC)8,8,5
```

```
    5 Q=P(J,I-1)/P(I-1,I-1)
      II=I-1
      DO 14 K2=II,M
   14 P(J,K2)=P(J,K2)-Q*P(I-1,K2)
C     USE THE BACK SUBSTITUTION TO COMPUTE X
      DO 15  I=2,N
      K=N-I+2
      IF(P(K,K) .EQ. 0.0) GO TO 8
      Q= P(K,M)/P(K,K)
      DO 15  J=I,N
      M2=N-J+1
   15 P(M2,M)= P(M2,M)-Q*P(M2,K)
      DO 16  I=1,N
   16 X(I)= P(I,M)/P(I,I)
      IF(KK-40) 32,32,36
   32 DO 30 I=1,N
   30 V(I,KK+1)= V(I,KK)+X(I)
      GO TO 1
   34 PRINT 28
   28 FORMAT(//,2X,*THE FINAL ANSWER IS*,/)
      DO 41 JJ=1,N
      PRINT 38,JJ,V(JJ,KK),P(JJ,M)
   38 FORMAT(10X,*V*,I2,10X,E15.8,5X,*P(JJ)*,5X,E15.8,/)
   41 CONTINUE
      GO TO 9
   36 PRINT 40
   40 FORMAT(5X,/,5X,* THE PRESENT INITIAL GUESS GIVES DIVERGENT
     +ITERATION*)
      DO 39 JJ=1,N
   39 PRINT 38,JJ,V(JJ,KK),P(JJ,M)
      GO TO 9
    8 PRINT 22
   22 FORMAT(10X,* MATRIX IS SINGULAR*)
    9 STOP
      END
```

```
      INITIAL GUESS IS
      V 1        .10000000E+01
      V 2        .10000000E+01


THE FINAL ANSWER IS

      V 1               .95966618E+00

      V 2               .11410472E+01
```

Let us now discuss a different but equally important method of computing the response of nonlinear resistive networks.

Piecewise Linear Approximation Method. The basic idea behind this method is to divide the v-i characteristic of each nonlinear element into several regions. In each region the nonlinear curve is then approximated by a straight line, and the problem is reduced to solving a group of linear equations corresponding to all possible segment combinations of nonlinear resistors. For example, consider the nonlinear resistive network shown in Fig. 7.4-6. The nonlinear resistors in this circuit have the *i-v* characteristics shown in Fig. 7.4-7 by the solid line. The linear approximation of these characteristics is shown with broken lines. The *i-v* characteristic of the first resistor can be divided into four regions, whereas three regions are sufficient to approximate the second resistor. Thus, to analyze this network by the piecewise linear

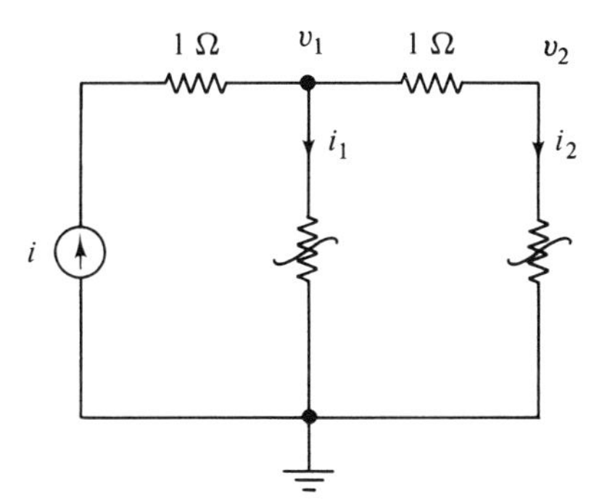

Figure 7.4-6 Nonlinear resistive network to be analyzed by piecewise linear approximation method.

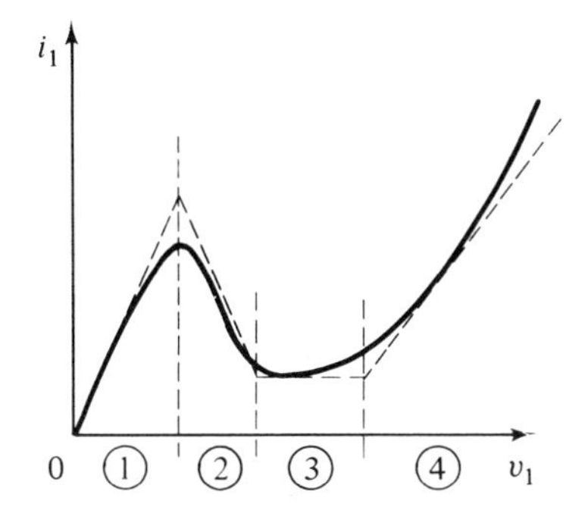

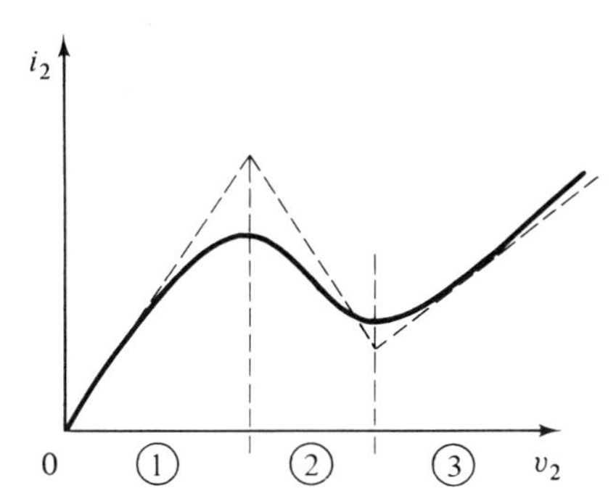

Figure 7.4-7 Piecewise linear approximation of the *v-i* characteristics of nonlinear resistors of the network of Figure 7.4-6.

approximation method, we should analyze $3 \times 4 = 12$ linear networks. In general, if the given network has n nonlinear resistors and if the jth resistor has m_j segments, the total number of linear resistive networks that should be solved will be

$$N = m_1 m_2 m_3 \dots m_n \tag{7.4-19}$$

For this reason, only networks with a relatively small number of nonlinearities can be analyzed by this method—computation time for large-scale nonlinear networks is prohibitive. Recently, however, a modified version of the original linear piecewise approximation has been developed that eliminates the need for solving the linear network for all possible combinations of the segments.

By doing so, the computation time is drastically reduced. A detailed treatment of this method is given by Chua [11].

7.5 COMPUTER SOLUTION OF STATE EQUATIONS

Derivation of the state equation of a large class of networks was considered in Chapter 5. We showed that different classes of networks can be represented by the following state equations:

Linear time invariant: $\dot{\mathbf{x}} = \mathbf{A}\mathbf{x} + \mathbf{B}\mathbf{u}$ (7.5-1)

Linear time varying: $\dot{\mathbf{x}} = \mathbf{A}(t)\mathbf{x} + \mathbf{B}(t)\mathbf{u}$ (7.5-2)

Nonlinear time varying: $\dot{\mathbf{x}} = \mathbf{f}(\mathbf{x}, \mathbf{u}, t)$ (7.5-3)

In Chapter 6 analytical solution of (7.5-1) was discussed in great detail. It was mentioned, however, that solutions of (7.5-2) and (7.5-3) can, in general, be obtained by numerical methods.

In this section we consider several methods of solving the most general of state representations, (7.5-3). These techniques can also be used to solve linear state equations. Before introducing any numerical solution to (7.5-3), let us write this equation in detail:

$$\dot{\mathbf{x}}(t) = \mathbf{f}[\mathbf{x}(t), \mathbf{u}(t), t], \qquad \mathbf{x}(t_0) = \mathbf{x}_0 \tag{7.5-4}$$

where $\mathbf{x}(t)$ is an n vector representing the state, $\mathbf{u}$ is an m vector representing the sources and possibly their derivatives, t denotes the dependence of $\mathbf{f}$ on time, and $\mathbf{f}(\cdot, \cdot, \cdot)$ is a vector-valued function. The initial time is t_0, and $\mathbf{x}_0$ represents the initial conditions.

Recall from Chapter 6 that by the solution of (7.5-4) we mean a vector-valued function $\mathbf{x}(t)$ which satisfies both the differential equation and the initial condition. Furthermore, $\mathbf{x}(t)$ must be defined on the semiinfinite interval $t \geq t_0$. In contrast to analytic solution of (7.5-4), due to finite computation time, a numerical solution of this equation is given only at discrete points in the *finite* interval $[t_0, t_0 + T]$, where t_0 and T are any two preassigned finite numbers. The general procedure for the numerical solution of (7.5-3) is to use $\mathbf{x}(t)$ and $\mathbf{f}(\mathbf{x}, \mathbf{u}, t)$ to extrapolate the state $\mathbf{x}(t + h)$, where h denotes the step size in the discretization of the time interval of interest. The computation time required for such analysis, therefore, is inversely proportional to the step size h. It is therefore essential to devise integration techniques that allow large step sizes without a loss of accuracy. These methods usually fall into two different categories, *explicit* and *implicit* integration methods. In explicit methods $\mathbf{x}(t + h)$ is expressed in terms of $\mathbf{x}(t)$ and past values of $\mathbf{x}(t)$, such as $\mathbf{x}(t - h)$, and so on. On the other hand, in an implicit method $\mathbf{x}(t + h)$ is defined in terms of $\mathbf{x}(t)$ and future values of $\mathbf{x}(t)$. In what follows we discuss several of these techniques; we start from the simplest method:

Euler's Method. In this method the solution of (7.5-4) at $t + h$ is approximated in terms of $\mathbf{x}(t)$ and h. To be more precise, let us divide the time interval of interest into n equal segments each of length h so that

$$h = \frac{T}{n} \tag{7.5-5}$$

To simplify the manipulations, let us adopt the following notations. Let

$$t_k = t_0 + kh \tag{7.5-6}$$

$$\mathbf{u}_k = \mathbf{u}(t_k) = \mathbf{u}(t_0 + kh) \tag{7.5-7}$$

$$\mathbf{x}_k = \mathbf{x}(t_k) = \mathbf{x}(t_0 + kh) \tag{7.5-8}$$

Now observe that the derivative of $\mathbf{x}$ at t_0 can be approximated by

$$\dot{\mathbf{x}}(t_0) \simeq \frac{1}{h}[\mathbf{x}(t_1) - \mathbf{x}(t_0)] \tag{7.5-9}$$

Using this equation together with (7.5-4), we get an approximate formula for computing $\mathbf{x}(t_1)$:

$$\mathbf{x}_1 \simeq \mathbf{x}_0 + h\mathbf{f}(\mathbf{x}_0, \mathbf{u}_0, t_0)$$

The geometrical interpretation of this formula can best be seen in Fig. 7.5-1. This figure shows the actual and the computed solution of a scalar differential equation.

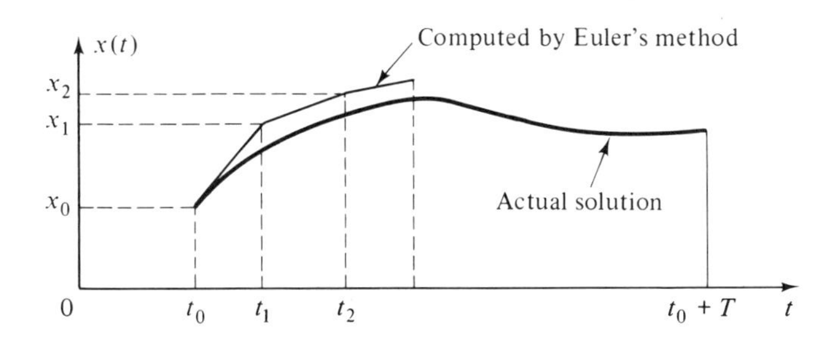

Figure 7.5-1 Result by Euler's method.

The general form of the preceding equation can be obtained by expanding $\mathbf{x}_{k+1}$ by a Taylor expansion around t_k; that is,

$$\mathbf{x}_{k+1} = \mathbf{x}_k + h\mathbf{f}(\mathbf{x}_k, \mathbf{u}_k, t_k) + \frac{h^2}{2}\ddot{\mathbf{x}}(\xi) \tag{7.5-10}$$

where $\ddot{\mathbf{x}}$ denotes the second derivative of $\mathbf{x}$, and ξ is a positive number in the interval $[t_k, t_{k+1}]$; that is,

$$t_k < \xi < t_{k+1} \tag{7.5-11}$$

For sufficiently small h, the last term in (7.5-10) can be neglected. The resulting approximate equation is called *Euler's forward integration formula* and is written as

$$\mathbf{x}_{k+1} = \mathbf{x}_k + h\mathbf{f}(\mathbf{x}_k, \mathbf{u}_k, t_k) \tag{7.5-12}$$

Obviously, the smaller the step size h, the smaller the truncation error will be. The last term in (7.5-10), of course, represents only the error at the kth step. The total error will be an accumulation of such terms. It can be shown [1] that the total truncation error in the Euler method is proportional to h. Euler's integration formula, although quite straightforward, is seldom used in the actual solution of differential equations. The reason for this is not only the lack of accuracy of this method, but also the numerical instability that might occur. We will discuss the numerical instability of a method later in this section. Let us now present an improvement of the Euler method just discussed.

Trapezoidal Rule. In (7.5-12) we used the slope of $\mathbf{x}(t)$ at t_k to approximate $\mathbf{x}(t_{k+1})$. If instead we choose the average of the slopes of $\mathbf{x}(t)$ at t_k and t_{k+1}, the following approximation formula will result:

$$\mathbf{x}_{k+1} = \mathbf{x}_k + \frac{h}{2}[\mathbf{f}(\mathbf{x}_k, \mathbf{u}_k, t_k) + \mathbf{f}(\mathbf{x}_{k+1}, \mathbf{u}_{k+1}, t_{k+1})] \tag{7.5-13}$$

This equation describes the trapezoidal rule of integration. Notice that (7.5-13) is an *implicit* equation in the sense that the unknown, $\mathbf{x}_{k+1}$, appears in both sides of the equation. This equation, therefore, cannot be solved directly; the Newton–Raphson method can be employed to solve (7.5-13). This procedure is called *implicit numerical integration*.

Alternatively, to solve (7.5-13) for $\mathbf{x}_{k+1}$, we can approximate $\mathbf{x}_{k+1}$ in the right-hand side by Euler's equation first (i.e., *predicting* $\mathbf{x}_{k+1}$), and then *correcting* this predicted value of $\mathbf{x}_{k+1}$ using (7.5-13). This is called the *predictor–corrector method*, which we shall discuss later.

To obtain an estimate of the error involved in (7.5-13), let us expand $\mathbf{x}_{k+1}$ around t_k by the following Taylor series:

$$\mathbf{x}_{k+1} = \mathbf{x}_k + h\dot{\mathbf{x}}_k + \frac{h^2}{2}\ddot{\mathbf{x}}_k + \frac{h^3}{6}\dddot{\mathbf{x}}(\xi) \tag{7.5-14}$$

where ξ is a real number so that $t_k < \xi < t_{k+1}$. If in (7.5-14) we approximate $\ddot{\mathbf{x}}_k$ by $(1/h)(\dot{\mathbf{x}}_{k+1} - \dot{\mathbf{x}}_k)$, we get

$$\mathbf{x}_{k+1} = \mathbf{x}_k + \frac{h}{2}[\dot{\mathbf{x}}_k + \dot{\mathbf{x}}_{k+1}] + \boldsymbol{\alpha}h^3 \tag{7.5-15}$$

where $\boldsymbol{\alpha}$ is a real column vector. In (7.5-15) we can replace $\dot{\mathbf{x}}_k$ and $\dot{\mathbf{x}}_{k+1}$ from (7.5-4) to get

$$\mathbf{x}_{k+1} = \mathbf{x}_k + \frac{h}{2}[\mathbf{f}(\mathbf{x}_k, \mathbf{u}_k, t_k) + \mathbf{f}(\mathbf{x}_{k+1}, \mathbf{u}_{k+1}, t_{k+1})] + \boldsymbol{\alpha} h^3$$

Comparison of the last equation to (7.5-13) shows that the error at each step is proportional to h^3. This is obviously an improvement over Euler's method, since typically $h \ll 1$. In this case, also, it can be shown that the total error is proportional to h^2.

General Integration Formulas. Euler's method and the trapezoidal rule discussed so far are special cases of a whole generation of equations called *numerical integration formulas*. Such equations are characterized by the following:

$$\mathbf{x}_{k+1} = \sum_{i=0}^{p} a_i \mathbf{x}_{k-i} + h \sum_{i=-1}^{p} b_i \dot{\mathbf{x}}_{k-i} \tag{7.5-16}$$

where p, i, and k are positive integers, and h is the integration step size. For example, if $p = 0$, $a_0 = b_0 = 1$, and $b_{-1} = 0$, Euler's equation will result. If $p = 0$, $a_0 = 1$, and $b_{-1} = b_0 = \frac{1}{2}$, the result is the trapezoidal rule. In general, if $b_{-1} = 0$, the resulting equations are called *explicit* integration schemes, and if $b_{-1} \neq 0$, the corresponding equations are called *implicit* formulas. For $p = 1$, (7.5-16) becomes $\mathbf{x}_{k+1} = a_0 \mathbf{x}_k + a_1 \mathbf{x}_{k-1} + h[b_{-1}\mathbf{f}(\mathbf{x}_{k+1}, \mathbf{u}_{k+1}, t_{k+1}) + b_0 \mathbf{f}(\mathbf{x}_k, \mathbf{u}_k, t_k) + b_1 \mathbf{f}(\mathbf{x}_{k-1}, \mathbf{u}_{k-1}, t_{k-1})]$. The coefficients a_i and b_i in (7.5-16) are determined by specific constraints on the accuracy and other computational properties of the formula ([1], p. 164).

The truncation error of (7.5-16) can also be determined in terms of the step size h and higher derivatives of $\mathbf{x}(t)$. This can be done by expanding $\mathbf{x}(t)$ around t_k and rearranging the equations so that the result can be compared to (7.5-16).

A basic criterion for usefulness of any of the numerical integration formulas given is their *stability*. The numerical stability of an integration method is defined next:

Definition 7.5-1 (Numerical Stability). A numerical integration method is said to be *stable* if the total error decreases as the step size h decreases.

Also, a numerical integration method is called *unstable* if the total error increases as h decreases. It should be mentioned at this time that the numerical instability defined is quite different from the physical instability of a network. If an unstable numerical integration is employed, a well-behaved network might result in an unstable response! Study of the stability of different integration methods can be carried out by considering the corresponding difference equation. We shall not, however, pursue this topic further and refer the interested reader to advanced texts in numerical analysis [2]. Let us now discuss some practical methods of solving nonlinear time-varying differential equations: the Runge–Kutta and predictor–corrector methods.

Runge-Kutta Method. A group of methods widely used in numerical solution of differential equations is known as Runge–Kutta methods. Their basic property is to maximize the accuracy of the result per unit of computation effort. The most popular among these is the fourth-order Runge–Kutta method, which we shall present shortly. Let us first outline the basic steps necessary for derivation of this method. Since the algebra involved in deriving the fourth-order formula is too involved to present in this chapter, we only show how a second-order Runge–Kutta formula can be obtained—the procedure will be the same for other Runge–Kutta methods.

To save space, let us assume that the scalar differential equation under consideration is

$$\dot{x} = f(x, t), \qquad x(t_0) = x_0 \tag{7.5-17}$$

In this equation the effect of the input $u(t)$ is incorporated in t. In a second-order Runge–Kutta method, we wish to obtain the scalar constants α, β, a, and b such that at the kth step the equation

$$x_{k+1} = x_k + aK_1 + bK_2 \tag{7.5-18}$$

where

$$K_1 = hf(x_k, t_k) \tag{7.5-19}$$

and

$$K_2 = hf(x_k + \beta K_1, t_k + \alpha h) \tag{7.5-20}$$

is the best approximation of the Taylor expansion of x_{k+1} around x_k: namely

$$x_{k+1} = x_k + h\dot{x}_k + \frac{h^2}{2}\ddot{x}_k + \cdots \tag{7.5-21}$$

To obtain these constants, let us use (7.5-17) and the chain rule to rewrite (7.5-21) as

$$x_{k+1} = x_k + hf(x_k, t_k) + \frac{h^2}{2!}\left[\frac{\partial f(x, t)}{\partial t} + \frac{\partial f(x, t)}{\partial x} f(x, t)\right]_{t=t_k} + \cdots \tag{7.5-22}$$

Also, substituting (7.5-19) in (7.5-20) and the resulting equation in (7.5-18), we get

$$x_{k+1} = x_k + ahf(x_k, t_k) + bhf[x_k + \beta hf(x_k, t_k), t_k + \alpha h] + \cdots \tag{7.5-23}$$

Expanding the last term in (7.5-23) by a Taylor expansion in both variables and comparing the result with (7.5-22), we obtain

$$a + b = 1, \qquad \alpha b = \tfrac{1}{2}, \quad \text{and} \quad \beta b = \tfrac{1}{2}$$

Choosing one of the constants arbitrarily, the rest of the constants will be specified. For example, if $a = \frac{1}{2}$, then $b = \frac{1}{2}$, and $\alpha = \beta = 1$. This results in the trapezoidal integration scheme discussed earlier.

By a similar, although much more tedious, procedure the celebrated fourth-order Runge–Kutta method can be obtained. The results for the general scalar differential equation

$$\dot{x}(t) = f[x(t), u(t), t], \qquad x(t_0) = x_0 \tag{7.5-24}$$

are

$$x_{k+1} = x_k + \tfrac{1}{6}(K_1 + 2K_2 + 2K_3 + K_4) \tag{7.5-25}$$

$$K_1 = hf(x_k, u_k, t_k) \tag{7.5-26}$$

$$K_2 = hf[x_k + \tfrac{1}{2}K_1, u_{k+(1/2)}, t_{k+(1/2)}] \tag{7.5-27}$$

$$K_3 = hf[x_k + \tfrac{1}{2}K_2, u_{k+(1/2)}, t_{k+(1/2)}] \tag{7.5-28}$$

$$K_4 = hf(x_k + K_3, u_{k+1}, t_{k+1}) \tag{7.5-29}$$

where

$$t_{k+(1/2)} \triangleq t_k + \tfrac{1}{2}h \tag{7.5-30}$$

and

$$u_{k+(1/2)} \triangleq u(t_k + \tfrac{1}{2}h) \tag{7.5-31}$$

This algorithm can easily be programmed on a digital computer to solve a given differential equation. A sample flow chart of such a program is illustrated on the following page.

Let us now work out an example to illustrate how the equations can be applied to solve a nonlinear and time-varying scalar differential equation.

Example 7.5-1. Use the fourth-order Runge–Kutta method to solve the following differential equation over the interval $0 \le t \le 10$ seconds.

$$\dot{x}(t) = -t^2x(t) - x^2(t) - \sin t, \qquad x(0) = 0.5 \tag{7.5-32}$$

Solution

Let us choose the step size h to be

$$h = \frac{10 - 0}{100} = 0.1$$

Runge–Kutta Flow Chart

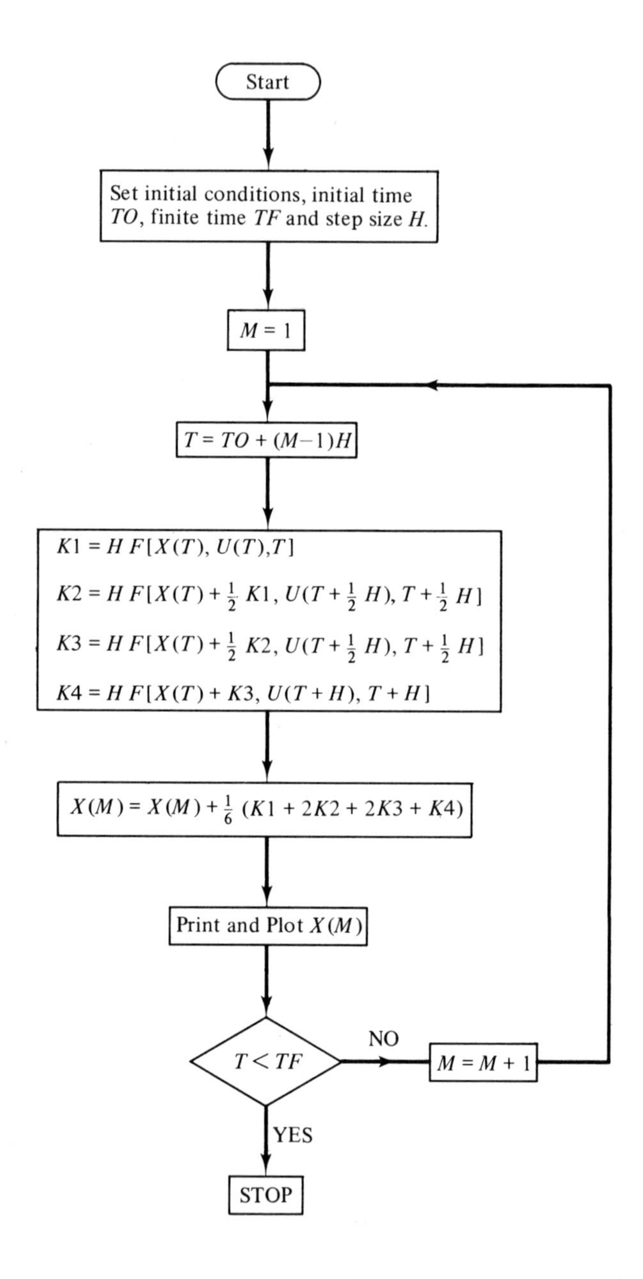

The following computer program can then be written to compute $x(k)$. (To plot $x(k)$, a subroutine GRAPHX is used.)

```
C     THIS PROGRAM IS FOR EXAMPLE 7.5-1
C     THIS PROGRAM USES A FOURTH ORDER RUNGE-KUTTA METHOD TO
C     SOLVE THE NONLINEAR DIFFERENTIAL EQUATION OF EXAMPLE 7.5-1
      DIMENSION X(401),DATA(210,3)
      REAL K1,K2,K3,K4
C     SET THE INITIAL CONDITIONS
      X(1)=.5
      T= 0.0
      PRINT 2
    2 FORMAT(22X,*T*,13X,*X*)
      N= 1
C     SET THE STEP SIZE
      H=.025
C     COMPUTE THE RUNGE-KUTTA COEFFICIENTS
    1 K1=H*(-T**2*X(N)-X(N)**2-SIN(T))
      K2=H*(-(T+H/2)**2*(X(N)+K1/2.)-(X(N)+K1/2.)**2-SIN(T+H/2.))
      K3=H*(-(T+.5*H)**2*(X(N)+K2/2.)-(X(N)+K2/2.)**2-SIN(T+H/2.))
      K4=H*(-(T+H)**2*(X(N)+K3)-(X(N)+K3)**2-SIN(T+H))
      X(N+1)=X(N)+(K1+2.*K2+2.0*K3+K4)/6.0
      T=T+H
      N=N+1
      IF(T .LT. 10.0) GO TO 1
      DO 5 N=1,401,8
      TIME=(N-1)*H
      M=N/8+1
      PRINT 3,TIME,X(N)
    3 FORMAT(19X,F6.2,5X,E15.8)
      DATA(M,1)=TIME
      DATA(M,2)=X(N)
    5 CONTINUE
C     PLOT
      CALL GRAPHX(DATA,M,1HT,5H  X  )
      END
```

The following subroutine is used to plot the dependent variable x versus the independent variable t. This subroutine plots up to three functions simultaneously. The functions are denoted by DATA(N,M), where M designates the function number; for example, if there are three functions X1(N), X2(N), and X3(N) to be plotted, we can write

$$X1(N)=DATA(N,1)$$

$$X2(N)=DATA(N,2)$$

$$X3(N)=DATA(N,3)$$

The integer N denotes the independent variable.

Subroutine GRAPHX

```
      SUBROUTINE GRAPHX(DATA,N,VINDEP,VARDEP)
      DIMENSION DATA(210,3),B(121)
      PRINT 300,VINDEP
  300 FORMAT(1H1,* THE INDEPENDENT VARIABLE IS *,2A10)
      PRINT 400,VARDEP
  400 FORMAT(1H,* THE DEPENDENT VARIABLE IS *,2A10//)
      BIGEST=DATA(1,2)
      SMAL = DATA(1,2)
      DO 1 I=2,N
      IF(DATA(I,2).GT.BIGEST)BIGEST=DATA(I,2)
      IF(DATA(I,2).LT.SMAL)SMAL=DATA(I,2)
    1 CONTINUE
      DO 2 I=2,N
      IF(DATA(I,3).GT.BIGEST)BIGEST=DATA(I,3)
      IF(DATA(I,3).LT.SMAL)SMAL=DATA(I,3)
    2 CONTINUE
      PRINT 200, SMAL, BIGEST
      BMINS = BIGEST-SMAL
      DO 3 I=1,61
    3 B(I) = 1H
      DO 4 I=1,N
      DATA(I,2)=(DATA(I,2)-SMAL)*61.0/BMINS+1.0
      DATA(I,3)=(DATA(I,3)-SMAL)*61.0/BMINS+1.0
      INDEX = DATA(I,2)
      JNDEX=DATA(I,3)
      B(INDEX) = 1H+
      B(JNDEX) = 1H*
      PRINT 100,DATA(I,1),(B(NN1),NN1=1,61)
      B(INDEX)=1H
    4 B(JNDEX) = 1H
  200 FORMAT(9X,E11.4,35X,1H,4X,E11.4,/10X,61(1H*))
  100 FORMAT(1HZ,F8.2,1X,61A1)
      RETURN
      END
```

The computer plot of $x(t)$ versus t is given in Fig. 7.5-2.

Quite clearly, if the differential equation under consideration is linear, a similar program can be written to solve it. However, if an nth-order differential equation is considered, it is much more economical to introduce a subroutine to solve the equations simultaneously. An example of such a subroutine is given next.

Example 7.5-2. Use the fourth-order Runge–Kutta method to solve the following set of coupled differential equations:

$$\begin{aligned} \dot{x}_1 &= -2x_1 + x_2^2 - e^{-t} & x_1(0) &= 1 \\ \dot{x}_2 &= -3x_2 - x_3 + 1 & x_2(0) &= 0 \\ \dot{x}_3 &= -x_1 - tx_3^3 + \sin t, & x_3(0) &= -1 \end{aligned}$$

Assume that $t_0 = 0$, $t_f = 5$ seconds, and take the step size h to be 0.05.

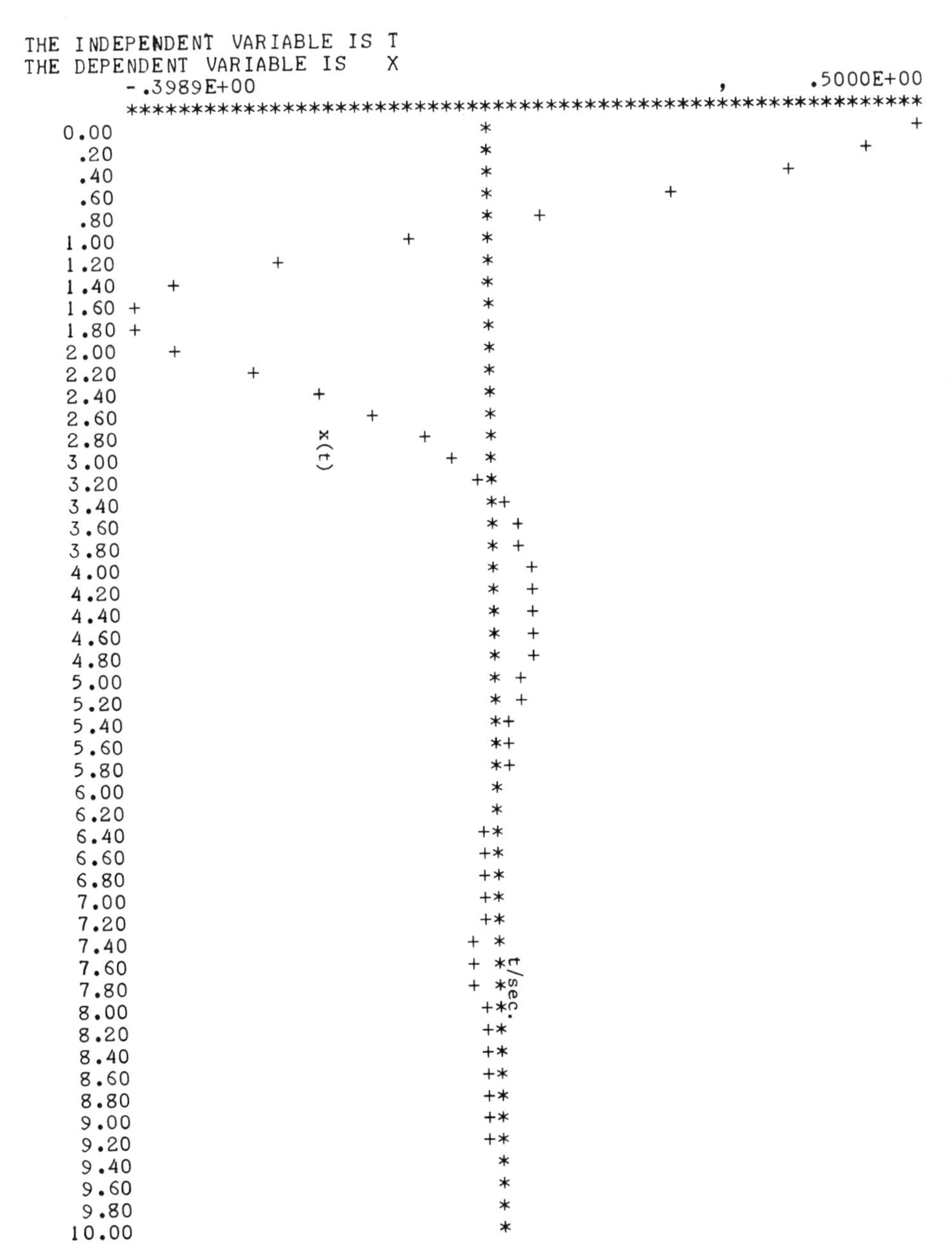

Figure 7.5-2 Computer plot of $x(t)$ versus t.

Solution

The following computer program and the accompanying subroutine can be used to provide the solution.

```
C     THIS PROGRAM IS FOR EXAMPLE 7.5-2
C     THIS PROGRAM USES A FOURTH ORDER RUNGE-KUTTA METHOD TO SOLVE
C     THE NONLINEAR TIME VARYING 3RD ORDER DIFFERENTIAL EQUATION OF
C     EXAMPLE 7.5-2 AND PLOTS X1 AND X2
      DIMENSION X(10),DX(10),DATA(210,3)
C     SET THE INITIAL CONDITIONS
      X(1)=1.0
      X(2)=0.0
      X(3)=-1.0
C     N IS NO. OF EQUATIONS
      N=3
C     SET INITIAL AND FINAL TIMES
      T=0.0
      TF=5.0
C     STORE INITIAL VALUES FOR PLOTTING.
      DATA(1,1)=T
      DATA(1,2)=X(1)
      DATA(1,3)=X(3)
C     SET STEP SIZE
      H=0.05
      PRINT 20
   20 FORMAT(15X,1HT,16X,2HX1,18X,2HX2,14X,2HX3)
      PRINT 30,T,(X(I),I=1,N)
   30 FORMAT(8X,6(F12.4,5X))
C     INITIALIZE K AND M
      K=0
      M=1
C     WRITE DOWN THE DIFFERENTIAL EQUATIONS
      H=.05
    1 DX(1)=-2.0*X(1)+X(2)**2-EXP(-T)
      DX(2)=-3.0*X(2)-X(3)+1.0
      DX(3)=-X(1)-T*(X(3)**3)+SIN(T)
      CALL RUNTA(N,K,I,X,DX,T,H)
      GO TO (1,2),I
C     PRINT AND PLOT X(1) AND X(3)
    2 NSTEP=2
      EPS=1.0E-04
      K1= T/H+EPS
      K2=NSTEP*AINT(T/H/NSTEP+EPS)
      IF(K1.NE.K2) GO TO 4
      PRINT 30, T,X(1),X(2),X(3)
      M= M+1
      DATA(M,1)= T
      DATA(M,2)= X(1)
      DATA(M,3)= X(3)
    4 IF(T.LE.(TF-EPS))GO TO 1
C     PLOT X(1) AND X(3)
      CALL GRAPHX(DATA,M,1HT,5HX1,X3)
      STOP
      END
```

The following subroutine employs the fourth-order Runge–Kutta method to solve a set of differential equations.

Subroutine RUNTA

```
      SUBROUTINE RUNTA(N,K,I,X,DX,T,H)
      DIMENSION Y(1000),Z(1000),X(10),DX(10)
      K=K+1
      GO TO (1,2,3,4,5),K
    2 DO 10 J=1,N
      Z(J)= DX(J)
      Y(J)= X(J)
   10 X(J)=Y(J)+.5*H*DX(J)
   25 T=T+.5*H
    1 I=1
      RETURN
    3 DO 15 J=1,N
      Z(J)= Z(J)+2.0*DX(J)
   15 X(J)=Y(J)+.5*H*DX(J)
      I=1
      RETURN
    4 DO 20 J=1,N
      Z(J)= Z(J)+2.0*DX(J)
   20 X(J)=Y(J)+.5*H*DX(J)
      GO TO 25
    5 DO 30 J=1,N
   30 X(J)=Y(J)+(Z(J)+DX(J))*H/6.0
      I=2
      K=0
      RETURN
      END
```

The computer plot of $x(1)$ and $x(3)$ appears in Fig. 7.5-3.

Remark 7.5-1. The main advantage of Runge–Kutta methods is that they are self-starting. This means that to compute x_{k+1} we only need to know x_k, in contrast to implicit methods in which x_{k+1} appears in both sides of the equation. The main disadvantage of Runge–Kutta methods is that the step size h should be chosen quite small because of the total truncation errors. Runge–Kutta methods are sometimes used as the starting equations for more efficient implicit integration methods. We now briefly discuss one such implicit method.

Predictor–Corrector Method. It was mentioned earlier in this section that to implement implicit integration formulas such as (7.5-13) we can use an explicit formula, say (7.5-12), to *predict* the value of x_{k+1} and then *correct* this value. These simple predictor–corrector formulas are repeated here for the scalar differential equation $\dot{x} = f(x, u, t)$.

Predictor

$$x_{k+1}^{*} = x_k + f(x_k, u_k, t_k) \tag{7.5-33}$$

Corrector

$$x_{k+1} = x_k + \frac{h}{2}[f(x_k, u_k, t_k) + f(x_{k+1}^{*}, u_{k+1}, t_{k+1})] \tag{7.5-34}$$

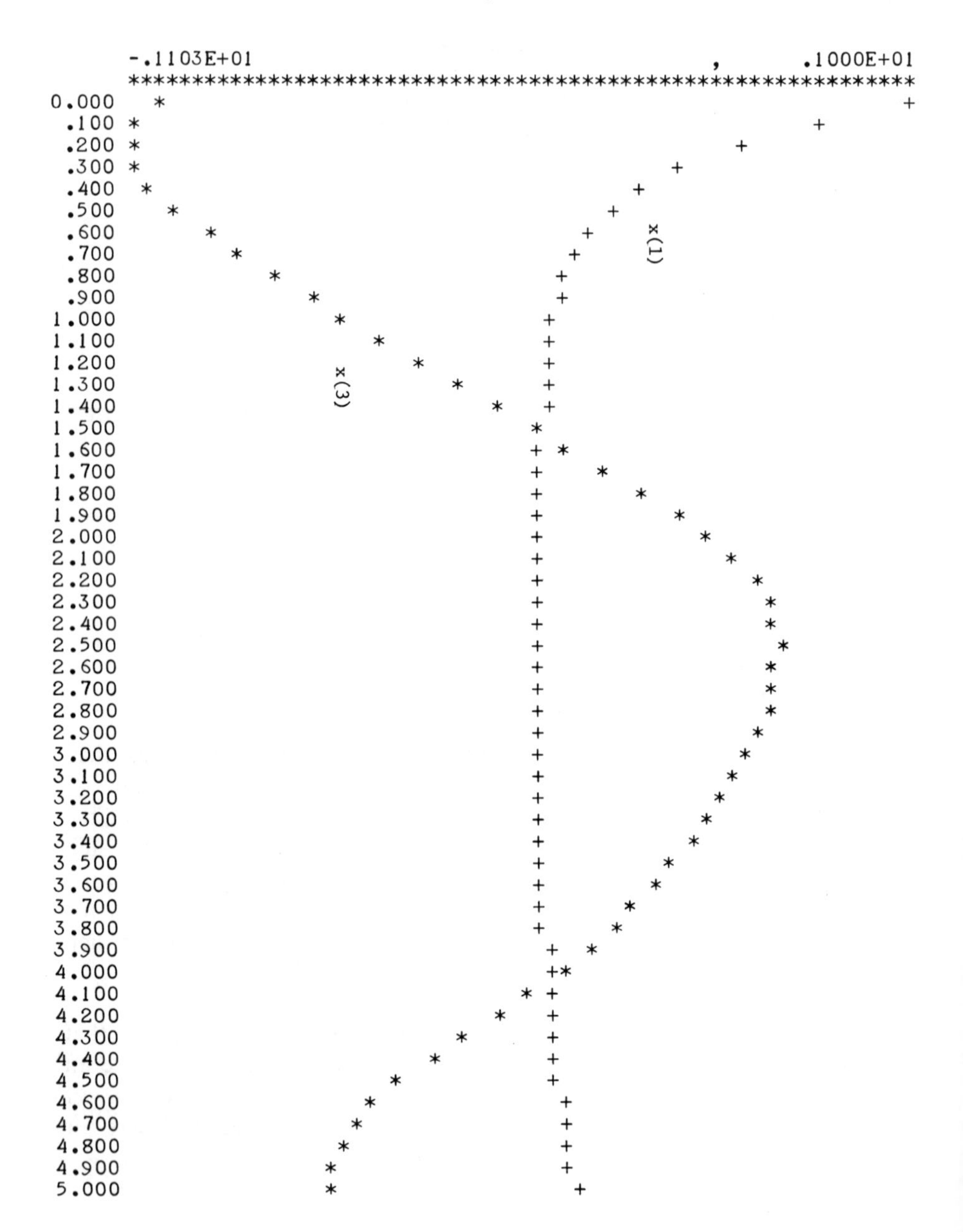

Figure 7.5-3 Computer plot of $x(1)$ and $x(3)$.

where x_{k+1}^* indicates the predicted and x_{k+1} represents the corrected value of $x(t_{k+1})$.

The general formula for the predictor–corrector methods is

Predictor

$$x_{k+1}^* = x_{k-q} + h \sum_{j=0}^{m} \beta_j f(x_{k-j}, u_{k-j}, t_{k-j}) \tag{7.5-35}$$

Corrector

$$x_{k+1} = x_{k-p} + h \sum_{i=1}^{n} \alpha_i f(x_{k+1-i}, u_{k+1-i}, t_{k+1-i}) + h\alpha_0 f(x_{k+1}^*, u_{k+1}, t_{k+1}) \tag{7.5-36}$$

For example, if $m = 0$, $q = 0$, $n = 1$, $p = 0$, $\beta_0 = 1$, and $\alpha_1 = \alpha_0 = \frac{1}{2}$, the Euler's predictor–corrector method given in (7.5-33) and (7.5-34) will result.

Some of the most popular predictor–corrector formulas are as follows:

Milne's Formula

$$x_{k+1}^* = x_{k-3} + \frac{4h}{3}[2f_k - f_{k-1} + 2f_{k-2}] \tag{7.5-37}$$

$$x_{k+1} = x_{k-1} + \frac{h}{3}[f_{k+1}^* + 4f_k + f_{k-1}] \tag{7.5-38}$$

where we have used the following shorthand notations:

$$f_k = f(x_k, u_k, t_k) \tag{7.5-39}$$

$$f_k^* = f(x_k^*, u_k, t_k) \tag{7.5-40}$$

The error involved in each step of Milne's method is proportional to h^5 (see, for example, [2]).

Another predictor–corrector formula suitable for digital computer application is

Adams–Moulton Method

$$x_{k+1}^* = x_k + \frac{h}{24}[55f_k - 59f_{k-1} + 37f_{k-2} - 9f_{k-3}] \tag{7.5-41}$$

$$x_{k+1} = x_k + \frac{h}{24}[9f_k^* + 19f_k - 5f_{k-1} + f_{k-2}] \tag{7.5-42}$$

As in the previous case, the truncation error in each step is proportional to h^5. Note that none of these methods is self-starting. This means to obtain x_{k+1} we not only need x_k, but also x_{k-1}, x_{k-2}, and so on. Hence, the initial condition given is not sufficient to start any of these programs. To overcome such deficiency, one can use a Runge–Kutta method or any other self-starting algorithm to obtain x_1, x_2, x_3, and then use Milne's or the Adams–Moulton method to obtain x_4, x_5, and so on. To conclude, let us work out an example to illustrate the application of predictor–corrector methods.

Example 7.5-3. Consider a third-order nonlinear network whose state equations are

$$\begin{aligned} \dot{x}_1 &= -x_1^4 + 2x_2 + x_3 & x_1(0) &= 0.5 \\ \dot{x}_2 &= -x_2^3 + 3x_3 + 1 & x_2(0) &= 0.5 \\ \dot{x}_3 &= -x_3^2 + \sin(2t) & x_3(0) &= 0.0 \end{aligned}$$

Use a predictor–corrector method to solve the equations for x_1, x_2, x_3 over the interval [0, 5] seconds.

Solution

Let $h = 0.1$, then the following subroutines together with subroutine RUNTA can be used to write the main program.

```
C     THIS PROGRAM IS FOR EXAMPLE 7.5-3
C     THIS PROGRAM SOLVES THE NONLINEAR DIFFERENTIAL EQUATION OF
C     EXAMPLE 7.5-3. THIS INTEGRATION METHOD USES A RUNGE-KUTTA
C     SUBROUTINE TO START THE PROGRAM AND THEN THE MILNES
C     PREDICTOR-CORRECTOR FORMULA  TO SOLVE THE EQUATIONS.
C     THE DIMENSION OF X IS FOUR TIMES THE NUMBER OF EQUATIONS
      DIMENSION X(12),DX(12),DATA(210,3)
C     SET THE INITIAL CONDITIONS
      T=0.
      TF=5.0
      H=.1
      X(1)= 0.5
      X(2)=0.5
      X(3)=0.0
C     STORE INITIAL VALUES FOR PLOTTING
      DATA(1,1)=T
      DATA(2,1)=X(1)
      DATA(3,1)=X(3)
C     INITIALIZE K
      K=0
      M=1
      PRINT 10
   10 FORMAT (8X,*T*,10X,*X1*,10X,*X2*,10X,*X3*)
      PRINT 11,T,X(1),X(2),X(3)
C     SPECIFY THE DIMENSION OF X
      N=12
      DX(1)=-X(1)**4+2*X(2)+X(3)
      DX(2)= -X(2)**3+3*X(3)+1
      DX(3)=-X(3)**2+SIN(2*T)
C     USE RUNGE-KUTTA TO START THE METHOD
      II=3
    1 IF(II)2,7,2
    2 DO 3 J=1,3
      X(J+3*II)=X(J)
    3 DX(J+3*II)=DX(J)
    4 CALL RUNTA(3,K,I,X,DX,T,H)
      GO TO (5,6),I
    5 DX(1)=-X(1)**4+2*X(2)+X(3)
      DX(2)= -X(2)**3+3*X(3)+1
      DX(3)=-X(3)**2+SIN(2*T)
```

```
      GO TO 4
    6 PRINT 11,T,X(1),X(2),X(3)
   11 FORMAT(5X,F6.3,3(5X,F7.3))
      II=II-1
      GO TO 1
    7 DX(1)=-X(1)**4+2*X(2)+X(3)
      DX(2)= -X(2)**3+3*X(3)+1
      DX(3)=-X(3)**2+SIN(2*T)
      CALL PMILNE(3,N,X,DX,T,H)
      DX(1)=-X(1)**4+2*X(2)+X(3)
      DX(2)= -X(2)**3+3*X(3)+1
      DX(3)=-X(3)**2+SIN(2*T)
      CALL CMILNE(3,N,X,DX,T,H)
      PRINT 11,T,X(1),X(2),X(3)
      M=M+1
      DATA(M,1)=T
      DATA(M,2)=X(1)
      DATA(M,3)=X(3)
      IF(T.LT.TF) GO TO 7
      CALL GRAPHX(DATA,M,1HT,5HX1,X3)
      END
```

Subroutine PMILNE

```
      SUBROUTINE PMILNE(M,N,X,DX,T,H)
C     THIS SUBROUTINE USES THE MLINE*S FORMULA TO PREDICT X(K+1)
      DIMENSION X(N),DX(N)
      DO 1 I=1,M
      XT=X(3*M+I)
      X(3*M+I)=X(2*M+I)
      X(2*M+I)=X(M+I)
      X(M+I)=X(I)
      X(I)=XT+4.0*H*(2.0*DX(I)-DX(M+I)+2.0*DX(2*M+I))/3.0
      DX(2*M+I)=DX(M+I)
    1 DX(M+I)=DX(I)
      T=T+H
      RETURN
      END
```

Subroutine CMILNE

```
      SUBROUTINE CMILNE(M,N,X,DX,T,H)
C     THIS SUBROUTINE USES THE MLINE*S FORMULA TO CORRECT X(K+1)
      DIMENSION X(N),DX(N)
      DO 2 I=1,M
    2 X(I)=X(2*M+I)+H*(DX(I)+4.0*DX(M+I)+DX(2*M+I))/3.0
      RETURN
      END
```

Subroutines RUNTA and GRAPHX are given in Examples 7.5-1 and 7.5-2 respectively. The computer plot of $x(1)$ and $x(3)$ are given in Fig. 7.5-4.

Computer Solution of Linear Time-Invariant State Equations. The techniques discussed so far in this section can be employed to solve linear time-invariant as well as nonlinear and time-varying state equations. Let us now explore the possibilities of using some of the properties of linear time-invariant state equations to obtain a more efficient computer method for their solution.

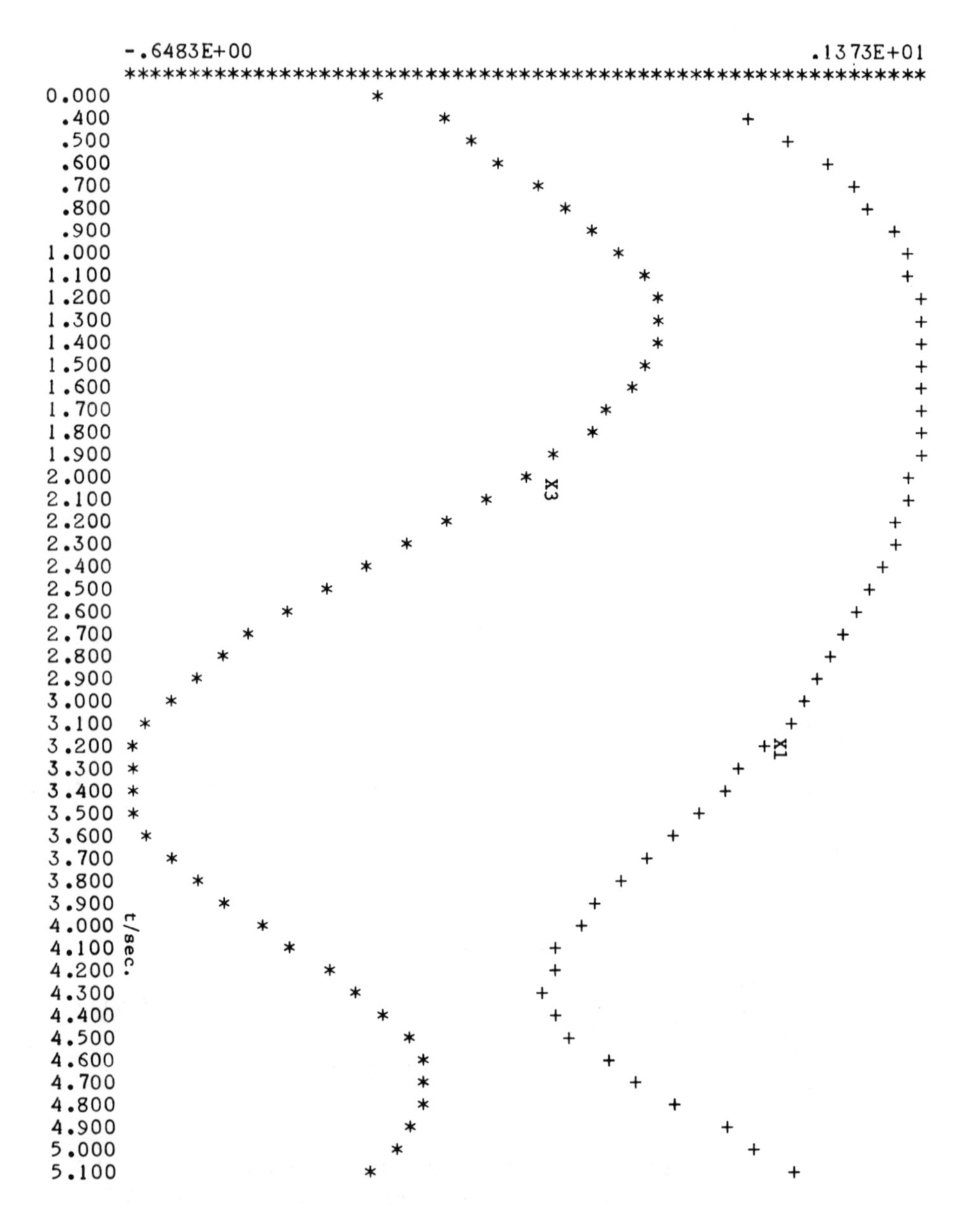

Figure 7.5-4 Computer plot of $x(1)$ and $x(3)$.

Recall that the state equation of such networks can be written as

$$\dot{\mathbf{x}}(t) = \mathbf{A}\mathbf{x}(t) + \mathbf{B}\mathbf{u}(t), \qquad \mathbf{x}(t_0) = \mathbf{x}_0 \tag{7.5-43}$$

In Chapter 6 we showed that the solution of this equation can be written as

$$\mathbf{x}(t) = e^{\mathbf{A}(t-t_0)}\mathbf{x}_0 + \int_{t_0}^{t} e^{\mathbf{A}(t-\tau)}\mathbf{B}\mathbf{u}(\tau)\,d\tau \tag{7.5-44}$$

In general, knowing $\mathbf{x}(t)$, we can obtain the state $\mathbf{x}(t+h)$ from

$$\mathbf{x}(t+h) = e^{\mathbf{A}h}\mathbf{x}(t) + \int_t^{t+h} e^{\mathbf{A}(t+h-\tau)}\mathbf{B}\mathbf{u}(\tau)\,d\tau$$

where h is a positive number denoting the step size in the numerical solution of (7.5-43). Using a simple change of variable on this equation, we obtain

$$\mathbf{x}(t+h) = e^{\mathbf{A}h}\mathbf{x}(t) + \int_0^{h} e^{\mathbf{A}(h-\tau)}\mathbf{B}\mathbf{u}(t+\tau)\,d\tau \tag{7.5-45}$$

The solution of (7.5-43) then amounts to computing $e^{\mathbf{A}h}$ and a simple numerical integration. Since h is typically a small number, a finite number of terms in the infinite series expansion can be used to approximate $e^{\mathbf{A}h}$. More precisely, consider the Taylor expansion of $e^{\mathbf{A}h}$:

$$e^{\mathbf{A}h} = \mathbf{1} + \mathbf{A}h + \frac{1}{2!}(\mathbf{A}h)^2 + \frac{1}{3!}(\mathbf{A}h)^3 + \cdots$$

If the magnitudes of the eigenvalues of $\mathbf{A}$ are relatively small, this infinite series converges quite rapidly. Consequently, $e^{\mathbf{A}h}$ can be approximated by the first few terms in this infinite series; that is, we have

$$e^{\mathbf{A}h} \cong \sum_{k=0}^{n} \frac{1}{k!}(\mathbf{A}h)^k \tag{7.5-46}$$

Equation (7.5-45) can then be written as

$$\mathbf{x}(t+h) = \sum_{k=0}^{n} \frac{1}{k!}\mathbf{A}^k\left[h^k\mathbf{x}(t) + \mathbf{B}\int_0^h (h-\tau)^k\mathbf{u}(t+\tau)\,d\tau\right] \tag{7.5-47}$$

For a given input function the integration can be performed using any of the standard integration routines, such as the trapezoidal rule, Newton–Cotes formula, and so on [1]. In many practical cases integration can be performed analytically using an integral table. For instance, if $\mathbf{u}(t)$ contains sinusoidal components or if it can be approximated by a finite polynomial in t, the integration can be carried out exactly. Consequently, the solution of (7.5-43) amounts to simple matrix multiplication and integration. However, as with most numerical techniques, the procedure outlined is not without drawbacks; one particularly significant shortcoming is discussed next.

State Equations with Very Large and Very Small Eigenvalues. If the eigenvalues of $\mathbf{A}$ are widely different from one another, the computer solution of linear time-invariant state equations as discussed, although conceptually simple, has some serious drawbacks. Generally speaking, in a numerical solution of (7.5-43) the largest eigenvalue of $\mathbf{A}$ dictates the upper bound on the step size h. More precisely, when $\mathbf{A}$ has an eigenvalue with large magnitude, the step size h must be kept small so that the infinite series expansion of $e^{\mathbf{A}h}$ converges rapidly and the approximation (7.5-46) remains valid. On the other hand, the smallest eigenvalue of **A** dictates the total length of time of integration. This is because the smaller the eigenvalue the longer it takes for the network to reach the steady state.

To satisfy the condition imposed by the largest eigenvalue, the step size h is usually taken as $h \leq \frac{1}{2}|\lambda_{max}|$, where λ_{max} denotes the largest eigenvalue of **A**. For example, in a given network, if the largest and the smallest eigenvalues of **A** are -500 and -1, respectively, the step size h should be chosen as $h = 0.001$. Furthermore, to study the transient behavior of this network, the total length of time over which the integration must be carried out should be anywhere between two to five times the largest time constant (reciprocal of the smallest eigenvalue). For the present example, this implies that equation (7.5-47) must be computed 2000 to 5000 times. This is indeed a slow and costly process. It should be mentioned that it is not uncommon for a network to have very large and very small eigenvalues. For example, consider the simple network shown in Fig. 7.5-5. Taking the capacitor charges as the state variables, the corresponding state equations are

$$\begin{bmatrix} \dot{x}_1 \\ \dot{x}_2 \end{bmatrix} = \begin{bmatrix} -10^3 & 0 \\ 100 & -2 \end{bmatrix} \begin{bmatrix} x_1 \\ x_2 \end{bmatrix} + \begin{bmatrix} 1 \\ 0 \end{bmatrix} u(t)$$

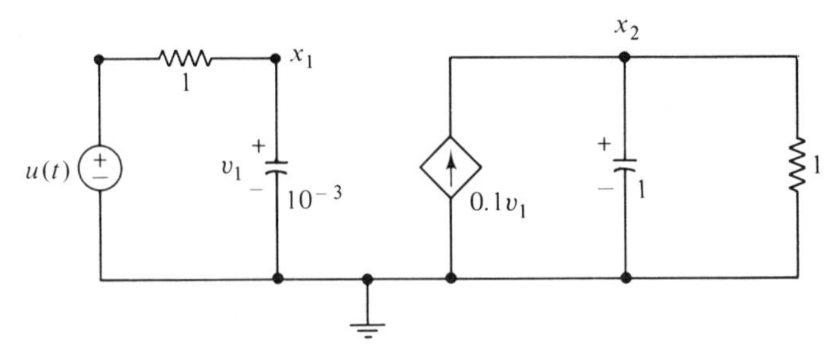

Figure 7.5-5 Network with widely different eigenvalues.

The eigenvalues are therefore -2 and -1000. In general, small energy-storing elements in the circuit generate the large eigenvalues and large energy-storing elements result in small eigenvalues.

To overcome this drawback, several modifications of the procedure are suggested. These modifications are generally based on decomposing **A** into the sum of two or more matrices so that the large and small eigenvalues of **A** can be dealt with separately. A detailed discussion of this topic, however, is beyond the intended scope of this book. The interested reader may refer to the literature [15, 16].

REFERENCES

[1] Ralston, A., *A First Course in Numerical Analysis*, McGraw-Hill Book Company, New York, 1965.

[2] Isaacson, E., and H. B. Keller, *Analysis of Numerical Methods*, John Wiley & Sons, Inc., New York, 1966.

[3] Sato, N., and W. F. Tinny, "Techniques for Exploiting the Sparsity of the Network Admittance Matrix," *IEEE Trans. Power Apparatus and Systems*, **82** (Dec. 1963), 944–950.

[4] Tinny, W. F., and J. W. Walker, "Direct Solution of Sparse Network Equations by Optimally Ordered Triangular Factorization," *Proc. IEEE*, **55**, No. 11 (1967), 1801–1809.

[5] Berry, R. D., "An Optimal Ordering of Electronic Circuit Equations for Sparse Matrix Solution," *IEEE Trans. Circuit Theory*, **CT-18**, No. 1 (1971), 40–50.

[6] Hatchel, G. D., R. K. Brayton, and F. G. Gustavson, "The Sparse Tableau Approach to Network Analysis and Design," *IEEE Trans. Circuit Theory*, **CT-18**, No. 1 (1971), 101–113.

[7] Katzenelson, J., and L. H. Seitelman, "An Iterative Method of Solution of Networks of Nonlinear Monotone Resistors," *IEEE Trans. Circuit Theory*, **CT-13** (Sept. 1966), 317–323.

[8] Sandberg, J. W., and A. N. Wilson, Jr., "Some Theorems on Properties of d.c. Equations of Nonlinear Networks," *Bell System Tech. J.* **48** (Jan. 1969), 1–34.

[9] Kuh, E. S., and I. N. Hajj, "Nonlinear Circuit Theory: Resistive Networks," *Proc. IEEE*, **59**, No. 3 (1971), 340–355.

[10] Katzenelson, J., "AEDENT: A Simulator for Nonlinear Networks," *Proc. IEEE*, **54** (Nov. 1966), 1536–1552.

[11] Chua, L. O., "Efficient Computer Algorithms for Piecewise-Linear Analysis of Resistive Nonlinear Networks," *IEEE Trans. Circuit Theory*, **CT-18**, No. 1 (1971), 73–85.

[12] Palais, R. S., "Natural Operations of Differential Forms," *Trans. Am. Math. Soc.*, **92** (July 1959), 125–141.

[13] Broyden, C. G., "A New Method of Solving Nonlinear Simultaneous Equations," *Computer J.*, **12** (Feb. 1969), 94–99.

[14] Brown, G. C., "DC Analysis of Nonlinear Networks," *Electronics Letters*, **5** (Aug. 1969), 374–375.

[15] Branin, F. H., Jr., "Computer Methods of Network Analysis," *Proc. IEEE*, **55**, No. 11 (1967), 1781–1801.

[16] Lee, H. B., "Matrix Filtering As an Aid to Numerical Integration," *Proc. IEEE*, **55**, No. 11 (1967), 1826–1831.

PROBLEMS

7.1 Show that to compute the determinant of an $n \times n$ matrix by expanding a row or a column requires $n! \sum_{k=1}^{n-1} [1/(n-k)!]$ multiplications. Also, show that for large n this number can be approximated by $(e-1)n!$.

7.2 Use the loop-analysis method to obtain a set of linear independent equations in the form $\mathbf{Px} = \mathbf{b}$, where the elements of $\mathbf{x}$ denote the loop currents. Specify $\mathbf{P}$ and $\mathbf{b}$ in terms of the matrices of the corresponding network graph.

7.3 Write the nodal equations of the network in Fig. P7.3 by inspection. (The conductances are in mhos and the currents are in amperes.)

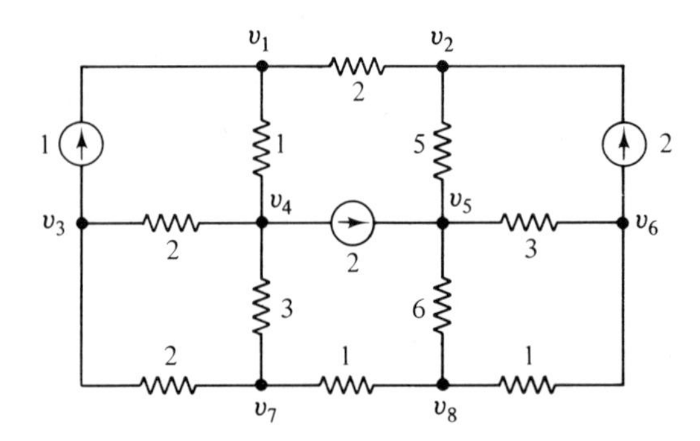

Figure P7.3

7.4 Use a Gaussian elimination method to solve the equations obtained in Problem 7.3.

7.5 Use a computer program similar to that of Example 7.2-2 to solve the equations obtained in Problem 7.3.

7.6 Consider the set of algebraic equations given in Example 7.2-2. Use an LU transformation to decompose $\mathbf{P}$. Specify $\mathbf{L}$ and $\mathbf{U}$.

7.7 Consider the networks shown in Fig. P7.7. Obtain the steady-state nodal equations of this network.

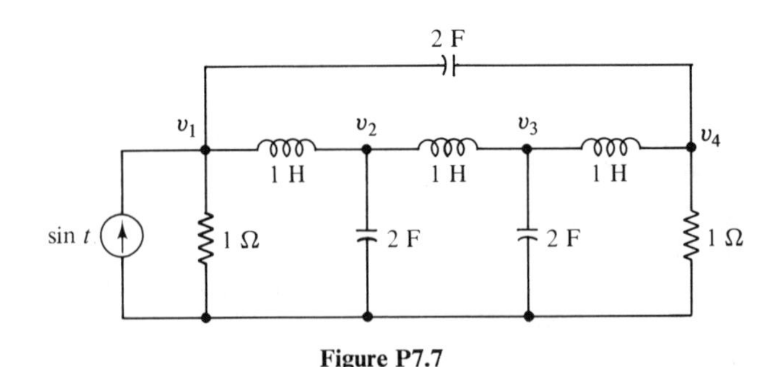

Figure P7.7

7.8 Use a computer program similar to that of Example 7.3-1 to solve the nodal equations obtained in Problem 7.7.

7.9 Use a Newton–Raphson method to solve the nonlinear equation

$$x^3 - \tfrac{1}{2}x^2 - \tfrac{1}{2}x = \tfrac{3}{2}$$

7.10 Use a Newton–Raphson method to solve the nonlinear coupled equations

$$x_1 + 3e^{x_2} - 2 = 0$$
$$-x_1^2 - 3x_1e^{x_2} + 1 = 0$$

7.11 Use a fourth-order Runge–Kutta method to solve the following first-order time-varying differential equation for $0 \leq t \leq 10$:

$$\dot{x} = -(1 + 0.5 \sin t)x + \cos t, \qquad x(0) = 1$$

7.12 Use a fourth-order Runge–Kutta method to solve the following first-order nonlinear differential equation for $0 \leq t \leq 5$:

$$\dot{x} = -x - x^3, \qquad x(0) = 1$$

7.13 Use a fourth-order Runge–Kutta method to solve the following nonlinear time-varying differential equation for $0 \leq t \leq 5$:

$$\dot{x} = -t^2x^2 - 0.5 \sin(2x), \qquad x(0) = 2$$

7.14 Use the fourth-order Runge–Kutta subroutine RUNTA of Example 7.5-2 to solve the following set of linear time-invariant differential equations for $0 \leq t \leq 5$:

$$\begin{aligned} \dot{x}_1 &= -x_1 + 2x_2 + 6x_3 & x_1(0) &= 1 \\ \dot{x}_2 &= -x_2 + 3x_3 + 2\sin t, & x_2(0) &= -1 \\ \dot{x}_3 &= -x_3 + t^2e^{-t} + \cos t, & x_3(0) &= 0 \end{aligned}$$

Use a graph subroutine to plot $x_1(t)$ and $x_2(t)$.

7.15 Solve the following linear time-varying differential equation using a fourth-order Runge–Kutta method for $0 \leq t \leq 5$:

$$\begin{aligned} \dot{x}_1 &= e^{-t}x_1 - 2x_2 + 3e^{-2t}x_2, & x_1(0) &= 0 \\ \dot{x}_2 &= \frac{1}{1+t}x_1 + 3x_2 - 6x_3, & x_2(0) &= 1 \\ \dot{x}_3 &= -x_1 + e^{-t}x_2 - x_3 + \sin t, & x_3(0) &= -1 \end{aligned}$$

7.16 Solve the following nonlinear differential equations using a fourth-order Runge–Kutta method for $0 \leq t \leq 5$:

$$\begin{aligned} \dot{x}_1 &= -x_1 + x_2^2 + 4x_3^3, & x_1(0) &= 1 \\ \dot{x}_2 &= -\sin x_1 - x_2 + 2x_3, & x_2(0) &= -1 \\ \dot{x}_3 &= -x_1^2 + 2x_2^2 - e^{-t} & x_3(0) &= 0 \end{aligned}$$

7.17 Use Milne's predictor–corrector to solve the differential equation of Problem 7.12 and compare the results.

7.18 Repeat Problem 7.17, but this time use the Adams–Moulton method.

7.19 Use the subroutine given in Example 7.5-3 to solve the differential equations of Problem 7.14 and compare the results.

Passive and Active Network Synthesis 8

8.1 INTRODUCTION

So far in this book we have concerned ourselves solely with the important topic of network analysis. That is, given a network and a set of inputs, we considered several methods for obtaining the corresponding responses. We now pose an equally important question: Given a set of input–output pairs (u, y), can we find the topology, element types, and element values of a network for which (u, y) would be an admissible signal pair? This is the question that a design engineer should be able to answer. As we shall see later on in this chapter, in contrast to the analysis problem, the synthesis problem does not have a unique solution. One can synthesize many networks to which a given input-output pair (u, y) is an admissible signal pair. The pair (u, y) may be specified either in the time domain or frequency domain. In the latter case, either the transfer function, driving-point impedance, or the driving-point admittance function is specified.

In this chapter we first outline the answer to the relatively simple problem of designing a linear time-invariant network with a prescribed driving-point impedance or admittance function using passive elements only. For this purpose we introduce the concept of positive real functions and use a method of classical network synthesis to obtain the desired network. Later, we take up active network synthesis and introduce some procedures for synthesizing network functions using active elements, such as negative impedance convertors, operational amplifiers, and the like.

8.2 POSITIVE REAL FUNCTIONS AND MATRICES

The problem of designing a network whose driving-point impedance or admittance function is a positive real function is commonly referred to as *network synthesis*. Before defining a positive real function, let us give a simple example of the basic idea behind network synthesis.

Suppose that the driving-point impedance of a linear time-invariant one-port network is given by

$$z(s) = \frac{2s^2 + 3s + 2}{s + 1}$$

The problem is to synthesize a network whose input impedance is $z(s)$. One approach would be to write $z(s)$ as the sum of several simple terms so that their corresponding networks could be recognized by inspection. For the function given, this can be accomplished by dividing the numerator by the denominator:

$$z(s) = 2s + 1 + \frac{1}{s + 1}$$

Thus, $z(s)$ can be synthesized by connecting three subnetworks whose driving-point functions are $2s$, 1, and $1/(s + 1)$ in series. The first two terms are immediately recognized as a 2**H** inductor and a 1**Ω** resistor, respectively. The last term can also be seen to be the impedance of a 1**Ω** resistor and a 1F capacitor connected in parallel. The desired network can then be drawn as in Fig. 8.2-1(a).

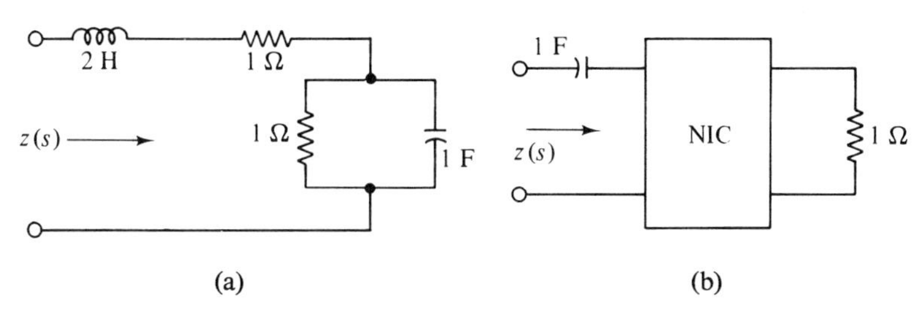

Figure 8.2-1 (a) Realization of a positive real driving point function; (b) synthesis of a nonpositive real function using active elements.

This example raises two important questions; first, can any given $z(s)$ be synthesized by linear time-invariant two-terminal passive *RLC* elements? Second, if the answer to the first question is positive, what is a systematic method for obtaining the corresponding network? The answer to the first question is obviously negative; this can be seen by the following counterexample. Suppose that the desired driving-point impedance is given by

$$z(s) = \frac{1 - s}{s}$$

This function can be written as

$$z(s) = \frac{1}{s} - 1$$

which corresponds to a network made up of a 1F capacitor in series with a -1Ω resistor. Instead of a -1Ω resistor we may use a $+1\Omega$ resistor in tandem with a negative impedance convertor, as in Fig. 8.2-1(b). (For the definition and properties of an NIC, see Chapter 1.) Consequently, the given function requires an active element and cannot be realized with passive elements. The class of driving-point functions that can be synthesized by linear time-invariant passive *RLC* elements is known as positive real functions, which we shall discuss next.

Positive Real Functions. Let $z(s)$ denote a scalar rational function of the complex variable $s = \sigma + j\omega$; then $z(s)$ can be written as the quotient of two polynomials $N(s)$ and $D(s)$:

$$z(s) = \frac{N(s)}{D(s)}$$

In order that $z(s)$ represent the driving-point impedance or admittance function of a linear time-invariant network with passive elements it must satisfy the conditions of the following definition:

Definition 8.2-1 (Positive Real Function). A rational function $z(s)$ is said to be *positive real* (p.r.) if and only if

(a) $z(s)$ is real when s is real.
(b) $\text{Re}\{z(s)\} \geq 0$, for all $\sigma \geq 0$.

Such functions were first defined by Brune [1]; for this reason they are sometimes called Brune functions. From a mathematical point of view, $z(s)$ is a function that maps the real axis of the complex s plane into the real axis of the complex $z(s)$ plane, and maps the right half of the s plane into the right half of the $z(s)$ plane. This fact is shown in Fig. 8.2-2.

Let us now state a fundamental theorem in passive network synthesis.

Theorem 8.2-1. A necessary and sufficient condition for a rational function $z(s)$ to be the driving-point function of a linear time-invariant network with passive two-terminal elements is that $z(s)$ be positive real.

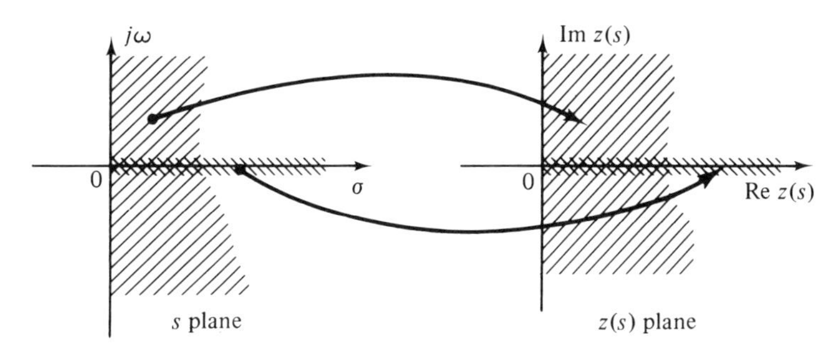

Figure 8.2-2 The positive real function $z(s)$ maps the real axis of the s plane into the real axis of the $z(s)$ plane and the right-half side of the s plane into the right-half side of the $z(s)$ plane.

An immediate consequence of this theorem is that if a function $z(s)$ is positive real then $y(s) = 1/z(s)$ is also a positive real function. This can be seen by observing that if $z(s)$ is the driving-point impedance of a network with linear time-invariant passive elements then $y(s)$ is the driving-point admittance of the same network. In Section 8.5 we prove the sufficiency condition of this theorem by giving a procedure for synthesizing any positive real function with passive *RLC* elements. A complete proof of this theorem can be found in [2].

Let us now present some simple rules for testing the positive realness of a given rational function. Using the maximum modulus theorem discussed in complex variable theory, it can be shown that (see, for example, reference [2], p. 82) the conditions stated in Definition 8.2-1 are equivalent to the conditions stated in the following theorem.

Theorem 8.2-2. A rational function $z(s)$ is positive real if and only if

(a) $z(s)$ has no poles in the right half of the complex s plane.
(b) The poles of $z(s)$ on the $j\omega$ axis are simple with real and positive residues.
(c) $\text{Re}\,\{z(j\omega)\} \geq 0$, for all $0 \leq \omega \leq \infty$.

These necessary and sufficient conditions are much easier to test than the conditions stated in Definition 8.2-1. Nevertheless, it is desirable to find a set of necessary conditions by which we can quickly single out those functions which are not positive real. A set of such necessary conditions is given in the following:

Theorem 8.2-3. Let $z(s)$ be written as the quotient of two polynomials in s:

$$z(s) = \frac{N(s)}{D(s)}$$

If $z(s)$ is positive real, the following conditions must hold:

(a) All the coefficients of $N(s)$ and $D(s)$ are real and positive.
(b) All the zeros of $D(s)$ *and* $N(s)$ are in the left half of the s-plane.
(c) The poles of $z(s)$ and $1/z(s)$ on the $j\omega$ axis are simple and have real and positive residues.
(d) The degrees of the polynomials $N(s)$ and $D(s)$ differ at most by 1.
(e) $z(s)$ has neither multiple poles nor multiple zeros at the origin.

Thus, if a given function $z(s)$ fails to satisfy any of these conditions, there is no need for further tests—the function is not positive real. If, however, a given function satisfies all the conditions, we should check to see whether it satisfies condition (c) of Theorem 8.2-2 or not.

Example 8.2-1. Test the following functions for positive realness:

(1) $z(s) = \dfrac{1}{s^2 + 1}$ (2) $z(s) = \dfrac{s^3 + 3s^2 - 2s + 1}{s^3 + s^2 + s + 6}$

(3) $z(s) = \dfrac{s^3 + 3s + 2}{s^4 + 3s^3 + s^2}$ (4) $z(s) = \dfrac{s^4 + 4s^3 + 3s^2 + 1}{s^4 + 2s^2 + 1}$

(5) $z(s) = \dfrac{s + 1}{s + 2}$ (6) $z(s) = \dfrac{(s + 1)(s + 2)}{(s + 3)(s + 4)}$

Solution

1. $z(s)$ given in this case fails condition (d) of Theorem 8.2-3 and hence is not positive real.
2. Since one of the coefficients of the numerator polynomial is negative, condition (a) of Theorem 8.2-3 is violated and therefore $z(s)$ is not positive real.
3. The function given in this case is not positive real since it has a multiple pole at the origin; therefore, condition (e) is violated.
4. $z(s)$ given in this case has a multiple pole on the $j\omega$ axis at $s = \pm j1$ and consequently fails to satisfy condition (c); hence, it is not positive real.
5. $z(s)$ given in this case satisfies all the conditions of Theorem 8.2-3. This, however, is not enough to conclude that $z(s)$ is positive real since these conditions are only necessary, but not sufficient. The necessary and sufficient conditions of Theorem 8.2-2 must, therefore, be utilized. Conditions (a) and (b) are obviously satisfied. Let us then check condition (c) for the problem in hand.

$$z(j\omega) = \frac{1 + j\omega}{2 + j\omega}$$

Thus,

$$\text{Re}\{z(j\omega)\} = \frac{2 + \omega^2}{4 + \omega^2}$$

which is clearly positive for all $0 \leq \omega \leq \infty$. The conclusion, therefore, is that $z(s)$ is positive real.

6. $z(s)$ given in this case also satisfies all the conditions of Theorem 8.2-3. It, therefore, remains to check condition (c) of Theorem 8.2-2:

$$z(j\omega) = \frac{(1 + j\omega)(2 + j\omega)}{(3 + j\omega)(4 + j\omega)} = \frac{(1 + j\omega)(2 + j\omega)(3 - j\omega)(4 - j\omega)}{(9 + \omega^2)(16 + \omega^2)}$$

Hence,

$$\text{Re}\{z(j\omega)\} = \frac{\omega^4 + 7\omega^2 + 24}{(9 + \omega^2)(16 + \omega^2)}$$

which is positive for all positive ω. Consequently, the function under study is positive real.

Remark 8.2-1. From the conclusion of Theorem 8.2-1 it is clear that to test the positive realness of a given rational function $z(s)$ one may try to synthesize a passive network whose driving-point impedance or admittance is $z(s)$. This procedure, however, is practical only if the corresponding network can be obtained by inspection. Let us now generalize the concept of positive real functions to matrices.

Positive Real Matrices. Let $\mathbf{H}(s)$ be an $n \times n$ matrix whose elements are rational functions of the complex variable s. $\mathbf{H}(s)$ may represent the transfer function, open-circuit impedance, or short-circuit admittance matrix of an $n \times n$ linear time-invariant network. It can be shown that to realize $\mathbf{H}(s)$ as the short-circuit impedance or the open-circuit admittance of an $n \times n$ network with passive elements it must satisfy certain conditions; more precisely, $\mathbf{H}(s)$ must be a positive real matrix.

Definition 8.2-2 (Positive Real Matrix). The matrix $\mathbf{H}(s)$ is called a *positive real (p.r.) matrix* if and only if the following conditions are satisfied:

(a) $\mathbf{H}(s)$ is real when s is real.
(b) The scalar function $\mathbf{x}^T\mathbf{H}(s)\mathbf{x}$ is a positive real function for any arbitrary real column vector $\mathbf{x}$.

For a given $\mathbf{H}(s)$, condition (a) is easy to check. Condition (b), however, is not so easy to verify since it must be satisfied for all possible $\mathbf{x}$. Fortunately, these conditions can be broken down into simpler tests that can be checked by inspection. The following theorem gives a set of necessary and sufficient conditions for $\mathbf{H}(s)$ to be positive real.

Theorem 8.2-4. The matrix $\mathbf{H}(s) = [h_{ij}(s)]$ is positive real (p.r.) if and only if

(a) All the coefficients of $h_{ij}(s)$ are real.
(b) $\mathbf{H}(s)$ has no poles in the right half of the s plane.

(c) All the poles of $\mathbf{H}(s)$ that are on the $j\omega$ axis are simple and the corresponding residue matrix is positive semidefinite.
(d) Re $\{\mathbf{H}(j\omega)\}$ is positive semidefinite for all $\omega \geq 0$.

Example 8.2-2. Test the following matrices for positive realness:

$$\mathbf{Y}(s) = \begin{bmatrix} \dfrac{-2s}{s^2+1} & \dfrac{s}{s^2+1} \\ \dfrac{s}{s^2+1} & \dfrac{3s}{s^2+1} \end{bmatrix}, \qquad \mathbf{Z}(s) = \begin{bmatrix} \dfrac{s^2+s+1}{s+1} & \dfrac{1}{s+1} \\ \dfrac{1}{s+1} & \dfrac{2s^2+2s+1}{s+1} \end{bmatrix}$$

Solution

$\mathbf{Y}(s)$ satisfies conditions (a) and (b) of the theorem, but it fails to satisfy condition (c), since the residue matrices corresponding to the $j\omega$ axis poles are not positive semidefinite. There is no need to proceed further; the matrix $\mathbf{Y}(s)$ is not positive definite. The matrix $\mathbf{Z}(s)$ clearly satisfies conditions (a), (b), and (c); to check condition (d) note that

$$\text{Re}\{\mathbf{Z}(j\omega)\} = \frac{1}{1+\omega^2}\begin{bmatrix} 1 & 1 \\ 1 & 1+\omega^2 \end{bmatrix}$$

which is positive definite. Consequently, $\mathbf{Z}(s)$ is a positive real matrix. A careful observation of $\mathbf{Z}(s)$ shows that it can be realized as the two-port network shown in Fig. 8.2-3.

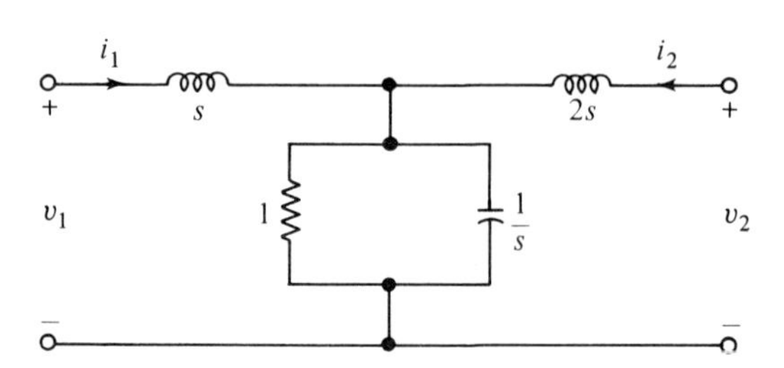

Figure 8.2-3 Realization of the p.r. matrix $Z(s)$ given in Example 8.2-2.

To answer the second basic question posed earlier, that is, how one can realize a positive real function by linear time-invariant elements, we start by considering the realization of a special class of positive real functions: reactance functions.

8.3 DRIVING-POINT SYNTHESIS OF REACTANCE FUNCTIONS

In this section we present several methods for synthesizing the driving-point impedance or admittance function of networks containing passive capacitors and inductors only. These networks are commonly referred to as *lossless* or *reactive* networks. The driving-point impedance and admittance of such networks are among a special class of positive real functions called the *reactance functions*. More precisely,

Definition 8.3-1 (Reactance Function). The rational function $z(s)$ is a *reactance function* if and only if

(a) All the poles of $z(s)$ are simple and lie on the $j\omega$ axis with real and positive residues.
(b) $\text{Re}\{z(j\omega)\} = 0$, for all $\omega \geq 0$.

From the definition it is easy to see that $z(s)$ can be written as

$$z(s) = K\frac{(s^2 + \omega_1^2)(s^2 + \omega_3^2)\cdots}{s(s^2 + \omega_2^2)(s^2 + \omega_4^2)\cdots} \tag{8.3-1}$$

or as

$$z(s) = K\frac{s(s^2 + \omega_1^2)(s^2 + \omega_3^2)\cdots}{(s^2 + \omega_2^2)(s^2 + \omega_4^2)\cdots} \tag{8.3-2}$$

Furthermore, it can be shown that the poles and zeros of a reactance function alternate on the $j\omega$ axis (see Problem 8.4). That is, in (8.3-1) we must have $0 < \omega_1 < \omega_2 < \omega_3 \ldots$ and in (8.3-2) we must have $0 < \omega_2 < \omega_1 < \omega_4 < \omega_3 \ldots$. Also, notice that $z(s)$ has either a pole or a zero at 0 and at ∞.

Since $z(s)$ is positive real, according to Theorem 8.2-1 there exists a passive network for which $z(s)$ is either the driving-point impedance or admittance function. Moreover, since $z(s)$ is a reactance function, it can be synthesized using LC elements only. To prove this fact, we shall next introduce several methods of realizing $z(s)$ by passive LC elements.

Foster's Synthesis Methods. Using a partial fraction expansion, $z(s)$ given in (8.3-1) or (8.3-2) can be written as

$$z(s) = K_\infty s + \frac{K_0}{s} + \frac{2K_2 s}{s^2 + \omega_2^2} + \frac{2K_4 s}{s^2 + \omega_4^2} + \cdots + \frac{2K_m s}{s^2 + \omega_m^2} \tag{8.3-3}$$

where all the constants are real and positive. If $z(s)$ is a driving-point impedance, the first term in (8.3-3) can be recognized to be the impedance of an inductor with inductance K_∞. Similarly, the second term is the impedance of a capacitor with capacitance $1/K_0$. Furthermore, each of the terms under the summation sign can be considered to be the impedance of a parallel connection of a capacitor with capacitance $1/2K_r$ and an inductor with inductance $2K_r/\omega_r^2$ where $r = 2, 4, 6, \ldots, m$. The realization of the final network will then be achieved by series connection of these terms. The resulting network is known as *first Foster form*. Figure 8.3-1 shows the final network.

To obtain the second Foster realization of the impedance function $z(s)$, we invert $z(s)$ to obtain the corresponding admittance function $y(s)$. Quite

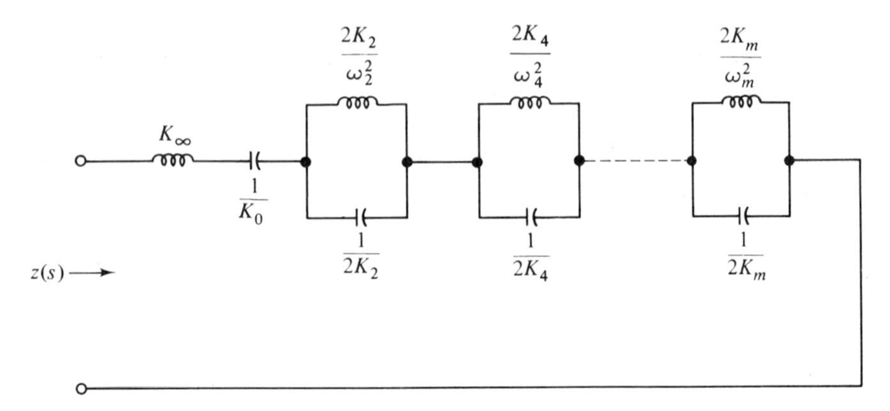

Figure 8.3-1 First Foster form realization of a reactance function.

clearly, $y(s)$ is also a reactance function and can be written as

$$y(s) = \frac{1}{z(s)} = K'_\infty s + \frac{K'_0}{s} + \frac{2K'_2 s}{s^2 + \omega_2^2} + \frac{2K'_4 s}{s^2 + \omega_4^2} + \cdots + \frac{2K'_m s}{s^2 + \omega_m^2} \tag{8.3-4}$$

where the K'_i are all real and positive constants. Thus, $y(s)$ can be realized as the parallel connection of several networks. In the right-hand side of (8.3-4), the first and second terms represent the admittance of a capacitor with capacitance K'_∞ and an inductor with inductance $1/K'_0$, respectively. The remaining terms are recognized as the series connection of an inductor with inductance $1/2K'_r$ and a capacitor with capacitance $2K'_r/\omega^2$. The final realization is known as *second Foster form* and is shown in Fig. 8.3-2.

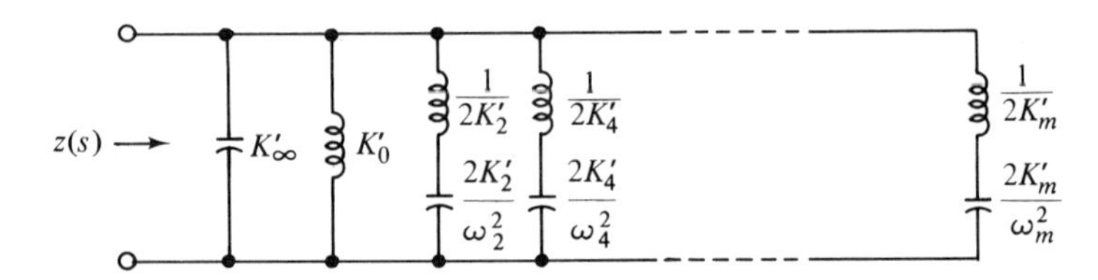

Figure 8.3-2 Second Foster form realization of a reactance function $z(s)$.

Let us now work out an example to illustrate this method.

Example 8.3-1. Consider the reactance function

$$z(s) = \frac{s^2 + 1}{s(s^2 + 2)}$$

Obtain the first and second Foster realization of this driving-point impedance.

Solution

$z(s)$ is indeed a reactance function, since it clearly satisfies the conditions of Definition 8.3-1. Let us then expand it by partial fraction expansion to get

$$z(s) = \frac{1/2}{s} + \frac{(1/2)s}{s^2 + 2}$$

Comparing this with equation (8.3-3), we get $K_\infty = 0$, $K_0 = \frac{1}{2}$, and $K_2 = \frac{1}{4}$. The resulting first Foster form is shown in Fig. 8.3-3(a). To obtain the

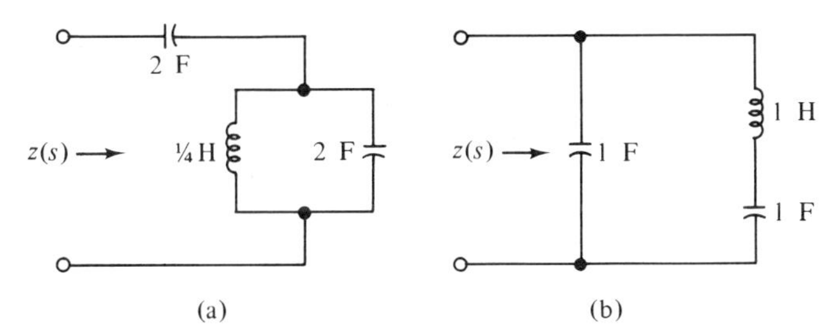

Figure 8.3-3 (a) First Foster realization and (b) second Foster realization of $z(s) = (s^2 + 1)/[s(s^2 + 2)]$.

second Foster realization, we must first invert $z(s)$ and then expand it; the result is

$$y(s) = \frac{1}{z(s)} = \frac{s(s^2 + 2)}{s^2 + 1} = s + \frac{s}{s^2 + 1}$$

Thus, $K'_\infty = 1$, $K'_0 = 0$, and $K'_2 = \frac{1}{2}$. The resulting network is shown in Fig. 8.3-3(b).

Cauer's Synthesis Methods. An alternative method of realizing a reactance function is by removing poles at zero and at infinity and realizing them as inductors or capacitors. More precisely, consider the reactance function given in (8.3-2). In this form of $z(s)$, the numerator polynomial is 1 degree higher than the denominator; that is, $z(s)$ has a pole at ∞. Thus, $z(s)$ can be written as

$$z(s) = L_1 s + z_1(s)$$

where $L_1 = K$ and $z_1(s)$ is another reactance function whose denominator polynomial is 1 degree higher than its numerator. It can therefore be inverted and written as

$$\frac{1}{z_1} = C_2 s + y_2(s)$$

where $y_2(s)$ is also a reactance function with a pole at ∞. This process can be continued until the remainder is simply K_r/s or $K_r s$. As a result, the reactance function $z(s)$ can be written as

$$z(s) = L_1 s + \cfrac{1}{C_2 s + \cfrac{1}{L_3 s + \cfrac{1}{C_4 s + \cdots}}} \tag{8.3-5}$$

Note that $L_1 s$ represents the impedance of an inductor with inductance L_1, $C_2 s$ is the admittance of a capacitor with capacitance C_2, and so on. Consequently, the resulting network, which is known as the *first Cauer's form*, can be synthesized as in Fig. 8.3-4.

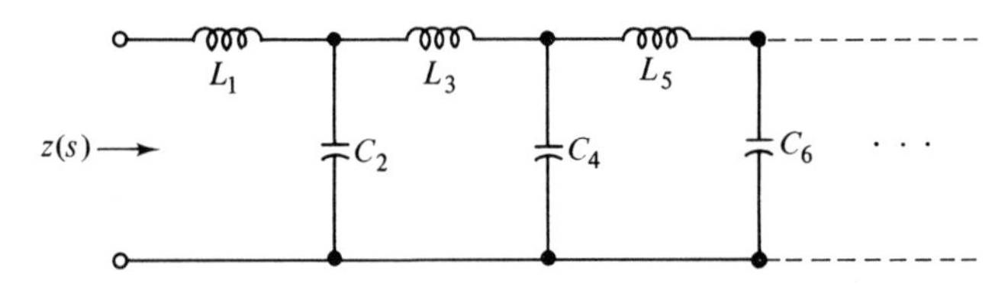

Figure 8.3-4 First Cauer realization of the reactance function that has a pole at ∞.

In the *second Cauer's form*, successive poles are removed from the origin. To clarify this statement, consider the reactance functions

$$z(s) = \frac{a_0 + a_1 s + a_2 s^2 + \cdots + a_n s^n}{b_1 s + b_2 s^2 + b_3 s^3 + \cdots + b_{n+1} s^{n+1}} = \frac{N(s)}{D(s)}$$

Note that the polynomials $N(s)$ and $D(s)$ are arranged in ascending order in s. Let us now divide $N(s)$ by $D(s)$, invert the remainder, and divide its numerator by its denominator again and continue. The resulting long division will then be

$$z(s) = \frac{1}{C_1 s} + \cfrac{1}{\cfrac{1}{L_2 s} + \cfrac{1}{\cfrac{1}{C_3 s} + \cfrac{1}{\cfrac{1}{L_4 s} + \cdots}}}$$

where the C_i and L_i are fixed constants obtained in the process of long division. From this equation it is clear that the resulting network can be synthesized as in Fig. 8.3-5.

Let us now conclude the discussion on reactance functions with an example.

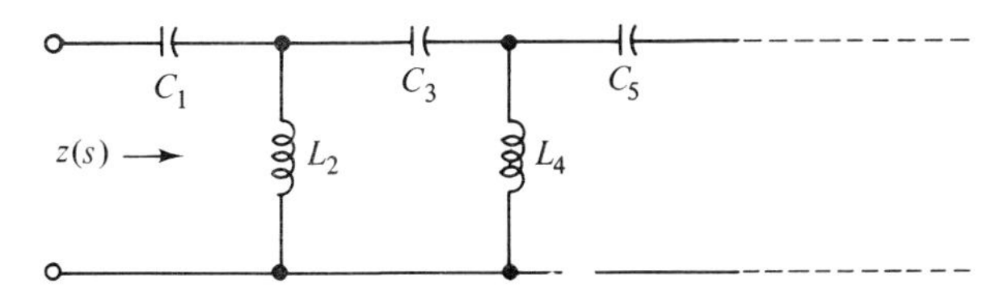

Figure 8.3-5 Second Cauer realization of a reactance function.

Example 8.3-2. Obtain the first and second Cauer realization of the reactance function

$$z(s) = \frac{(s^2 + 1)(s^2 + 3)}{s(s^2 + 2)}$$

Solution

Removing consecutive poles at ∞, by dividing, inverting, and so on, $z(s)$ can be written as

$$z(s) = s + \cfrac{1}{\frac{1}{2}s + \cfrac{1}{4s + \cfrac{1}{\frac{1}{6}s}}}$$

The corresponding first Cauer form can then be obtained as in Fig. 8.3-6(a).

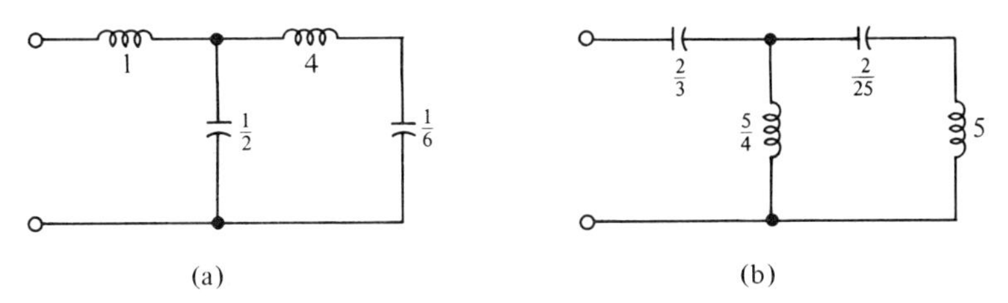

Figure 8.3-6 (a) First Cauer form and (b) second Cauer form realization of the reactance function $[(s^2 + 1)(s^2 + 3)]/[s(s^2 + 2)]$.

To obtain the second Cauer form, first arrange the numerator and denominator polynomials in ascending order and then remove consecutive poles at the origin by the dividing–inverting process to get

$$z(s) = \frac{3 + 4s^2 + s^4}{2s + s^3} = \frac{3}{2s} + \cfrac{1}{\frac{4}{5s} + \cfrac{1}{\frac{25}{2s} + \cfrac{1}{\cfrac{1}{5s}}}}$$

The resulting network is shown in Fig. 8.3-6(b).

Before proceeding to the synthesis of general *RLC* networks, let us discuss the realization of *RC* and *RL* networks.

8.4 SYNTHESIS OF *RC* AND *RL* NETWORKS

Another special class of positive real functions is the one that can be realized as the driving-point impedance and/or admittance of passive *RL* or *RC* networks. These functions can generally be written in one of the following forms:

$$F_1(s) = K\frac{(s+\sigma_1)(s+\sigma_3)\dots(s+\sigma_m)}{(s+\sigma_2)(s+\sigma_4)\dots(s+\sigma_n)} \tag{8.4-1}$$

or

$$F_2(s) = K\frac{(s+\sigma_2)(s+\sigma_4)\dots(s+\sigma_m)}{(s+\sigma_1)(s+\sigma_3)\dots(s+\sigma_n)} \tag{8.4-2}$$

where K and the σ_i are positive numbers so that

$$0 \leq \sigma_1 < \sigma_2 < \sigma_3 \dots \tag{8.4-3}$$

and

$$m = n - 1 \quad \text{or} \quad m = n + 1 \tag{8.4-4}$$

In what follows we shall show that these functions can be synthesized as the driving-point impedance or admittance of *RC* and/or *RL* networks. For this purpose let us consider two distinct cases:

I. Synthesis of F_1 (s)

1. If $m = n - 1$, the degree of the numerator polynomial will be equal to the degree of the denominator polynomial. In this case, the function $F_1(s)/s$ can be expanded as the following partial fraction expansion:

$$\frac{F_1(s)}{s} = \frac{K_1}{s} + \frac{K_2}{s+\sigma_2} + \frac{K_4}{s+\sigma_4} + \dots + \frac{K_n}{s+\sigma_n} \tag{8.4-5}$$

or, equivalently,

$$F_1(s) = K_1 + \frac{K_2 s}{s+\sigma_2} + \frac{K_4 s}{s+\sigma_4} + \dots + \frac{K_n s}{s+\sigma_n} \tag{8.4-6}$$

where all the coefficients are real and positive [show this by using equation (8.4-3)]. This function can now be synthesized either as the driving-point impedance of an *RL* network, known as the *first Foster realization* [Fig. 8.4-1 (a)], or as the driving-point admittance of an *RC* network, known as the *second Foster realization* [Fig. 8.4-1(b)].

Example 8.4-1. Synthesize the following as the driving-point impedance or admittance of a network.

$$F(s) = \frac{(s+2)(s+4)}{(s+3)(s+5)}$$

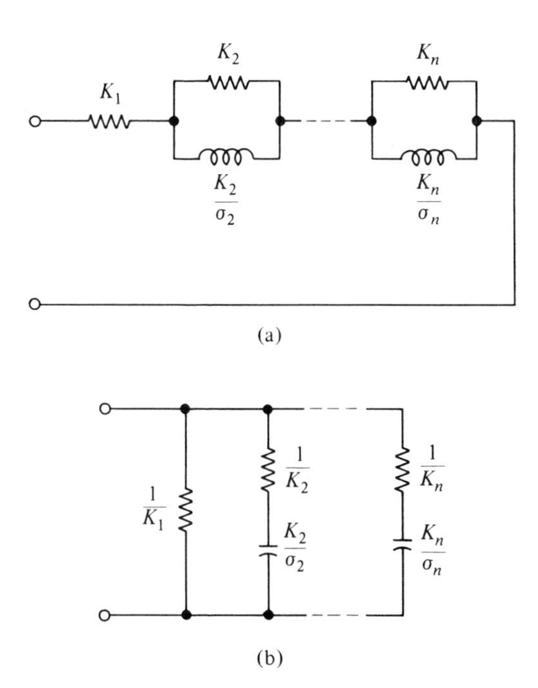

Figure 8.4-1 (a) First Foster form *RL* realization; (b) second Foster form *RC* realization of $F_1(s)$ when $m = n - 1$.

Solution

We can write

$$\frac{F(s)}{s} = \frac{(s+2)(s+4)}{s(s+3)(s+5)} = \frac{8/15}{s} + \frac{1/6}{s+3} + \frac{3/10}{s+5}$$

Thus,

$$F(s) = \frac{8}{15} + \frac{(1/6)s}{s+3} + \frac{(3/10)s}{s+5}$$

The desired realizations are given in Fig. 8.4-1 with corresponding coefficients: $K_1 = \frac{8}{15}$, $K_2 = \frac{1}{6}$, and $K_4 = \frac{3}{10}$

2. If $m = n + 1$, the degree of the numerator polynomial will be 1 greater than the degree of the denominator polynomial. This implies that $F_1(s)$ has a pole at ∞, which can be removed. More precisely, in this case $F_1(s)$ can be written as

$$F_1(s) = K_\infty s + \hat{F}(s) \tag{8.4-7}$$

A quick reference to Theorem 8.2-2 shows that $\hat{F}_1(s)$ is a positive real function (why?) Furthermore, the degree of the numerator polynomial of $\hat{F}_1(s)$ is equal to the degree of its denominator polynomial. Consequently, it can

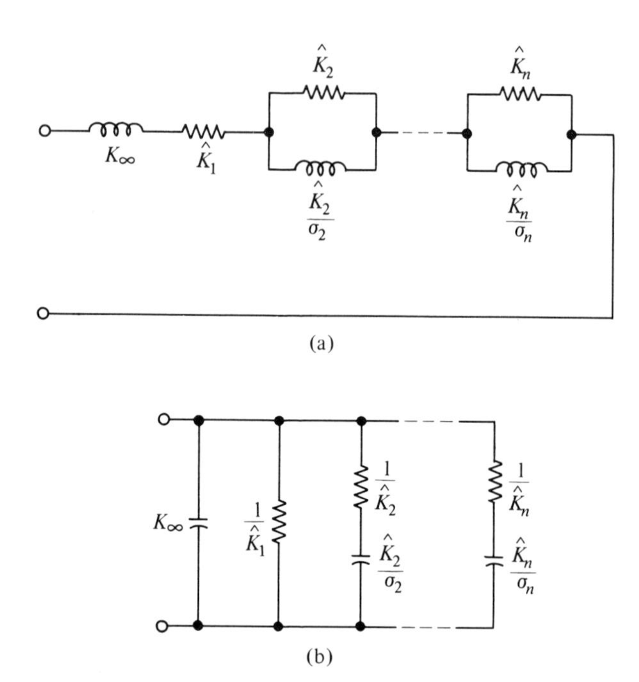

Figure 8.4-2 First and second Foster form of $F_1(s)$ when $m = n + 1$.

be synthesized as in case 1 just discussed. The resulting realizations are then as in Fig. 8.4-2.

Example 8.4-2. Synthesize the positive real function

$$F_1(s) = \frac{(s+1)(s+3)}{s+2}$$

Solution
Extracting the pole at ∞ yields

$$F_1(s) = s + \frac{2s+3}{s+2}$$

If the second term in the right-hand side of this equation is expanded as in case 1, we obtain

$$F_1(s) = s + \frac{3}{2} + \frac{(1/2)s}{s+2}$$

Hence, the corresponding constants are $K_\infty = 1$, $\hat{K}_1 = \frac{3}{2}$, and $\hat{K}_2 = \frac{1}{2}$. The resulting network then will be as those in Fig. 8.4-2 with appropriate element values.

II. Synthesis of $F_2(s)$

1. If $m = n - 1$, the degree of the numerator polynomial will be less than the degree of the denominator. If, on the other hand, $m = n + 1$, the degree of the numerator and denominator polynomials will be the same. In either case, $F_2(s)$ can be written as

$$F_2(s) = K_\infty + \frac{K_1}{s+\sigma_1} + \frac{K_3}{s+\sigma_3} + \frac{K_5}{s+\sigma_5} + \cdots + \frac{K_n}{s+\sigma_n} \tag{8.4-8}$$

where the K_i are the residues of the corresponding poles, which can be obtained from

$$K_i = \frac{(\sigma_2 - \sigma_i)(\sigma_4 - \sigma_i)\ldots(\sigma_{i-1} - \sigma_i)(\sigma_{i-1} - \sigma_{i+1})\ldots(\sigma_m - \sigma_i)}{(\sigma_1 - \sigma_i)(\sigma_3 - \sigma_i)\ldots(\sigma_{i-2} - \sigma_i)(\sigma_{i+2} - \sigma_i)\ldots(\sigma_n - \sigma_i)} \tag{8.4-9}$$

Comparing (8.4-9) with the inequality (8.4-3) it is easy to see that the K_i are all positive; either all the terms $\sigma_k - \sigma_i$ are positive, or, if they are negative, there is an even number of them. Thus, $K_i > 0$ for $i = 1, 3, 5, \ldots, n$. Similarly, K_∞ is the residue of the pole at ∞ and can be obtained from the equation

$$K_\infty = \lim_{s\to\infty} F_2(s) \tag{8.4-10}$$

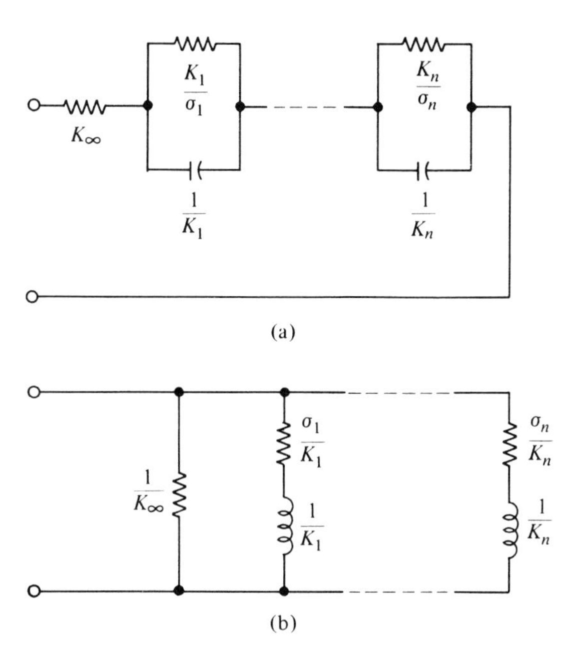

Figure 8.4-3 (a) Realization of $F_2(s)$ as a driving-point impedance; (b) realization of $F_2(s)$ as a driving-point admittance.

Quite clearly, since $F_2(s)$ is positive real, K_∞ is always positive. The circuit realization of $F_2(s)$ given in (8.4-8) can then be obtained as in Fig. 8.4-3(a) or (b).

So far in this chapter we have considered the synthesis of special classes of positive real functions. We shall next discuss the synthesis of a general positive real function.

8.5 SYNTHESIS OF GENERAL *RLC* NETWORKS (BRUNE SYNTHESIS)

Let $z(s)$ be a given positive real function; then, according to Theorem 8.2-1, it can be realized as the driving-point impedance or admittance function of a network comprised of linear time-invariant passive *RLC* elements. For simplicity, we assume that $z(s)$ is to be realized as a driving-point impedance throughout the rest of this section. The procedure for realizing $z(s)$ as a driving-point admittance function is similar and can, therefore, be omitted.

The basic idea behind synthesizing a general positive real function $z(s)$ is to write it as the sum of several simple positive real functions so that each function can be recognized as the driving-point impedance of a simple *RLC* network by inspection. The implementation of this method obviously depends on the form of $z(s)$. Several different possibilities are discussed.

Removal of a Pole at Infinity. If the numerator polynomial of $z(s)$ is 1 degree higher than its denominator, then $z(s) \longrightarrow \infty$ as $s \longrightarrow \infty$. In this case $z(s)$ is said to have a pole at ∞. The general form of $z(s)$ in this case is

$$z(s) = \frac{a_{n+1}s^{n+1} + a_n s^n + \cdots + a_1 s + a_0}{b_n s^n + b_{n-1}s^{n-1} + \cdots + b_1 s + b_0} \tag{8.5-1}$$

The first step in synthesizing $z(s)$ is to "extract" the pole at ∞. This can be done by dividing the numerator of $z(s)$ by its denominator; the result is

$$z(s) = sL_1 + \frac{a'_n s^n + a'_{n-1}s^{n-1} + \cdots + a'_1 s + a'_0}{b_n s^n + b_{n-1}s^{n-1} + \cdots + b_1 s + b_0} \tag{8.5-2}$$

where L_1 is a real positive constant given by

$$L_1 = \frac{a_{n+1}}{b_n} \tag{8.5-3}$$

If we now let

$$z_1(s) = \frac{a'_n s^n + a'_{n-1}s^{n-1} + \cdots + a'_1 s + a'_0}{b_n s^n + b_{n-1}s^{n-1} + \cdots + b_1 s + b_0} \tag{8.5-4}$$

then $z(s) = sL_1 + z_1(s)$. Thus, $z(s)$ can be considered as the driving-point impedance of an inductor with inductance L_1 in series with another network whose driving-point impedance is $z_1(s)$. This is shown in Fig. 8.5-1(a). We must now show that $z_1(s)$ obtained in this fashion is also positive real. Otherwise, it cannot be synthesized as the driving-point impedance of a linear time-invariant network with passive elements. To prove that $z_1(s)$ is positive real, we check the necessary and sufficient conditions of Theorem 8.2-2:

(a) The poles of $z_1(s)$ are the same as the poles of $z(s)$. Since $z(s)$ is positive real, all the poles of $z_1(s)$ are in the left half-plane.
(b) Removal of a pole at ∞ will not affect the residues of the poles on the $j\omega$ axis. This can be seen by a partial fraction expansion of $z(s)$. Thus, $j\omega$ axis poles of $z_1(s)$ are simple and have real and positive residues.
(c) The real part of $z_1(j\omega)$ can be obtained from

$$\text{Re}\{z(j\omega)\} = \text{Re}\{j\omega L_1\} + \text{Re}\{z_1(j\omega)\}$$

and since $\text{Re}\{j\omega L_1\} = 0$, we get

$$\text{Re}\{z_1(j\omega)\} = \text{Re}\{z(j\omega)\} \geq 0, \qquad 0 \leq \omega \leq \infty \tag{8.5-5}$$

Consequently, $z_1(s)$ is positive real and, hence, it can be realized by linear time-invariant passive elements.

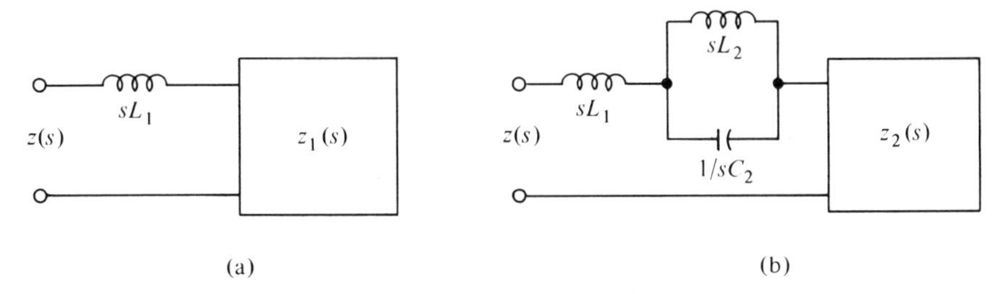

Figure 8.5-1 (a) Removing a pole at ∞ amounts to extracting an inductor from the network; (b) removal of a pair of poles from the $j\omega$ axis leaves the remaining function positive real.

The next subnetwork that can be extracted from $z_1(s)$ may be obtained by removing poles of $z_1(s)$ which lie on the $j\omega$ axis. We discuss this topic next.

Removal of the Poles from the Imaginary Axis. Suppose that $z(s)$ contains a pair of poles on the $j\omega$ axis, say, at $\omega = \pm\omega_1$. Then $z_1(s)$ can be written as

$$z_1(s) = \frac{N'(s)}{(s^2 + \omega_1^2)D'(s)} \tag{8.5-6}$$

where $N'(s)$ and $D'(s)$ are polynominals with appropriate degrees in s. Using a partial fraction expansion, $z_1(s)$ can be written as

$$z_1(s) = \frac{k_1}{s + j\omega_1} + \frac{k_2}{s - j\omega_1} + z_2(s) \tag{8.5-7}$$

where k_1 is a constant denoting the residue of the pole at $s = -j\omega_1$ and k_2 denotes the residue of the pole at $s = j\omega_1$. Since $z_1(s)$ is positive real, then, according to Theorem 8.2-2, k_1 and k_2 are both real and positive. Furthermore, it is an easy exercise to show that

$$k_1 = k_2$$

Rearranging (8.5-7) and using this equation, we get

$$z_1(s) = \frac{2k_1 s}{s^2 + \omega_1^2} + z_2(s) \tag{8.5-8}$$

The first term can be recognized as the driving-point impedance of a capacitor and an inductor connected in parallel. From this observation and the previous discussions, $z(s)$ can be decomposed as in Fig. 8.5-1(b) in which C_2 and L_2 are given by

$$C_2 = \frac{1}{2k_1}, \qquad L_2 = \frac{2k_1}{\omega_1^2}$$

By the same method used in the previous case, it can be shown that $z_2(s)$ given in (8.5-7) is positive real.

Exercise 8.5-1. Show that $z_2(s)$ defined in (8.5-8) is positive real.

If $z_2(s)$ has other poles on the $j\omega$ axis, they can be removed as previously and realized as parallel LC subnetworks. Thus, the remaining subnetwork will have no poles on the $j\omega$ axis or at ∞. Before discussing a method for synthesizing this subnetwork, let us consider a case where the given network has a pole at the origin instead of one at ∞.

Removal of a Pole at the Origin. If the given positive real function has a pole at the origin, it can be written as

$$z(s) = \frac{a_n s^n + a_{n-1}s^{n-1} + \cdots + a_1 s + a_0}{s(b_n s^n + b_{n-1}s^{n-1} + \cdots + b_1 s + b_0)} \tag{8.5-9}$$

Using a partial fraction expansion, this can be written as

$$z(s) = \frac{1}{sC_1} + z_1(s) \tag{8.5-10}$$

where
$$z_1(s) = \frac{a'_n s^n + a'_{n-1}s^{n-1} + \cdots + a'_1 s + a'_0}{b_n s^n + b_{n-1}s^{n-1} + \cdots + b_1 s + b_0} \tag{8.5-11}$$

and
$$C_1 = \frac{b_0}{a_0}$$

Consequently, a capacitor with capacitance C_1 can be extracted from $z(s)$. The remaining subnetwork, $z_1(s)$, can then be shown to be positive real. This process is illustrated in Fig. 8.5-2.

Let us now go back to the case where the positive real subnetwork under study has no poles on the $j\omega$ axis or at ∞. It is assumed that such poles have already been removed. The next step in this case is to extract a resistor from the remaining subnetwork.

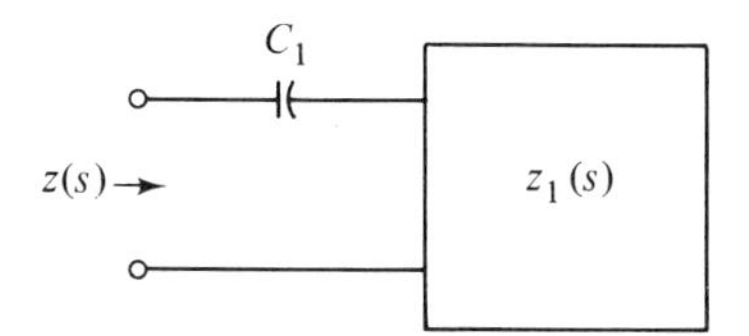

Figure 8.5-2 The pole at the origin can be removed as a series capacitor.

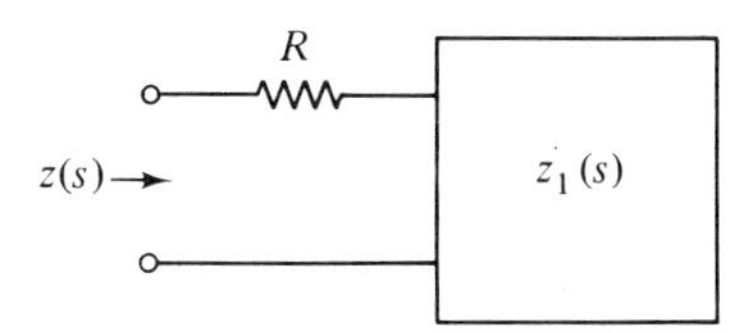

Figure 8.5-3 A constant can be removed as a series resistor.

Removal of a Constant. Consider a positive real function $z(s)$, which has no poles on the $j\omega$ axis. Let us write $z(s)$ as

$$z(s) = R + z_1(s)$$

where R is a real and positive number that can be realized as a resistor with resistance R, and $z_1(s)$ is a positive real function. The function $z(s)$ can then be decomposed as in Fig. 8.5-3. Notice that R must be chosen such that the function $z_1(s)$ remains positive real. According to Theorem 8.2-2, this requirement imposes the following constraint on the value of the resistance R:

$$R \leq \min_{\omega}[\text{Re}\{z(j\omega)\}] \tag{8.5-12}$$

In words, the value of the resistance R must be less than or equal to the minimum of the real part of $z(j\omega)$ for all positive ω. Let us now work out an example to illustrate this procedure.

Example 8.5-1. Remove the largest possible resistance from the following positive real function so that the remaining function is positive real.

$$z(s) = \frac{3s^2 + 3s + 9}{s^2 + s + 4}$$

Solution

This function has obviously no poles on the $j\omega$ axis, at the origin, or at ∞. To remove a constant R, we must first determine the minimum of $\text{Re}\{z(j\omega)\}$;

$$\text{Re}\{z(j\omega)\} = \text{Re}\left\{\frac{9 - 3\omega^2 + j3\omega}{4 - \omega^2 + j\omega}\right\} = \frac{3\omega^4 - 18\omega^2 + 36}{\omega^4 - 7\omega^2 + 16}$$

The minimum of the above function can be obtained by putting its derivative with respect to ω equal to zero:

$$\frac{d}{d\omega}\text{Re}\{z(j\omega)\} = 0 \Longrightarrow \omega = \sqrt{2}$$

Then, we choose

$$R = \min_{\omega}[\text{Re}\{z(j\omega)\}] = \text{Re}\{z(j\omega)\}|_{\omega=\sqrt{2}} = 2$$

Thus, $z(s)$ can be written as

$$z(s) = 2 + \frac{s^2 + s + 1}{s^2 + s + 4}$$

Clearly, the last term in the right-hand side of the equation is positive real and can be realized as the driving-point impedance of an RLC network.

Let us now consider the final step in the realization of a general positive real function.

Brune Synthesis. Having removed all the poles at 0, ∞, and the $j\omega$ axis, we use the procedure discussed to remove a resistor whose resistance is equal to the minimum of the $\text{Re}\{z(j\omega)\}$. The remaining function is a positive real function known as minimum function. More precisely,

Definition 8.5-1 (Minimum Function). A positive real function $z(s)$ is called a *minimum function* if

(a) $z(s)$ has no poles or zeros on the $j\omega$ axis, 0, or at ∞.
(b) $\text{Re}\{z(j\omega_0)\} = 0$ for some $\omega_0 \geq 0$.

Let us denote the imaginary part of the minimum function $z(j\omega)$ by $x(\omega)$; that is, let

$$x(\omega) = \text{Im}\{z(j\omega)\} \tag{8.5-13}$$

Since $\text{Re}\{z(j\omega_0)\} = 0$, we can write

$$z(j\omega_0) = jx(\omega_0) \tag{8.5-14}$$

We now consider two distinct cases: $x(\omega_0) < 0$ and $x(\omega_0) > 0$. Note that $x(\omega_0) \neq 0$; otherwise, $z(j\omega)$ will have a zero on the $j\omega$ axis, which contradicts the assumption that $z(s)$ is minimum function.

Case 1. If $x(\omega_0) < 0$, we remove a *negative inductor* with inductance $L_1 \triangleq x(\omega_0)/\omega_0$. That is, the given positive real minimum function can be written as

$$z(s) = z_1(s) + sL_1$$

or, equivalently,

$$z_1(s) = z(s) - sL_1 \tag{8.5-15}$$

Since $L_1 < 0$, then $-sL_1$ is positive real and, consequently, $z_1(s)$, which is the sum of two positive real functions, is positive real. The resulting network is shown in Fig. 8.5-4. We shall soon show that this negative inductor can be replaced by a positive inductor with mutual coupling. The important fact to recognize here is that $z_1(s)$ *has a pair of zeros on the* $j\omega$ *axis* at $\omega = \pm\omega_0$

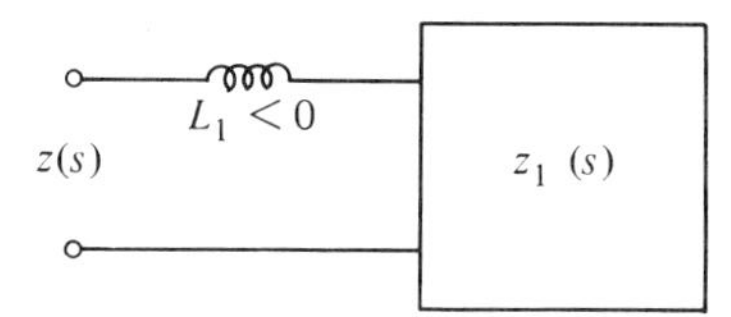

Figure 8.5-4 Removal of a negative inductor $L_1 = (1/\omega_0)x(j\omega_0)$ leaves $z_1(s)$ positive real.

(why?). As a result, $y_1(s) = 1/z_1(s)$ has a pair of poles on the $j\omega$ axis at $\omega = \pm\omega_0$ that can be removed as a series *LC* network:

$$y_1(s) = \frac{1}{z_1(s)} = \frac{N(s)}{(s^2 + \omega_0^2)D(s)} = \frac{2k_1 s}{s^2 + \omega_0^2} + y_2(s) \qquad (8.5\text{-}16)$$

Since $y_1(s)$ is positive real, using Theorem 8.2-2, it becomes clear that $y_2(s)$ is also positive real, and its inverse, $z_2(s) = 1/y_2(s)$, can then be realized as the driving-point impedance of a network with linear time-invariant passive elements. Furthermore, a rearrangement of the last three equations together with the fact that $z(s)$ is a minimum function shows that $z_2(s)$ has a pole at ∞. Thus, an inductor with inductance $L_3 > 0$ can be removed from $z_2(s)$, leaving the remaining function, $z_3(s)$, positive real. The realization discussed so far can be summarized in Fig. 8.5-5.

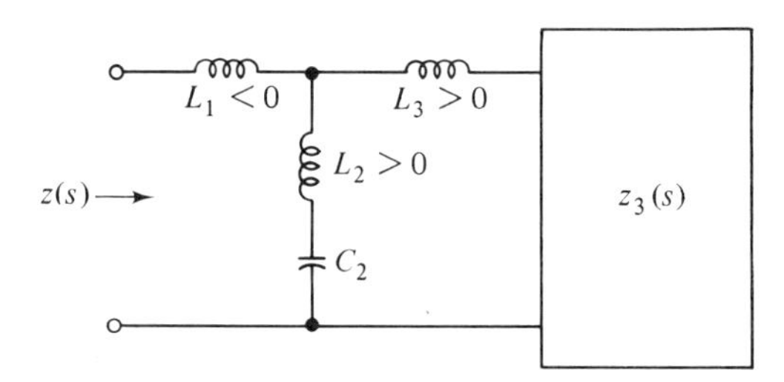

Figure 8.5-5 Realization of a minimum function for which $x(\omega_0) < 0$.

It now remains to replace the T-shape inductors in Fig. 8.5-5 by a subnetwork all of whose elements are positive. By doing so we shall have disposed of the negative inductance L_1. This can readily be achieved by replacing the T-shape inductors of Fig. 8.5-5 by a pair of coupled inductors, as in Fig. 8.5-6, where the values of the self-inductances L_a and L_b and the mutual

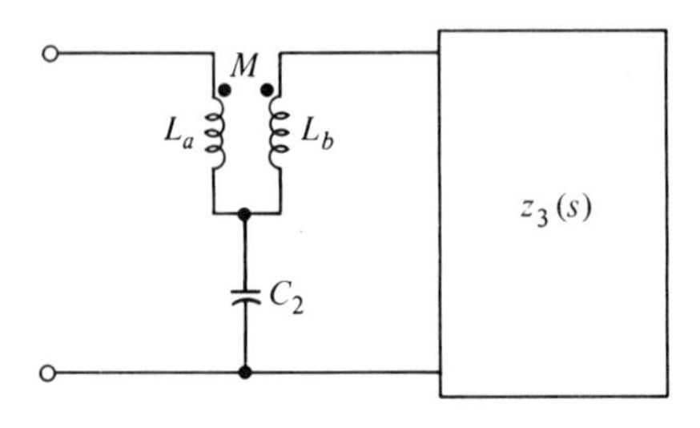

Figure 8.5-6 Negative inductance of Fig. 8.5-5 can be replaced with coupled inductors.

inductance M are given by

$$L_a = L_1 + L_2 \tag{8.5-17}$$

$$L_b = L_2 + L_3 \tag{8.5-18}$$

$$M = L_2 \tag{8.5-19}$$

Exercise 8.5-2. Derive equations (8.5-17), (8.5-18), and (8.5-19), and show that L_a and L_b are both positive.

Case 2. If $x(\omega_0) > 0$, an inductor with positive inductance $L_1 = x(\omega_0)/\omega_0$ can be removed. The remaining function $z_1(s)$, however, will not be positive real, since it will have a zero in the right half-plane. Nevertheless, we proceed with the synthesis, allowing the presence of negative inductors. Note that as in the previous case $z_1(s)$ has a pair of zeros on the $j\omega$ axis at $\omega = \pm\omega_0$. Consequently, $y_1(s) \triangleq 1/z_1(s)$ will have a pair of poles at $\pm j\omega_0$, which can be removed as a series LC network. The remaining function, $z_2(s)$, will then have a pole at ∞, which can be removed as a negative inductor. The resulting network is shown in Fig. 8.5-7. As in case 1, the T-shape inductor subnetwork can be replaced by a pair of coupled inductors with positive self-and mutual inductances.

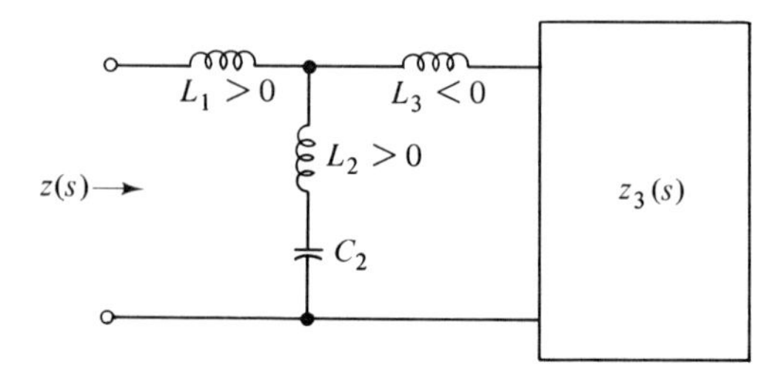

Figure 8.5-7 Realization of a minimum function for which $x(\omega_0) > 0$.

Introducing negative inductors for synthesizing a minimum function was first presented by Brune [1]. Since Brune's original work, numerous methods of realizing a positive real function have been developed. An excellent presentation of these methods can be found in [2] and [3].

Let us now conclude this section by working out a simple example to illustrate the synthesis procedure presented.

Example 8.5-2. Synthesize a network with linear time-invariant passive elements whose driving-point impedance is given by the positive real function

$$z(s) = \frac{s^5 + 2s^4 + 7s^3 + 4s^2 + 9s + 1}{s^4 + s^3 + 5s^2 + s + 4}$$

Solution

Clearly, $z(s)$ has a pole at ∞; hence, an inductor can be removed from the network, leaving the remaining network positive real:

$$z(s) = s + z_1(s)$$

where

$$z_1(s) = \frac{s^4 + 2s^3 + 3s^2 + 5s + 1}{s^4 + s^3 + 5s^2 + s + 4}$$

The function $z_1(s)$ can also be seen to have a pair of poles on the $j\omega$ axis at $\omega = \pm 1$; it can then be written as

$$z_1(s) = \frac{s}{s^2 + 1} + z_2(s)$$

where the first term in the right-hand side of the equation can be realized as the impedance of a parallel LC subnetwork with $L = 1$ H and $C = 1$ F. The second term is then

$$z_2(s) = \frac{s^2 + s + 1}{s^2 + s + 4}$$

The next step is to realize $z_2(s)$. This function, however, has no poles at ∞, 0, or on the $j\omega$ axis. We might then try to remove a constant from it by first computing its real part:

$$\text{Re}\{z_2(j\omega)\} = \text{Re}\left\{\frac{1 - \omega^2 + j\omega}{4 - \omega^2 + j\omega}\right\} = \frac{\omega^4 - 4\omega^2 + 4}{(4 - \omega^2)^2 + \omega^2}$$

But,

$$\min[\text{Re}\{z_2(j\omega)\}] = 0 \quad \text{at } \omega = \pm\sqrt{2}$$

Then $z_2(s)$ is a minimum function and the Brune synthesis discussed previously can be used to realize it. The imaginary part of $z_2(s)$ is found to be

$$x(\sqrt{2}) = \text{Im}\{z_2(j\sqrt{2})\} = \frac{\sqrt{2}}{2} > 0$$

This means that case 2 should be used. The value of the first inductor to be removed from $z_2(s)$ is therefore

$$L_1 = \frac{x(\sqrt{2})}{\sqrt{2}} = \frac{1}{2}\text{H}$$

The resulting $z_3(s)$ can then be found from

$$z_3(s) = \frac{s^2 + s + 1}{s^2 + s + 4} - \frac{s}{2} = \frac{-s^3 + s^2 - 2s + 2}{2(s^2 + s + 4)}$$

Notice that $z_3(s)$ has a pair of zeros at $\omega = \pm\sqrt{2}$; hence, its inverse, $y_3(s)$, will have a pair of poles on the $j\omega$ axis that can be removed:

$$y_3(s) = \frac{2(s^2 + s + 4)}{-s^3 + s^2 - 2s + 2} = \frac{2(s^2 + s + 4)}{(s^2 + 2)(1 - s)} = \frac{2s}{s^2 + 2} + \frac{4}{1 - s}$$

The function $2s/(s^2 + 2)$ can be realized as the driving-point admittance of a capacitor with a capacitance of 1 F in series with an inductor with an inductance of $\frac{1}{2}$ H. Also, the function $4/(1 - s)$ can be realized as the driving point of a negative inductor with an inductance $-\frac{1}{4}$ H in series with a resistor with a resistance of $\frac{1}{4}$ Ω. The resulting network is shown in Fig. 8.5-8.

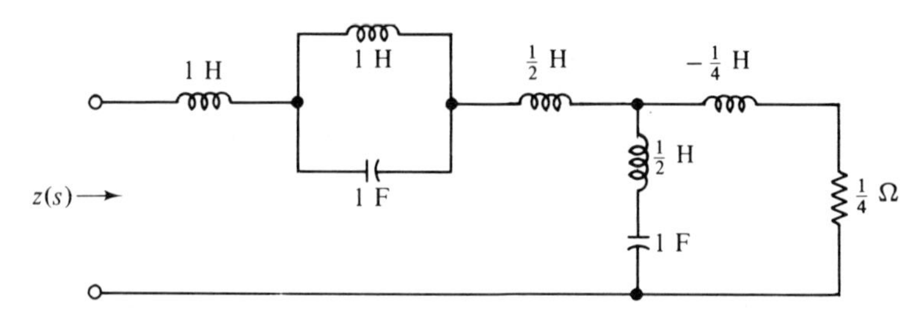

Figure 8.5-8 Brune realization of the p.r. function of Example 8.5-2.

To eliminate the negative inductor, we can use equations (8.5-17) through (8.5-19) to replace the T-shape inductors by a pair of inductors with mutual coupling:

$$L_a = L_1 + L_2 = 1 \text{ H}$$
$$L_b = L_2 + L_3 = \tfrac{1}{4} \text{ H}$$
$$M = \tfrac{1}{2} \text{ H}$$

The final network is as shown in Fig. 8.5-9.

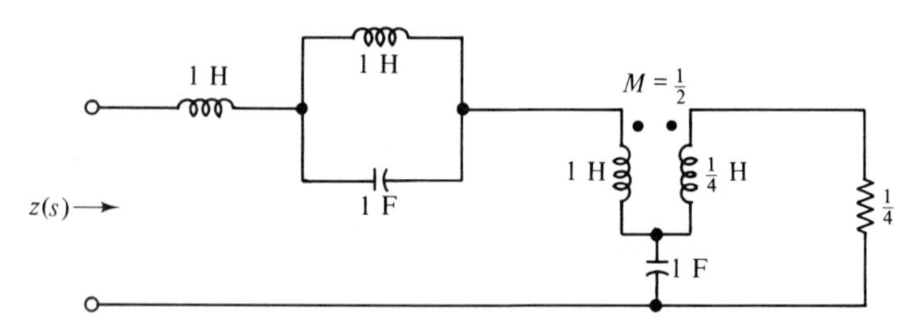

Figure 8.5-9 Final network of Example 8.5-2.

8.6 BOTT–DUFFIN SYNTHESIS (TRANSFORMERLESS SYNTHESIS)

The general synthesis procedure developed in the last section requires transformers (or coupled inductors) in order to avoid the presence of non-passive elements in the final network. In many practical cases, due to cost, bulkiness, and construction difficulties, it is desirable to obtain a transformerless synthesis of a given network function. In this section we discuss an alternative procedure for synthesizing minimum functions without transformers; this method was first introduced by Bott and Duffin [7].

Richards [8] has shown that if $z(s)$ is a positive real function, then $R(s)$, defined by

$$R(s) = \frac{kz(s) - sz(k)}{kz(k) - sz(s)} \tag{8.6-1}$$

is also positive real for all real and positive values of k. Furthermore, the degree of the numerator or denominator of $R(s)$ is no higher than that of $z(s)$. In other words, the *complexity* of $R(s)$ is not greater than the complexity of $z(s)$. We shall next show that the value of k can be chosen so that $R(s)$ can be synthesized without the use of transformers.

Solving (8.6-1) for $z(s)$, we obtain

$$z(s) = \frac{kz(k)R(s)}{k + sR(s)} + \frac{sz(k)}{k + sR(s)}$$

or, equivalently;

$$z(s) = \frac{1}{\dfrac{1}{z(k)R(s)} + \dfrac{s}{kz(k)}} + \frac{1}{\dfrac{k}{sz(k)} + \dfrac{R(s)}{z(k)}} \tag{8.6-2}$$

The first step in the Bott–Duffin synthesis is to realize the impedance function of (8.6-2) as the network of Fig. 8.6-1, where $z_1(s)$ and $z_2(s)$ are both

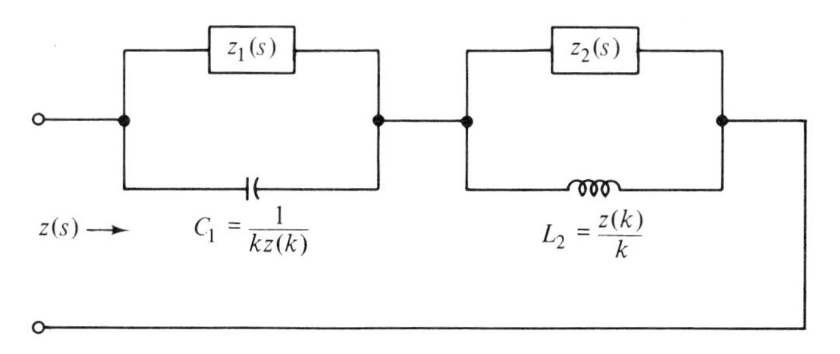

Figure 8.6-1 First step in the Bott-Duffin synthesis.

positive real functions defined by

$$z_1(s) = z(k)R(s) \tag{8.6-3}$$

and

$$z_2(s) = z(k)/R(s) \tag{8.6-4}$$

Note that by specifying k the values of the capacitor and the inductor in Fig. 8.6-1 are specified. It now remains to show that $z_1(s)$ and $z_2(s)$ can be synthesized without the use of transformers. *This will be done by choosing k such that R(s) will have a pair of zeros or poles on the imaginary axis.*

Let us assume that $z(s)$ is a minimum function (if not, we can first remove all the poles and zeros from the $j\omega$ axis, zero, and infinity and reduce it to a minimum function). Consequently, for some ω_0 we must have

$$z(j\omega_0) = jx(\omega_0) \tag{8.6-5}$$

where $x(\omega_0)$ denotes the imaginary part of $z(j\omega_0)$, which may be positive or negative. We consider each case separately.

Case 1 $\boldsymbol{x(\omega_0) > 0}$**.** In this case we choose the constant k such that $R(s)$ given by (8.6-1) will have a pair of zeros at $s = \pm j\omega_0$. More precisely, choose k such that

$$kz(j\omega_0) - j\omega_0 z(k) = 0$$

Using (8.6-5), this equation can be written as

$$kx(\omega_0) - \omega_0 z(k) = 0 \tag{8.6-6}$$

Note that since ω_0 is known equation (8.6-6) represents a polynomial in k, which can be solved for the *real and positive* value of k. Furthermore, comparing (8.6-5) and (8.6-6) it is clear that $k = +j\omega_0$ and $k = -j\omega_0$ are two roots of this polynomial; observation of this fact often simplifies the solution of (8.6-6). Having obtained k, then C_1 and L_2 of Fig. 8.6-1 can be obtained from

$$C_1 = \frac{1}{kz(k)} \quad \text{and} \quad L_2 = \frac{z(k)}{k}$$

Observe that since $k > 0$ and $z(s)$ is p.r., both C_1 and L_2 are positive. With this choice of k, $R(s)$ will have a pair of zeros on the $j\omega$ axis at $s = \pm j\omega_0$. Thus, (8.6-3) and (8.6-4) can be written as

$$y_1(s) = \frac{1}{z_1(s)} = \frac{1}{z(k)R(s)} = \frac{2K_0 s}{s^2 + \omega_0^2} + y_3(s) \tag{8.6-7}$$

and

$$z_2(s) = \frac{z(k)}{R(s)} = \frac{2K_0' s}{s^2 + \omega_0^2} + z_4(s) \tag{8.6-8}$$

The first terms in the extreme right-hand side of these equations can be synthesized as parallel and series LC elements, as shown in Fig. 8.6-2, where

$$L_3 = \frac{1}{2k_0}, \quad C_3 = \frac{2K_0}{\omega_0^2}, \quad L_4 = \frac{2K_0'}{\omega_0^2}, \quad \text{and} \quad C_2 = \frac{1}{2K_0'} \tag{8.6-9}$$

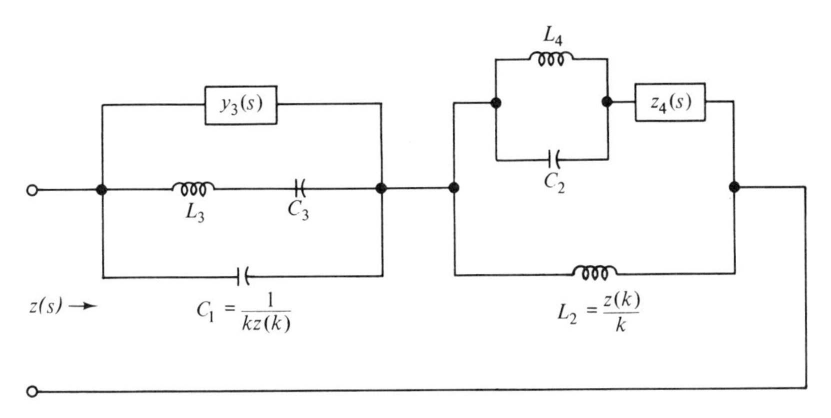

Figure 8.6-2 Completion of a Bott-Duffin cycle.

Note that since $z_1(s)$ and $z_2(s)$ are p.r., residues K_0 and K_0' are real and positive. Furthermore, $y_3(s)$ and $z_4(s)$ are p. r. functions with polynomials of lower degree than those of $y_1(s)$ and $z_2(s)$. The procedure discussed so far is called a *Bott–Duffin cycle.* This cycle can be repeated until the synthesis is complete. The important fact is that the complexity of the p.r. function in each cycle will be reduced by 2.

Case 2 $\boldsymbol{x(\omega_0) < 0}$**.** If $x(\omega_0) < 0$, the choice of k as in case 1 results in negative C_1 and L_2. To avoid this, we choose k such that $R(s)$ will have a pair of *poles* at $s = \pm j\omega_0$. Thus, from (8.6-1) we have

$$kz(k) - j\omega_0 z(j\omega_0) = 0$$

which, using (8.6-5), yields

$$kz(k) = -\omega_0 x(\omega_0)$$

But since $x(\omega_0)$ is, by assumption, negative, both $C_1 = 1/kz(k)$ and $L_2 = z(k)/k$ will be positive. Hence, $z(s)$ can be synthesized as the network of Fig. 8.6-1. The p.r. functions $z_1(s)$ and $1/z_2(s)$ will, in this case, have a pair of poles at $s = \pm j\omega_0$, which can be removed. Removal of these poles completes the Bott–Duffin cycle for this case. This cycle can then be repeated to synthesize the entire network. Let us now work out an example to illustrate the procedure discussed so far.

Example 8.6-1. Synthesize the following minimum function using the Bott–Duffin method:

$$z(s) = \frac{s^2 + s + 1}{s^2 + s + 4}$$

Solution

As we have already seen in Example 8.5-2, $z(s)$ is indeed a minimum function and its real part vanishes at $\omega_0 = \pm\sqrt{2}$. We then have

$$x(\omega_0) = \text{Im}\{z(j\sqrt{2})\} = \frac{\sqrt{2}}{2} > 0$$

Since $x(\omega_0) > 0$, we must choose the synthesis procedure outlined in case 1. Thus, the first step is to choose k from (8.6-6); this yields

$$k^3 - k^2 + 2k - 2 = 0$$

Knowing that $k^2 + 2$ can be factored in this equation, we get

$$(k^2 + 2)(k - 1) = 0$$

The real and positive root of this equation is $k = 1$. We therefore have

$$z(k) = \frac{1}{2}, \qquad C_1 = \frac{1}{kz(k)} = 2, \quad \text{and} \quad L_2 = \frac{z(k)}{k} = \frac{1}{2}$$

The Richard's function $R(s)$ is obtained from

$$R(s) = \frac{kz(s) - sz(k)}{kz(k) - sz(s)} = \frac{(1 - s)(s^2 + 2)}{(1 - s)(2s^2 + 3s + 4)} = \frac{s^2 + 2}{2s^2 + 3s + 4}$$

As expected, $R(s)$ has a pair of zeros at $s = \pm j\sqrt{2}$. Using (8.6-7) and (8.6-8), we get

$$y_1(s) = \frac{4s^2 + 6s + 8}{s^2 + 2} = \frac{6s}{s^2 + 2} + 4$$

and
$$z_2(s) = \frac{s^2 + \frac{3}{2}s + 2}{s^2 + 2} = \frac{\frac{3}{2}s}{s^2 + 2} + 1$$

Hence, $\quad K_0 = 3, \quad K_0' = \frac{3}{4}, \quad y_3 = 4, \quad \text{and} \quad z_4 = 1$

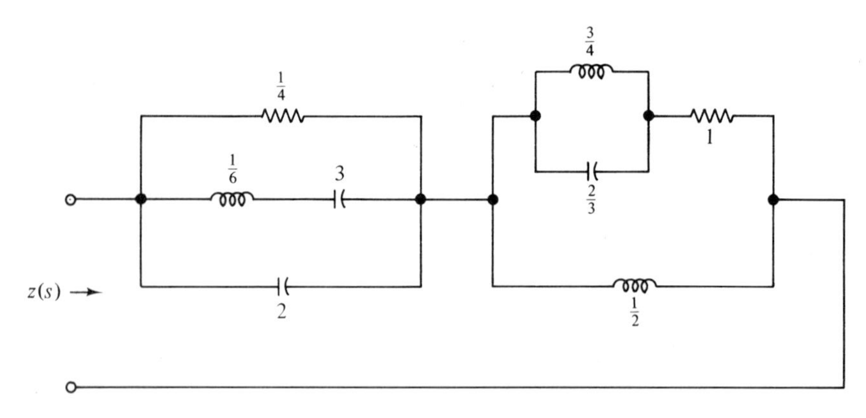

Figure 8.6-3 Bott-Duffin realization of $z(s) = (s^2 + s + 1)/(s^2 + s + 4)$.

The complete realization of $z(s)$ can then be obtained using Fig. 8.6-2 and equation (8.6-9). The resulting network is shown in Fig. 8.6-3.

8.7 ACTIVE NETWORK SYNTHESIS

So far in this chapter we have discussed the synthesis of passive networks. As we have already seen, only positive real network functions can be realized using passive *RLC* elements. In this section we broaden our scope and permit the inclusion of active elements, such as transistors, negative impedance convertors, operational amplifiers, and so on, in the synthesis procedure. The advantages of doing so are manifold. First, inclusion of active elements allows the design engineer to synthesize nonpositive real functions as well as positive real ones. He can design an active filter, which will do the job of a conventional passive filter and at the same time amplify the signal to any desired degree.

However, the main advantage of the active synthesis of a given network function is that the entire synthesis can be carried out without the use of any inductors; only resistors, capacitors, and active elements such as operational amplifiers are needed. This implies that an active network can be built using integrated circuits (due to practical difficulties inductors cannot, as yet, be included in integrated circuits). These and other features of active networks and filters (such as the low production cost, small weight, reliability, low sensitivity, etc.) have caused them to become more and more popular during the past decade.

There are numerous books and articles available on the subject of active network synthesis [11–17]. In the remainder of this chapter, however, we confine ourselves to the fundamental concepts and encourage the interested reader to consult the cited references.

In the previous sections we only discussed the synthesis of a driving-point impedance or admittance. Here we take up the synthesis of transfer functions, in particular that of voltage transfer functions.

Synthesis of Active Filters. Before discussing the synthesis of active filters, let us introduce electrical filters and define some of the terminology used in filter analysis and design. Electrical filters have numerous applications in electronic and communication systems. They are mainly used to discriminate against the input frequencies of a system—to pass the desirable frequencies and reject the unwanted noise. Recall from Chapter 4 that if a network with transfer function $H(s)$ is excited by a sinusoidal function $u(t) = \sin \omega t$, the steady-state response $y(t)$ will be in the form

$$y(t) = |H(j\omega)| \sin[\omega t + \phi(\omega)]$$

where $|H(j\omega)|$ and $\phi(\omega)$ are, respectively, the magnitude and the phase of the transfer function. Depending upon the plot of $|H(j\omega)|$ versus ω, filters are divided into three basic categories: *low-pass*, *band-pass*, and *high-pass* filters. In each case the ω axis is divided into two separate regions: *passband* and *stopband.* Ideally, all the signals whose frequencies fall in the passband will "pass" through the filter with little or no attenuation. On the other hand, all the signals whose frequencies fall in the stopband will be "stopped" by the filter. A graphical representation of these terms is shown in Fig. 8.7-1. For a second-order filter the transfer functions of different categories are

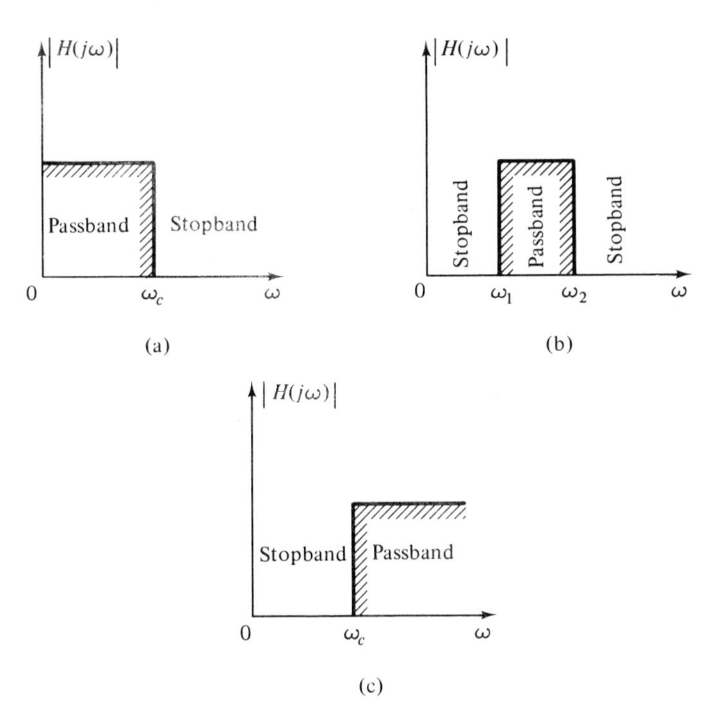

Figure 8.7-1 Various filter classifications: (a) low pass; (b) band pass; (c) high pass.

Low Pass

$$H(s) = H_0 \frac{1}{s^2 + 2\xi\omega_0 s + \omega_0^2} \tag{8.7-1}$$

where ω_0, ξ, and H_0/ω_0^2 are, respectively, called the *damped natural frequency*, *damping ratio*, and *dc gain*. Note that, for $0 < \xi < 1$, $H(s)$ will have complex roots.

Band Pass

$$H(s) = H_0 \frac{2\xi\omega_0 s}{s^2 + 2\xi\omega_0 s + \omega_0^2} \tag{8.7-2}$$

In this particular case, the *quality factor* Q of the filter is defined as

$$Q = \frac{1}{2\xi} \tag{8.7-3}$$

Note that the dc gain in this case is zero.

High Pass

$$H(s) = H_0 \frac{s^2}{s^2 + 2\xi\omega_0 s + \omega_0^2} \tag{8.7-4}$$

In this case also the dc gain is zero; H_0 denotes the gain of the filter at $\omega \longrightarrow \infty$.

Let us now proceed with the synthesis of active filters. Perhaps, the best way to discuss this topic is to start with a simple example.

Example 8.7-1. Consider the transfer function

$$H(s) = \frac{y(s)}{u(s)} = \frac{b_1 s + b_0}{s^2 + a_1 s + a_0} \tag{8.7-5}$$

where $y(s)$ and $u(s)$ represent the Laplace transforms of the output and the input, respectively. Let us introduce an auxiliary variable x, and write $H(s)$ as

$$\frac{y}{x} \cdot \frac{x}{u} = (b_1 s + b_0)\frac{1}{s^2 + a_1 s + a_0} \tag{8.7-6}$$

The terms in the left- and the right-hand side of (8.7-6) can be identified as

$$\frac{x}{u} = \frac{1}{s^2 + a_1 s + a_0}$$

and

$$\frac{y}{x} = b_1 s + b_0$$

or, upon cross multiplication and rearranging, as

$$(s + a_1)sx = u - a_0 x \tag{8.7-7}$$

and

$$y = b_1 sx + b_0 x \tag{8.7-8}$$

Let us next use a block diagram to represent these equations. A *block diagram* is a pictorial representation of a transfer function; it consists of three basic components: *integrators*, *adders* (or summers), and *scalar multipliers*. The symbolic representation of each of these elements is shown in Fig. 8.7-2.

As we have already seen in Chapter 1, these components can be constructed using operational amplifiers, resistors, and capacitors only (see Section 1.5).

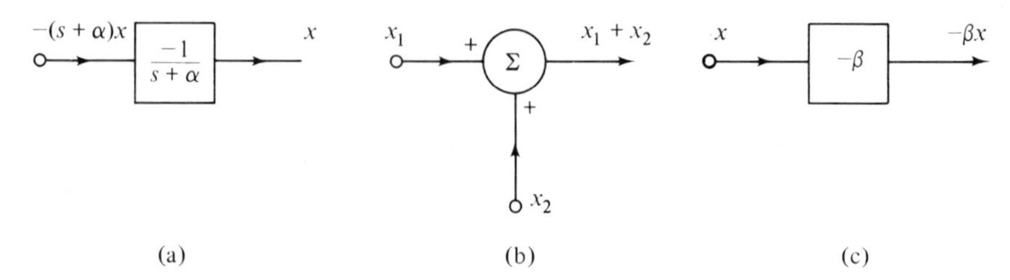

Figure 8.7-2 Symbolic representation of (a) an integrator, (b) an adder, and (c) a scalar multiplier.

For convenience, the network realization of these components is shown in Fig. 8.7-3. Note that operational amplifiers are assumed to have infinite input impedance so that the current through them may be considered negligible. Furthermore, their open-loop gain is assumed to be infinity ($A \longrightarrow \infty$) so that the input voltage can be considered zero.

The element values shown in the figure are normalized values; they may be adjusted if variations of the above devices are needed. For example, if an

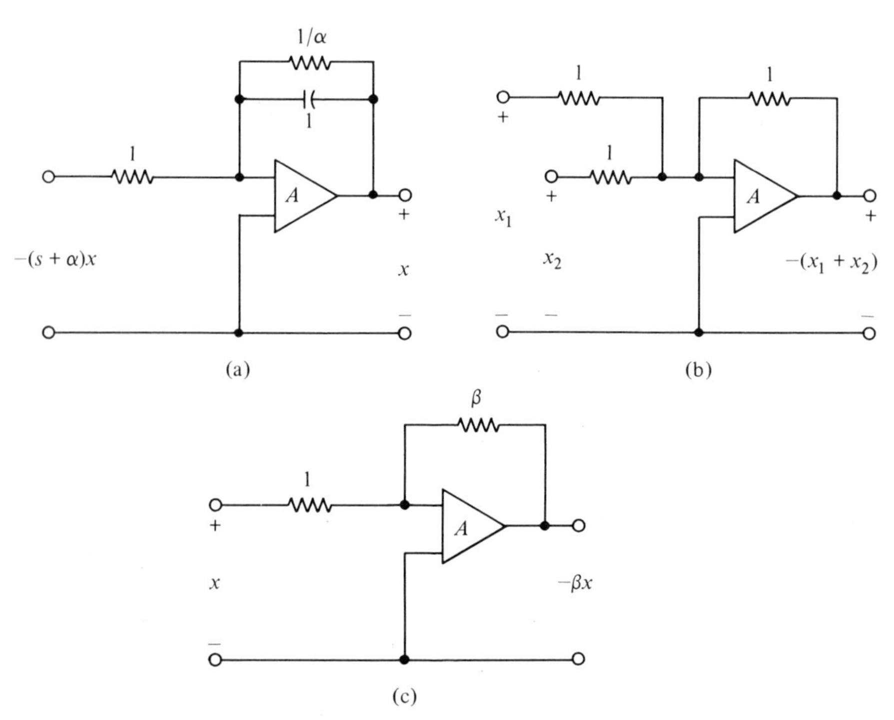

Figure 8.7-3 Op-amp realization of (a) integrator, (b) adder, and (c) scalar multiplier.

integrator with a transfer function of $-(1/s)$ is required, the resistor in the feedback of the op-amp of Fig. 8.7-3(a) may be open circuited to yield the desired integrator. Or if an adder is to multiply the signals x_1 and x_2 by β_1 and β_2 before adding them, the resistors in the forward path of the op-amp of Fig. 8.7-3(b) can be replaced by resistors with resistances $1/\beta_1$ and $1/\beta_2$, respectively.

Let us now return to equations (8.7-7) and try to obtain the corresponding block-diagram representation. Since the transfer function under consideration is of second order, we need two integrators. Let us take the output of the second integrator to be x and its transfer function to be $-(1/s)$; the input of this integrator will be $-sx$. Take the transfer function of the first integrator to be $-1/(s + a_1)$ and its output to be $-sx$; the input to it must then be $(s + a_1)sx$. But from equation (8.7-7) this term is equal to $u - a_0x$, which can be generated by adding u and $-a_0x$ by an adder. The term $-a_0x$ is obtained by "feeding back" the output of the second integrator. From (8.7-8) it is clear that the output y may be obtained by appropriately scaling and adding the output of the integrators. The overall block diagram is shown in Fig. 8.7-4. Using this block diagram and the components of Fig. 8.7-3, the resulting network realization of the given transfer function may be obtained as in

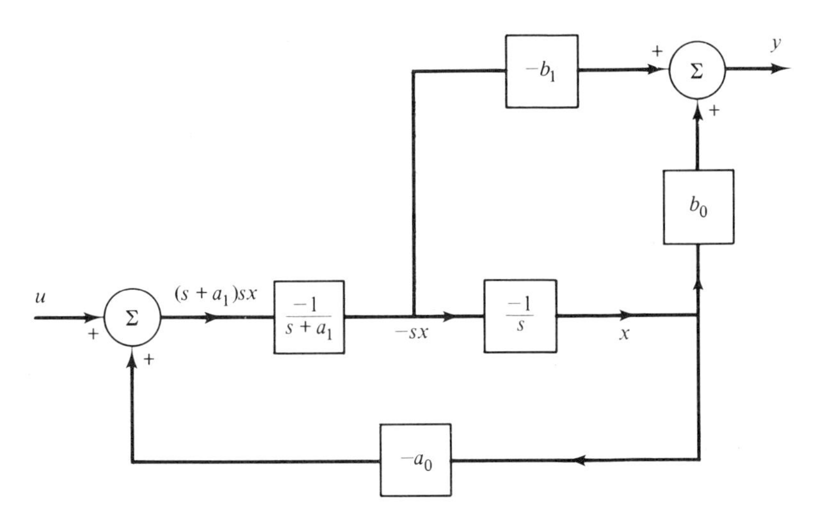

Figure 8.7-4 Block-diagram representation of $[y(s)]/[u(s)] = (b_1s + b_0)/(s^2 + a_1s + a_0)$.

Fig. 8.7-5. Note that the op-amps A_3 and A_4 are used for sign inversion only. (If either of the coefficients a_0 and b_1 are negative, the corresponding op-amp should be eliminated.) The op-amp A_5 is used as a scalar multiplier and adder. Furthermore, the input adder has been incorporated in A_1. In an actual design configuration the normalized resistors and capacitors of Fig. 8.7-5

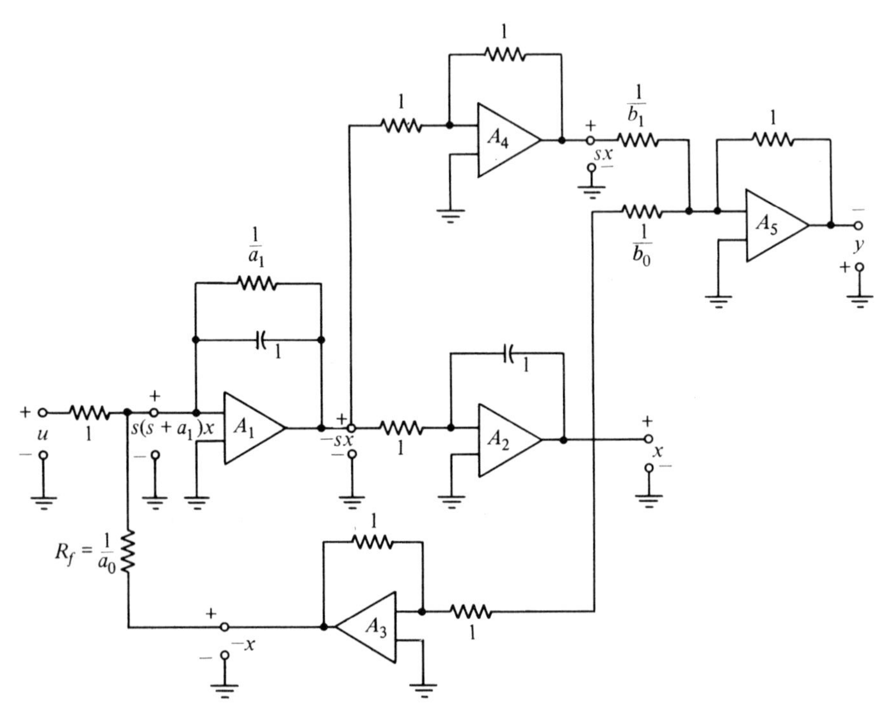

Figure 8.7-5 Op-amp realization of the voltage transfer function $y/u = (b_1 s + b_0)/(s^2 + a_1 s + a_0)$.

will be scaled appropriately to match the characteristics of the op-amps. For example, the 1Ω resistors in the feedback and the forward path of the sign-inverting op-amps A_3 and A_4 are replaced by 10kΩ resistors.

In general, consider a transfer function $H(s)$ defined by

$$H(s) = \frac{y(s)}{u(s)} = \frac{b_n s^n + b_{n-1}s^{n-1} + \cdots + b_1 s + b_0}{s^n + a_{n-1}s^{n-1} + \cdots + a_1 s + a_0} \tag{8.7-9}$$

where the coefficients a_i and b_i may be positive or negative. To obtain a block diagram for this general case, let us introduce the auxiliary variable x and proceed as in Example 8.71. Thus,

$$\frac{y}{x}\cdot\frac{x}{u} = \frac{b_n s^n + b_{n-1}s^{n-1} + \cdots + b_1 s + b_0}{1} \cdot \frac{1}{s^n + a_{n-1}s^{n-1} + \cdots + a_1 s + a_0} \tag{8.7-10}$$

which yields

$$s^n x = u - a_{n-1}s^{n-1}x - a_{n-2}s^{n-2}x - \cdots - a_1 sx - a_0 x \tag{8.7-11}$$

and $$y = b_n s^n x + b_{n-1}s^{n-1}x + \cdots + b_1 sx + b_0 x \tag{8.7-12}$$

We therefore need n integrators each with transfer function $-(1/s)$; if the output of the nth integrator is to be x, the output of the $(n-1)$st integrator must be $-sx$, and so on. The term $s^n x$ is then generated by feeding back $-a_0 x, -a_1 sx, \cdots, -a_{n-1}s^{n-1}x$ and adding them to u. Similarly, the output is obtained by adding $b_0 x, b_1 sx, \cdots, b_n s^n x$ together. The resulting block diagram is shown in Fig. 8.7-6.

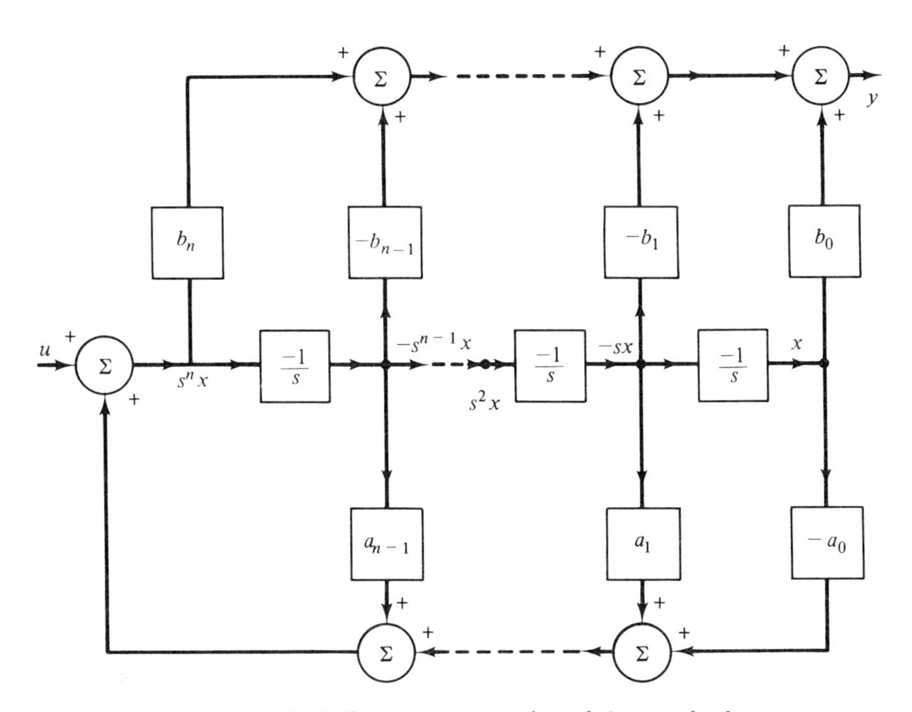

Figure 8.7-6 Block-diagram representation of the transfer function

$$\frac{y}{u} = (b_n s^n + b_{n-1}s^{n-1} + \cdots + b_1 s + b_0)/(s^n + a_{n-1}s^{n-1} + \cdots + a_1 s + a_0).$$

Using the integrators and adders of Fig. 8.7-3, a network realization of the transfer function under consideration can be obtained. In practice, however, due to sensitivity considerations, cost, and ease of tuning, only transfer functions with $n \leq 2$ (biquadratic or *biquad* transfer functions) are constructed with the procedure described. Transfer functions with $n > 2$ are usually written as the product of several biquad transfer functions so that each biquad can be realized as a separate unit; overall transfer function is then realized as the tandem connection of these units [17]. Let us work out an example to illustrate this procedure.

Example 8.7-2 (Synthesis of a Low-Pass Butterworth Filter). Consider a low-pass fifth-order Butterworth filter whose transfer function is given as

$$H(s) = \frac{y}{u} = \frac{1}{s^5 + 3.236s^4 + 5.236s^3 + 5.236s^2 + 3.236s + 1} \tag{8.7-13}$$

The denominator polynomial has two pairs of complex roots and one real root. Using a table on the Butterworth polynomials [11], this transfer function can be written as

$$H(s) = \frac{1}{s^2 + 0.618s + 1} \cdot \frac{1}{s + 1} \cdot \frac{1}{s^2 + 1.618s + 1} \tag{8.7-14}$$

Thus, H can be written as the product of three transfer functions; that is,

$$H(s) = H_1(s) \cdot H_2(s) \cdot H_3(s) \tag{8.7-15}$$

where

$$H_1(s) = \frac{1}{s^2 + 0.618s + 1}, \quad H_2(s) = \frac{1}{s + 1},$$

and

$$H_3(s) = \frac{1}{s^2 + 1.618s + 1}$$

The transfer functions $H_1(s)$ and $H_3(s)$ can be synthesized as biquad sections similar to that shown in Fig. 8.7-5. More precisely, comparing the numerators of $H_1(s)$ and $H_3(s)$ with that of (8.7-5), it becomes clear that to obtain the realization of $H_1(s)$ and $H_3(s)$ it is sufficient to remove the op-amps A_4 and A_5, and choose a_0, b_0, and a_1 appropriately. Similarly, $H_2(s)$ can be realized from Fig. 8.7-3(a) by letting $\alpha = 1$. The cascade connection of these sections results in the network of Fig. 8.7-7, which has the desired transfer function. Note that due to sign inversion the polarity of the output is reversed. Furthermore, since ideal op-amps are assumed to have zero output impedance and infinite input impedance, the loading effect in cascade connection of H_1, H_2, and H_3 is negligible.

It should be mentioned at this juncture that the biquad realization of active filters has a basic drawback—it requires too many operational amplifiers. In some cases this may not be a desirable characteristic of such filters. As we shall soon see, there are special design configurations that require much fewer op-amps than biquad realization. However, since the price of op-amps is decreasing as new methods of production develop, this might not be a serious drawback after all.

Let us now introduce some special configurations used in the design of active filters.

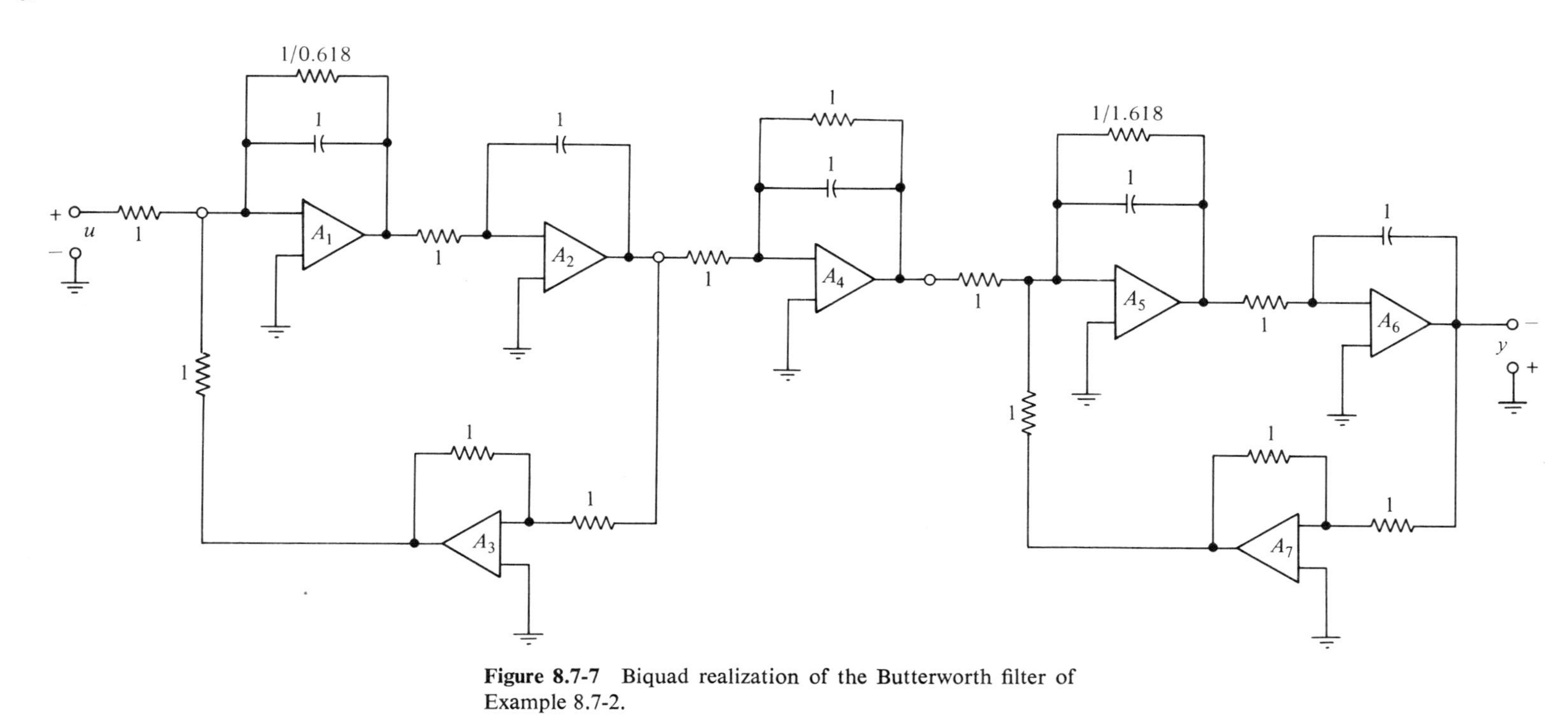

Figure 8.7-7 Biquad realization of the Butterworth filter of Example 8.7-2.

Special Configurations. Aside from the general procedure outlined, there are several special cofigurations for synthesizing a given filter. The main feature of these configurations is that fewer op-amps are needed for the synthesis of the corresponding filter. In what follows we introduce some of the most popular schemes that have been developed during the last two decades. In each case the topology and the element types are specified; the element values can then be chosen by the design engineer to match the coefficients of the desired filter. As in the previous case, only second-order building blocks are presented here. The more complex filters can be synthesized by cascade connection of these units.

Consider the general configuration shown in Fig. 8.7-8(a), where y_1, y_2,

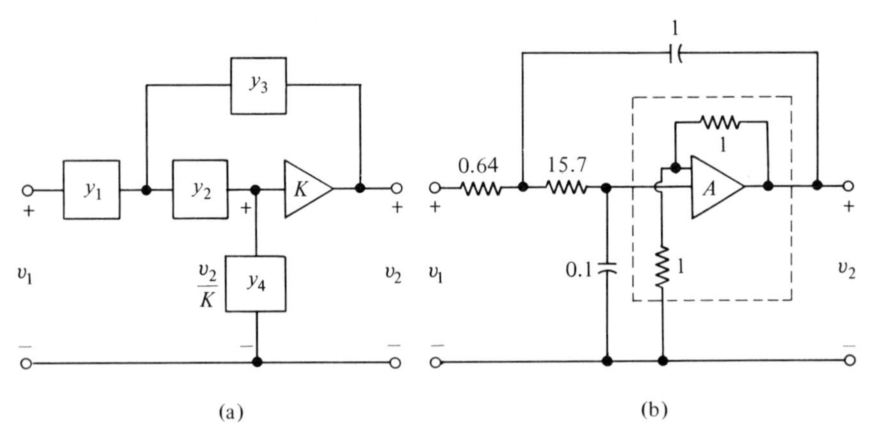

Figure 8.7-8 (a) General second-order filter; (b) realization of a low-pass second-order filter with transfer function $y/u = 2/(s^2 + s + 1)$.

y_3, and y_4 denote the admittance of single resistor or capacitor elements, and K denotes the gain of a *noninverting* op-amp (see problem 8.25). Such a non-inverting op-amp is shown in Fig. 8.7-8(b) by the broken line. Using a simple nodal analysis, the transfer function of this filter can be seen to be (due to high input impedance the current through the noninverting op-amp is assumed to be zero)

$$H(s) = \frac{v_2(s)}{v_1(s)} = \frac{Ky_1y_2}{y_3y_4 + (1 - K)y_2y_3 + y_2y_4 + y_1y_2 + y_1y_4} \tag{8.7-16}$$

By appropriately choosing the y_i various types of filters known as Sallen and Key [16] filters can be realized. Some of these are discussed next.

Low Pass. Let us choose the admittances in Fig. 8.7-8(a) as follows:

$$y_1 = G_1, \quad y_2 = G_2 \quad y_3 = sC_1 \quad \text{and} \quad y_4 = sC_2$$

Equation (8.7-16) can then be written as

$$H(s) = \frac{y(s)}{u(s)} = \frac{K\dfrac{G_1G_2}{C_1C_2}}{s^2 + \left[(1-K)\dfrac{G_2}{C_2} + \dfrac{G_1}{C_1} + \dfrac{G_2}{C_1}\right]s + \dfrac{G_1G_2}{C_1C_2}} \tag{8.7-17}$$

which, clearly, represents a low-pass RC filter. The element values C_1, C_2, G_1, G_2, and K are chosen by comparing the numerator and the denominator of (8.7-17) with that of the prescribed transfer function. This procedure yields fewer equations than unknowns. This is to the advantage of the design engineer, since he can arbitrarily choose some of the element values to satisfy other constraints, such as the size, cost, sensitivity, and availability of components. Let us now work out an example to illustrate this procedure.

Example 8.7-3. Use the topology of the network of Fig. 8.7-8(a) to synthesize a second-order low-pass filter whose voltage transfer function is given by

$$H(s) = \frac{2}{s^2 + s + 1} \tag{8.7-18}$$

Solution

Comparing $H(s)$ in (8.7-18) with (8.7-17), we obtain the following relations:

$$\frac{G_1G_2}{C_1C_2} = 1, \qquad K = 2, \qquad -\frac{G_2}{C_2} + \frac{G_1}{C_1} + \frac{G_2}{C_1} = 1$$

Let us choose $C_1 = 1$ F and $C_2 = 0.1$ F; the equations then yield

$$G_1G_2 = 0.1 \quad \text{and} \quad G_1 - 9G_2 = 1$$

which solved simultaneously result in

$$R_2 = \frac{1}{G_2} \cong 15.7\ \Omega \quad \text{and} \quad R_1 = \frac{1}{G_1} \cong 0.64\ \Omega$$

The resulting filter is shown in Fig. 8.7-8(b).

High Pass. To obtain a high-pass second-order filter from the general configuration shown in Fig. 8.7-8, it is sufficient to choose

$$y_1 = sC_1, \quad Y_2 = sC_2, \quad y_3 = G_3, \quad \text{and} \quad y_4 = G_4$$

The corresponding transfer function becomes

$$H(s) = \frac{Ks^2}{s^2 + \left[\dfrac{G_4}{C_1} + (1-K)\dfrac{G_3}{C_1}\right]s + \dfrac{G_3G_4}{C_1C_2}} \tag{8.7-19}$$

Example 8.7-4. Use the general configuration of Fig. 8.7-8(a) to design an active RC second-order Butterworth filter whose transfer function is given by

$$H(s) = \frac{2s^2}{s^2 + \sqrt{2}\,s + 1} \tag{8.7-20}$$

Solution

Comparison of (8.7-19) and (8.7-20) yields

$$K = 2, \qquad G_3G_4 = C_1C_2, \qquad G_4 - G_3 = \sqrt{2}\,C_1$$

Choosing $C_1 = 1$ and $C_2 = 0.1$ these equations become

$$G_3G_4 = 0.1 \quad \text{and} \quad G_4 - G_3 = \sqrt{2}$$

The simultaneous solution of these equations is

$$R_3 = \frac{1}{G_3} = 14.7\,\Omega \quad \text{and} \quad R_4 = \frac{1}{G_4} = 0.68\,\Omega$$

The resulting filter is shown in Fig. 8.7-9.

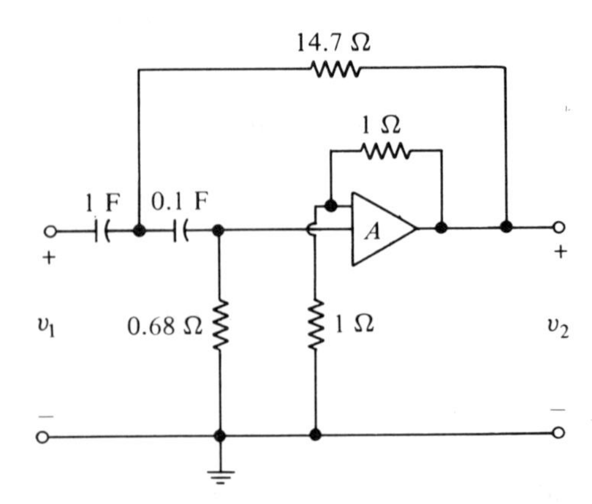

Figure 8.7-9 Active *RC* realization of a second-order high-pass Butterworth filter.

Band Pass. There are several configurations for realization of second-order band-pass building blocks. One of the most popular ones in shown in Fig. 8.7-10(a). As in the previous case, the transfer function of this network can

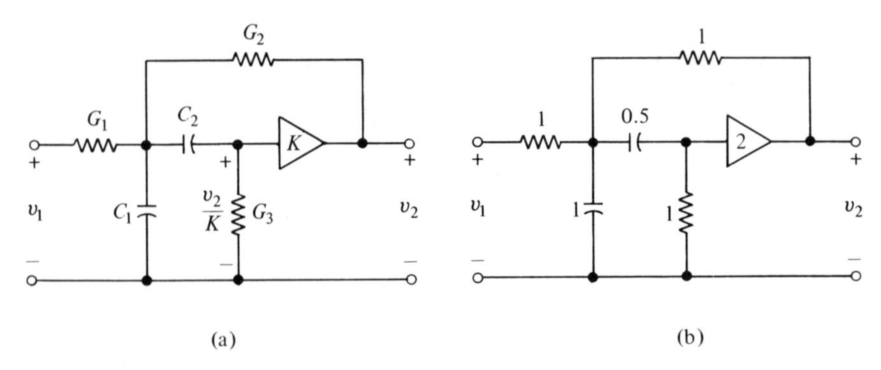

Figure 8.7-10 (a) General configuration for a second-order band-pass filter; (b) active *RC* realization of the band-pass filter whose voltage transfer function is $v_2/v_1 = 2s/(s^2 + 3s + 4)$.

be written as

$$H(s) = \frac{v_2(s)}{v_1(s)} = \frac{K\dfrac{G_1}{C_1}s}{s^2 + \left[\dfrac{G_1}{C_1} + \dfrac{G_3}{C_1} + \dfrac{G_3}{C_2} + \dfrac{1-K}{C_1}G_2\right]s + \dfrac{G_3(G_1+G_2)}{C_1C_2}} \tag{8.7-21}$$

By appropriately choosing the element values, a prescribed filter may be synthesized. One such case is given in the following example.

Example 8.7-5. Obtain an active *RC* realization of the band-pass voltage transfer function

$$H(s) = \frac{v_2(s)}{v_1(s)} = \frac{2s}{s^2 + 3s + 4}$$

Solution

Comparing the transfer function under study with (8.7-21), we obtain

$$K\frac{G_1}{C_1} = 2, \quad \frac{G_1 + G_2 + (1-K)G_2}{C_1} + \frac{G_3}{C_2} = 3, \quad \text{and} \quad \frac{G_3(G_1+G_2)}{C_1C_2} = 4$$

Let us choose $K = 2$, $C_1 = 1$, and $G_1 = G_2 = 1$. Then the preceding relations yield

$$G_3 = 1 \quad \text{and} \quad C_2 = 0.5$$

The resulting filter is shown in Fig. 8.7-10(b).

Filters with Finite $j\omega$ Axis Zeros. None of the special configuration filters discussed so far can provide finite zeros on the $j\omega$ axis. To obtain such zeros, the networks of Fig. 8.7-11 were introduced by Kerwin and Huelsman ([10], Chapter 1).

The transfer function of the filter of Fig. 8.7-11(a) can be written as

$$H(s) = \frac{K(s^2 + \alpha^2)}{s^2 + \alpha(k+1)[1/R + (2-K)/k]s + \alpha^2[1 + (k+1)/R]} \tag{8.7-22}$$

And the transfer function of the filter of Fig. 8.7-11(b) is

$$H(s) = \frac{\dfrac{K}{(k+1)C+1}(s^2 + \alpha^2)}{s^2 + \alpha\dfrac{[C + (2-K)/k](k+1)}{(k+1)C+1}s + \dfrac{\alpha^2}{(k+1)C+1}} \tag{8.7-23}$$

where the scalar constants α, k, C, K, and R are chosen so that the prescribed characteristics of the filter are met. The transfer functions given are both in the form of

$$H(s) = H_0\frac{s^2 + \alpha^2}{s^2 + 2\xi\omega_0 s + \omega_0^2}$$

with the major difference that (8.7-22) can only realize transfer functions for which $\alpha \leq \omega_0$, whereas (8.7-23) can realize only those for which $\alpha \geq \omega_0$.

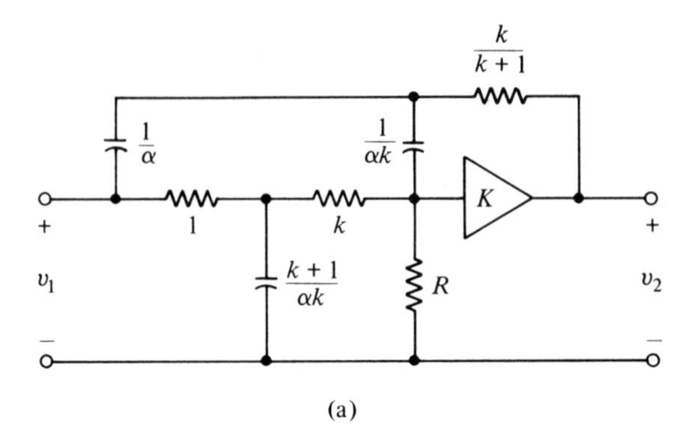

(a)

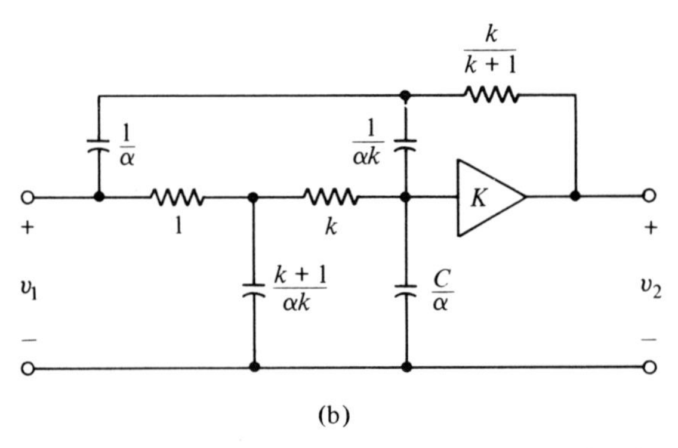

(b)

Figure 8.7-11 Two different realizations of second-order *RC* active filters with $j\omega$ axis zeros.

Exercise 8.7-1. Show that (8.7-22) and (8.7-23) are indeed the transfer functions of the *RC* active filters of Figs. 8.7-11(a) and (b), respectively.

8.8 SENSITIVITY CONSIDERATIONS IN NETWORK DESIGN

In the synthesis procedures discussed so far we have assumed that the network elements used are ideal in the sense that their value will remain constant under all operating conditions. In reality, however, the element values change due to a variety of causes, such as temperature, humidity, pressure, aging, and fluctuation in the bias sources. For example, the operating characteristics of a transistor or an op-amp are usually specified under room temperature and normal humidity. Consequently, a circuit constructed with such components may be *sensitive* to temperature and humidity; that is, its behavior will change with these parameters. Likewise, passive elements used in actual circuits usually have a tolerance of 5 to 10 per cent of their specified value. Such fluctuations in element values will affect the network function and consequently its response.

A basic criterion in network design is then to choose the topology and the element type so that the resulting network is least sensitive to element tolerance and fluctuations. Sensitivity analysis is extensively discussed in various books and articles (see, for example, [11], Appendix C). In this section we merely define sensitivity and discuss some of the most often used sensitivity functions.

Generally speaking, the sensitivity of a network function with respect to a particular parameter is the ratio of the percentage of change in the network function to the percentage of change of that parameter. More specifically, let $H(s, p)$ represent the network function. This may be the transfer function, driving-point function, or the like. The parameter p may represent the value of a particular element or an independent variable, such as the temperature. The following definition was originally introduced by Bode [18].

Definition 8.8-1 (Sensitivity). The *sensitivity function* S_p^H of $H(s, p)$ with respect to the parameter p is defined as

$$S_p^H = \frac{dH(s, p)/H(s, p)}{dp/p} = \frac{p}{H(s, p)} \cdot \frac{dH(s, p)}{dp} \tag{8.8-1}$$

where $dH(s, p)$ and dp represent the incremental changes in $H(s, p)$ and p, respectively. An alternative form of (8.8-1) is

$$S_p^H = \frac{d[\log H(s, p)]}{d[\log p]} \tag{8.8-2}$$

Let us now illustrate the application of this definition by an example.

Example 8.8-1. Consider the band-pass RC active filter of Fig. 8.7-11(a). The transfer function of this filter is given in equation (8.7-22). Let us compute the sensitivity of the dc gain of this filter with respect to the resistor R, assuming that $K = 2$ and $k = 1$.

Solution

The dc gain of this filter can be obtained by putting $s = 0$ in (8.7-22). Using the values of K and k given, we get

$$H(0, R) = \frac{2R}{2 + R}$$

The sensitivity of $H(0, R)$ with respect to R can then be computed using (8.8-1); the result is

$$S_R^{H(0, R)} = \frac{R(2 + R)}{2R} \cdot \frac{d}{dR}\left(\frac{2R}{2 + R}\right) = \frac{2}{2 + R}$$

Thus, to obtain minimum sensitivity, R must be chosen as large as other constraints on the network permit.

In the actual design of filters, usually the sensitivity of particular network functions such as the quality factor Q, damped natural frequency ω_0, damping ratio ξ, gain $|H(j\omega)|$, and phase $\phi(\omega)$ are specified. This is because these functions determine the behavior of the network response. The performance of the network, therefore, depends on the sensitivity of these functions with respect to the network elements. Among various design possibilities, the design engineer then chooses the configuration that is least sensitive. Let us now work out an example to illustrate the derivation of a sensitivity function for a particular second-order low-pass filter; derivation of sensitivity functions for other filters can then be obtained in a similar manner.

Example 8.8-2. Consider the second-order low-pass active RC filter whose transfer function is given in (8.7-17). For this filter we have

$$\omega_0 = \frac{\sqrt{G_1 G_2}}{\sqrt{C_1 C_2}}$$

$$\xi = \frac{1}{2}\left[(1-K)\frac{\sqrt{C_1 G_2}}{\sqrt{G_1 C_2}} + \frac{\sqrt{G_1 C_2}}{\sqrt{C_1 G_2}} + \frac{\sqrt{C_2 G_2}}{\sqrt{C_1 G_1}}\right]$$

Let us now compute the sensitivity of ω_0 and ξ with respect to, say, G_1. Using (8.8-1), we readily get

$$S_{G_1}^{\omega_0} = \frac{1}{2}$$

Similarly,

$$S_{G_1}^{\xi} = \frac{1}{2}\cdot\frac{C_2 + C_2 G_2 - (1-K)C_1 G_2}{G_1 C_2 + C_2 G_2 + (1-K)C_1 G_2}$$

The first equation implies that a 1 percent change in G_1 results in a $\frac{1}{2}$ percent change in the damped natural frequency ω_0.

Effect of Feedback on Sensitivity. One of the most important applications of feedback in network design is the reduction of the sensitivity of the overall network with respect to the fluctuations in the forward path. To give a quantitative account of this effect, let us consider the feedback network whose block-diagram representation is shown in Fig. 8.8-1. The transfer function of the overall network can be computed by observing that

$$y(s) = G(s, p)e(s)$$

and

$$e(s) = u(s) - K(s)y(s)$$

Thus,

$$H(s, p) = \frac{y(s)}{u(s)} = \frac{G(s, p)}{1 + K(s)G(s, p)} \tag{8.8-3}$$

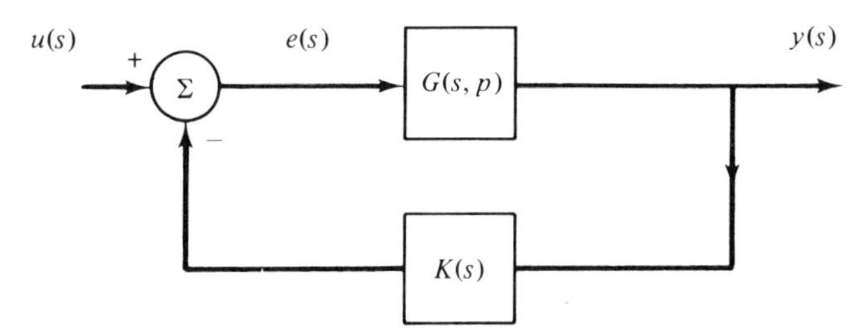

Figure 8.8-1 Block-diagram representation of a feedback network whose overall sensitivity is much smaller than the sensitivity of $G(s, p)$.

where we have denoted the overall transfer function by $H(s, p)$. To compare the sensitivity of $H(s, p)$ to that of $G(s, p)$, let us use (8.8-1) to write

$$S_p^{H(s,p)} = \frac{p[1 + K(s)G(s, p)]}{G(s, p)} \cdot \frac{d}{dp}\left[\frac{G(s, p)}{1 + K(s)G(s, p)}\right] \tag{8.8-4}$$

which simplifies to

$$S_p^{H(s,p)} = \frac{p}{G(s, p)} \cdot \frac{dG(s, p)}{dp} \cdot \frac{1}{1 + K(s)G(s, p)} \tag{8.8-5}$$

But, by Definition (8.8-1), we have

$$S_p^{G(s,p)} = \frac{p}{G(s, p)} \cdot \frac{dG(s, p)}{dp} \tag{8.8-6}$$

Thus, (8.8-5) and (8.8-6) yield

$$S_p^{H(s,p)} = S_p^{G(s,p)} \cdot \frac{1}{1 + K(s)G(s, p)} \tag{8.8-7}$$

In an actual design, usually $|1 + K(s)G(s, p)| \gg 1$; consequently, the sensitivity function $S_p^{H(s,p)}$ is much smaller than $S_p^{G(s,p)}$. This low sensitivity, of course, has been obtained at the price of a loss in the magnitude. This can be seen from an example.

Example 8.8-3. Consider the filter whose block diagram is shown in Fig. 8.8-1. In practice, $G(s, p)$ is usually made up of the cascade connection of an amplifier and a filter. Assume that at a particular frequency ω_0 the forward and feedback gains are

$$G(\omega_0, p) = 100 \quad \text{and} \quad K(\omega_0) = 1$$

From (8.8-7) we have

$$S_p^{H(\omega_0,p)} \cong \frac{1}{100} S_p^{G(\omega_0,p)}$$

That is, the sensitivity has been reduced 100 times. The overall gain, however, is also reduced to

$$H(\omega_0, p) \cong 1$$

Thus, the amplifier in the forward path is mainly used for reducing the sensitivity of the filter rather than amplifying the input.

REFERENCES

[1] BRUNE, O., "Synthesis of a Finite Two-Terminal Network Whose Driving-Point Impedance Is a Prescribed Function of Frequency," *J. Math. Phys.*, **10** (1931), 191–236.

[2] VAN VALKENBURG, M. E., *Introduction to Modern Network Synthesis*, John Wiley & Sons, Inc., New York, 1962.

[3] HAZONY, D., *Elements of Network Synthesis*, Van Nostrand Reinhold Company, New York, 1963.

[4] NEWCOMB, R. W., *Linear Multiport Synthesis*, McGraw-Hill Book Company, New York, 1966.

[5] KARNI, S., *Network Theory: Analysis and Synthesis*, Allyn and Bacon, Inc., Boston, 1966.

[6] CHEN, W. H., *Linear Network Design and Synthesis*, McGraw-Hill Book Company, New York, 1964.

[7] BOTT, R., and R. J. DUFFIN, "Impedance Synthesis Without the Use of Transformers," *J. Appl. Phys.* **20** (1949), 816.

[8] RICHARDS, P. I., "A Special Class of Functions with Positive Real Part in Half-Plane," *Duke Math. J.* **14** (1947), 777–786.

[9] HUMPHREYS, D. S., *The Analysis, Design and Synthesis of Electrical Filters*, John Wiley & Sons, Inc., New York, 1969.

[10] HUELSMAN, L. P. (Ed.), *Active Filters: Lumped, Distributed, Integrated, Digital and Parametric*, McGraw-Hill Book Company, New York, 1970.

[11] TOBEY, G. E., J. G. GRAEME, and L. P. HUELSMAN, *Operational Amplifiers: Design and Applications*, McGraw-Hill Book Company, New York, 1971.

[12] HUELSMAN, L. P., *Theory and Design of Active RC Circuits*, McGraw-Hill Book Company, New York, 1968.

[13] NEWCOMB, R. W., *Active Integrated Network Synthesis*, Prentice-Hall, Inc., Englewood Cliffs, N.J., 1968.

[14] GHAUSI, M. S., *Principles and Design of Linear Active Circuits*, McGraw-Hill Book Company, New York, 1965.

[15] MITRA, S. K., *Analysis and Synthesis of Linear Active Networks*, John Wiley & Sons, Inc., New York, 1969.

[16] SALLEN, R. P., and E. L. KEY, "A Practical Method of Designing *RC* Active Filters," *IRE Trans. Circuit Theory*, **CT-2**, No. 1 (1955), 74–85.

[17] Tow, J., "A Step by Step Active-Filter Design," *IEEE Spectrum*, 6 (Dec. 1969), 64–68.

[18] Bode, H., *Network Analysis and Feedback Amplifier Design*, Van Nostrand Reinhold Company, New York, 1950.

PROBLEMS

8.1 Synthesize the following driving-point admittance functions by linear time-invariant passive *RLC* elements.

(a) $y(s) = \dfrac{s+2}{s^2+3s+3}$

(b) $y(s) = \dfrac{3s^2+11s+6}{s+3}$

8.2 Determine whether the following rational functions are positive real:

(a) $z(s) = s^2 + 2s + 1$ (b) $z(s) = s + \dfrac{1}{s} + 1$

(c) $y(s) = \dfrac{s^2+s+2}{s^3+s+2}$ (d) $y(s) = \dfrac{3s^2+2s+3}{2s^2+s+1}$

8.3 Show that the following matrices are positive real:

(a) $\mathbf{Z}(s) = \begin{bmatrix} 2+s & \dfrac{s}{2} \\ \dfrac{s}{2} & 1+2s \end{bmatrix}$

(b) $\mathbf{Z}(s) = \begin{bmatrix} \dfrac{3s^2+s+1}{s} & s+1 \\ s+1 & \dfrac{2s^2+s+2}{s} \end{bmatrix}$

(c) $\mathbf{Z}(s) = \begin{bmatrix} \dfrac{s(2s+1)}{s^2+2s+1} & \dfrac{s}{s^2+2s+1} \\ \dfrac{s}{s^2+2s+1} & \dfrac{s+2}{s^2+2s+1} \end{bmatrix}$

8.4 Show that the poles and zeros of a reactance function alternate on the $j\omega$ axis. [Hint: If $z(s)$ is a reactance function, note that $(d/d\omega)z(j\omega) > 0$ for all ω.]

8.5 Obtain the first Foster realization of the reactance function

$$z(s) = \frac{2s^4+5s^2+2}{s^3+s}$$

8.6 Obtain the second Foster realization of the driving-point impedance function

$$z(s) = \frac{s^5+3s^3+2s}{2s^6+11s^4+16s^2+6}$$

8.7 Determine the first Cauer realization of the reactance function considered in Problems 8.5 and 8.6.

8.8 Determine the second Cauer realization of the reactance functions considered in Problems 8.5 and 8.6.

8.9 Obtain the first Foster realization of the driving-point impedance function

$$z(s) = \frac{10s^2 + 10s + 2}{3s^2 + 4s + 1}$$

8.10 Synthesize the positive real function given in Problem 8.9 as a driving-point admittance function.

8.11 Synthesize the following positive real functions:

(a) $z(s) = \dfrac{3s^2 + 8s + 2}{3s + 1}$

(b) $z(s) = \dfrac{4s^2 + 25s + 38}{2s^2 + 11s + 15}$

8.12 Determine the positive realness of the following driving-point functions and synthesize them by linear time-invariant passive *RLC* elements.

(a) $y(s) = \dfrac{s + 3}{2s^2 + 7s + 4}$ (b) $z(s) = \dfrac{3s + 4}{s^2 + 3s + 2}$

(c) $y(s) = \dfrac{s^2 + 1}{2s^3 + 2s + 1}$ (d) $z(s) = \dfrac{3s^2 + 12s + 11}{s^3 + 6s^2 + 11s + 6}$

8.13 Show that the removal of a pair of $j\omega$ axis poles from a positive real function will leave the remaining function positive real.

8.14 Synthesize the following positive real function by first removing a constant:

$$z(s) = \frac{s^2 + 4s + 3}{s^2 + 6s + 8}$$

8.15 Repeat Problem 8.14 for the function

$$y(s) = \frac{s^2 + s + (3/2)}{s^2 + s + 1}$$

8.16 Give a Bott–Duffin realization of the driving-point impedance functions given in Problems 8.14 and 8.15.

8.17 Obtain the topology and element values of an active *RC* second-order low-pass filter with a damped natural frequency of 1 radian per second, a damping factor of 0.707, and a dc gain of 10.

8.18 Obtain the block-diagram representation and an active *RC* realization of the filters whose transfer functions are

(a) Chebyshev filter:

$$H(s) = \frac{1}{(s + 0.348)(s + 0.697 + j0.868)(s + 0.697 - j0.868)}$$

(b) Gaussian filter:

$$H(s) = \frac{1}{(s + 1.374)(s + 1.511 + j0.94)(s + 1.511 - j0.94)}$$

(c) Butterworth filter:

$$H(s) = \frac{1}{(s + 1)(s + 0.5 + j0.866)(s + 0.5 - j0.866)}$$

8.19 Determine the block-diagram representation and op-amp realization of the following low-pass filters:

(a) $H(s) = \dfrac{1}{s^3 + 2s^2 + 2s + 1}$

(b) $H(s) = \dfrac{1}{s^4 + 2.61s^3 + 3.61s^2 + 2.61s + 1}$

8.20 Draw the block-diagram representation of the second-order filter

$$H(s) = \frac{a_2 s^2 + a_1 s + a_0}{b_2 s^2 + b_1 s + b_0}$$

8.21 Obtain the biquad active *RC* realization of the filters whose transfer functions are given in Problem 8.19.

8.22 Consider the biquad realization of the fifth-order filter discussed in Example 8.7-2. Show that $H_2(s)$ given in (8.7-15) can be synthesized using a simple *RC* circuit. Assume that the op-amps used in the realizations of $H_1(s)$ and $H_3(s)$ are ideal. Obtain the values of R and C.

8.23 Use the topology of the second-order filter given in Fig. 8.7-8(a) to realize the second-order low-pass Butterworth filter whose transfer function is

$$H(s) = \frac{1}{s^2 + \sqrt{2}\,s + 1}$$

8.24 Use the configurations of Fig. 8.7-11 to realize the following transfer functions:

(a) $H(s) = \dfrac{s^2 + 1}{s^2 + 2s + 2}$

(b) $H(s) = \dfrac{s^2 + 4}{s^2 + \sqrt{2}\,s + 1}$

8.25 Consider the noninverting op-amp shown in Fig. P8.25.

(a) Show that the input impedance of this op-amp is infinite.

(b) Show that

$$\frac{v_2}{v_1} = \frac{Z_1 + Z_2}{Z_1}$$

[Hint: The current in and out of the open loop op-amp A is assumed to be zero.]

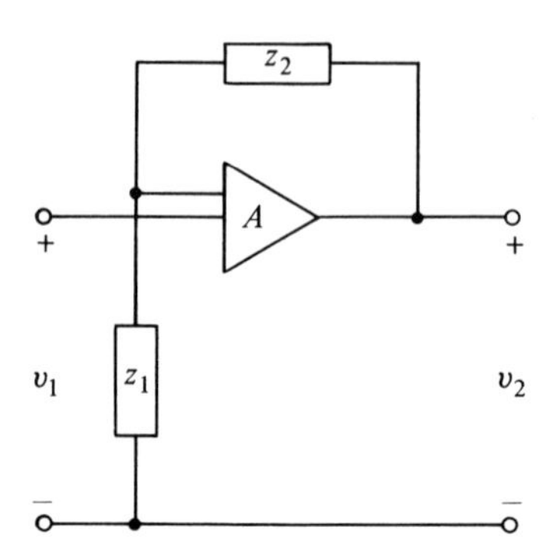

Figure P8.25

8.26 The voltage transfer function of a second-order low-pass filter is

$$H(s) = \frac{-1/R_1R_2C}{s^2 + s(1/R_1 + 1/R_2 + 1/R_3)/C + 1/R_2R_3C}$$

Show that

$$S_{R_2}^{\omega_0} = S_{R_3}^{\omega_0} = S_C^{\omega_0} = -\frac{1}{2}$$

$$S_{R_1}^{\xi} = \frac{1}{-2\xi\omega_0R_1C}$$

$$S_{R_2}^{\xi} = \frac{1}{2} - \frac{1}{\xi\omega_0R_2C}$$

Computer-Aided Network Design

9

9.1 INTRODUCTION

In the previous chapter we discussed the analytical synthesis of passive and active networks. We showed how a given network function can be exactly and systematically synthesized if certain conditions were satisfied. In many practical situations, however, the network function is not available analytically; rather it is given as a set of data obtained experimentally or as a result of the simulation of another network on a digital computer. Furthermore, analytical synthesis procedures discussed earlier do not take into account constraints imposed by element value, noise level, cost, sensitivity, and the like. These drawbacks together with the lack of a general analytical method for the design of time-varying and nonlinear networks have necessitated a new approach to the problem of network synthesis. In this chapter we discuss an iterative method that can be successfully implemented to design linear time-invariant as well as nonlinear and time-varying networks. This method uses a digital computer to carry out the approximation and/or the optimization processes involved. Since the introduction of digital computers in network design, computer-aided design techniques have emerged as an indispensable tool in designing large-scale and nonlinear time-varying networks [1–10].

The basic advantage of computer-aided network design over conventional network synthesis is that various constraints imposed on the final network can, in most cases, be satisfied simultaneously. In many practical cases, the specific constraints might contradict one another; under these circumstances, usually a compromise is reached among various constraints and the *optimum* network is obtained. Virtually all the computer-aided design techniques are based on iterative methods. In such methods, usually, a topology for the desired network is chosen and tentative values are assigned to the elements. Then in each iteration these values are altered in such a way that a preassigned error function is reduced. This iterative process is continued until the error function is minimized. A major disadvantage of computer-aided design is that generally there is no guarantee that the iterations converge to the optimal network. As we shall see later in this chapter, this convergence depends on the convexity (or unimodality) of the error function (see Problems 9.1 and 9.2).

A detailed discussion of iterative methods is given in the next section, and various optimization techniques currently used in network design are considered in Sections 9.3 and 9.4. Section 9.5 is devoted to the application of the adjoint computer-aided design. The remainder of the chapter is then devoted to a discussion on the design of linear time-varying networks.

9.2 ITERATIVE METHODS IN COMPUTER-AIDED DESIGN

In this section we discuss the general iterative methods used in almost all computer-aided design techniques. To introduce the basic idea underlying these methods, we consider a particular design problem, the design of linear time-invariant networks operated under steady-state sinusoidal excitation. The extension of this method to designing linear time-varying networks is postponed until Section 9.6.

Frequency-Domain Synthesis. Let us consider the design of a network whose transfer function $H(j\omega)$ is to approximate a prescribed transfer function $\hat{H}(j\omega)$ over the frequency range of interest $[\omega_0, \omega_1]$. Suppose that the topology and element type of the desired network are specified. From a design point of view this information may be gathered by considering various constraints, such as the cost, size, reliability, sensitivity, and availability of the elements. A design engineer can generally choose a reasonable topology and element type for the desired network, although in recent automated design techniques this requirement is removed. All one has to specify are the port characteristics of the network; completely automated programs choose the optimal topology and "grow" elements on the network [3]. In this chapter, however, we assume that the structure and the element types are given. The design problem can then be stated as follows: Obtain the element values of the network so that its transfer function is *as close as possible* to the prescribed transfer function $\hat{H}(j\omega)$ over the frequency range of interest $[\omega_0, \omega_1]$.

Error Criteria. To give a precise meaning to the phrase "as close as possible," the design engineer may choose one of the several error criteria commonly used. The most popular one is the *mean squared error function* defined by

$$E = \frac{1}{2}\int_{\omega_0}^{\omega_1} |H(j\omega) - \hat{H}(j\omega)|^2 \, d\omega \tag{9.2-1}$$

or, more generally, the *weighted mean squared error function* defined by

$$E = \frac{1}{2}\int_{\omega_0}^{\omega_1} W(\omega)\,|H(j\omega) - \hat{H}(j\omega)|^2 \, d\omega \tag{9.2-2}$$

where $|\cdot|$ represents the absolute value of a complex function and $W(\omega)$ is a real nonnegative weighting function. In what follows, however, we use the simpler error function defined in (9.2-1) with the understanding that the results can easily be extended to the case where E is chosen as in (9.2-2). Quite clearly, $H(j\omega)$ depends on the value of the network elements; hence, a detailed description of the arguments in $H(j\omega)$ is given by

$$H(j\omega) = H(R_1, R_2, \ldots, R_m; C_{m+1}, C_{m+2}, \ldots, C_n; L_{n+1}, L_{n+2}, \ldots, L_q; j\omega) \tag{9.2-3}$$

where we have assumed that the network under consideration is to be realized by resistors, capacitors, and inductors only. Furthermore, we have numbered the resistors from 1 to m, capacitors from $m + 1$ to n, and so on.

The problem is then to obtain the value of all the resistors, capacitors, and inductors such that E is minimized. To simplify the notations, let

$$\mathbf{p} \triangleq [R_1, R_2, \ldots, R_m, C_{m+1}, C_{m+2}, \ldots, C_n, L_{n+1}, L_{n+2}, \ldots, L_q]^T \tag{9.2-4}$$

Then (9.2-1) can be rewritten as

$$E(\mathbf{p}) = \frac{1}{2}\int_{\omega_0}^{\omega_1} |H(\mathbf{p}, j\omega) - \hat{H}(j\omega)|^2 \, d\omega \tag{9.2-5}$$

The problem is then to obtain the elements of the parameter vector $\mathbf{p}$ such that $E(\mathbf{p})$ is minimized. Before going through the mechanics of the iterative solution of this problem, let us digress for a while and reformulate the error function in terms of the response of the network.

If the desired network is a one-port network, usually either its input impedance or admittance is specified. Let us assume that such a network is excited by the current source phasor $I(j\omega)$. The design problem would then be to obtain the parameter vector $\mathbf{p}$ such that its port voltage phasor $V(\mathbf{p}, j\omega)$ is as close as possible to a prescribed voltage phasor $\hat{V}(j\omega)$. If the mean squared error function is adopted, the error function will be

$$E(\mathbf{p}) = \frac{1}{2}\int_{\omega_0}^{\omega_1} |V(\mathbf{p}, j\omega) - \hat{V}(j\omega)|^2 \, d\omega \tag{9.2-6}$$

Remark 9.2-1. It should be noted that the error functions discussed are just the squared L_2 norm of the difference between the desired and the actual network function. Henceforth, a general error function can be defined in terms of the L_r norm of the error; more specifically, we can have

$$E(\mathbf{p}) = \frac{1}{r} \int_{\omega_0}^{\omega_1} |V(\mathbf{p}, j\omega) - \hat{V}(j\omega)|^r \, d\omega \tag{9.2-7}$$

where r is an integer. For $r \longrightarrow \infty$ the error function will become the L_∞ norm of the difference. In most engineering applications, however, the mean squared error functions defined in (9.2-5) and (9.2-6) are quite satisfactory.

Remark 9.2-2. In actual computation of the error function $E(\mathbf{p})$, the integral is approximated by a finite summation of weighted points using the trapezoidal rule, Newton–Cotes formula, or the like. We can then write

$$E(\mathbf{p}) = \tfrac{1}{2} \sum_{k=1}^{n} W_k | V(\mathbf{p}, j\omega_k) - \hat{V}(j\omega_k)|^2 \tag{9.2-8}$$

where W_k is a weighting factor and n represents the number of sample points chosen in the interval $[\omega_0, \omega_1]$. The number n depends on the function $V(\mathbf{p}, j\omega)$ and $\hat{V}(j\omega)$; if these functions are rapidly varying with ω, then the number of the sample points, n, must be chosen to be large. If, on the other hand, these functions are relatively smooth, n is chosen to be small. In the latter case choosing n large will increase the computation time and the round-off errors.

Let us now proceed with a description of the iterative design of networks and present a step-by-step algorithm for finding the parameter vector $\mathbf{p}$ such that the error function is minimized. In what follows we assume a general error function $E(\mathbf{p})$. The special cases of this general function are those defined in (9.2-5) through (9.2-8).

Iteration Algorithm. The following steps are involved in a typical iteration process.

Step 1. Make a reasonable guess on the element values; that is, choose some initial values for R_k, C_k, and L_k. This initial guess need not be close to the optimal values. However, if through experience or practical considerations the design engineer is able to make a reasonable guess, the number of iterations (and hence the computation time) will be reduced. With this choice of element values, the analysis techniques of Chapter 4 and the numerical methods of Chapter 7 can be used to compute $H(\mathbf{p}, j\omega)$ or $V(\mathbf{p}, j\omega)$, depending on whether the error function of equation (9.2-1) or that of equation (9.2-8) is chosen.

Step 2. Use the element values of step 1 and an analysis program to compute the error function $E(\mathbf{p})$. If $E(\mathbf{p}) \leq \epsilon$, where ϵ is a small preassigned positive number, the problem is solved; the element values chosen in step 1 are satisfactory. This situation, however, is not likely to happen for a relatively complicated network. Usually we get $E(\mathbf{p}) \gg \epsilon$. In this case we must proceed to step 3.

Step 3. Introduce a small perturbation on the element values. That is, replace R_k by $R_k + \delta R_k$, C_k by $C_k + \delta C_k$, and L_k by $L_k + \delta L_k$, where δR_k, δC_k, and δL_k represent the small change on the element values R_k, C_k, and L_k, respectively. This means that the parameter vector $\mathbf{p}$ is replaced by $\mathbf{p} + \delta\mathbf{p}$. Using these adjusted element values, the corresponding transfer function $H(\mathbf{p} + \delta\mathbf{p}, j\omega)$ and the error function $E(\mathbf{p} + \delta\mathbf{p})$ can be computed.

We then choose $\delta\mathbf{p}$ so that the following two conditions are satisfied:

1. $E(\mathbf{p} + \delta\mathbf{p}) < E(\mathbf{p})$.
2. $|E(\mathbf{p} + \delta\mathbf{p}) - E(\mathbf{p})|$ is *maximized.*

The first condition guarantees that the adjustment of the element values is in the right direction in the sense that the new network is one step closer to the desired one. The second condition assures that $\delta\mathbf{p}$ is chosen in an optimal fashion in the sense that the difference between the original error function and the adjusted one is as large as possible. In Section 9.3 we discuss an optimization technique by which $\delta\mathbf{p}$ can be chosen such that conditions 1 and 2 are satisfied simultaneously.

Step. 4. Use the new element values, $\mathbf{p} + \delta\mathbf{p}$, to compute $H(\mathbf{p} + \delta\mathbf{p}, j\omega)$ and $E(\mathbf{p} + \delta\mathbf{p})$. If $E(\mathbf{p} + \delta\mathbf{p}) \leq \epsilon$, stop the iterations; $\mathbf{p} + \delta\mathbf{p}$ is the optimal parameter vector. If $E(\mathbf{p} + \delta\mathbf{p}) > \epsilon$, use $\mathbf{p} + \delta\mathbf{p}$ as the new element values, go back to step 3, and continue the process until one of the following two conditions is satisfied:

1. $E \leq \epsilon$, or
2. $|E_{i+1} - E_i| \leq \eta$,

where η is a preassigned positive number and E_i denotes the error function at the ith iteration. The first condition implies that the error function has reached the desired limit and the iterations can be terminated. The second condition means that the error function has reached a minimum point (or a saddle point) and further iterations will not reduce E appreciably. This minimum, however, might be a local minimum rather than the global minimum. In the case where E has reached a local minimum, a new initial guess might be chosen, and the iterative process may be repeated until the global minimum is found and the optimum parameter values are obtained.

As was mentioned earlier, the lack of certainty in reaching the global minimum is the main drawback of such iterative procedures. In most practical problems, however, the advantages of this technique heavily outweigh its disadvantages.

From the outlined procedure it is now clear that the heart of the proposed iterative technique is to compute the optimal $\delta\mathbf{p}$. This is discussed in the next section.

9.3 MEAN SQUARED ERROR OPTIMIZATION

In this section we use a perturbation technique to compute the optimum change $\delta\mathbf{p}$ on the parameter vector $\mathbf{p}$ that satisfies conditions 1 and 2 of step 3 of the previous algorithm.

Perturbation Analysis. To introduce the basic idea behind the perturbation analysis, let us first consider the mean squared error function of (9.2-1). The generalization to any error criterion will then follow.

Changing the parameter vector $\mathbf{p}$ by the "small" vector $\delta\mathbf{p}$ will cause a change δH on the transfer function $H(\mathbf{p}, j\omega)$, which in turn will result in a perturbation δE on the error function $E(\mathbf{p})$.

Using (9.2-1), the change on $E(\mathbf{p})$ is

$$E(\mathbf{p}+\delta\mathbf{p}) - E(\mathbf{p}) = \frac{1}{2}\int_{\omega_0}^{\omega_1} [|H(\mathbf{p}, j\omega) + \delta H(\mathbf{p}, j\omega) - \hat{H}(j\omega)|^2 - |H(\mathbf{p}, j\omega) - \hat{H}(j\omega)|^2]\,d\omega \tag{9.3-1}$$

which can be written as

$$E(\mathbf{p}+\delta\mathbf{p}) - E(\mathbf{p}) = \frac{1}{2}\int_{\omega_0}^{\omega_1} [(H + \delta H - \hat{H})(\bar{H} + \delta\bar{H} - \bar{\hat{H}}) - (H - \hat{H})(\bar{H} - \bar{\hat{H}})]\,d\omega \tag{9.3-2}$$

where $\bar{H}$, $\delta\bar{H}$, and $\bar{\hat{H}}$ denote the complex conjugates of H, δH, and $\hat{H}$, respectively. (Since it is understood that the arguments of the functions are $\mathbf{p}$ and $j\omega$, we have omitted these arguments to save space in the following manipulations.)

Simplifying (9.3-2) yields

$$E(\mathbf{p}+\delta\mathbf{p}) - E(\mathbf{p}) = \frac{1}{2}\int_{\omega_0}^{\omega_1} [(H - \hat{H})\delta\bar{H} + (\bar{H} - \bar{\hat{H}})\,\delta H + |\delta H|^2]\,d\omega \tag{9.3-3}$$

Note that since by assumption $\delta\mathbf{p}$ is small and $H(\mathbf{p}, j\omega)$ is considered to be a smooth continuous function of $\mathbf{p}$ then δH is small. Consequently, $|\delta H|^2$ is a second-order term in comparison with the other terms of the integrand,

and hence can be neglected in this equation. We then have

$$E(\mathbf{p} + \delta\mathbf{p}) - E(\mathbf{p}) \cong \delta E(\mathbf{p}) = \tfrac{1}{2}\int_{\omega_0}^{\omega_1} [(H - \hat{H})\,\delta\bar{H} + (\bar{H} - \bar{\hat{H}})\,\delta H]\,d\omega \tag{9.3-4}$$

where $\delta E(\mathbf{p})$ is the approximate change in $E(\mathbf{p})$ after neglecting the second-order term. Using the properties of complex functions, the equation simplifies to

$$\delta E(\mathbf{p}) = \int_{\omega_0}^{\omega_1} \mathrm{Re}\,\{[\bar{H}(\mathbf{p}, j\omega) - \bar{\hat{H}}(j\omega)]\,\delta H(\mathbf{p}, j\omega)\}\,d\omega \tag{9.3-5}$$

Observe that $\delta H(\mathbf{p}, j\omega)$ can be expanded in terms of the perturbations on the element values as follows:

$$\delta H = \sum_{k=1}^{m} \frac{\partial H}{\partial R_k}\delta R_k + \sum_{k=m+1}^{n} \frac{\partial H}{\partial C_k}\delta C_k + \sum_{k=n+1}^{q} \frac{\partial H}{\partial L_k}\delta L_k \tag{9.3-6}$$

where the partial derivatives are taken with respect to adjustable elements. If there are other parameters in the network that can be adjusted (such as the gain of the amplifiers, gyration constant, etc.), the corresponding partial derivatives will be added to (9.3-6). Using a vector notation, this equation can be written as

$$\delta H(\mathbf{p}, j\omega) = \mathbf{g}^T(\mathbf{p}, j\omega)\,\delta\mathbf{p} \tag{9.3-7}$$

where we have used the following notations:

$$\mathbf{g}^T(\mathbf{p}, j\omega) \triangleq \left[\frac{\partial H}{\partial R_1}, \ldots, \frac{\partial H}{\partial R_m}, \frac{\partial H}{\partial C_{m+1}}, \ldots, \frac{\partial H}{\partial C_n}, \frac{\partial H}{\partial L_{n+1}}, \ldots, \frac{\partial H}{\partial L_q}\right] \tag{9.3-8}$$

and

$$\delta\mathbf{p} = [\delta R_1, \ldots, \delta R_m, \delta C_{m+1}, \ldots, \delta C_n, \delta L_{n+1}, \ldots, \delta L_q]^T \tag{9.3-9}$$

The function $\mathbf{g}(\mathbf{p}, j\omega)$ as defined is commonly referred to as the *gradient* of H with respect to $\mathbf{P}$. Replacing $\delta H(\mathbf{p}, j\omega)$ from (9.3-7) into (9.3-5) and observing that $\delta\mathbf{p}$ is independent of ω, we get

$$\delta E = \left(\int_{\omega_0}^{\omega_1} \mathrm{Re}\,\{[\bar{H}(\mathbf{p}, j\omega) - \bar{\hat{H}}(j\omega)]\mathbf{g}^T(\mathbf{p}, j\omega)\}\,d\omega\right)\delta\mathbf{p} \tag{9.3-10}$$

We now make the following important observation. Since $\hat{H}(j\omega)$ is given and at each iteration the parameter vector $\mathbf{p}$ is known (by choosing), then $\bar{H}(\mathbf{p}, j\omega)$ and $\mathbf{g}^T(\mathbf{p}, j\omega)$ can be computed for each ω. Consequently, the coefficient of $\delta\mathbf{p}$ in (9.3-10) can be computed in terms of the element values of the previous iterations. Let us, therefore, rewrite (9.3-10) in compact form:

$$\delta E(\mathbf{p}) = \boldsymbol{\beta}^T(\mathbf{p})\,\delta\mathbf{p} \tag{9.3-11}$$

where $\boldsymbol{\beta}(\mathbf{p})$ is a column vector called the *gradient of* $E(\mathbf{p})$ and is defined by

$$\boldsymbol{\beta}^T(\mathbf{p}) = \int_{\omega_0}^{\omega_1} \mathrm{Re}\,\{[\bar{H}(\mathbf{p}, j\omega) - \bar{\hat{H}}(j\omega)]\mathbf{g}^T(\mathbf{p}, j\omega)\}\,d\omega \tag{9.3-12}$$

The optimization problem now is reduced to obtaining $\delta\mathbf{p}$ such that, for a given $\boldsymbol{\beta}^T(\mathbf{p})$, the perturbation on the error function δE satisfies the following two conditions:

1. $\delta E < 0$.
2. $|\delta E|$ is *maximized.*

Note that to obtain an efficient and rapidly convergent algorithm it is necessary to satisfy condition 2.

To find the solution to our optimization problem, we must make certain assumptions regarding the "smallness" of the element perturbation vector $\delta\mathbf{p}$. Depending on whether we put an upper bound on the l_2 norm of $\delta\mathbf{p}$ or a bound on the magnitude of each element of $\delta\mathbf{p}$, we obtain two entirely different answers. For this reason, each of these cases is considered separately in the following theorems.

Case 1 (l_2 Norm of δp Is Bounded)

Theorem 9.3-1. Let the l_2 norm of $\delta\mathbf{p}$ be bounded by a small positive number η; that is,

$$|\delta\mathbf{p}|_2^2 = \delta\mathbf{p}^T\delta\mathbf{p} \leq \eta^2 \tag{9.3-13}$$

where $|\cdot|_2$ denotes the l_2 norm of a vector. Then $\delta\mathbf{p}$ given in (9.3-11) satisfies conditions (1) $\delta E < 0$ and (2) $|\delta E|$ is maximized if and only if

$$\delta\mathbf{p} = -\mu\boldsymbol{\beta}(\mathbf{p}) \tag{9.3-14}$$

where μ is a small positive number that depends on η and $\boldsymbol{\beta}(\mathbf{p})$.

Proof

Since $\mathbf{p}$, $\boldsymbol{\beta}(\mathbf{p})$, and μ are real, the following inequality is always true:

$$[\delta\mathbf{p} - \mu\boldsymbol{\beta}(\mathbf{p})]^T[\delta\mathbf{p} - \mu\boldsymbol{\beta}(\mathbf{p})] \geq 0 \tag{9.3-15}$$

or, equivalently,

$$\delta\mathbf{p}^T\,\delta\mathbf{p} + \mu^2\boldsymbol{\beta}^T(\mathbf{p})\boldsymbol{\beta}(\mathbf{p}) - 2\mu\boldsymbol{\beta}^T(\mathbf{p})\delta\mathbf{p} \geq 0$$

where μ is an arbitrary positive number to be determined later. Using (9.3-11) in this equation and rearranging the result, we get

$$\delta E \leq \frac{1}{2\mu}[\delta\mathbf{p}^T\,\delta\mathbf{p} + \mu^2\boldsymbol{\beta}^T(\mathbf{p})\boldsymbol{\beta}(\mathbf{p})]$$

Upon using (9.3-13), we get

$$\delta E \leq \frac{1}{2\mu}[\eta^2 + \mu^2\boldsymbol{\beta}^T(\mathbf{p})\boldsymbol{\beta}(\mathbf{p})]$$

Since the right-hand side of this inequality is independent of $\delta\mathbf{p}$, the left-hand side is maximized if and only if it is equal to the right-hand side. This can be achieved if and only if (9.3-15) is satisfied with the *equality* sign; that is,

$$[\delta\mathbf{p} - \mu\boldsymbol{\beta}(\mathbf{p})]^T[\delta\mathbf{p} - \mu\boldsymbol{\beta}(\mathbf{p})] = 0$$

which implies that

$$\delta\mathbf{p} = \mu\boldsymbol{\beta}(\mathbf{p})$$

This choice of $\delta\mathbf{p}$ yields the maximum δE:

$$\delta E = \mu\boldsymbol{\beta}^T(\mathbf{p})\boldsymbol{\beta}(\mathbf{p})$$

Since the right-hand side of this equation is always positive, we get

$$|\delta E|_{\max} = \delta E = \mu\boldsymbol{\beta}^T(\mathbf{p})\boldsymbol{\beta}(\mathbf{p})$$

Finally, to satisfy condition 1, it is necessary and sufficient to replace μ by $-\mu$; this yields

$$\delta E = -\mu\boldsymbol{\beta}^T(\mathbf{p})\boldsymbol{\beta}(\mathbf{p}) < 0$$

and

$$|\delta E| = |-\mu|\boldsymbol{\beta}^T(\mathbf{p})\boldsymbol{\beta}(\mathbf{p}) = |\delta E|_{\max}$$

Thus, we must have

$$\delta\mathbf{p} = -\mu\boldsymbol{\beta}(\mathbf{p})$$

This proves the theorem.

Consequently, if we impose an upper bound on the l_2 norm of $\delta\mathbf{p}$, at each iteration the optimum perturbation of the *RLC* elements is given by

$$\delta\mathbf{p} = -\mu\int_{\omega_0}^{\omega_1} \operatorname{Re}\{[\bar{H}(\mathbf{p}, j\omega) - \bar{\hat{H}}(j\omega)]\mathbf{g}^T(\mathbf{p}, j\omega)\}\, d\omega \qquad (9.3\text{-}16)$$

The positive number μ is chosen small enough so that the inequality (9.3-13) is satisfied. Obtaining the relationship between η and μ is left as an exercise for the reader. (See Problem 9.6.)

Remark 9.3-1. The scalar constant μ is called the *step size* and is usually chosen small enough so that the approximation (9.3-4) remains valid. In actual design problems the step size μ is chosen through a group of techniques known as *one-dimensional searches*. One of these techniques is to assume a particular geometry (usually a third-order polynomial) for the error function at each iteration and obtain μ such that $E(\mathbf{p})$ is minimized. A detailed discussion on various one-dimensional search techniques is given in [4] and [6].

It should be mentioned at this point that the step size for obtaining each component of $\delta\mathbf{p}$ may be chosen independently; more precisely, we may choose δp_k as

$$\delta p_k = -\mu_k\beta_k(\mathbf{p})$$

where μ_k denotes the step size of δp_k, and $\beta_k(\mathbf{p})$ represents the kth component of $\boldsymbol{\beta}^T(\mathbf{p})$. Before going any further, let us give a geometrical interpretation of the conclusion of Theorem 9.3-1. Note from (9.3-14) that the optimum $\boldsymbol{\delta}\mathbf{p}$ is proportional to $\boldsymbol{\beta}(\mathbf{p})$. For this reason $-\boldsymbol{\beta}(\mathbf{p})$ is called the *steepest-descent direction*, which is related to the gradient of $H(\mathbf{p}, j\omega)$ through (9.3-10). For a two-parameter error function the iteration procedure can be shown schematically as in Figs. 9.3-1(a) and (b). In Fig. 9.3-1(a) a convex error function is shown. Starting with the initial guess $\mathbf{p}^1$ and the corresponding error function E^1, we proceed "downhill" along the steepest-descent direction $-\beta(\mathbf{p}^1)$ to get to E^2. Repeating this process several times, the optimum parameter vector $\mathbf{p}_{\text{opt}}$ can be found that corresponds to the minimum error

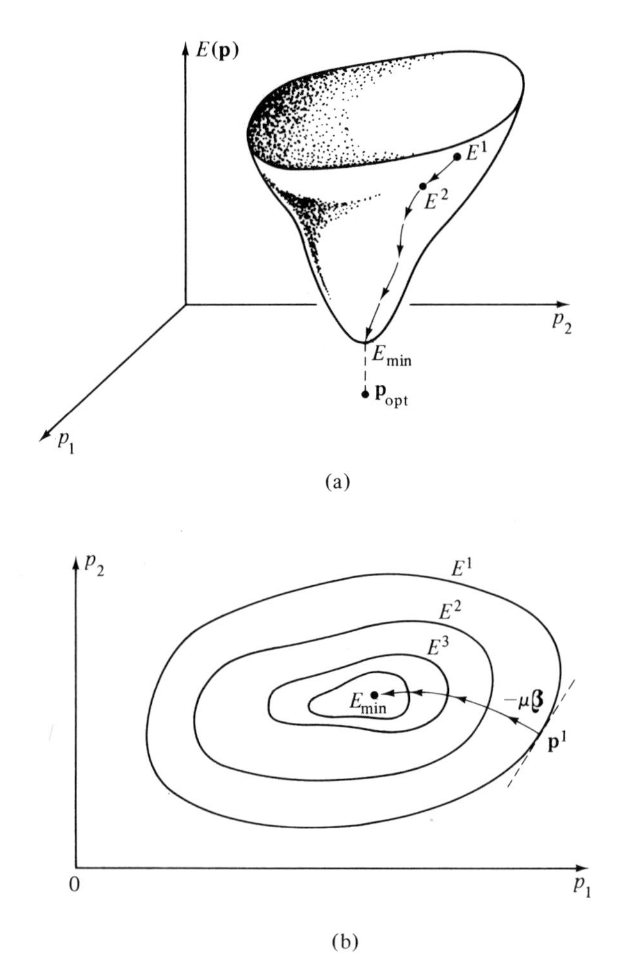

Figure 9.3-1 Schematic representation of the steepest-descent method.

function $E_{\min}$. The projection of $E(\mathbf{p})$ on the p_1-p_2 plane and the corresponding optimum path are shown in Fig. 9.3-1(b).

Let us now work out a simple example to illustrate the steepest-descent criterion discussed so far.

Example 9.3-1. Consider the design of a second-order active low-pass Butterworth filter whose transfer function is given by

$$\hat{H}(s) = \frac{2}{s^2 + \sqrt{2}\,s + 1} \tag{9.3-17}$$

The desired topology is shown in Fig. 9.3-2. The only constraint on the element values is that capacitors C_1 and C_2 must be equal; that is,

$$C_1 = C_2 = C \tag{9.3-18}$$

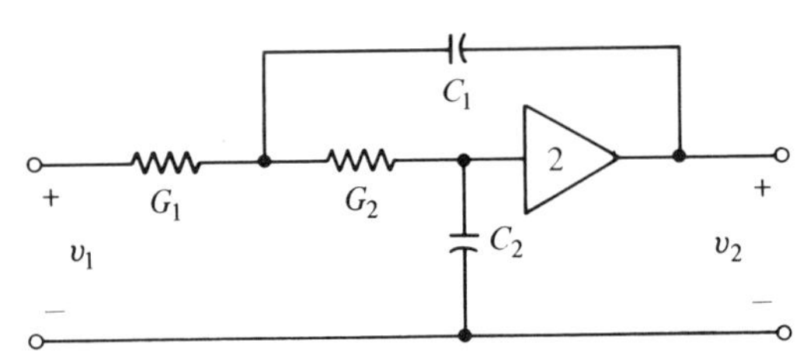

Figure 9.3-2 Topology and the element type of the second-order low-pass Butterworth filter to be designed using the steepest-descent technique.

It is desired that the resulting transfer function be as close as possible (in the mean squared error sense) to the transfer function given in (9.3-17) under the steady-state sinusoidal excitation over the frequency range of [0, 5] radians per second.

Solution

The transfer function of the network shown in Fig. 9.3-2 can be written as

$$H(\mathbf{p}, s) = \frac{2G_1G_2}{C_1C_2s^2 + [C_2G_1 + (C_2 - C_1)G_2]s + G_1G_2}$$

Under the steady-state sinusoidal excitation and (9.3-18), this transfer function becomes

$$H(\mathbf{p}, j\omega) = \frac{2G_1G_2}{G_1G_2 - C^2\omega^2 + jCG_1\omega} \tag{9.3-19}$$

where

$$\mathbf{p} = [G_1 \quad G_2 \quad C]^T \tag{9.3-20}$$

The desired transfer function $\hat{H}(j\omega)$ can also be obtained from (9.3-17) as

$$\hat{H}(j\omega) = \frac{2}{1 - \omega^2 + j\sqrt{2}\,\omega} \tag{9.3-21}$$

The design problem is then to choose G_1, G_2, and C such that the error function

$$E(\mathbf{p}) = \tfrac{1}{2}\int_0^5 |H(\mathbf{p}, j\omega) - \hat{H}(j\omega)|^2 \, d\omega \tag{9.3-22}$$

is minimized. The gradient vector $\mathbf{g}(\mathbf{p}, j\omega)$ can be computed from (9.3-8) and (9.3-19) to be

$$\mathbf{g}^T(\mathbf{p}, j\omega) = \begin{bmatrix} \dfrac{-2G_2C^2\omega^2}{D^2} & \dfrac{-2G_1C^2\omega^2 + j2CG_1^2\omega}{D^2} & \dfrac{4G_1G_2C\omega^2 - j2G_1^2G_2\omega}{D^2} \end{bmatrix} \tag{9.3-23}$$

where the denominator D^2 is

$$D^2 = (G_1G_2 - C^2\omega^2 + jCG_1\omega)^2$$

Let us choose some arbitrary initial values for the parameter vector $\mathbf{p}$, say,

$$\mathbf{p}^0 = [2 \quad 1 \quad 0.5]^T \tag{9.3-24}$$

The optimum adjustment vector at each iteration can then be computed from

$$\delta G_1 = -\mu_1\beta_1(\mathbf{p}), \qquad \delta G_2 = -\mu_2\beta_2(\mathbf{p}), \qquad \delta C = -\mu_3\beta_3(\mathbf{p}) \tag{9.3-25}$$

where $\beta_1(\mathbf{p})$, $\beta_2(\mathbf{p})$, and $\beta_3(\mathbf{p})$ are the components of $\boldsymbol{\beta}(\mathbf{p})$ defined by

$$\boldsymbol{\beta}(\mathbf{p}) = \tfrac{1}{2}\int_0^5 \mathrm{Re}\{[\bar{H}(\mathbf{p}, j\omega) - \bar{\hat{H}}(j\omega)]\mathbf{g}^T(\mathbf{p}, j\omega)\} \, d\omega \tag{9.3-26}$$

To monitor the error at each iteration, the corresponding error function can also be computed from (9.3-22). The result of each iteration is summarized in Table 9.3-1.

Table 9.3-1

RESULT OF THE ITERATIONS FOR THE NETWORK OF EXAMPLE 9.3-1

Iteration No.	G_1	G_2	C	E
0	2.0000	1.0000	0.5000	4.1215
1	1.8900	0.6544	0.8639	0.1216
2	1.7676	0.5987	0.8962	0.0362
4	1.5938	0.5912	0.9125	0.0188
8	1.4571	0.6122	0.9152	0.0070
12	1.3761	0.6275	0.9153	0.0021
20	1.3120	0.6415	0.9144	0.0001

The corresponding computer program and the associated subroutine are given next. Note that the iterations are stopped when the error function is less than 0.0001. To achieve this objective with the preceding choice of initial values 20 iterations were needed. The computer program for the design of the active filter of Fig. 9.3-2 is shown on the next two pages.

```
C     THIS PROGRAM IS FOR EXAMPLE 9.3-1.
C     DESIGN OF SECOND ORDER ACTIVE LOW PASS BUTTERWORTH FILTER
      DIMENSION S1(100),S2(100),S3(100),E(100),SE(100),DATA(210,3)
      COMPLEX D,NA,NB,NC,DD,H1,H2,HH,B1,B2,B3,DH1
      REAL MU1,MU2,MU3
C     INITIAL GUESS
      G1=2.0
      G2=1.0
      C=.5
      W=0.0
      N=51
      HW=5.0/(N-1)
      PRINT 6,G1,G2,C
      DO 2 ITER=1,30
      DO 4 I=1,N
      D=CMPLX(G1*G2-(C*W)**2,C*G1*W)
      DD=D**2
      NA=CMPLX(-2.0*G2*(C*W)**2,0.0)
      NB=CMPLX(-2.0*G1*(C*W)**2,2.0*C*W*G1*G1)
      NC=CMPLX(4.0*C*G1*G2*W*W,-2.0*G2*W*G1*G1)
      DH1=CMPLX(1.0-W**2,W*SQRT(2.0))
C     DESIRED TRANSFER FUNCTION IS H1,ACTUAL TRANSFER FUNCTION IS H2
      H1=2.0/DH1
      H2=2.0*G1*G2/D
      DATA(I,1)=W
      DATA(I,2)=CABS(H1)
      DATA(I,3)=CABS(H2)
      HH=H2-H1
      B1=CONJG(HH)*(NA/DD)/2.0
      B2=CONJG(HH)*(NB/DD)/2.0
      B3=CONJG(HH)*(NC/DD)/2.0
      S1(I)=REAL(B1)
      S2(I)=REAL(B2)
      S3(I)=REAL(B3)
      E(I)=((CABS(HH))**2)/2.0
      W=W+HW
    4 CONTINUE
      CALL GRAPHX(DATA,N,1HN,5HH1,H2)
C     CHOOSE THE FOLLOWING STEP SIZES
      MU1=1.2
      MU2=.1
      MU3=.05
C     USE THE SIMPSONS RULE OF INTEGRATION TO COMPUTE THE GRADIENT
      CALL SMPSN(S1,HW,N,ANS)
      G1=-MU1*ANS+G1
      CALL SMPSN(S2,HW,N,ANS)
      G2=-MU2*ANS+G2
      CALL SMPSN(S3,HW,N,ANS)
      C=-MU3*ANS+C
      CALL SMPSN(E,HW,N,EP)
      SE(ITER)=EP
      PRINT 6,G1,G2,C,EP
    6 FORMAT(/,5X,4F14.4)
C     STOP THE ITERATION IF THE ERROR  EP IS LESS THAN .0001
      IF(EP .LE. .0001) GO TO 12
      W=0.0
    2 CONTINUE
   12 DO 10 I=1,30
      DATA(I,1)=I
      DATA(I,2)=SE(I)
   10 DATA(I,3)=0.0
      CALL GRAPHX(DATA,30,1HN,5HERROR)
      STOP
      END
```

```
      SUBROUTINE SMPSN(Y,H,N,ANS)
      DIMENSION Y(1)
      ODD=0.0
      EVEN=0.0
      NN=N-3
      DO 1 I=2,NN,2
      EVEN= EVEN+Y(I)
    1 ODD=ODD+Y(I+1)
      ANS=(H/3.0)*(Y(1)+4.0*(EVEN+Y(N-1))+2.0*ODD+Y(N))
      RETURN
      END
```

The error function $E(\mathbf{p})$ is plotted versus the number of the iterations in Fig. 9.3-3. The resulting transfer functions at several iterations are shown in Fig. 9.3-4. From these figures it is clear that the iterations converge quite rapidly.

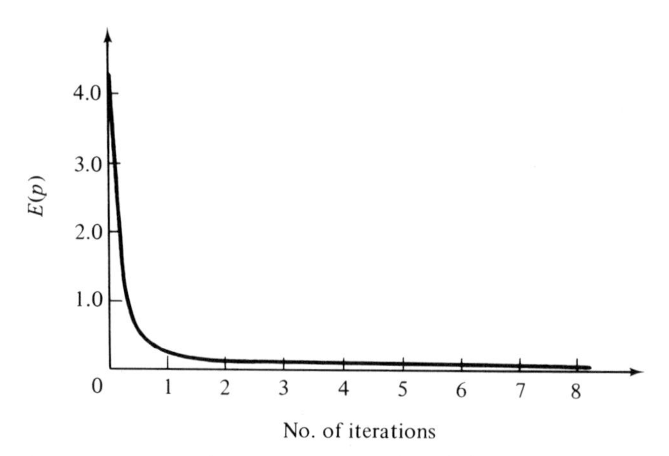

Figure 9.3-3 Plot of the error function of Example 9.3-1 versus the number of iterations.

Let us now consider the case where the size of each element of $\delta\mathbf{p}$ is bounded.

Case 2 (Each Element of δp is Bounded). In some cases each element of the parameter vector **p** is allowed to change at most by a certain amount. In such cases the conclusion of Theorem 9.3-1 should be modified as in the following theorem.

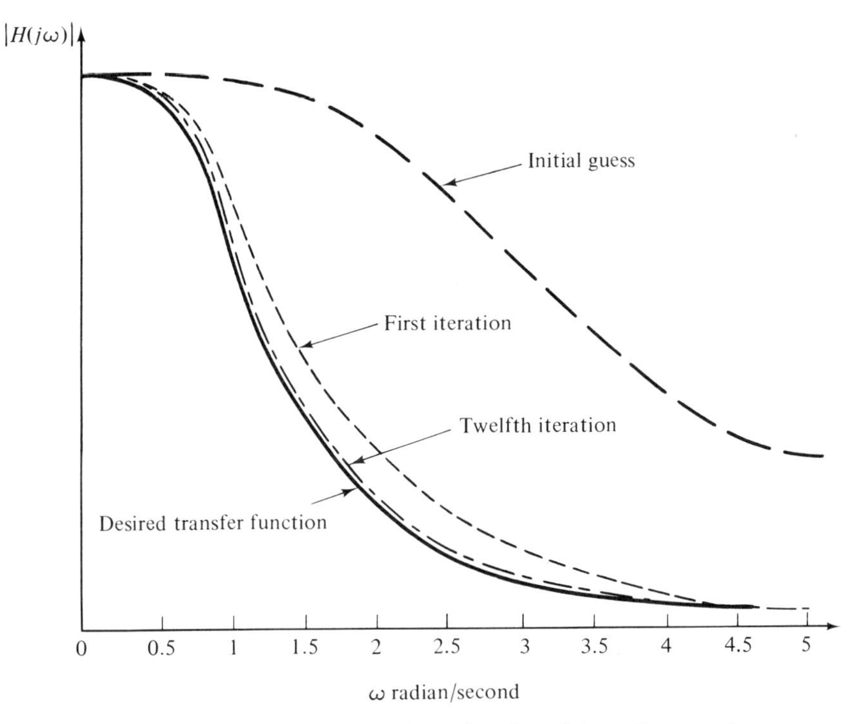

Figure 9.3-4 Plot of the magnitude function of the active second-order low-pass Butterworth filter at various iterations.

Theorem 9.3-2. Let the kth element of $\boldsymbol{\delta p}$ be bounded by a small positive number η_k; that is,

$$|\delta p_k| \leq \eta_k \tag{9.3-27}$$

Then $\boldsymbol{\delta p}$ given in (9.3-11) satisfies conditions 1 and 2 stated previously if and only if

$$\delta p_k = -\eta_k \operatorname{sgn}(\beta_k) \tag{9.3-28}$$

where the function $\operatorname{sgn}(\beta_k)$ is defined as

$$\operatorname{sgn}(\beta_k) = +1 \quad \text{if } \beta_k \geq 0$$

and

$$\operatorname{sgn}(\beta_k) = -1 \quad \text{if } \beta_k < 0$$

Equation (9.3-28) implies that in each iteration δp_k must be chosen as its maximum allowable value (with appropriate sign). This procedure is sometimes referred to as *bang-bang* optimization.

Proof

From (9.3-11) we can write

$$\delta E = \sum_{k=1}^{q} \beta_k \, \delta p_k \tag{9.3-29}$$

or, using Schwartz's inequality, we get

$$|\delta E| \leq \sum_{k=1}^{q} |\beta_k| \cdot |\delta p_k|$$

Using (9.3-27), this inequality becomes

$$|\delta E| \leq \sum_{k=1}^{q} \eta_k |\beta_k| \tag{9.3-30}$$

Since β_k and η_k are fixed numbers, $|\delta E|$ achieves its maximum if and only if the equality sign in (9.3-30) is assumed; that is,

$$|\delta E|_{\max} = \sum_{k=1}^{q} \eta_k |\beta_k|$$

Comparing this last equation to (9.3-29), to maximize $|\delta E|$, we must have

$$\delta p_k = \eta_k \operatorname{sgn}(\beta_k)$$

On the other hand, since δE must be negative to satisfy condition 1, the optimal choice of δp_k will be

$$\delta p_k = -\eta_k \operatorname{sgn}(\beta_k)$$

The proof is, therefore, complete.

Equation (9.3-28) can be written in the following detailed form:

$$\delta p_k = -\eta_k \operatorname{sgn} \int_{\omega_0}^{\omega_1} \operatorname{Re}\{[\bar{H}(\mathbf{p}, j\omega) - \bar{\hat{H}}(j\omega)] g_k(\mathbf{p}, j\omega)\} \, d\omega \tag{9.3-31}$$

where g_k denotes the kth element of $\mathbf{g}$. Consequently, in this case, to obtain the perturbation on each network element, it is sufficient to determine the sign of the integral in (9.3-31) and use the maximum allowable change on the element with appropriate sign.

The optimization method discussed so far was directed toward the mean squared error function of (9.2-1). Extension of this method to other mean squared error functions, such as those given in (9.2-6) and (9.2-7), is trivial and hence will be omitted. Instead, let us develop a general perturbation analysis that applies to any form of error function $E(\mathbf{p})$.

9.4 GENERAL OPTIMIZATION METHODS

Let $E(\mathbf{p})$ denote a general error function that depends on the parameter vector $\mathbf{p}$. Suppose that during the iteration process $\mathbf{p}$ is perturbed by

$\delta\mathbf{p}$. The corresponding $E(\mathbf{p} + \delta\mathbf{p})$ can then be expanded by a Taylor expansion around $\mathbf{p}$ as

$$E(\mathbf{p} + \delta\mathbf{p}) = E(\mathbf{p}) + \nabla^T E\delta\mathbf{p} + \tfrac{1}{2}\delta\mathbf{p}^T\mathbf{H}\delta\mathbf{p} + \cdots \tag{9.4-1}$$

where $\nabla^T E$ is a row vector called the *gradient* of $E(\mathbf{p})$ and is defined as

$$\nabla^T E = \begin{bmatrix} \dfrac{\partial E}{\partial p_1} & \dfrac{\partial E}{\partial p_2} & \cdots & \dfrac{\partial E}{\partial p_q} \end{bmatrix} \tag{9.4-2}$$

$\mathbf{H}$ is a symmetric matrix called the *Hessian* matrix of E. The elements of $\mathbf{H}$ are second derivatives of E with respect to the elements of $\mathbf{p}$; that is,

$$\mathbf{H} = \begin{bmatrix} \dfrac{\partial^2 E}{\partial p_1^2} & \dfrac{\partial^2 E}{\partial p_1\, \partial p_2} & \cdots & \dfrac{\partial^2 E}{\partial p_1\, \partial p_n} \\ \dfrac{\partial^2 E}{\partial p_2\, \partial p_1} & \dfrac{\partial^2 E}{\partial p_2^2} & \cdots & \dfrac{\partial^2 E}{\partial p_2\, \partial p_n} \\ \vdots & \vdots & & \vdots \\ \dfrac{\partial^2 E}{\partial p_n\, \partial p_1} & \dfrac{\partial^2 E}{\partial p_n\, \partial p_2} & \cdots & \dfrac{\partial^2 E}{\partial p_n^2} \end{bmatrix} \tag{9.4-3}$$

In theory, all the terms in the Taylor expansion (9.4-1) should be calculated. In actual computations, however, since it is assumed that the norm of $\delta\mathbf{p}$ is small and E is a smooth function of $\mathbf{p}$, the higher-order terms can be neglected and only the first three terms retained. The approximate change on the error function $\delta E(\mathbf{p})$ can then be written as

$$E(\mathbf{p} + \delta\mathbf{p}) - E(\mathbf{p}) = \delta E(\mathbf{p}) = \nabla^T E\, \delta\mathbf{p} + \tfrac{1}{2}\delta\mathbf{p}^T\mathbf{H}\, \delta\mathbf{p} \tag{9.4-4}$$

In some cases the last term in (9.4-4) is also dropped and only the linear portion is used. This, indeed, was the case in the optimization procedure discussed in the last section.

Note that $E(\mathbf{p})$ is minimized if $\nabla^T E = \mathbf{0}$ and $\mathbf{H}$ is positive definite. In an iteration process, therefore, the objective is to find $\delta\mathbf{p}$ such that at each cycle the following two conditions are satisfied:

1. The error function is reduced; that is, $\delta E(\mathbf{p}) \leq 0$.
2. The absolute value of the change in the error function is maximized; that is, $|\delta E(\mathbf{p})|$ is maximized.

There are several methods of obtaining the optimum $\delta\mathbf{p}$. In what follows we only consider two of these methods, which have proven successful in most engineering problems. For a discussion on the other methods, the interested reader may consult the literature [4, 6, 7].

Steepest Descent Method. The perturbation $\delta\mathbf{p}$ just discussed is a vector in the $\mathbf{p}$ space. Let us denote the direction of this vector by the unit vector $\mathbf{s}$

and its magnitude by the positive scalar α. Thus, $\delta\mathbf{p}$ can be written as

$$\delta\mathbf{p} = \alpha\mathbf{s} \tag{9.4-5}$$

The objective is to obtain $\mathbf{s}$ such that the absolute value of the perturbation of the error function $|\delta E|$ is maximized. To do so, let us use (9.4-5) in (9.4-1) to get

$$\delta E(\mathbf{p}) = \alpha\nabla^T E\mathbf{s} + \frac{\alpha^2}{2}\mathbf{s}^T\mathbf{H}\mathbf{s} + \cdots$$

Let $\delta E_{\max}$ denote the desired maximum; then we must have

$$\delta E_{\max} = \frac{\partial}{\partial\alpha}\delta E(\mathbf{p})|_{\alpha=0} = \nabla^T E\mathbf{s} + \alpha\mathbf{s}^T\mathbf{H}\mathbf{s} + \cdots|_{\alpha=0} = \nabla^T E\mathbf{s} \tag{9.4-6}$$

Note that the second- and higher-order terms in (9.4-6) vanish at $\alpha = 0$. Since $\mathbf{s}$ is the unit vector, we have

$$|\mathbf{s}|_2^2 = \mathbf{s}^T\mathbf{s} = 1 \tag{9.4-7}$$

Therefore, by an argument similar to that given in Theorem 9.3-1, we can show that the optimal $\mathbf{s}$ is in the direction of the gradient ∇E. Moreover, since $\mathbf{s}$ is a unit vector and $\delta E_{\max} \leq 0$, we must have

$$\mathbf{s} = -\frac{1}{|\nabla E|_2}\nabla E \tag{9.4-8}$$

where $|\nabla E|_2$ denotes the l_2 norm of ∇E. Finally, from (9.4-8) and (9.4-5) we get

$$\delta\mathbf{p} = -\mu\nabla E \tag{9.4-9}$$

where μ is a positive scalar called the step size and is

$$\mu = \frac{\alpha}{|\nabla E|_2} \tag{9.4-10}$$

Conceptually, μ can be chosen as large as possible. In practice, however, it must be chosen small enough so that the linearization of the error function remains valid. One popular method of determining the step size is to approximate $E(\mathbf{p})$ by a polynomial (usually of degree 3) and then choose μ such that this polynomial is minimized. Details of this and other methods for obtaining the step size μ are given in [6–8].

Note that (9.4-9) is a generalization of (9.3-14), which is derived for the arbitrary error function $E(\mathbf{p})$.

If the magnitude of the elements of $\delta\mathbf{p}$ are bounded, the optimum $\delta\mathbf{p}$ is determined from the sign of the elements of ∇E. More precisely, let

$$\delta p_k \leq \eta_k \tag{9.4-11}$$

Then using a procedure similar to that in Theorem 9.3-2, the bang-bang optimization results in

$$\delta p_k = -\eta_k \operatorname{sgn}\{\nabla_k E\} \tag{9.4-12}$$

where $\nabla_k E$ denotes the kth component of the gradient ∇E and sgn$\{\cdot\}$ is defined in Theorem 9.3-2.

The main advantage of the steepest-descent criterion is that it always converges to a minimum. This minimum, however, may not be the global minimum. The convergence of the iterations to the global minimum depends on the shape of the error function in the parameter space. One major disadvantage of the steepest-descent method is that if the shape of the error function is relatively flat the convergence will be rather slow. This disadvantage is particularly evident near the minimum point. In many optimization processes, steepest descent is used as a starting strategy and then other optimization methods, such as the Fletcher–Powell technique, are used to obtain the minimum.

Fletcher–Powell Method. The rate of convergence in the steepest-descent method can be greatly improved if the Hessian matrix, **H**, defined in (9.4-3) is used. To see how **H** can be implemented, let us expand the gradient of the error function at the $(j+1)$st iteration, $\nabla E(\mathbf{p}^{j+1} + \delta\mathbf{p}^{j+1})$, about $\mathbf{p}^{j+1}$ by a Taylor expansion:

$$\nabla E(\mathbf{p}^{j+1} + \delta\mathbf{p}^{j+1}) = \nabla E(\mathbf{p}^{j+1}) + \mathbf{H}^{j+1}\,\delta\mathbf{p}^{j+1} + \cdots$$

Retaining the linear terms only, we get

$$\nabla E(\mathbf{p}^{j+1} + \delta\mathbf{p}^{j+1}) \cong \nabla E(\mathbf{p}^{j+1}) + \mathbf{H}^{j+1}\,\delta\mathbf{p}^{j+1}$$

If the minimum is reached at the $(j+1)$st iteration, the corresponding gradient must be equal to zero; that is,

$$\delta\mathbf{p}^{j+1} = -(\mathbf{H}^{j+1})^{-1}\nabla E^{j+1} \tag{9.4-13}$$

where we have used the following notations:

$$\delta\mathbf{p}^{j+1} = \mathbf{p}^{j+1} - \mathbf{p}^{j} \quad \text{and} \quad \nabla E^{j+1} = \nabla E(\mathbf{p}^{j+1})$$

Conceptually, the element values can be determined by iterating (9.4-13). In practice, however, inverting $\mathbf{H}^{j+1}$ at each iteration can easily amount to an expensive and time-consuming process, particularly if the number of adjustable parameters is large. To avoid this difficulty, Fletcher and Powell have suggested a simple method of estimating $(\mathbf{H}^{j+1})^{-1}$. In the following we introduce the basic equations for the Fletcher–Powell method without proof. The interested reader can then consult the original work of Fletcher and Powell [9] or other sources [4] for a more detailed discussion.

Let us denote the approximate $(\mathbf{H}^{j+1})^{-1}$ by a positive definite matrix $\mathbf{G}^{j+1}$; that is,

$$\mathbf{G}^{j+1} \cong (\mathbf{H}^{j+1})^{-1} \tag{9.4-14}$$

The Fletcher–Powell method can then be summarized as follows: Let

$$\mathbf{d}^{j} = \nabla E^{j+1} - \nabla E^{j} \tag{9.4-15}$$

and

$$\mathbf{s}^{j} = -\mathbf{G}^{j}\nabla E^{j} \tag{9.4-16}$$

then
$$\mathbf{G}^{j+1} = \mathbf{G}^j + \frac{\mathbf{s}^j(\mathbf{s}^j)^T}{(\mathbf{s}^j)^T\mathbf{d}^j} - \frac{\mathbf{G}^j\mathbf{d}^j(\mathbf{d}^j)^T\mathbf{G}^j}{(\mathbf{d}^j)^T\mathbf{G}^j\mathbf{d}^j} \tag{9.4-17}$$

Note that since $\mathbf{d}^j$ and $\mathbf{s}^j$ are column vectors $\mathbf{s}^j(\mathbf{s}^j)^T$ and $\mathbf{d}^j(\mathbf{d}^j)^T$ represent square matrices with the same dimension as $\mathbf{G}^j$. Then it can be shown [9] that the optimum $\delta\mathbf{p}^{j+1}$ is

$$\delta\mathbf{p}^{j+1} = -\mathbf{G}^{j+1}\nabla E^{j+1} \tag{9.4-18}$$

To start the iterations, $\mathbf{G}^0$ is usually taken to be the identity matrix. This process is then repeated until the minimum is reached. This iteration has proved to be a powerful and rapidly convergent scheme for obtaining the optimum element values. The main drawback of this method is that it requires a large amount of memory core storage if the number of adjustable elements is large. This deficiency can be overcome by adopting the *conjugate-gradient method* developed by Fletcher and Reeves. A detailed discussion of this method can be found in [10]. Let us now turn our attention to an interesting and useful method of computing the gradient of the error function.

9.5 APPLICATION OF ADJOINT NETWORK IN COMPUTING THE GRADIENT

In all the optimization techniques discussed so far, the gradient of the error function plays an essential role in the total process. Since at each iteration the gradient must be computed, the computation time could be quite large, even in moderate design problems. To circumvent this problem Director and Rohrer [5] have developed a simple method of computing the gradient by implementing the properties of the adjoint network. To present the basic idea underlying this technique, let us consider the one-port network of Fig. 9.5-1(a). Let the input to this network be the current source phasor $I(j\omega)$ and the desired output be a given voltage phasor $\hat{V}(j\omega)$. The design problem is to choose the element values such that the output

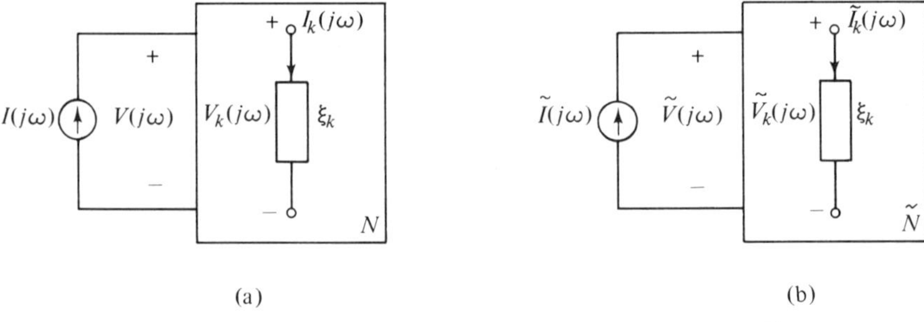

Figure 9.5-1 (a) Original network N; (b) adjoint network $\tilde{N}$.

voltage phasor $V(j\omega)$ is as close as possible to desired $\hat{V}(j\omega)$. Choosing the mean squared error function

$$E(p) = \frac{1}{2}\int_{\omega_0}^{\omega_1} |V(\mathbf{p}, j\omega) - \hat{V}(j\omega)|^2 \, d\omega$$

and using a perturbation analysis similar to that discussed in Section 9.2, we can show that the optimal perturbation on the element values is given by

$$\delta\mathbf{p} = -\mu\boldsymbol{\beta}(\mathbf{p}) \tag{9.5-1}$$

where $\boldsymbol{\beta}(\mathbf{p})$ is the corresponding gradient and is given as

$$\boldsymbol{\beta}^T(\mathbf{p}) = \int_{\omega_0}^{\omega_1} \operatorname{Re}\{[\bar{V}(\mathbf{p}, j\omega) - \bar{\hat{V}}(j\omega)]\mathbf{g}^T(\mathbf{p}, j\omega)\} \, d\omega \tag{9.5-2}$$

where $\mathbf{g}^T(\mathbf{p}, j\omega)$ is a row vector defined as

$$\mathbf{g}^T(\mathbf{p}, j\omega) = \left[\frac{\partial V}{\partial R_1} \quad \frac{\partial V}{\partial R_2} \quad \cdots \quad \frac{\partial V}{\partial C_k}, \quad \ldots, \quad \frac{\partial V}{\partial L_q}\right] \tag{9.5-3}$$

Furthermore, the perturbation on the voltage can be written as

$$\delta V(\mathbf{p}, j\omega) = \mathbf{g}^T(\mathbf{p}, j\omega)\,\delta\mathbf{p} \tag{9.5-4}$$

To use the adjoint network for computing the gradient $\boldsymbol{\beta}^T(\mathbf{p})$, let us first define an adjoint network in conjunction with the network under study. The adjoint of the *RLC* network of Fig. 9.5-1(a) is a network with the same topology and the same element values, but driven with the current source phasor $\tilde{I}(j\omega)$, which is different from the driving source of the original network $I(j\omega)$. The relationship between $\tilde{I}(j\omega)$ and $I(j\omega)$ will be determined later. Thus, the adjoint of the network N of Fig. 9.5-1(a) is the network $\tilde{N}$ as shown in Fig. 9.5-1(b). According to Tellegen's theorem (Theorem 3.5-2), the branch voltages and currents of N and $\tilde{N}$ are related by

$$\sum_{k=1}^{q} V_k(j\omega)\tilde{I}_k(j\omega) + V(j\omega)\tilde{I}(j\omega) = 0 \tag{9.5-5}$$

and

$$\sum_{k=1}^{q} I_k(j\omega)\tilde{V}_k(j\omega) + I(j\omega)\tilde{V}(j\omega) = 0 \tag{9.5-6}$$

If the element values of the original network are changed by $\delta\mathbf{p}$, the change on $V_k(j\omega)$ will be $\delta V_k(j\omega)$ and that of I_k will be δI_k. Applying Tellegen's theorem to the perturbed network will, therefore, yield

$$\sum_{k=1}^{q} (V_k + \delta V_k)\tilde{I}_k + (V + \delta V)\tilde{I} = 0 \tag{9.5-7}$$

and

$$\sum_{k=1}^{q} (I_k + \delta I_k)\tilde{V}_k + I\tilde{V} = 0 \tag{9.5-8}$$

where the dependence of V_k and I_k on $\mathbf{p}$ and $j\omega$ is understood. Notice that the perturbation on $I(j\omega)$ is zero since by assumption the driving function is an ideal current source; thus, its value is not affected by element changes.

Comparing (9.5-7) and (9.5-5), we get

$$\sum_{k=1}^{q} \delta V_k \tilde{I}_k + \delta V \tilde{I} = 0 \tag{9.5-9}$$

Similarly, comparing (9.5-8) and (9.5-6), we obtain

$$\sum_{k=1}^{q} \delta I_k \tilde{V}_k = 0 \tag{9.5-10}$$

Subtracting (9.5-10) from (9.5-9) yields

$$\tilde{I}\,\delta V = \sum_{k=1}^{q} (\delta I_k \tilde{V}_k - \delta V_k \tilde{I}_k) \tag{9.5-11}$$

Let us now consider the branch voltage relationships of the perturbed network. For the kth perturbed resistor the v-i relation is

$$V_k + \delta V_k = (R_k + \delta R_k)(I_k + \delta I_k)$$

Comparing this with the v-i relation of the original network

$$V_k = R_k I_k$$

and neglecting the second-order term $\delta R_k\,\delta I_k$, we get

$$\delta V_k = R_k\,\delta I_k + I_k\,\delta R_k, \qquad k = 1, 2, \ldots, m \tag{9.5-12}$$

In a similar manner, the perturbations δV_k and δI_k for the inductors and capacitors can be found to be

$$\delta V_k = j\omega L_k\,\delta I_k + j\omega I_k\,\delta L_k, \qquad k = n+1, \ldots, q \tag{9.5-13}$$

and

$$\delta I_k = j\omega C_k\,\delta V_k + j\omega V_k\,\delta C_k, \qquad k = m+1, \ldots, n$$

Regrouping terms in the right-hand side of (9.5-11) and using the last three equations, we get

$$\begin{aligned} \tilde{I}\,\delta V = \sum_{k=1}^{m} [(\tilde{V}_k - R_k \tilde{I}_k)\,\delta I_k - \tilde{I}_k I_k\,\delta R_k] + \sum_{k=n+1}^{q} [(\tilde{V}_k - j\omega L_k \tilde{I}_k)\,\delta I_k \\ - j\omega \tilde{I}_k I_k\,\delta L_k] + \sum_{k=m+1}^{n} [-(\tilde{I}_k - j\omega C_k \tilde{V}_k)\,\delta V_k + j\omega \tilde{I}_k I_k\,\delta C_k] \end{aligned} \tag{9.5-14}$$

Since the RCL elements of the adjoint network $\tilde{N}$ are the same as those of the original network N, the following equations hold:

$$\tilde{V}_k = R_k \tilde{I}_k, \qquad k = 1, 2, \ldots, m$$

$$\tilde{I}_k = j\omega C_k \tilde{V}_k, \qquad k = m+1, \ldots, n$$

and

$$\tilde{V}_k = j\omega L_k \tilde{I}_k, \qquad k = n+1, \ldots, q$$

Using these three equations in (9.5-14), we get

$$\tilde{I}\,\delta V = -\sum_{k=1}^{m} \tilde{I}_k I_k\,\delta R_k - \sum_{k=n+1}^{q} j\omega \tilde{I}_k I_k\,\delta L_k + \sum_{k=m+1}^{n} j\omega \tilde{I}_k I_k\,\delta C_k$$

or, equivalently,

$$-\tilde{I}(j\omega)\,\delta V(\mathbf{p}, j\omega) = \mathbf{h}^T(\mathbf{p}, j\omega)\,\delta\mathbf{p} \tag{9.5-15}$$

where $\delta\mathbf{p} = [\delta R_1 \quad \delta R_2 \ldots \delta C_{m+1} \ldots \delta L_q]^T$

and $$\mathbf{h}^T(\mathbf{p}, j\omega) \triangleq [\tilde{I}_{R_1} I_{R_1}, \ldots, \tilde{I}_{R_m} I_{R_m}, -j\omega \tilde{I}_{C_{m+1}} I_{C_{m+1}}, \ldots, j\omega \tilde{I}_{L_q} I_{L_q}] \tag{9.5-16}$$

We have now arrived at the crux of this presentation: Let us *choose the input current source of the adjoint network as*

$$\tilde{I}(j\omega) = -[\bar{V}(\mathbf{p}, j\omega) - \bar{\hat{V}}(j\omega)] \tag{9.5-17}$$

In words, choose the excitation of the adjoint network to be the negative of the complex conjugate of the error between the actual and the desired response. Then from (9.5-4), (9.5-15), and (9.5-17) we obtain

$$[\bar{V}(\mathbf{p}, j\omega) - \bar{\hat{V}}(j\omega)]\mathbf{g}^T(\mathbf{p}, j\omega)\,\delta\mathbf{p} = \mathbf{h}^T(\mathbf{p}, j\omega)\,\delta\mathbf{p} \tag{9.5-18}$$

Taking the real part of (9.5-18) and integrating with respect to ω, we get

$$\left(\int_{\omega_0}^{\omega_1} \operatorname{Re}\{\bar{V}(\mathbf{p}, j\omega) - \bar{\hat{V}}(j\omega)]\mathbf{g}^T(\mathbf{p}, j\omega)\}\,d\omega\right)\delta\mathbf{p} = \left(\int_{\omega_0}^{\omega_1} \operatorname{Re}\{\mathbf{h}^T(\mathbf{p}, j\omega)\}\,d\omega\right)\delta\mathbf{p} \tag{9.5-19}$$

Since (9.5-19) must hold for *any* $\delta\mathbf{p}$, we can conclude that

$$\int_{\omega_0}^{\omega_1} \operatorname{Re}\{[\bar{V}(\mathbf{p}, j\omega) - \bar{\hat{V}}(j\omega)]\mathbf{g}^T(\mathbf{p}, j\omega)\}\,d\omega = \int_{\omega_0}^{\omega_1} \operatorname{Re}\{\mathbf{h}^T(\mathbf{p}, j\omega)\}\,d\omega \tag{9.5-20}$$

Comparing (9.5-20) with the gradient given in (9.5-2), we obtain the following important result:

$$\boldsymbol{\beta}(\mathbf{p}) = \int_{\omega_0}^{\omega_1} \operatorname{Re}\{\mathbf{h}(\mathbf{p}, j\omega)\}\,d\omega \tag{9.5-21}$$

Thus, from (9.5-1) we obtain the optimal change on the parameter vector as

$$\delta\mathbf{p} = -\mu \int_{\omega_0}^{\omega_1} \operatorname{Re}\{\mathbf{h}(\mathbf{p}, j\omega)\}\,d\omega \tag{9.5-22}$$

Consequently, the bulk of the work in computing $\delta\mathbf{p}$ reduces to computing $\mathbf{h}^T(\mathbf{p}, j\omega)$ as given in (9.5-16). This, in contrast to computing $[\bar{V}(\mathbf{p}, j\omega) - \bar{\hat{V}}(j\omega)]\mathbf{g}^T(\mathbf{p}, j\omega)$, is quite straightforward and less time consuming, particularly if the number of network elements is large.

To obtain $\mathbf{h}^T(\mathbf{p}, j\omega)$, it is sufficient to excite the adjoint network by the current source given in (9.5-17) and the original network by $I(j\omega)$, and compute all the branch currents in both networks (i.e., $\tilde{I}_k$ and I_k). Equation (9.5-16) can then be used to compute $\mathbf{h}^T(\mathbf{p}, j\omega)$.

To appreciate the advantage of using the adjoint network in computing the gradient, let us compare the effort required for computing $\mathbf{g}(\mathbf{p}, j\omega)$ to that of $\mathbf{h}(\mathbf{p}, j\omega)$. In computing $\mathbf{h}(\mathbf{p}, j\omega)$ one needs to analyze only two networks at each iteration (the original network to obtain I_k and the adjoint network to obtain $\tilde{I}_k$), whereas in computing $\mathbf{g}(\mathbf{p}, j\omega)$ one analysis is needed for each adjustable element [see equation (9.5-3)]. Thus, if the number of adjustable elements is large, the method just discussed is more efficient.

Remark 9.5-1. If the network under consideration contains gyrators, or controlled sources, the corresponding elements in the adjoint network are not quite the same. For example, if the gyration constant of a gyrator in the original network is α, the corresponding gyrator in the adjoint network has a gyration constant of $-\alpha$. A complete list of relationships between the elements of the original and the adjoint network is given in [5].

So far in this chapter we have discussed the design of linear time-invariant networks operated under the sinusoidal steady-state case. The methods discussed can be applied, with minor adjustments, to design n-port *RLC* networks or realize a given driving-point admittance matrix. A detailed discussion of these topics, however, would be beyond the intended scope of this book. Interested readers are referred to the references [1, 3, 5, 7]. Let us now give a brief discussion of the design of linear time-varying networks.

9.6 COMPUTER-AIDED DESIGN OF LINEAR TIME-VARYING NETWORKS

Since linear time-varying networks can be best represented by their state equations, their design problems are usually formulated in the time domain. As in Section 9.2, to design a linear time-varying network, a suitable error function is chosen and the characteristics of the network elements are determined so that the error function is minimized. In this section we first introduce a general approach to the design of nth-order linear time-varying networks and then consider a specific design problem.

Consider a linear time-varying network whose input–output state equations are

$$\dot{\mathbf{x}} = \mathbf{A}(\mathbf{p})\mathbf{x} + \mathbf{b}(\mathbf{p})u \tag{9.6-1}$$

$$y = \mathbf{c}^T(\mathbf{p})\mathbf{x} \tag{9.6-2}$$

where $\mathbf{x}(t)$ is an n vector representing the state, and $u(t)$ and $y(t)$ are scalar time functions representing the input and the output, respectively. The parameter vector $\mathbf{p}$ is defined by

$$\mathbf{p} \triangleq [R_1(t), \ldots, R_m(t), C_{m+1}(t), \ldots, C_n(t), L_{n+1}(t), \ldots, L_q(t)]^T \tag{9.6-3}$$

and matrices $\mathbf{A}$, $\mathbf{b}$, and $\mathbf{c}$ are assumed to have appropriate dimensions. The design problem is to obtain the elements of the vector $\mathbf{p}$ as defined so that the resulting network approximates a desired property, such as a prescribed step response, time delay, or steady-state behavior. For example, we may require that the response of the network to a step function applied at τ, $s(t, \tau; \mathbf{p})$ be "as close as possible" to a given time function $\alpha(t, \tau)$ over a finite domain $T \geq t \geq \tau \geq 0$. Again, to specify the problem more clearly, we may choose the quadratic error function

$$E(\mathbf{p}) = \int_0^T \int_0^t [s(t, \tau; \mathbf{p}) - \alpha(t, \tau)]^2 \, d\tau \, dt \tag{9.6-4}$$

where $s(t, \tau; \mathbf{p})$ can be obtained from the state equations by specifying the input $u(t)$, using any of the numerical methods discussed in Chapter 7. We next apply the following iterative technique to obtain the elements of $\mathbf{p}(t)$. As in the previous case, for simplicity, we assume that the topology and element types of the desired network are known.

Step 1. Make a reasonable guess on the vector function $\mathbf{p}(t)$ and use the input–output state equations to obtain $s(t, \tau; \mathbf{p})$ for $T \geq t \geq \tau \geq 0$. Also, use (9.6-4) to compute E. If $E \leq \epsilon$ (where ϵ is a preassigned small positive number), there is no need to proceed further; the initial guess is satisfactory. In an actual design problem, however, this situation is unlikely to happen. In that case, we proceed to step 2.

Step 2. Perturb $\mathbf{p}(t)$ by a small function $\delta\mathbf{p}(t)$; that is, let the new resistors be $R_k(t) + \delta R_k(t)$, the new capacitors be $C_k(t) + \delta C_k(t)$, and the new inductors be $L_k(t) + \delta L_k(t)$. The perturbation $\delta\mathbf{p}(t)$ will cause a perturbation δs on the step response and a change δE on the error so that

$$E(\mathbf{p} + \delta\mathbf{p}) - E(\mathbf{p}) = \tfrac{1}{2} \int_0^T \int_0^t [(s + \delta s - \alpha)^2 - (s - \alpha)^2] \, d\tau \, dt$$

Simplifying and neglecting the second-order terms, we get

$$\delta E(\mathbf{p}) = \int_0^T \int_0^t [s(t, \tau; \mathbf{p}) - \alpha(t, \tau)] \, \delta s(t, \tau; \mathbf{p}) \, d\tau \, dt \tag{9.6-5}$$

Step 3. Choose $\delta\mathbf{p}(t)$ such that the following two conditions are satisfied:

1. $\delta E < 0$.
2. $|\delta E|$ is maximized.

This objective can be achieved by a steepest-descent technique similar to the previous case. This will be demonstrated in a particular case, which we discuss soon.

Step 4. Choose $\mathbf{p} + \delta\mathbf{p}$ as the new element value vector and go to step 2. Continue the iterations until E is sufficiently small or E is minimized.

To discuss step 3 of the iteration method in detail, we consider the following special case.

Design of Networks with a Prescribed Impulse Response. To keep the analysis simple, we only discuss the design of a first-order network; the same techniques can be applied to nth-order networks. Consider a network whose state equation is

$$\dot{x} = a(t)x + u(t) \tag{9.6-6}$$

where $x(t)$ and $u(t)$ are scalar functions representing the state and the input of the network, respectively. The function $a(t)$ is to be determined so that the impulse response of the resulting network is as close as possible to a desired impulse response. More precisely, denote the impulse response of (9.4-6) by $h(t, \tau)$, where τ denotes the time of the application of the impulse and t denotes the time of the observation of the response; then

$$h(t, \tau) = \int_0^t \exp\left[\int_{t'}^t a(\xi)\, d\xi\right]\delta(t' - \tau)\, dt' \tag{9.6-7}$$

where $\delta(t' - \tau)$ denotes a unit impulse function applied at time τ. Using the sifting property of the impulse function, (9.6-7) can be simplified as

$$h(t, \tau) = \exp\left[\int_\tau^t a(\xi)\, d\xi\right], \qquad t \geq \tau \tag{9.6-8}$$

The problem is to choose $a(t)$ such that $h(t, \tau)$ is a "good" approximation of a given function $\hat{h}(t, \tau)$. For this purpose the following error function is chosen:

$$E(a) = \frac{1}{2}\int_0^T \int_0^t [h(t, \tau) - \hat{h}(t, \tau)]^2\, d\tau\, dt \tag{9.6-9}$$

where T is a finite constant. To start the iteration, we choose a reasonable function $a(t)$ as the initial guess and use (9.6-8) to compute $h(t, \tau)$ and (9.6-9) to compute the error function. If we perturb $a(t)$ by a small function $\delta a(t)$, the corresponding perturbation on the impulse response will be $\delta h(t, \tau)$ and the change in the error function is

$$E(a + \delta a) - E(a) = \frac{1}{2}\int_0^T \int_0^t [(h + \delta h - \hat{h})^2 - (h - \hat{h})^2]\, d\tau\, dt$$

Simplifying and neglecting the second-order term, we get

$$\delta E = \int_0^T \int_0^t [h(t, \tau) - \hat{h}(t, \tau)]\, \delta h(t, \tau)\, d\tau\, dt \tag{9.6-10}$$

To obtain $\delta h(t, \tau)$ in terms of $\delta a(t)$, we use (9.6-8):

$$h(a + \delta a) - h(a) = \exp\left[\int_\tau^t (a + \delta a)\, d\xi\right] - \exp\left(\int_\tau^t a\, d\xi\right)$$

or

$$h(a + \delta a) - h(a) = \left[\exp\left(\int_\tau^t \delta a\, d\xi\right) - 1\right]\exp\left(\int_\tau^t a\, d\xi\right)$$

But for sufficiently small $\delta a(t)$, we can approximate the exponential function as

$$\exp\left[\int_\tau^t \delta a(\xi)\,d\xi\right] \simeq 1 + \int_\tau^t \delta a(\xi)\,d\xi \tag{9.6-11}$$

Thus,

$$\delta h(t, \tau) = \exp\left[\int_\tau^t a(\xi)\,d\xi\right] \int_\tau^t \delta a(\xi)\,d\xi$$

Using δh from this equation in (9.6-10), we get

$$\delta E = \int_0^T \int_0^t \int_\tau^t g(t, \tau)\,\delta a(\xi)\,d\xi\,d\tau\,dt \tag{9.6-12}$$

where $g(t, \tau)$ is defined as

$$g(t, \tau) \triangleq [h(t, \tau) - \hat{h}(t, \tau)] \exp\left[\int_\tau^t a(\xi)\,d\xi\right] \tag{9.6-13}$$

Note that since $a(t)$ is specified $g(t, \tau)$ can be considered a known function at each iteration, because it can be computed using (9.6-13) and (9.6-8). To be able to use a steepest-descent technique similar to that used in Section 9.2, we should try to "pull out" δa from under the first two integrals in (9.6-12). This can be done by changing the order of integration twice and noting that $T \geq t \geq \tau \geq 0$; the result is

$$\delta E = \int_0^T \left[\int_\xi^T \int_0^\xi g(t, \tau)\,d\tau\,dt\right] \delta a(\xi)\,d\xi \tag{9.6-14}$$

which can be put in the form

$$\delta E = \int_0^T l(\xi)\,\delta a(\xi)\,d\xi \tag{9.6-15}$$

where we have used the following notation:

$$l(\xi) \triangleq \int_\xi^T \int_0^\xi g(t, \tau)\,d\tau\,dt \tag{9.6-16}$$

Notice that since $g(t, \tau)$ is known from (9.6-13), $l(\tau)$ is completely specified and can be computed from (9.6-16). At this point, to obtain the optimum δa, we must bound the norm of $\delta a(t)$. As in the frequency-domain case, there are two common ways of restricting δa; the first is to impose a bound on the L_2 norm of $\delta a(t)$ and the second is to impose a bound on the L_∞ norm of $\delta a(t)$. In either case we must choose $\delta a(t)$ such that $\delta E < 0$ and $|\delta E|$ is maximized. The following two theorems will provide a method for computing optimum $\delta a(t)$. The proofs of these theorems are analogous to Theorems 9.3-1 and 9.3-2.

Theorem 9.6-1. Let the L_2 norm of $\delta a(t)$ be bounded by η; that is, let

$$\int_0^T \delta a^2(t)\,dt \leq \eta^2 \tag{9.6-17}$$

Then δE given in (9.6-15) satisfies conditions (a) $\delta E < 0$, and (b) $|\delta E|$ is maximized, if and only if

$$\delta a(t) = -\mu l(t) \tag{9.6-18}$$

where μ is the step size. As before, the step size μ is chosen small enough so that the approximation (9.6-11) remains valid.

Consequently, if we impose a bound on the L_2 norm of $\delta a(t)$, the desired perturbation is "proportional" to $l(t)$, which can be computed using (9.6-16).

Theorem 9.6-2. Let the L_∞ norm of $\delta a(t)$ be bounded by δ; that is, let

$$\sup_{0 \leq t \leq T} |\delta a(t)| \leq \eta \tag{9.6-19}$$

Then δE satisfies conditions (a) and (b) stated in Theorem 9.6-1 if and only if

$$\delta a(t) = -\mu \operatorname{sgn}[l(t)] \tag{9.6-20}$$

where $\operatorname{sgn}[l(t)] = 1 \quad \text{for all } l(t) \geq 0$

and $\operatorname{sgn}[l(t)] = -1 \quad \text{for all } l(t) < 0$

Hence, in this case only the sign of $l(t)$ needs to be determined; $\delta a(t)$ is either $+\mu$ or $-\mu$, a discontinuous "bang-bang" solution. However, if a continuous $\delta a(t)$ is required, the previous theorem must be used. Having computed $\delta a(t)$, we choose $a(t) + \delta a(t)$ as the new function and continue the iterations until the cost function is minimized.

Since the field of computer-aided design of linear time-varying and nonlinear networks is relatively unexplored, there are many questions concerning efficiency, convergence, sensitivity, and so on, that still remain to be answered. These questions will eventually be answered as the research in this area continues [12–15]. The design of a special class of nonlinear networks will be considered in Chapter 10, after the small–signal behavior of these networks is discussed.

REFERENCES

[1] DESOER, C. A., and S. K. MITRA, "Design of Lossy Ladder Filters by Digital Computer," *IRE Trans. Circuit Theory*, **CT-8** (Sept. 1961), 192–201.

[2] CALAHAN, D. A., "Computer Design of Linear Frequency Selective Networks," *Proc. IEEE*, **53** (Nov. 1965), 1701–1706.

[3] ROHRER, R. A., "Fully Automated Network Design by Digital Computer: Preliminary Consideration," *Proc. IEEE*, **55** (Nov. 1967), 1929–1939.

[4] TEMES, G. C., and D. A. CALAHAN, "Computer-Aided Network Optimization —The State of The Art," *Proc. IEEE*, **55** (Nov. 1967), 1832–1864.

[5] DIRECTOR, S. W., and R. A. ROHRER, "Automated Network Design—The Frequency Domain Case," *IEEE Trans. Circuit Theory*, **CT-16**, No. 3 (Aug. 1969), 330–337.

[6] DIRECTOR, S. W., "Survey of Circuit-Oriented Optimization Techniques," *IEEE Trans. Circuit Theory*, **CT-18**, No. 1 (1971), 3–10.

[7] KUO, F. F., and W. G. MAGNUSON, JR., *Computer Oriented Circuit Design*, Prentice-Hall Inc., Englewood Cliffs, N. J., 1969.

[8] WILDE, D. J., and C. S. BEIGHTLER, *Foundations of Optimization*, Prentice-Hall Inc., Englewood Cliffs, N.J., 1967.

[9] FLETCHER, R. M., and J. C. POWELL, "A Rapidly Convergent Descent Method for Minimization," *Computer J.*, **6**, No. 4 (1963), 163–168.

[10] FLETCHER, R. M., and C. M. REEVES, "Function Minimization by Conjugate Gradient," *Computer J.*, **7** (July 1964), 149–154.

[11] CALAHAN, D. A., *Computer-Aided Network Design*, Preliminary Edition, McGraw-Hill Book Company, New York, 1968.

[12] DESOER, C. A., and B. PEIKARI, "Design of Linear Time-Varying and Nonlinear Time-Invariant Networks," *IEEE Trans. Circuit Theory*, **CT-17** (May 1970), 232–240.

[13] PEIKARI, B., "On the Synthesis of Nonlinear Time-Invariant Networks," *IEEE Trans. Circuit Theory*, **CT-17** (Nov. 1970), 657–659.

[14] PEIKARI, B., "Design of Nonlinear Networks with a Prescribed Small-Signal Behavior," *IEEE Trans. Circuit Theory*, **CT-19**, No. 4 (1972), 389–391.

[15] CHUA, L. O., "Synthesis of Nonlinear Systems with Prescribed Singularities," *IEEE Trans. Circuit Theory*, **CT-18** (May 1971), 375–382.

PROBLEMS

9.1 An error function $E(\mathbf{p})$ is said to be *convex* if for any two points $\mathbf{p}^1$ and $\mathbf{p}^2$ in the parameter space the following inequality holds:

$$E[\mathbf{p}^1 + \lambda(\mathbf{p}^2 - \mathbf{p}^1)] < E(\mathbf{p}^1) + \lambda[E(\mathbf{p}^2) - E(\mathbf{p}^1)] \quad \text{for all } 0 < \lambda < 1$$

Give a geometric interpretation of this definition of convexity and show that a convex function has at most one extremum (minimum or maximum).

9.2 An error function $E(\mathbf{p})$ is said to be *unimodal* if for any two points $\mathbf{p}^1$ and $\mathbf{p}^2$, $E(\mathbf{p}^2) < E(\mathbf{p}^1)$ implies that there exists a monotonically decreasing path on the surface of $E(\mathbf{p})$ connecting $E(\mathbf{p}^2)$ to $E(\mathbf{p}^1)$. Give a geometrical interpretation of this definition and convince yourself that a unimodal error function has but one extremum.

9.3 Determine which of the following functions are convex and/or unimodal:

(a) $E(\mathbf{p}) = p_1^2 + p_2^2 + 2$

(b) $E(\mathbf{p}) = p_1^3 + p_2^3 - p_1 - p_2$

(c) $E(\mathbf{p}) = \sin p_1 + \sin p_2 + 2(p_1 + p_2)$

9.4 Consider the design of a linear time-invariant network whose impulse response $h(t)$ is to be as close as possible to a prescribed function $\hat{h}(t)$. Let the error function be

$$E(\mathbf{p}) = \frac{1}{2}\int_0^T [h(\mathbf{p}, t) - \hat{h}(t)]^2\, dt$$

Use a perturbation analysis similar to that discussed in Section 9.3 to show that if the L_2 norm of $\delta\mathbf{p}$ is bounded then the optimal change on the parameter vector is

$$\delta\mathbf{p} = -\mu \int_0^T [h(\mathbf{p}, t) - \hat{h}(t)]\mathbf{g}(\mathbf{p}, t)\, dt$$

where
$$\mathbf{g}^T(\mathbf{p}, t) = \left[\frac{\partial h}{\partial p_1}, \frac{\partial h}{\partial p_2}, \ldots, \frac{\partial h}{\partial p_q}\right]$$

9.5 Repeat Problem 9.5, but this time assume that $|\delta p_k| \leq \eta_k$. Show that the optimum δp_k is

$$\delta p_k = -\mu \operatorname{sgn}\left\{\int_0^T [h(\mathbf{p}, t) - \hat{h}(t)]g_k(\mathbf{p}, t)\right\} dt$$

9.6 Find the optimum value of the step size μ that satisfies conditions (9.3-13) and (9.3-14) of Theorem 9.3-1.
[Ans.: $\mu_{\text{opt}} = \eta/\sqrt{\boldsymbol{\beta}^T\boldsymbol{\beta}}$]

9.7 Consider the active RC filter of Fig. 9.3-2. Let the desired transfer function be as in (9.3-17). By simple coefficient matching, show that the optimal element values of the filter are

$$G_1 = \sqrt{2}, \quad G_2 = \frac{\sqrt{2}}{2}, \quad \text{and} \quad C_1 = C_2 = 1$$

and show that these values correspond to $E(\mathbf{p}) = 0$.

9.8 Choose different initial values in the iterative solution of the design problem discussed in Example 9.3-1. Are the resulting parameters different? Why?

9.9 Choose an appropriate topology for a third-order active low-pass Butterworth filter whose desired transfer function is given by

$$\hat{H}(s) = \frac{1}{s^3 + 2s^2 + 2s + 1}$$

Let all the resistors be 1Ω resistors and use an iterative method to obtain the values of the capacitors such that the resulting transfer function is a good approximation of $\hat{H}(s)$ over the frequency range of [1, 5] radians per second.

9.10 Choose one of the configurations of Fig. 8.7-11, and use an iterative method to obtain the values of the parameters α, k, and R [or C if 8.7-11(b) is chosen] so that the resulting transfer function is a good approximation of

$$\hat{H}(s) = \frac{2(s^2 + 1)}{s^2 + 2s + 3}$$

Assume that the noninverting operational amplifiers have a gain of 2.

9.11 Show that a Hessian matrix $\mathbf{H}$ is always symmetric.

9.12 Show that the Hessian matrix of a convex error function is positive definite.

9.13 Use the Fletcher–Powell method to design the active *RC* network of Example 9.3-1. Plot the error function with respect to the number of iterations and compare it with the plot of Fig. 9.3-2.

9.14 Consider the iterative design of a two-port *RLC* network in the frequency domain using the adjoint network.
(a) Choose an appropriate error function.
(b) Use an analysis similar to that developed in Section 9.5 to determine the appropriate excitations for the adjoint network.

9.15 Derive equation (9.6-14) from (9.6-12). Note that $T \geq t \geq \tau \geq 0$.

9.16 Give a rigorous proof for Theorem 9.6-2.

Small-Signal Analysis and Design of Nonlinear Networks

10

10.1 INTRODUCTION

In this chapter we present a detailed discussion on the small-signal behavior of nonlinear networks and introduce a method for designing those nonlinear networks whose small-signal equivalent is a desired linear time-varying network. Implementation of the small-signal properties of nonlinear networks has become a standard method of designing modern electronic circuits and devices, particularly those with adjustable or adaptive characteristics, such as filters with variable bandwidth or delay, automatic tuning circuits, and the like.

The small-signal behavior of individual nonlinear capacitors, inductors, and resistors was extensively discussed in Chapter 1. We showed that if the nonlinear element is used with a dc bias source it behaves as a linear time-invariant element as far as the input and the output signals are concerned, provided that the amplitude of the input signal is sufficiently small. The value of the resulting linear element was shown to be equal to the slope of the corresponding nonlinear element at the operating point generated by the dc source. We also showed that if the bias source is time varying then the corresponding operating point and the resulting linear element are time varying. In this chapter we extend these results to networks that contain an arbitrary number of nonlinear elements. We show that the corresponding *linearized*, *incremental*, or *small-signal equivalent* network is a linear network with the same topology and element type except that all the nonlinear elements are replaced by their small-signal equivalent. A mathematical representation of this fundamental idea is given in the next two sections; the state equations of the linearized network are derived and a nonlinear differential equation is obtained that can be solved for the time-varying operating points.

If the time-varying bias sources are *slowly varying*, the frozen-operating-point concept can be used to derive a simple method of computing the operating points (Section 10.4). Finally, a systematic method of designing nonlinear networks whose small-signal equivalent is a desired linear time-varying network will be presented in Section 10.5. It will be shown that any linear time-varying network satisfying certain conditions can be designed using nonlinear time-invariant elements and variable bias sources.

10.2 SMALL-SIGNAL ANALYSIS OF AUTONOMOUS NETWORKS

Consider a nonlinear time-invariant network whose state equation can be written in the form

$$\dot{\mathbf{x}}(t) = \mathbf{f}[\mathbf{x}(t), \mathbf{E}] \tag{10.2-1}$$

where $\mathbf{x}(t)$ denotes the state vector, $\mathbf{f}[\mathbf{x}(t), \mathbf{E}]$ is a vector-valued function, and $\mathbf{E}$ is a vector denoting the time-invariant bias sources of the network. An example of such networks is illustrated in Fig. 10.2-1, in which the energy-storing elements are nonlinear time invariant and the resistors are linear time invariant. The bias sources are constant voltage sources e_1 and e_2. Equations of the form of (10.2-1) describe a special class of nonlinear networks known as autonomous networks; more precisely,

Definition 10.2-1 (Autonomous Versus Nonautonomous Networks). A network is said to be *autonomous* if it is comprised of time-invariant elements (linear

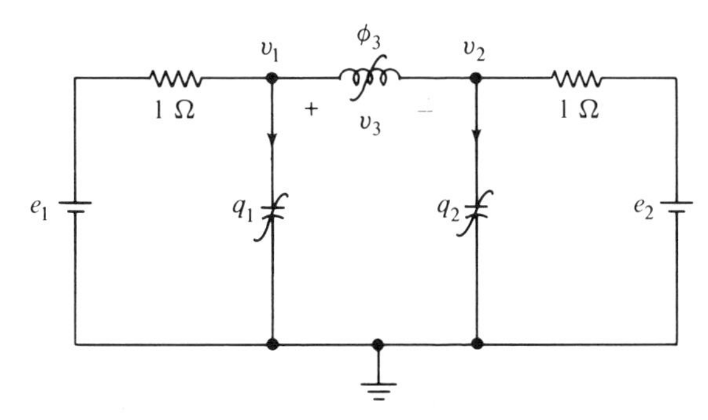

Figure 10.2-1 Autonomous network containing linear resistors, nonlinear energy-storing elements, and dc sources.

or nonlinear) and dc voltage and/or current sources only. A network that is not autonomous is called *nonautonomous.*

Before discussing the small-signal behavior of autonomous networks, let us introduce the concept of equilibrium point. Since autonomous networks are usually linearized around one of their equilibrium points, this concept plays a crucial role in our later discussions. Generally speaking, an equilibrium point is a point in the state space at which the state of the network $\mathbf{x}$ becomes a constant vector $\mathbf{x}_e$; that is, $\dot{\mathbf{x}}_e(t) = \mathbf{0}$. More specifically,

Definition 10.2-2 (Equilibrium Point). The point $\mathbf{x}_e$ in the state space is said to be an *equilibrium point* or *equilibrium state* of (10.2-1) if

$$\dot{\mathbf{x}}_e = \mathbf{f}(\mathbf{x}_e, \mathbf{E}) = \mathbf{0} \quad \text{for all } t \geq t_0 \tag{10.2-2}$$

where t_0 denotes the time at which the state $\mathbf{x}(t)$ first reaches the equilibrium point.

Note that if the autonomous network under consideration is stable, the equilibrium state will be the same as the steady state.

Example 10.2-1. The state equation of a nonlinear first-order circuit is

$$\dot{x} = -x + x^3$$

Obtain the equilibrium points of this network.

Solution

By definition, the equilibrium points of the network under study are the solutions of

$$x_e^3 - x_e = 0$$

which are

$$x_e = 0 \quad \text{and} \quad x_e = \pm 1$$

Example 10.2-2. Consider the network shown in Fig. 10.2-1. Let the capacitors be charge controlled with v-q relation

$$v_1 = f_1(q_1), \qquad v_2 = f_2(q_2)$$

and the inductor be flux controlled with i-φ relation

$$i_3 = f_3(\varphi_3)$$

Obtain the equilibrium points of this network.

Solution

Take the capacitor charges and the inductor flux as the state variables and write the state equations as follows.

$$\dot{q}_1 = -f_1(q_1) - f_3(\varphi_3) + e_1$$
$$\dot{q}_2 = -f_2(q_2) + f_3(\varphi_3) + e_2$$
$$\dot{\varphi}_3 = f_1(q_1) - f_2(q_2)$$

The equilibrium points, therefore, are the solutions of

$$-f_1(q_1) - f_3(\varphi_3) + e_1 = 0$$
$$-f_2(q_2) + f_3(\varphi_3) + e_2 = 0$$
$$f_1(q_1) - f_2(q_2) = 0$$

A simple manipulation of these equations yields

$$f_1(q_1) = \frac{e_1 + e_2}{2}, \qquad f_2(q_2) = \frac{e_1 + e_2}{2}$$

and $$f_3(\varphi_3) = \frac{e_1 - e_2}{2}$$

Then the equilibrium point is

$$q_1 = f_1^{-1}\left(\frac{e_1 + e_2}{2}\right), \qquad q_2 = f_2^{-1}\left(\frac{e_1 + e_2}{2}\right), \quad \text{and} \quad \varphi_3 = f_3^{-1}\left(\frac{e_1 - e_2}{2}\right)$$

where $f^{-1}(\cdot)$ denotes the inverse function of $f(\cdot)$. As a numerical example, let $f_1(q_1) = \sqrt{q_1}$, $f_2(q_2) = e^{q_2}$, $f_3(\varphi_3) = \varphi_3$, $e_1 = e_2 = 1\text{ V}$

Hence, $q_1 = 1^2 = 1$, $q_2 = \log 1 = 0$, and $\varphi_3 = 0$

Remark 10.2-1. As was shown in the first example, a nonlinear network may have more than one equilibrium point. A linear autonomous network, however, has either a unique equilibrium point or infinitely many of them. To clarify this statement, consider a linear autonomous network whose state equations are

$$\dot{\mathbf{x}}(t) = \mathbf{A}\mathbf{x}(t) + \mathbf{E} \tag{10.2-3}$$

where $\mathbf{x}(t)$ is the state vector, $\mathbf{A}$ is an $n \times n$ constant matrix, and E is an n-column vector. The equilibrium point $\mathbf{x}_e$ is then the solution of

$$\mathbf{A}\mathbf{x}_e = -\mathbf{E} \tag{10.2-4}$$

If $\mathbf{A}$ is nonsingular, the network under consideration has but one equilibrium point, $\mathbf{x}_e$, given by

$$\mathbf{x}_e = -\mathbf{A}^{-1}\mathbf{E} \tag{10.2-5}$$

On the other hand, if $\mathbf{A}$ is singular, (10.2-4) has infinitely many solutions. In fact, in this case (10.2-4) describes an *equilibrium line* or an *equilibrium plane* in the state space.

Alternative Method of Obtaining the Equilibrium State. Let us now present an alternative method of obtaining the equilibrium state of an autonomous network without writing the state equations. Consider a nonlinear autonomous network whose state equations are given as in (10.2-1). In this equation, $\mathbf{x}(t)$ denotes either the capacitor charges or voltages and inductor fluxes or currents. Hence, putting $\dot{\mathbf{x}}_e = \mathbf{0}$ implies that *at the equilibrium state* $\mathbf{x}_e$ *the voltages across all the inductors and the currents through all the capacitors are equal to zero*. Consequently, the equilibrium state $\mathbf{x}_e$ can be obtained by solving a purely resistive network, which is obtained from the original network by

1. Short circuiting all the inductors and
2. Open circuiting all the capacitors.

Let us now work out an example to illustrate this method.

Example 10.2-3. Obtain the equilibrium points (or state) of the network discussed in Example 10.2-2 using the method just outlined.

Solution

Short circuiting the inductor and open circuiting the capacitors, the network shown in Fig. 10.2-1 becomes as shown in Fig. 10.2-2. Solving this simple

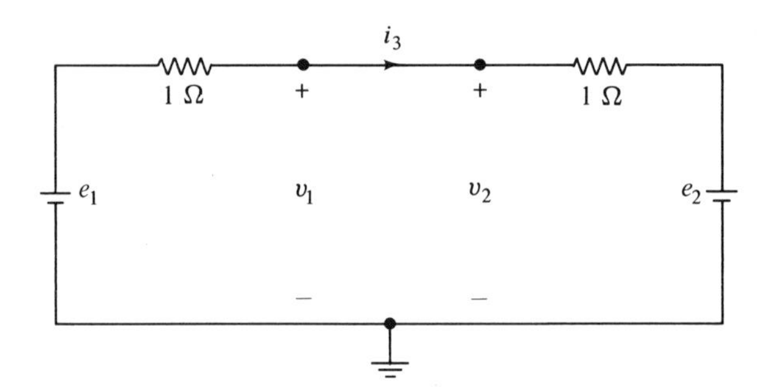

Figure 10.2-2 Solving this resistive network yields the equilibrium points of the network shown in Fig. 10.2-1.

resistive network for i_3, v_1, and v_2, we get

$$i_3 = \frac{e_1 - e_2}{2}, \qquad v_1 = v_2 = \frac{e_1 + e_2}{2}$$

Hence, from the v-q and i-φ characteristics of the nonlinear energy-storing elements given in Example 10.2-2, we can write

$$f_1(q_1) = \frac{e_1 + e_2}{2}, \qquad f_2(q_2) = \frac{e_1 + e_2}{2}, \quad \text{and} \quad f_3(\varphi_3) = \frac{e_1 - e_2}{2}$$

or, as in the previous example, the equilibrium point(s) are given by

$$q_1 = f_1^{-1}\left(\frac{e_1 + e_2}{2}\right), \qquad q_2 = f_2^{-1}\left(\frac{e_1 + e_2}{2}\right), \quad \text{and} \quad \varphi_3 = f_3^{-1}\left(\frac{e_1 - e_2}{2}\right)$$

Consequently, to obtain the equilibrium point of an autonomous network, the state equations of the network are not needed. These equilibrium points can be obtained by solving the corresponding resistive network. Let us now proceed with the small-signal analysis.

Perturbation Analysis. To understand the basic idea behind the small-signal behavior of a network, it is best to consider a specific example, such as the network shown in Fig. 10.2-1. This network can be considered as a low-pass filter whose energy-storing elements are nonlinear. The signal to be filtered usually is superimposed with one of the bias sources, say e_1(see Fig. 10.2-3).

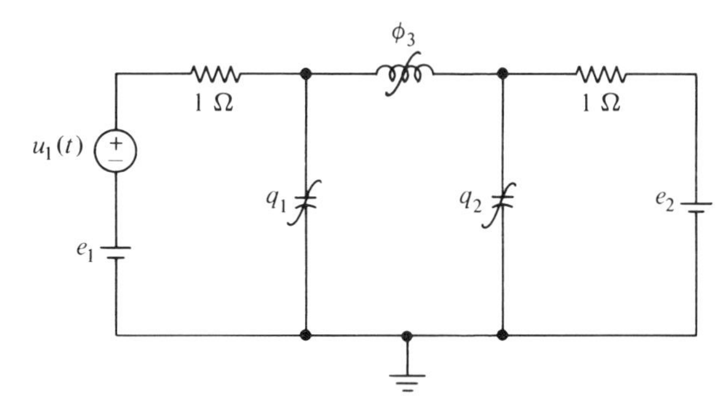

Figure 10.2-3 Small signal $u_1(t)$ superimposed on the bias e_1 will cause a perturbation $\xi(t)$ on the equilibrium point.

Let us denote this signal by $u_1(t)$. If the state of the network under study was at an equilibrium point prior to applying the signal $u_1(t)$, it will be perturbed by $\xi(t)$ due to the application of $u_1(t)$. If the signal $u_1(t)$ is small, that is, if $|u_1(t)|_\infty$ is a small number for each $t \geq 0$, the corresponding perturbation $\xi(t)$ is generally small. Henceforth, the perturbation vector $\xi(t)$ is referred to as the *small-signal response* of the network due to $u_1(t)$.

With this background let us now perform a precise small-signal analysis on a given autonomous network. Consider a nonlinear autonomous network whose state equations are given as (10.2-1). Assume that to the bias vector **E** we have added a small-signal input $\mathbf{u}(t)$; then this equation becomes

$$\dot{\mathbf{x}}(t) = \mathbf{f}[\mathbf{x}(t), \mathbf{E} + \mathbf{u}(t)]$$

where $\mathbf{x}(t)$ is an **n**-column vector, $\mathbf{u}(t)$ is an **m**-column vector, and $\mathbf{f}(\cdot, \cdot)$ is a vector-valued function whose first argument represents the state and whose second argument represents the bias and the small signals. In the absence of the input $\mathbf{u}(t)$, the equilibrium $\mathbf{x}_e$ is the solution of

$$\dot{\mathbf{x}}_e = \mathbf{f}(\mathbf{x}_e, \mathbf{E}) = \mathbf{0} \tag{10.2-6}$$

The small input signal $\mathbf{u}(t)$ will cause a perturbation $\xi(t)$ on $\mathbf{x}_e$; then $\mathbf{x}_e + \xi(t)$

must satisfy the original state equations. That is, we must have

$$\dot{\mathbf{x}}_e + \dot{\boldsymbol{\xi}}(t) = \mathbf{f}[\mathbf{x}_e + \boldsymbol{\xi}(t), \mathbf{E} + \mathbf{u}(t)]$$

Since by (10.2-6), $\dot{\mathbf{x}}_e = \mathbf{0}$, this equation becomes

$$\dot{\boldsymbol{\xi}}(t) = \mathbf{f}[\mathbf{x}_e + \boldsymbol{\xi}(t), \mathbf{E} + \mathbf{u}(t)] \tag{10.2-7}$$

To proceed with the analysis, we must make certain assumptions concerning the nonlinear function $\mathbf{f}(\cdot, \cdot)$ and the norm of the input signal $\mathbf{u}(t)$. These are

1. First and second derivatives of $\mathbf{f}(\cdot, \cdot)$ with respect to both arguments exist and are continuous.
2. The L_∞ norm of $\mathbf{u}(t)$, (i.e. $\|\mathbf{u}\|_\infty$), is a "small" number.

Then, since by assumption 1 $\mathbf{f}(\cdot, \cdot)$ is twice continuously differentiable, we can use a Taylor expansion to expand the right-hand side of (10.2-7) around the fixed equilibrium point $(\mathbf{x}_e, \mathbf{E})$; this yields

$$\dot{\boldsymbol{\xi}}(t) = \mathbf{f}(\mathbf{x}_e, \mathbf{E}) + \mathbf{f}'_x(\mathbf{x}_e, \mathbf{E})\boldsymbol{\xi}(t) + \mathbf{f}'_u(\mathbf{x}_e, \mathbf{E})\mathbf{u}(t) + \mathbf{g}(\boldsymbol{\xi}, \mathbf{u}) \tag{10.2-8}$$

This expression requires some explanation. The coefficient of $\boldsymbol{\xi}(t)$, that is, $\mathbf{f}'_x(\mathbf{x}_e, \mathbf{E})$, is an $n \times n$ constant matrix whose elements are the derivative of $\mathbf{f}(\cdot, \cdot)$ with respect to its first argument evaluated at $(\mathbf{x}_e, \mathbf{E})$; that is,

$$\mathbf{f}'_x(\mathbf{x}_e, \mathbf{E}) \triangleq \begin{bmatrix} \frac{\partial f_1}{\partial x_1} & \frac{\partial f_1}{\partial x_2} & \cdots & \frac{\partial f_1}{\partial x_n} \\ \frac{\partial f_2}{\partial x_1} & \frac{\partial f_2}{\partial x_2} & \cdots & \frac{\partial f_2}{\partial x_n} \\ \cdot & \cdot & & \\ \cdot & \cdot & & \\ \cdot & \cdot & & \\ \frac{\partial f_n}{\partial x_1} & \frac{\partial f_n}{\partial x_2} & \cdots & \frac{\partial f_n}{\partial x_n} \end{bmatrix}\Bigg|_{(\mathbf{x}_e, \mathbf{E})} \tag{10.2-9}$$

where $f_1, f_2, \ldots, f_n$ are the components of $\mathbf{f}$, and $x_1, x_2, \ldots, x_n$ are the components of $\mathbf{x}$. Similarly, $\mathbf{f}'_u(\mathbf{x}_e, \mathbf{E})$ is an $n \times m$ constant matrix whose elements are the derivative of $\mathbf{f}(\cdot, \cdot)$ with respect to its second argument evaluated at $(\mathbf{x}_e, \mathbf{E})$:

$$\mathbf{f}'_u(\mathbf{x}_e, \mathbf{E}) \triangleq \begin{bmatrix} \frac{\partial f_1}{\partial u_1} & \frac{\partial f_1}{\partial u_2} & \cdots & \frac{\partial f_1}{\partial u_m} \\ \frac{\partial f_2}{\partial u_1} & \frac{\partial f_2}{\partial u_2} & \cdots & \frac{\partial f_2}{\partial u_m} \\ \cdot & \cdot & & \\ \cdot & \cdot & & \\ \cdot & \cdot & & \\ \frac{\partial f_n}{\partial u_1} & \frac{\partial f_n}{\partial u_2} & \cdots & \frac{\partial f_n}{\partial u_m} \end{bmatrix}\Bigg|_{(\mathbf{x}_e, \mathbf{E})} \tag{10.2-10}$$

The first term in the right-hand side of equation (10.2-8) is zero by (10.2-6);

hence,

$$\dot{\boldsymbol{\xi}}(t) = \mathbf{f}'_x(\mathbf{x}_e, \mathbf{E})\boldsymbol{\xi}(t) + \mathbf{f}'_u(\mathbf{x}_e, \mathbf{E})\mathbf{u}(t) + \mathbf{g}(\boldsymbol{\xi}, \mathbf{u}) \tag{10.2-11}$$

The function $\mathbf{g}(\boldsymbol{\xi}, \mathbf{u})$ in (10.2-8) denotes the second- and higher-order terms of the Taylor expansion. It is intuitively clear and has been rigorously shown [1] that if conditions 1 and 2 are satisfied, $\|\mathbf{g}(\boldsymbol{\xi}, \mathbf{u})\|_\infty$ is small compared to the norm of the rest of the terms and hence can be neglected. The *approximate* small-signal response of the network, therefore, is the solution of the following linear time-invariant differential equation:

$$\dot{\boldsymbol{\xi}}(t) = \mathbf{A}\boldsymbol{\xi}(t) + \mathbf{B}\mathbf{u}(t) \tag{10.2-12}$$

where we have chosen the following simplified notations:

$$\mathbf{A} \triangleq \mathbf{f}'_x(\mathbf{x}_e, \mathbf{E}) \tag{10.2-13}$$

and

$$\mathbf{B} \triangleq \mathbf{f}'_u(\mathbf{x}_e, \mathbf{E}) \tag{10.2-14}$$

Remark 10.2-2. Notice that (10.2-12) can be considered as the state equation of a linear time-invariant network whose input is $\mathbf{u}(t)$ and whose response is $\boldsymbol{\xi}(t)$. Hence, *so far as the small signal* $\mathbf{u}(t)$ *is concerned, the nonlinear network whose state equation is given by* (10.2-1) *behaves as a linear time-invariant network.*

Let us now illustrate the procedure by a simple example.

Example 10.2-4. Consider the nonlinear time-invariant network shown in Fig. 10.2-4. Let the inductors be flux controlled with i-φ characteristics

$$i_1 = \varphi_1^{1/3}, \qquad i_2 = \varphi_2^{1/3}$$

and the capacitor be charge controlled with v-q characteristic

$$v_3 = e^{q_3}$$

Obtain the state equations of the small-signal equivalent of this network.

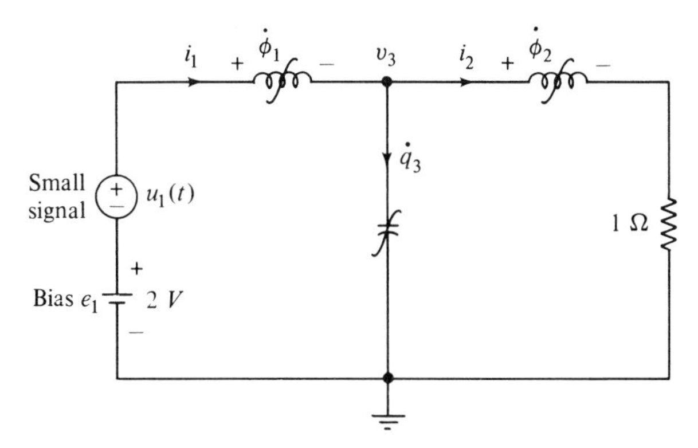

Figure 10.2-4 Nonlinear time-invariant network that behaves as a linear time-invariant network.

Solution

Let us first write the state equations of this network. Taking the capacitor charge and inductors fluxes as the state variables, we can write

$$\dot{\varphi}_1 = -e^{q_3} + u_1(t) + 2$$
$$\dot{\varphi}_2 = -\varphi_2^{1/3} + e^{q_3}$$
$$\dot{q}_3 = \varphi_1^{1/3} - \varphi_2^{1/3}$$

Next, the equilibrium point $\mathbf{x}_e$ can either be obtained from the state equations or directly from the network by short circuiting the inductors and open circuiting the capacitor. In either case, it is

$$\varphi_{1e} = 2^3, \qquad \varphi_{2e} = 2^3, \quad \text{and} \quad q_{3e} = \log 2$$

The matrices **A** and **B** can then be obtained by differentiating the state equations:

$$\mathbf{A} = \begin{bmatrix} 0 & 0 & e^{q_3} \\ 0 & -\frac{1}{3}\varphi_1^{-2/3} & e^{q_3} \\ \frac{1}{3}\varphi_1^{-2/3} & -\frac{1}{3}\varphi_1^{-2/3} & 0 \end{bmatrix}_{\substack{\varphi_1=\varphi_2=8 \\ q_3=\log 2}} = \begin{bmatrix} 0 & 0 & 2 \\ 0 & \frac{-1}{3\sqrt{2}} & 2 \\ \frac{1}{3\sqrt{2}} & \frac{-1}{3\sqrt{2}} & 0 \end{bmatrix}$$

and

$$\mathbf{B} = [1 \quad 0 \quad 0]^T$$

The state equations of the small-signal equivalent network are, therefore,

$$\begin{bmatrix} \dot{\xi}_1 \\ \dot{\xi}_2 \\ \dot{\xi}_3 \end{bmatrix} = \begin{bmatrix} 0 & 0 & 2 \\ 0 & \frac{-1}{3\sqrt{2}} & 2 \\ \frac{1}{3\sqrt{2}} & \frac{-1}{3\sqrt{2}} & 0 \end{bmatrix} \begin{bmatrix} \xi_1 \\ \xi_2 \\ \xi_3 \end{bmatrix} = \begin{bmatrix} 1 \\ 0 \\ 0 \end{bmatrix}^e$$

where ξ_1, ξ_2, and ξ_3 are the small-signal responses of φ_1, φ_2 and q_3, respectively, due to $u_1(t)$.

If the nonlinear network under consideration has more than one equilibrium point, the constant matrices **A** and **B** will in general, be different for each operating point. Furthermore, these matrices depend on the value of the bias sources **E**. *It is, therefore, possible to generate many small-signal equivalents to a given nonlinear network by simply changing the bias* **E.** This idea has been successfully used in designing electronically adjustable filters. We shall discuss this technique in detail when we discuss the nonlinear network synthesis in Section 10.5. Let us now develop a similar small-signal analysis on nonlinear time-varying networks.

10.3 SMALL-SIGNAL ANALYSIS OF NONAUTONOMOUS NETWORKS

If a nonlinear nonautonomous network is operated on a small region of its entire characteristic, it behaves as a *linear* time-varying network. To prove this interesting fact, let us start by defining the *operating point* of a nonautonomous network, which is the counterpart of the equilibrium point in autonomous networks.

Consider a nonautonomous network whose state equation is

$$\dot{\mathbf{x}}(t) = \mathbf{f}[\mathbf{x}(t), \hat{\mathbf{u}}(t), t] \tag{10.3-1}$$

where $\mathbf{x}(t)$ is the state vector, $\hat{\mathbf{u}}(t)$ is an m vector denoting the time-varying bias sources and the "small" input sources, and $\mathbf{f}(\cdot, \cdot, t)$ is a vector-valued function. If we denote the bias sources by $\bar{\mathbf{u}}(t)$ and the input signals by $\mathbf{u}(t)$, then $\hat{\mathbf{u}}(t)$ can be written as

$$\hat{\mathbf{u}}(t) = \bar{\mathbf{u}}(t) + \mathbf{u}(t) \tag{10.3-2}$$

where we assume that the L_∞ norm of $\mathbf{u}(t)$ is a small number. The state $\mathbf{x}(t)$ in (10.3-1) can then be written as the sum of two vectors $\bar{\mathbf{x}}(t)$ and $\boldsymbol{\xi}(t)$:

$$\mathbf{x}(t) = \bar{\mathbf{x}}(t) + \boldsymbol{\xi}(t) \tag{10.3-3}$$

where $\bar{\mathbf{x}}(t)$ corresponds to the bias source $\bar{\mathbf{u}}(t)$ and $\boldsymbol{\xi}(t)$ corresponds to the small input signal $\mathbf{u}(t)$. If $\boldsymbol{\xi}(t) = \mathbf{0}$, (10.3-1) can be written as

$$\dot{\bar{\mathbf{x}}}(t) = \mathbf{f}[\bar{\mathbf{x}}(t), \bar{\mathbf{u}}(t), t] \tag{10.3-4}$$

We can then define the operating point of a nonautonomous network as follows:

Definition 10.3-1. (Operating Point). The vector function $\bar{\mathbf{x}}(t)$ defined by (10.3-4) is called the *operating point* of the nonautonomous network (10.3-1) corresponding to the bias source $\bar{\mathbf{u}}(t)$.

If the nonlinear network under consideration is time invariant, then (10.3-4) reduces to

$$\dot{\bar{\mathbf{x}}}(t) = \mathbf{f}[\bar{\mathbf{x}}(t), \bar{\mathbf{u}}(t)] \tag{10.3-5}$$

In either case, if the bias sources $\bar{\mathbf{u}}(t)$ are time varying, the operating point is in fact a "point" in *function space* rather than the Euclidean space. The operating point as defined is sometimes referred to as the *generating solution* of the nonautonomous network under study [2].

Perturbation Analysis of Nonautonomous Networks. Consider a nonlinear time-varying network whose state equations are given as in (10.3-1). The operating point of this network corresponding to $\bar{\mathbf{u}}(t)$ is given by (10.3-4). If a small input signal $\mathbf{u}(t)$ is superimposed on the source $\mathbf{u}(t)$, the operating point $\bar{\mathbf{x}}(t)$ will be perturbed by a function $\boldsymbol{\xi}(t)$, so that from (10.3-3) and (10.3-4) we can write

$$\dot{\bar{\mathbf{x}}} + \dot{\boldsymbol{\xi}} = \mathbf{f}(\bar{\mathbf{x}} + \boldsymbol{\xi}, \bar{\mathbf{u}} + \mathbf{u}, t) \tag{10.3-6}$$

To justify the expansion of (10.3-6) by a Taylor expansion and disregarding the second- and higher-order terms, we must make the following assumptions:

1. $\mathbf{f}(\cdot, \cdot, t)$ has continuous, bounded, second partial derivatives with respect to its first and second arguments.
2. $\bar{\mathbf{u}}(t)$ is continuous.
3. $\|\mathbf{u}\|_\infty$ is a "small" number.

Under these assumptions, the right-hand side of (10.3-4) can be expanded by a Taylor expansion as

$$\dot{\bar{\mathbf{x}}} + \dot{\boldsymbol{\xi}} = \mathbf{f}(\bar{\mathbf{x}}, \bar{\mathbf{u}}, t) + \mathbf{f}'_x(\bar{\mathbf{x}}, \bar{\mathbf{u}}, t)\boldsymbol{\xi} + \mathbf{f}'_{\hat{u}}(\bar{\mathbf{x}}, \bar{\mathbf{u}}, t)\mathbf{u} + \mathbf{g}(\boldsymbol{\xi}, \mathbf{u}, t)$$

This equation together with (10.3-4) yields

$$\dot{\boldsymbol{\xi}} = \mathbf{f}'_x(\bar{\mathbf{x}}, \bar{\mathbf{u}}, t)\boldsymbol{\xi} + \mathbf{f}'_{\hat{u}}(\bar{\mathbf{x}}, \bar{\mathbf{u}}, t)\mathbf{u} + \mathbf{g}(\boldsymbol{\xi}, \mathbf{u}, t) \tag{10.3-7}$$

where $\mathbf{f}'_x(\bar{\mathbf{x}}, \bar{\mathbf{u}}, t)$ is an $n \times n$ time-varying matrix whose elements are the derivatives of $\mathbf{f}(\cdot, \cdot, t)$ with respect to its first argument evaluated at $(\bar{\mathbf{x}}, \bar{\mathbf{u}}, t)$. Similarly, $\mathbf{f}'_{\hat{u}}(\bar{\mathbf{x}}, \bar{\mathbf{u}}, t)$ is an $n \times m$ time-varying matrix whose elements are the derivatives of $\mathbf{f}(\cdot, \cdot, t)$ with respect to its second argument evaluated at $(\bar{\mathbf{x}}, \bar{\mathbf{u}}, t)$. More specifically,

$$\mathbf{f}'_x(\bar{\mathbf{x}}, \bar{\mathbf{u}}, t) = \begin{bmatrix} \dfrac{\partial f_1}{\partial x_1} & \dfrac{\partial f_1}{\partial x_2} & \cdots & \dfrac{\partial f_1}{\partial x_n} \\ \dfrac{\partial f_2}{\partial x_1} & \dfrac{\partial f_2}{\partial x_2} & \cdots & \dfrac{\partial f_2}{\partial x_n} \\ \vdots & \vdots & & \\ \dfrac{\partial f_n}{\partial x_1} & \dfrac{\partial f_n}{\partial x_2} & \cdots & \dfrac{\partial f_n}{\partial x_n} \end{bmatrix}_{(\bar{\mathbf{x}}, \bar{\mathbf{u}}, t)} \triangleq \mathbf{A}(t) \tag{10.3-8}$$

where f_i and x_i represent the ith component of $\mathbf{f}$ and $\mathbf{x}$, respectively. Also,

$$\mathbf{f}'_{\hat{u}}(\bar{\mathbf{x}}, \bar{\mathbf{u}}, t) = \begin{bmatrix} \dfrac{\partial f_1}{\partial \hat{u}_1} & \dfrac{\partial f_1}{\partial \hat{u}_2} & \cdots & \dfrac{\partial f_1}{\partial \hat{u}_m} \\ \dfrac{\partial f_2}{\partial \hat{u}_1} & \dfrac{\partial f_2}{\partial \hat{u}_2} & \cdots & \dfrac{\partial f_2}{\partial \hat{u}_m} \\ \vdots & & & \\ \dfrac{\partial f_n}{\partial \hat{u}_1} & \dfrac{\partial f_n}{\partial \hat{u}_2} & \cdots & \dfrac{\partial f_n}{\partial \hat{u}_m} \end{bmatrix}_{(\bar{\mathbf{x}}, \bar{\mathbf{u}}, t)} \triangleq \mathbf{B}(t) \tag{10.3-9}$$

where $\hat{u}_i$, $i = 1, 2, \ldots, m$, represents the ith component of $\hat{\mathbf{u}}$. We can then write (10.3-7) as

$$\dot{\boldsymbol{\xi}}(t) = \mathbf{A}(t)\boldsymbol{\xi}(t) + \mathbf{B}(t)\mathbf{u}(t) + \mathbf{g}(\boldsymbol{\xi}, \mathbf{u}, t) \qquad (10.3\text{-}10)$$

The last term in (10.3-10) represents the second- and higher-order terms in $\boldsymbol{\xi}$ and $\mathbf{u}$. Since $\boldsymbol{\xi}$ and $\mathbf{u}$ are essentially small, it can be shown [1] that under conditions 1, 2, and 3 the norm of $\mathbf{g}(\boldsymbol{\xi}, \mathbf{u}, t)$ is much smaller than the norm of the rest of the terms in (10.3-10). Consequently, $\mathbf{g}(\boldsymbol{\xi}, \mathbf{u}, t)$ can be ignored and the state equations of the *approximate small-signal equivalent* network become

$$\dot{\boldsymbol{\xi}}(t) = \mathbf{A}(t)\boldsymbol{\xi}(t) + \mathbf{B}(t)\mathbf{u}(t) \qquad (10.3\text{-}11)$$

Hence, *as far as the small input signal* $\mathbf{u}(t)$ *is concerned, the nonlinear time-varying network of* (10.3-1) *behaves as a linear time-varying network whose state equations are given in* (10.3-11). This *linearized* network is called the *small-signal equivalent* of (10.3-1) around the operating point $\bar{\mathbf{x}}(t)$. Let us now consider an example that illustrates this procedure.

Example 10.3-1. The state equations of a specific nonautonomous network are

$$\begin{aligned}
\dot{x}_1 &= 2x_1 + 3x_2^3 + t \sin x_3 + \hat{u}_1 + \hat{u}_2^2, & \bar{x}_1(0) &= 0 \\
\dot{x}_2 &= x_1^2 + 2x_2 + 3x_3^2 + \hat{u}_2 & \bar{x}_2(0) &= 0 \\
\dot{x}_3 &= 2x_2^2 + t^2 x_3^2 + \hat{u}_3 + \hat{u}_1^2 & \bar{x}_3(0) &= 0
\end{aligned}$$

and the time-varying bias sources are

$$\bar{\mathbf{u}}(t) = [\sin t \quad \cos t \quad 0]^T$$

Find the state equations of its small-signal equivalent network.

Solution

Let $\mathbf{u}(t)$ be the small signal and $\boldsymbol{\xi}(t)$ be the small-signal response; then the state equations of the small-signal equivalent network are given as in (10.3-11), where $\mathbf{A}(t)$ and $\mathbf{B}(t)$ can be obtained from (10.3-8) and (10.3-9) as

$$\mathbf{A}(t) = \begin{bmatrix} 2 & 9\bar{x}_2^2 & t\cos\bar{x}_3 \\ 2\bar{x}_1 & 2 & 6\bar{x}_3 \\ 0 & 4\bar{x}_2 & 2t^2\bar{x}_3 \end{bmatrix}$$

and

$$\mathbf{B}(t) = \begin{bmatrix} 1 & 2\cos t & 0 \\ 0 & 1 & 0 \\ 2\sin t & 0 & 1 \end{bmatrix}$$

and $\bar{x}_1$, $\bar{x}_2$, and $\bar{x}_3$ are the solutions of the following differential equations:

$$\begin{aligned}
\dot{\bar{x}}_1 &= 2\bar{x}_1 + 3\bar{x}_2^3 + t\sin\bar{x}_3 + \sin t + \cos^2 t \\
\dot{\bar{x}}_2 &= \bar{x}_1^2 + 2\bar{x}_2 + 3\bar{x}_3^2 + \cos t \\
\dot{\bar{x}}_3 &= 2\bar{x}_2^2 + t^2\bar{x}_3^2 + \sin^2 t
\end{aligned}$$

with the initial conditions

$$\bar{\mathbf{x}}(0) = \mathbf{0}$$

Example 10.3-2. Consider the nonlinear time-invariant network shown in Fig. 10.3-1. Let the bias source be a time-varying voltage source $\bar{u}_1(t) = 1 + 0.5 \sin t$. The inductor is flux controlled with i-φ relation

$$i_2 = \varphi_2^2$$

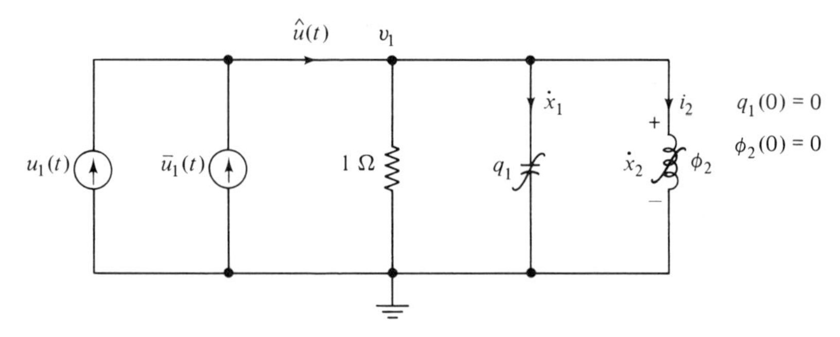

Figure 10.3-1 Nonlinear network whose small-signal equivalent behaves as a linear time-varying network.

and the capacitor is charge controlled with v-q relation

$$v_1 = \frac{1}{q_1}$$

Obtain the small-signal equivalent of this network.

Solution

Taking the charge on the capacitor and the flux through the inductor as the state variables, the state equations can be written as

$$\hat{u} = v_1 + \dot{x}_1 + i_2$$
$$v_1 = \dot{x}_2$$

or, equivalently,

$$\dot{x}_1 = \frac{-1}{x_1} - x_2^2 + \hat{u}(t)$$

$$\dot{x}_2 = \frac{1}{x_1}$$

The small-signal equivalent network, therefore, is given by

$$\begin{bmatrix} \dot{\xi}_1 \\ \dot{\xi}_2 \end{bmatrix} = \begin{bmatrix} \dfrac{1}{\bar{x}_1^2(t)} & -2\bar{x}_2(t) \\ \dfrac{-1}{\bar{x}_1^2(t)} & 0 \end{bmatrix} \begin{bmatrix} \xi_1 \\ \xi_2 \end{bmatrix} + \begin{bmatrix} 1 \\ 0 \end{bmatrix} u(t)$$

where $\bar{x}_1(t)$ and $\bar{x}_2(t)$ can be obtained by solving

$$\dot{\bar{x}}_1(t) = \frac{-1}{\bar{x}_1} - \bar{x}_2^2 + 1 + 0.5 \sin t$$

$$\dot{\bar{x}}_2(t) = \frac{1}{\bar{x}_1}$$

with the initial conditions $\bar{\mathbf{x}}(t) = [0 \quad 0]^T$.

Remark 10.3-1. It should be clear by now that, in contrast to autonomous networks, to obtain the small-signal equivalent of a nonautonomous network we must first solve a set of nonlinear differential equations for the operating point $\bar{\mathbf{x}}(t)$. Finding the solution of such equations usually is not an easy task; in many cases we must resort to numerical solutions. To extend the idea of the equilibrium point to nonautonomous networks, we next consider the concept of a frozen operating point. We show that in many cases with practical importance, instead of using the operating point discussed previously, we can linearize the network around the frozen operating point. This will simplify the calculations since the frozen operating point can be obtained by solving a set of nonlinear algebraic equations.

10.4 FROZEN-OPERATING-POINT METHOD OF SMALL-SIGNAL ANALYSIS

In the last two sections we showed that a given nonlinear network can be linearized around an equilibrium point or around an operating point, depending on whether the network was autonomous or nonautonomous. The main disadvantage in linearizing a network about an operating point was the amount of effort involved in solving a set of nonlinear differential equations for the operating point. In this section we show that the calculations can be greatly simplified if the bias source is *slowly varying*. The procedure is based on linearizing the network around a neighboring operating point called the *frozen operating point*.

Before going through the analysis, let us give precise definition of the frozen operating point. For the sake of simplicity consider a nonlinear time-invariant network with time-varying bias sources:

$$\dot{\mathbf{x}}(t) = \mathbf{f}[\mathbf{x}(t), \hat{\mathbf{u}}(t)] \tag{10.4-1}$$

where $\mathbf{x}(t)$ is an n-column vector representing the state, $\hat{\mathbf{u}}(t)$ is an m-column vector representing the inputs and bias sources, and $\mathbf{f}(\cdot, \cdot)$ is a vector-valued function. Let $\bar{\mathbf{u}}(t)$ denote the bias sources; then

Definition 10.4-1 (Frozen Operating Point). The *frozen operating point* $\mathbf{x}^\circ$ of the nonautonomous network (10.4-1) with respect to the bias source $\bar{\mathbf{u}}(t)$ is defined to be the solution of

$$\mathbf{f}[\mathbf{x}^\circ(t), \bar{\mathbf{u}}(t)] = \mathbf{0} \tag{10.4-2}$$

In contrast to the operating point $\bar{\mathbf{x}}(t)$ defined in the previous section, the frozen operating point is the solution of a nondynamical equation, which can be solved using a Newton–Raphson or any other numerical method. We shall show that in some cases of practical interest the frozen operating point can be obtained by inspection. We illustrate this fact by some examples later on in this section. Let us now proceed with the perturbation analysis.

Perturbation Analysis. Since the frozen operating point $\mathbf{x}^\circ(t)$ and the operating point $\mathbf{x}(t)$ are presumed to be close "points" in the function space, an arbitrary component of each, say $x_1(t)$ and $x_1^\circ(t)$, can be plotted as in Fig. 10.4-1. As was mentioned earlier, the input vector $\hat{\mathbf{u}}(t)$ is composed of the

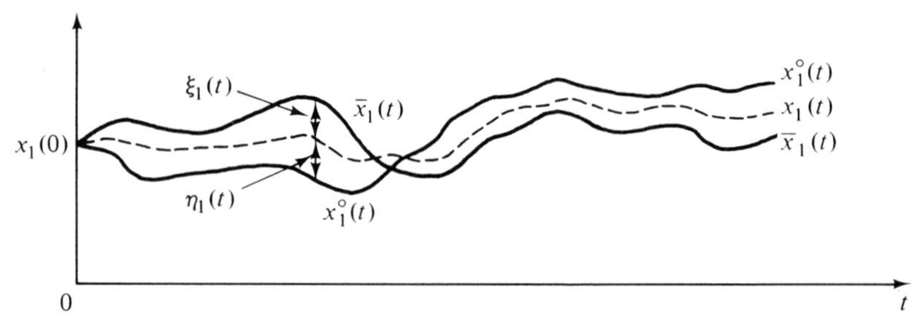

Figure 10.4-1 Operating points $x_1^0(t)$ and $\bar{x}_1(t)$, and their corresponding perturbations $\eta_1(t)$ and $\xi_1(t)$.

sum of two vectors, the bias source vector $\bar{\mathbf{u}}(t)$ and the small-signal vector $\mathbf{u}(t)$. Thus, we can write

$$\hat{\mathbf{u}}(t) = \bar{\mathbf{u}}(t) + \mathbf{u}(t) \tag{10.4-3}$$

where $\mathbf{u}(t)$ is assumed to be slowly varying and $||\mathbf{u}||_\infty$ is assumed to be small. The small signal $\mathbf{u}(t)$ will cause a small perturbation $\boldsymbol{\eta}(t)$ on the frozen operating point $\mathbf{x}^\circ(t)$ so that

$$\mathbf{x}(t) = \mathbf{x}^\circ(t) + \boldsymbol{\eta}(t) \tag{10.4-4}$$

Replacing $\mathbf{x}(t)$ and $\hat{\mathbf{u}}(t)$ from these equations in (10.4-1) and using a Taylor expansion, we get

$$\dot{\mathbf{x}}^\circ + \dot{\boldsymbol{\eta}} = \mathbf{f}(\mathbf{x}^\circ, \bar{\mathbf{u}}) + \mathbf{A}^\circ(t)\boldsymbol{\eta} + \mathbf{B}^\circ(t)\mathbf{u} + \mathbf{g}(\boldsymbol{\eta}, \mathbf{u}) \tag{10.4-5}$$

The first term in the right-hand side of the equation is zero by (10.4-2), and $\mathbf{A}^\circ(t)$ and $\mathbf{B}^\circ(t)$ are defined by

$$\mathbf{A}^\circ(t) \triangleq \begin{bmatrix} \frac{\partial f_1}{\partial x_1} & \frac{\partial f_1}{\partial x_2} & \cdots & \frac{\partial f_1}{\partial x_n} \\ \frac{\partial f_2}{\partial x_1} & \frac{\partial f_2}{\partial x_2} & \cdots & \frac{\partial f_2}{\partial x_n} \\ \vdots & \vdots & & \\ \frac{\partial f_n}{\partial x_1} & \frac{\partial f_n}{\partial x_2} & \cdots & \frac{\partial f_n}{\partial x_n} \end{bmatrix}_{(\mathbf{x}^\circ(t),\, \bar{\mathbf{u}}(t))} \tag{10.4-6}$$

$$\mathbf{B}^\circ(t) \triangleq \begin{bmatrix} \frac{\partial f_1}{\partial u_1} & \frac{\partial f_1}{\partial u_2} & \cdots & \frac{\partial f_1}{\partial u_m} \\ \frac{\partial f_2}{\partial u_1} & \frac{\partial f_2}{\partial u_2} & \cdots & \frac{\partial f_2}{\partial u_m} \\ \vdots & & & \\ \frac{\partial f_n}{\partial u_1} & \frac{\partial f_n}{\partial u_2} & \cdots & \frac{\partial f_n}{\partial u_m} \end{bmatrix}_{(\mathbf{x}^\circ(t),\, \bar{\mathbf{u}}(t))} \tag{10.4-7}$$

Rearranging (10.4-5) and using (10.4-2) we get

$$\dot{\boldsymbol{\eta}}(t) = \mathbf{A}^\circ(t)\boldsymbol{\eta}(t) + \mathbf{B}^\circ(t)\mathbf{u}(t) + \mathbf{g}(\eta, u) - \dot{\mathbf{x}}^\circ(t) \tag{10.4-8}$$

As in the previous cases, provided that

1. $\mathbf{f}(\cdot, \cdot)$ has continuous bounded second derivatives,
2. $\bar{\mathbf{u}}(t)$ is continuous
3. $\|\mathbf{u}\|_\infty$ is sufficiently small,

it can be shown [1] that the norm of $\mathbf{g}(\eta, u)$ is much smaller than the norm of the rest of the terms in the right-hand side of (10.4-8) and hence can be neglected; then

$$\dot{\boldsymbol{\eta}}(t) = \mathbf{A}^\circ(t)\boldsymbol{\eta}(t) + \mathbf{B}^\circ(t)\mathbf{u}(t) - \dot{\mathbf{x}}^\circ(t) \tag{10.4-9}$$

This equation represents the small-signal equivalent of (10.4-1) with respect to the frozen operating point $\mathbf{x}^\circ$.

Remark 10.4-1. Note that the only difference between this small-signal equivalent and the one given in (10.3-11) is the term $\dot{\mathbf{x}}^\circ(t)$. Having computed $\mathbf{x}^\circ(t)$ from (10.4-2), it is then easy to obtain $\dot{\mathbf{x}}^\circ(t)$, which acts as a correction term in (10.4-9) since we have linearized the network about $\mathbf{x}^\circ(t)$ rather than $\bar{\mathbf{x}}(t)$.

Next we present two theorems concerning the relationship between $\mathbf{x}^\circ(t)$ and $\bar{\mathbf{x}}(t)$. These theorems essentially imply that if the bias source $\bar{\mathbf{u}}(t)$ is slowly varying, then $\bar{\mathbf{x}}(t)$ and $\mathbf{x}^\circ(t)$ are "close" together; furthermore, $\dot{\mathbf{x}}^\circ(t)$ in (10.4-9) can be neglected. Before stating the theorems, let us make certain assumptions concerning the network under consideration. For future reference we denote these assumptions as A1 through A5.

A1. The function $\mathbf{f}(\cdot, \cdot)$ is twice continuously differentiable with respect to both arguments.

A2. The norms of matrices $\mathbf{A}(t)$ and $\mathbf{B}(t)$ are bounded; that is,

$$\|\mathbf{A}\|_\infty < \infty \quad \text{and} \quad \|\mathbf{B}\|_\infty < \infty$$

A3. For each $t \geq 0$, all the eigenvalues of $\mathbf{A}(t)$ have negative real parts.

A4. Second derivatives of $\mathbf{f}(\cdot, \cdot)$ with respect to its first and second arguments are continuous and bounded.

A5. The derivative of the bias source $\mathbf{u}(t)$ is continuous.

Theorem 10.4-1. Suppose that assumptions A1 through A5 hold and that $\|\dot{\bar{\mathbf{u}}}\|_\infty$ is small; then $\|\mathbf{x}^\circ - \bar{\mathbf{x}}\|_\infty$ is also small, and

$$\|\dot{\bar{\mathbf{u}}}\|_\infty \longrightarrow 0 \Longrightarrow \|\mathbf{x}^\circ - \bar{\mathbf{x}}\|_\infty \longrightarrow 0 \tag{10.4-10}$$

This means that the smaller the derivative of $\bar{\mathbf{u}}(t)$, the closer $\bar{\mathbf{x}}(t)$ is to the frozen operating point $\mathbf{x}^\circ$.

The proof of this theorem is quite lengthy; hence, we omit it. The interested reader is encouraged to consult [3].

Theorem 10.4-2. Suppose that assumptions A1 through A5 are satisfied and that $\|\mathbf{u}\|_\infty$ and $\|\dot{\bar{u}}\|_\infty$ are small. Then $\boldsymbol{\xi}(t)$ and $\boldsymbol{\eta}(t)$ defined by (10.3-11) and (10.4-9), respectively, are related by

$$\boldsymbol{\xi}(t) = \boldsymbol{\eta}(t) + \mathbf{h}(\|\mathbf{u}\| + \|\dot{\bar{\mathbf{u}}}\|) \tag{10.4-11}$$

where

$$\mathbf{h}(\|\mathbf{u}\| + \|\dot{\bar{\mathbf{u}}}\|) \longrightarrow \mathbf{0} \quad \text{as} \quad (\|\mathbf{u}\| + \|\dot{\bar{\mathbf{u}}}\|) \longrightarrow 0$$

The proof of this theorem can also be found in [3]. An important consequence of this theorem is that if $\|\dot{\bar{\mathbf{u}}}\|$ is sufficiently small, that is, if $\bar{\mathbf{u}}(t)$ is slowly varying, then we can perform the small-signal analysis around the frozen operating point $\mathbf{x}^\circ$ instead of $\bar{\mathbf{x}}$. The penalty that we have to pay for doing this is the extra term $\dot{\mathbf{x}}^\circ$ in (10.4-9). However, Theorem 10.4-2 guarantees that if $\bar{\mathbf{u}}(t)$ is sufficiently slowly varying, this penalty will reduce to zero. Consequently, if $\|\dot{\bar{\mathbf{u}}}\|$ is sufficiently small, $\dot{\mathbf{x}}^\circ$ can be neglected and (10.4-9) can be written as

$$\dot{\boldsymbol{\eta}}(t) = \mathbf{A}^\circ(t)\boldsymbol{\eta}(t) + \mathbf{B}^\circ(t)\mathbf{u}(t) \tag{10.4-12}$$

Let us now work out an example to show how, in certain cases, the frozen operating point can be obtained by inspection.

Example 10.4-1. Consider the nonlinear time-invariant network shown in Fig. 10.4-2. Let the capacitors be charge controlled with v-q characteristics

$$v_1 = q_1^3, \qquad v_2 = q_2^3$$

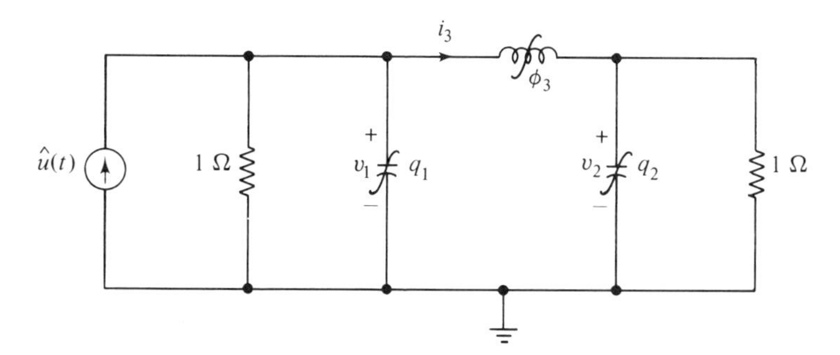

Figure 10.4-2 Nonlinear time-invariant network whose small-signal equivalent behaves as a linear time-varying low-pass filter.

and let the inductor be flux controlled with i-φ characteristic

$$i_3 = \tanh \varphi_3$$

Use the frozen-operating-point method to obtain the small-signal equivalent of this network. Assume that the bias $\bar{u}(t)$ is

$$\bar{u}(t) = 2 \sin \omega t$$

Solution

Taking the capacitor charges and inductor flux as the state variables, the state equations of this network can be written as

$$\dot{x}_1 = -x_1^3 - \tanh x_3 + \hat{u}$$
$$\dot{x}_2 = -x_2^3 + \tanh x_3$$
$$\dot{x}_3 = x_1^3 - x_2^3$$

where we have chosen the following notations:

$$x_1 = q_1, \qquad x_2 = q_2, \quad \text{and} \quad x_3 = \varphi_3$$

The frozen operating point is, therefore, the solution of

$$-(x_1^\circ)^3 - \tanh(x_3^\circ) + 2 \sin \omega t = 0$$
$$\tanh(x_3^\circ) - (x_2^\circ)^3 = 0$$
$$(x_1^\circ)^3 - (x_2^\circ)^3 = 0$$

A simple manipulation of these equations yields

$$(x_1^\circ)^3 = (x_2^\circ)^3 = \tanh(x_3^\circ) = \sin \omega t$$

or by inverting each function we get

$$x_1^\circ(t) = (\sin \omega t)^{1/3}, \qquad x_2^\circ(t) = (\sin \omega t)^{1/3}, \quad \text{and} \quad x_3^\circ(t) = \tanh^{-1}(\sin \omega t)$$

Then $\mathbf{A}^\circ(t)$ and $\mathbf{B}^\circ(t)$ matrices can be found to be

$$\mathbf{A}^\circ(t) = \begin{bmatrix} -3x_1^2 & 0 & \tanh^2 x_3 - 1 \\ 0 & -3x_2^2 & 1 - \tanh^2 x_3 \\ 3x_1^2 & -3x_2^2 & 0 \end{bmatrix}\Bigg|_{\substack{x_1 = x_2 = (\sin \omega t)^{1/3} \\ x_3 = \tanh^{-1}(\sin \omega t)}}$$

and $\qquad \mathbf{B}^\circ(t) = [1 \quad 0 \quad 0]^T$

Using (10.4-12), the small-signal equivalent network will have the following state equations:

$$\begin{bmatrix} \dot{\eta}_1 \\ \dot{\eta}_2 \\ \dot{\eta}_3 \end{bmatrix} = \begin{bmatrix} -3(\sin \omega t)^{2/3} & 0 & -1 + (\sin \omega t)^2 \\ 0 & -3(\sin \omega t)^{2/3} & 1 - (\sin \omega t)^2 \\ 3(\sin \omega t)^{2/3} & -3(\sin \omega t)^{2/3} & 0 \end{bmatrix} \begin{bmatrix} \eta_1 \\ \eta_2 \\ \eta_3 \end{bmatrix} + \begin{bmatrix} 1 \\ 0 \\ 0 \end{bmatrix} u(t)$$

Exercise 10.4-1. Determine the small-signal equivalent capacitances and the small-signal equivalent inductance of the previous example and draw the schematic of the small-signal equivalent network.

Remark 10.4-2. As was mentioned earlier, the main advantage of the frozen-operating-point method of small-signal analysis over the usual operating-point method is the simplicity of calculation of the frozen operating point. In fact, the frozen operating point $\mathbf{x}^\circ(t)$ can be calculated from the network without writing the state equations. To clarify this point, recall that the operating point $\bar{\mathbf{x}}(t)$ is the solution of

$$\dot{\bar{\mathbf{x}}}(t) = \mathbf{f}[\bar{\mathbf{x}}(t), \bar{\mathbf{u}}(t)]$$

where $\bar{\mathbf{x}}(t)$ denotes either the capacitor charges or voltages and inductor fluxes or currents. Hence, $\dot{\bar{\mathbf{x}}}(t)$ represents either the *capacitor currents* or *inductor voltages*. Comparing this equation with the equation defining the frozen operating point,

$$\mathbf{0} = \mathbf{f}[\mathbf{x}^\circ(t), \bar{\mathbf{u}}(t)]$$

we realize that $\mathbf{x}^\circ(t)$ can be computed by solving the resistive network, which is obtained from the original network by

1. Short circuiting all the inductors and
2. Open circuiting all the capacitors.

The resulting network will then be a purely resistive network, which can be solved using the numerical methods discussed in Chapter 7.

Example 10.4-2. Compute the frozen operating point of the network shown in Fig. 10.4-2 using the method outlined.

Solution

Open circuiting the capacitors and short circuiting the inductor, the network becomes as shown in Fig. 10.4-3. Solving this network for v_1, v_2, and i_3, we get

$$v_1(t) = v_2(t) = i_3(t) = \sin \omega t$$

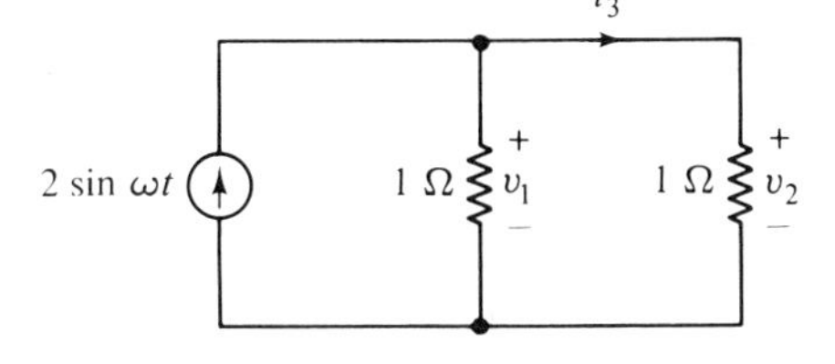

Figure 10.4-3 Resistive network used for computing the frozen operating point of the network shown in Fig. 10.4-2.

Using the v-q and i-φ characteristics given in Example 10.4-1, we get

$$v_1 = x_1^{\circ 3}, \qquad v_2 = x_2^{\circ 3}, \quad \text{and} \quad i_3 = \tanh x_3^{\circ}$$

Consequently,

$$x_1^{\circ}(t) = (\sin \omega t)^{1/3}, \qquad x_2^{\circ}(t) = (\sin \omega t)^{1/3}, \quad \text{and} \quad x_3^{\circ}(t) = \tanh^{-1}(\sin \omega t)$$

These results agree, as they should, with the results obtained in Example 10.4-1.

Example 10.4-3. Obtain the frozen operating point of the network shown in Fig. 10.4-4(a) where $v_1 = \sqrt{q_1}$, $v_2 = \sqrt{q_2}$, $i_3 = \varphi_3^5$, $v_4 = \frac{1}{2}i_4^3$ and $v_5 = \frac{1}{2}i_5^3$.

Solution

Let

$$x_1 = q_1, \qquad x_2 = q_2, \qquad x_3 = \varphi_3$$

Then open circuiting the capacitors and short circuiting the inductor, Fig. 10.4-4(a) becomes as shown in Fig. 10.4-4(b), (see next page).
Then

$$e^{-\alpha t} = \tfrac{1}{2}i_4^3 + \tfrac{1}{2}i_5^3$$

But since the current through the capacitor is zero, we get

$$i_5 = i_4$$

Inserting this in the previous equation and solving for i_4 and i_5, we obtain

$$i_4 = i_5 = e^{-\alpha t/3}$$

Consequently,

$$v_1 = e^{-\alpha t} - v_4 = e^{-\alpha t} - \tfrac{1}{2}e^{-\alpha t} = \tfrac{1}{2}e^{-\alpha t}$$

$$v_2 = v_4 + v_5 = \tfrac{1}{2}e^{-\alpha t} + \tfrac{1}{2}e^{-\alpha t} = e^{-\alpha t}$$

$$i_3 = i_4 = i_5 = e^{-\alpha t/3}$$

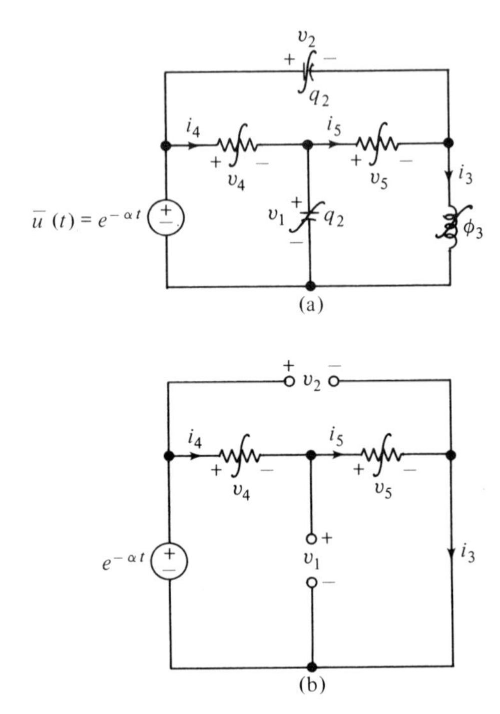

Figure 10.4-4 (a) Nonlinear time-invariant network with slowly varying bias source; (b) resistive network used for obtaining the frozen operating point of the network in (a).

Using the nonlinear characteristics, we get

$$x_1^\circ(t) = \tfrac{1}{4}e^{-2\alpha t}$$

$$x_2^\circ(t) = e^{-2\alpha t}$$

and

$$x_3^\circ(t) = e^{-\alpha t/15}$$

To show that $\|\dot{\mathbf{x}}^\circ(t)\| \longrightarrow 0$ as $\|\dot{\bar{u}}\| \longrightarrow 0$, let us take the derivative of $\mathbf{x}^\circ(t)$ with respect to t:

$$\dot{x}_1^\circ(t) = -\frac{\alpha}{2}e^{-2\alpha t}$$

$$\dot{x}_2^\circ(t) = -2\alpha e^{-2\alpha t}$$

$$\dot{x}_3^\circ(t) = -\frac{\alpha}{15}e^{-\alpha t/15}$$

Hence, as the variation of $\bar{u}(t)$ becomes slower and slower (i.e., α in $e^{-\alpha t}$ becomes smaller and smaller), the derivative of the frozen operating point $\dot{\mathbf{x}}^\circ$ tends to zero. This, once again, reaffirms our claim that for slowly varying bias sources the small-signal response around the frozen operating point,

$\boldsymbol{\eta}(t)$, is a good approximation of the small-signal response around the usual operating point, $\boldsymbol{\xi}(t)$.

An interesting problem, which makes use of the analysis discussed in the previous sections, is the design of a nonlinear network whose small-signal equivalent is identical to a prescribed linear network. This problem is considered in the next section.

10.5 DESIGN OF NONLINEAR NETWORKS

The problem of designing nonlinear networks, in general, is extremely difficult and as yet unsolved. In special cases, however, nonlinear networks can easily be designed to perform a particular function [4–6]. Indeed, for many years, electronic engineers have been designing special-purpose nonlinear circuits, such as wave shapers, clippers, limiters, frequency multipliers, detection devices, and the like. In this section we discuss a method for designing a special class of nonlinear networks, those whose small-signal equivalent realizes a desired linear time-varying network. Such networks have applications in various modern electronic devices, such as electronically adjustable filters, parametric amplifiers, and electronic tuning devices. In what follows we first consider the design of first-order nonlinear networks and derive a set of conditions under which a given linear time-varying network can be realized by nonlinear time-invariant elements, and then use the frozen-operating-point method to introduce a design criterion for nth-order uncoupled networks.

Design of First-Order Nonlinear Networks. In Section 10.3 we showed that a nonlinear time-invariant network operated in the small-signal mode behaves as a linear time-varying network as far as the input–output signals are concerned. The design problem to be considered here can be stated as: Design a nonlinear time-invariant network whose small-signal equivalent is identical to a desired linear time-varying network.

In this subsection we consider a first-order nonlinear network whose state equation can be written as

$$\dot{x}(t) = f[x(t)] + \hat{u}(t) \tag{10.5-1}$$

where $x(t)$ is a scalar function representing the state, and $\hat{u}(t)$ is the input of the network, which consists of a bias source $\bar{u}(t)$ and the "small" input signal $u(t)$; thus, $\hat{u}(t)$ can be written as

$$\hat{u}(t) = \bar{u}(t) + u(t) \tag{10.5-2}$$

As was mentioned earlier, the state $x(t)$ can be written as the sum of the operating point $\bar{x}(t)$ and the small-signal response $\xi(t)$; that is,

$$x(t) = \bar{x}(t) + \xi(t) \tag{10.5-3}$$

where $\xi(t)$ is the solution of

$$\dot{\xi}(t) = f'[\bar{x}(t)]\xi(t) + u(t), \qquad \xi(0) = 0 \tag{10.5-4}$$

and $x(t)$ must satisfy the original state equations; that is,

$$\dot{\bar{x}}(t) = f[\bar{x}(t)] + \bar{u}(t), \qquad \bar{x}(0) = 0 \tag{10.5-5}$$

Intuitively, it is clear that if the nonlinear function $f(\cdot)$ is appropriately chosen, equation (10.5-4) can represent a desired linear time-varying network. In what follows we utilize this intuitive idea to present a systematic method of obtaining $f(\cdot)$.

Suppose that the state equation of the desired linear time-varying network is given by

$$\dot{\eta}(t) = a(t)\eta(t) + u(t), \qquad \eta(0) = 0 \tag{10.5-6}$$

where $\eta(t)$ is the state, $u(t)$ is the input, and $a(t)$ is a scalar function representing the time behavior of the desired network.

The design problem can then be stated in the following terms: Obtain the nonlinear function $f(\cdot)$ such that the behaviors of the linear networks whose state equations are given in (10.5-4) and (10.5-6) are the same. To achieve this objective, it is clear that we must have

$$f'[\bar{x}(t)] = a(t) \quad \text{for all } t \tag{10.5-7}$$

But upon multiplication of both sides by $\dot{\bar{x}}$, integration, and noting that $\bar{x}(0) = 0$, we obtain

$$f[\bar{x}(t)] = p(t) \tag{10.5-8}$$

where $p(t)$ is a scalar function defined by

$$p(t) = \int_0^t a(\sigma)\dot{\bar{x}}(\sigma)\, d\sigma + f(0) \tag{10.5-9}$$

$f(0)$ can be assigned arbitrarily. From (10.5-8) and (10.5-9) it is clear that to obtain $f(\cdot)$ we need to compute $\bar{x}(t)$ and $\dot{\bar{x}}(t)$. To compute these functions, let us differentiate (10.5-5) with respect to t to get

$$\ddot{\bar{x}}(t) = f'[\bar{x}(t)]\dot{\bar{x}}(t) + \dot{\bar{u}}(t) \tag{10.5-10}$$

Comparing (10.5-10) with (10.5-7) yields the following second-order differential equation:

$$\ddot{\bar{x}}(t) = a(t)\dot{\bar{x}}(t) + \dot{\bar{u}}(t) \tag{10.5-11}$$

with the initial conditions

$$\bar{x}(0) = 0 \quad \text{and} \quad \dot{\bar{x}}(0) = f(0) + \bar{u}(0) \tag{10.5-12}$$

Since $\bar{u}(t)$ and $a(t)$ are given, equation (10.5-11) can be solved for $\dot{\bar{x}}(t)$ and $\bar{x}(t)$.

Let us now return to (10.5-8) and note that for given $\bar{x}(t)$ and $p(t)$ the nonlinear function $f(\cdot)$ can be obtained by eliminating the independent variable t between the two functions. This can be done either by a graphical method (see [8], p. 923) or by numerical means ([6] and [9]). Unfortunately, unless stringent conditions are imposed on $\bar{x}(t)$ and $p(t)$, the function $f(\cdot)$ will not be single valued. (Strictly speaking, $f(\cdot)$ will be a *relation* rather than a function, which implies single valuedness.) From a practical design point of view, multivalued nonlinear elements are quite undesirable due to difficulties involved in their manufacturing. For this reason it is essential to choose an appropriate bias source $\bar{u}(t)$ such that the resulting $f(\cdot)$ is single valued. In what follows we use the frozen-operating-point concept to obtain a restriction on the choice of bias sources such that the resulting nonlinear characteristics are single valued. Furthermore, we generalize the design problem to the nth-order case and give a step-by-step method of obtaining the desired nonlinear network.

Design of *N*th-Order Nonlinear Networks. Consider a nonlinear time-invariant *RLC* network whose state equations can be written as

$$\dot{\mathbf{x}}(t) = \mathbf{A}\mathbf{f}[\mathbf{x}(t)] + \hat{\mathbf{u}}(t) \tag{10.5-13}$$

where $\mathbf{x}$ is an n-column vector representing the state, $\mathbf{A}$ is an $n \times n$ constant matrix, and

$$\mathbf{f}(\mathbf{x}) = [f_1(x_1), f_2(x_2), \ldots, f_n(x_n)]^T \tag{10.5-14}$$

The n vector $\hat{\mathbf{u}}(t)$ represents the input of the network, which consists of a bias source vector $\bar{\mathbf{u}}(t)$ and a small-signal vector $\mathbf{u}(t)$; more precisely,

$$\hat{\mathbf{u}}(t) = \bar{\mathbf{u}}(t) + \mathbf{u}(t) \tag{10.5-15}$$

Note that the class of nonlinear networks defined by (10.5-13) is a special case of general nonlinear networks, whose state equations are given in (10.3-1), in the sense that in (10.5-13) the network is time invariant and the input appears as an additive term. Furthermore, the form of the nonlinear function defined in (10.5-14) implies that the network under consideration must not contain any dependent sources or coupling elements such as gyrators or coupled inductors.

If the bias source $\bar{\mathbf{u}}(t)$ is sufficiently slowly varying, then, as was mentioned in Section 10.4, we can define a frozen operating point $\mathbf{x}^\circ(t)$ for (10.5-13) by

$$\mathbf{A}\mathbf{f}[\mathbf{x}^\circ(t)] + \bar{\mathbf{u}}(t) = \mathbf{0} \tag{10.5-16}$$

The small-signal equivalent of the network under study can then be represented by the following state equation:

$$\dot{\boldsymbol{\xi}} = \mathbf{A}\mathbf{f}'[\mathbf{x}^\circ(t)]\boldsymbol{\xi} + \mathbf{u} \tag{10.5-17}$$

where $\boldsymbol{\xi}$ represents the small-signal response and $\mathbf{f}'[\mathbf{x}^\circ(t)]$ denotes the Jacobian of $\mathbf{f}(\mathbf{x})$ evaluated at $\mathbf{x}^\circ$. Notice that since elements of $\mathbf{f}(\mathbf{x})$ given in (10.5-14) are uncoupled, $\mathbf{f}'[\mathbf{x}^\circ(t)]$ is an $n \times n$ *diagonal* matrix.

Suppose that the state equation of the desired linear time-varying network N_l is given by ·

$$\dot{\boldsymbol{\eta}} = \mathbf{A}\mathbf{D}(t)\boldsymbol{\eta}(t) + \mathbf{u}(t) \tag{10.5-18}$$

where $\mathbf{A}$ is the same as in (10.5-13), $\boldsymbol{\eta}(t)$ is the state vector, $\mathbf{u}(t)$ is the input vector, and $\mathbf{D}(t)$ is an $n \times n$ diagonal matrix whose ith diagonal element $d_i(t)$ is positive for all $t \geq 0$.

The design problem is to obtain a nonlinear time-invariant network with time-varying bias sources so that its small-signal equivalent network [described by (10.5-17)] is identical to the linear time-varying network described by (10.5-18). We assume that the topology and element types of the linear network N_l are given. This information may be obtained by considering the linear time-invariant counterpart of the network and the design constraints. For instance, if a time-varying low-pass filter is to be designed, the topology and element type of one of the known time-invariant low-pass filters, such as Butterworth, Chebyshev, or maximally flat, may be used. For simplicity, we assume that all the resistors in the desired linear network are time invariant. The following procedure can then be used to obtain the corresponding nonlinear network N_n.

Design Algorithm

1. Replace the kth linear time-varying capacitor in N_l by a nonlinear charge-controlled capacitor whose v-q characteristic is
$$v_k = f_k(q_k) \tag{10.5-19}$$
in *parallel* with a current source $\bar{u}_k(t)$ [functions $f_k(\cdot)$ and $\bar{u}_k(\cdot)$ will be specified later]. Repeat this for every capacitor in N_l.
2. Replace the kth linear time-varying inductor in N_l by a nonlinear flux-controlled inductor whose i-ϕ characteristic is
$$i_k = f_k(\phi_k) \tag{10.5-20}$$
in *series* with a voltage source $\bar{u}_k(t)$. Repeat this for every inductor in N_l.
3. Leave all the resistive elements unchanged.

The network so constructed will be the desired nonlinear network, N_n. Before proving this claim, let us make two assumptions concerning the networks N_l and N_n.

Assumption 1. The $n \times n$ matrix $\mathbf{A}$ is nonsingular.

To state the second assumption, consider (10.5-16); since by Assumption 1 the matrix $\mathbf{A}$ is nonsingular, we can write

$$\mathbf{f}[\mathbf{x}^\circ(t)] = -\mathbf{A}^{-1}\bar{\mathbf{u}}(t)$$

Thus,

$$f_k(x_k^\circ) = -\sum_{j=1}^{n} b_{kj}\bar{u}_j \triangleq \alpha_k(t) \tag{10.5-21}$$

where b_{kj} denotes the kjth entry of $\mathbf{A}^{-1}$. Equation (10.5-21) defines $\alpha_k(t)$; for a given bias vector $\bar{\mathbf{u}}$ and $\mathbf{A}^{-1}$, $\alpha_k(t)$ is completely specified. In fact, $\alpha_k(t)$ as given denotes a linear combination of the bias sources $\bar{u}_j$. We then make the following assumption:

Assumption 2. The bias sources $\bar{u}_j(t)$ are chosen so that

$$h_k[\alpha_k(t)] = \frac{1}{d_k(t)} \quad \text{for } t \geq 0 \tag{10.5-22}$$

where $h(\cdot)$ is an arbitrary continuous function. This assumption implies that the bias sources should be related, through a continuous function, to the diagonal elements of $\mathbf{D}(t)$.

In an actual design problem, for a given $d_k(t)$ we choose $h_k(\cdot)$ such that the bias source $\bar{u}_k(t)$ is a function that can easily be generated. For example, if $d_k(t)$ is a periodic function with period T, $h_k(\cdot)$ is chosen so that $\bar{u}_k(t)$ is also a periodic function whose period is a multiple of T. Having determined $\bar{\mathbf{u}}(t)$ [and, consequently, $h_k(\cdot)$], we can now use the following theorem to determine the characteristics of the nonlinear elements $f_k(\cdot)$.

Theorem 10.5-1. Let the nonlinear network N_n be described by (10.5-13), and assumptions 1 and 2 hold. Then the small-signal equivalent of N_n [described by (10.5-17)] is identical to the desired linear time-varying N_l [described by (10.5-18)] if and only if

$$(f_k^{-1})'(x_k) = h_k(x_k) \tag{10.5-23}$$

Proof

Since $\mathbf{A}$ is nonsingular, (10.5-17) and (10.5-18) are identical if and only if

$$\mathbf{D}(t) = \mathbf{f}'[\mathbf{x}^\circ(t)] \quad \text{for all } t \geq 0 \tag{10.5-24}$$

Note that from (10.5-21) we obtain

$$x_k^\circ(t) = f_k^{-1}[\alpha_k(t)] \tag{10.5-25}$$

where $f_k^{-1}(\cdot)$ denotes the inverse function of $f(\cdot)$, and $x_k^\circ(t)$ is the kth element of $\mathbf{x}^\circ(t)$. Replacing $x_k^\circ(t)$ from (10.5-25) into (10.5-24), we get for each diagonal element

$$d_k(t) = f_k'[f_k^{-1}(\alpha_k(t))] \tag{10.5-26}$$

Now, using the inverse-function theorem (see, for example, [7], p. 34), equation (10.5-26) can be written as

$$(f_k^{-1})'[\alpha_k(t)] = \frac{1}{d_k(t)} \tag{10.5-27}$$

Then from (10.5-22) we get

$$(f_k^{-1})'(x_k) = h_k(x_k)$$

where x_k is a dummy variable replacing α_k. The proof of the theorem is therefore complete.

Notice that from (10.5-23) we obtain $(\mathbf{f}^{-1})'(\mathbf{x})$, which upon integration yields $\mathbf{f}^{-1}(\mathbf{x})$; hence, $\mathbf{f}(\mathbf{x})$ can be determined by inversion. It should be mentioned that the function $\mathbf{f}(\cdot)$ obtained in this fashion is not unique. An arbitrary constant will result from integrating $(\mathbf{f}^{-1})'(\mathbf{x})$, which can be used to our advantage for a more flexible design. Let us now work out an example to illustrate this design procedure.

Example 10.5-1. Consider the linear time-varying low-pass Butterworth filter shown in Fig. 10.5-1. Let the terminating resistor be linear time-invariant and the time-varying capacitor and inductors be defined as

$$C_1(t) = C_2(t) = (1 + 0.5\,|\sin \omega t\,|)^{-2} \tag{10.5-28}$$

and

$$L_3(t) = 2(1 + 0.5\,|\sin \omega t\,|)^{-2} \tag{10.5-29}$$

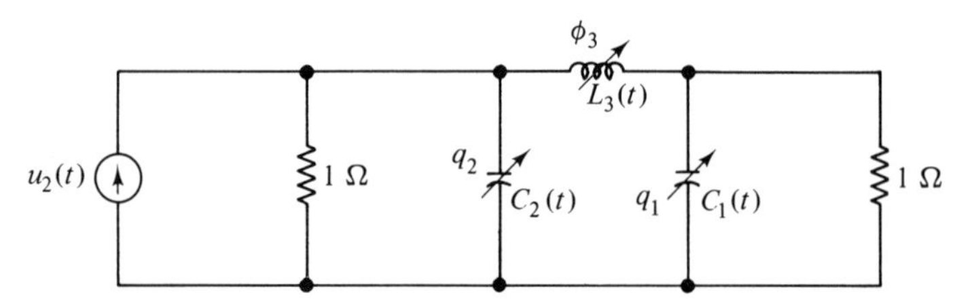

Figure 10.5-1 Prescribed linear time-varying Butterworth filter.

The design problem is to obtain a nonlinear time-invariant network whose small-signal equivalent is identical to the filter shown in Fig. 10.5-1.

Solution

Choosing the capacitor charges and the inductor flux as the state variables, the state equations of the linear network can then be written as

$$\begin{bmatrix} \dot{\eta}_1 \\ \dot{\eta}_2 \\ \dot{\eta}_3 \end{bmatrix} = (1 + 0.5\,|\sin \omega t\,|)^2 \begin{bmatrix} -1 & 0 & \frac{1}{2} \\ 0 & -1 & -\frac{1}{2} \\ -1 & 1 & 0 \end{bmatrix} \begin{bmatrix} \eta_1 \\ \eta_2 \\ \eta_3 \end{bmatrix} + \begin{bmatrix} 0 \\ u_2 \\ 0 \end{bmatrix} \tag{10.5-30}$$

where $\eta_1 = q_1$, $\eta_2 = q_2$, and $\eta_3 = \phi_3$. Decomposing the 3×3 matrix in

(10.5-30) into **A** and **D**(t), we get

$$\mathbf{A} = \begin{bmatrix} -1 & 0 & 1 \\ 0 & -1 & -1 \\ -1 & 1 & 0 \end{bmatrix}, \qquad \mathbf{D}(t) = (1 + 0.5\,|\sin \omega t\,|)^2 \begin{bmatrix} 1 & 0 & 0 \\ 0 & 1 & 0 \\ 0 & 0 & \frac{1}{2} \end{bmatrix}$$

Following steps 1, 2, and 3 of the design algorithm, the nonlinear network of Fig. 10.5-2 will result. In this particular example, we choose the bias sources as

$$\bar{u}_1(t) = 0, \qquad \bar{u}_2(t) = \sin \omega t, \qquad \bar{u}_3(t) = 0 \tag{10.5-31}$$

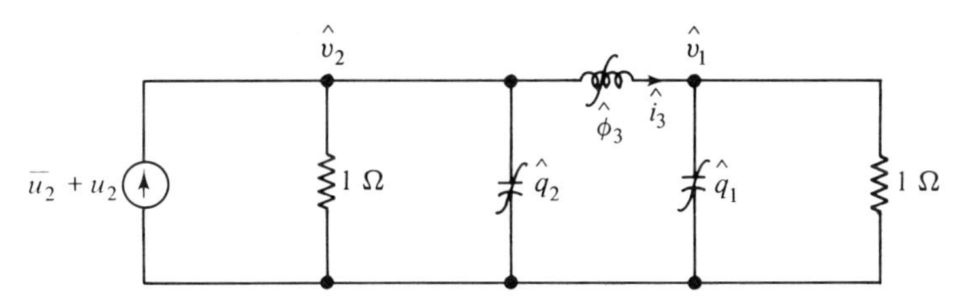

Figure 10.5-2 Nonlinear time-invariant network whose small-signal equivalent behavior is as that of the network of Fig. 10.5-1.

This choice of $\bar{\mathbf{u}}(t)$ complies with Assumption 2. The nonlinear elements are characterized by

$$\hat{v}_1 = f_1(\hat{q}_1), \qquad \hat{v}_2 = f_2(\hat{q}_2), \qquad \hat{i}_3 = f_3(\hat{\varphi}_3) \tag{10.5-32}$$

where $f_1(\cdot)$, $f_2(\cdot)$, and $f_3(\cdot)$ are yet to be determined. The corresponding state equation of the nonlinear network is

$$\begin{bmatrix} \dot{x}_1 \\ \dot{x}_2 \\ \dot{x}_3 \end{bmatrix} = \begin{bmatrix} -1 & 0 & 1 \\ 0 & -1 & -1 \\ -1 & 1 & 0 \end{bmatrix} \begin{bmatrix} f_1(x_1) \\ f_2(x_2) \\ f_3(x_3) \end{bmatrix} + \begin{bmatrix} 0 \\ \hat{u}_2(t) \\ 0 \end{bmatrix} \tag{10.5-33}$$

where $x_1 = \hat{q}_1$, $x_2 = \hat{q}_2$, and $x_3 = \hat{\varphi}_3$. Applying (10.5-21) to the present problem and using the matrix **A** obtained previously, we get

$$\alpha_k(t) = 0.5 \sin \omega t, \qquad k = 1, 2, 3 \tag{10.5-34}$$

Comparing (10.5-34) with (10.5-28) and (10.5-31) and using (10.5-22), we get

$$h_1(x) = h_2(x) = \frac{1}{(1 + |x|)^2}$$

and

$$h_3(x) = \frac{1/2}{(1 + |x|)^2}$$

Hence, by Theorem 10.5-1, we can write

$$(f_1^{-1})'(x) = (f_2^{-1})'(x) = \frac{1}{(1 + |x|)^2}$$

and
$$(f_3^{-1})'(x) = \frac{1/2}{(1+|x|)^2}$$

which upon integration and inversion yields

$$f_1(x_1) = \frac{x_1}{1-|x_1|}, \qquad f_2(x_2) = \frac{x_2}{1-|x_2|}, \qquad f_3(x_3) = \frac{x_3}{2-|x_3|}$$

This completely specifies the nonlinear network.

The design procedure outlined is one of several methods of designing nonlinear networks. The interested reader is encouraged to consult the references [4, 5, 6, and 8].

REFERENCES

[1] DESOER, C. A., and K. K. WONG, "Small-Signal Behavior of Nonlinear Lumped Networks," *Proc. IEEE*, vol 56 (Jan. 1968), 14–22.

[2] STERN, T. E., *Theory of Nonlinear Networks and Systems*, Addison-Wesley Publishing Company, Inc., Reading, Mass., 1965.

[3] DESOER, C. A., and B. PEIKARI, "The 'Frozen Operating' Point Method of Small-Signal Analysis," *IEEE Trans. Circuit Theory*, **CT-17** (May 1970), 259–261.

[4] DESOER, C. A., and B. PEIKARI, "Design of Linear Time-Varying and Nonlinear Time-Invariant Networks," *IEEE Trans. Circuit Theory*, **CT-17**, No. 2 (1970), 232–240.

[5] PEIKARI, B., "On the Design of Nonlinear Time-Invariant Networks," *IEEE Trans. Circuit Theory*, **CT-17**, No. 4 (1970), 657–659.

[6] PEIKARI, B., "Design of Nonlinear Networks with a Prescribed Small-Signal Behavior," *IEEE Trans. Circuit Theory*, **CT-19**, No. 4 (1972), 389–391.

[7] SPIVAK, M., *Calculus on Manifolds*, W. A. Benjamin, Inc., Menlo Park, Calif., 1965.

[8] CHUA, L. O., *Introduction to Nonlinear Network Theory*, McGraw-Hill Book Company, New York, 1969.

[9] White, J. V. and B. PEIKARI, "Comments on Design of Networks with a Prescribed Small-Signal Behavior," *IEEE Trans. Circuit Theory*, **CT-20**, No. 5 (Sept. 1973), 611–612.

PROBLEMS

10.1 Consider an autonomous network whose state equations are

$$\dot{x}_1 = x_1^2 - x_2^2$$
$$\dot{x}_2 = x_1^2 + 6x_2 + 5$$

(a) Find all the equilibrium points of this network.
(b) Find the small-signal equivalent equations around each equilibrium point.

10.2 Find the small-signal equivalent of a network whose state equations are

$$\dot{x}_1 = x_1 + 2x_1x_2 - x_3^3$$
$$\dot{x}_2 = -x_2 + x_1x_3 + x_2^3$$
$$\dot{x}_3 = -2x_1 + x_2$$

around the equilibrium point $\mathbf{x}_e = \mathbf{0}$.

10.3 Consider the nonlinear network shown in Fig. P10.3(a). The i-v characteristic of the tunnel diode is given in Fig. P10.3(b). Use a graphical technique to obtain the approximate location of the equilibrium points.

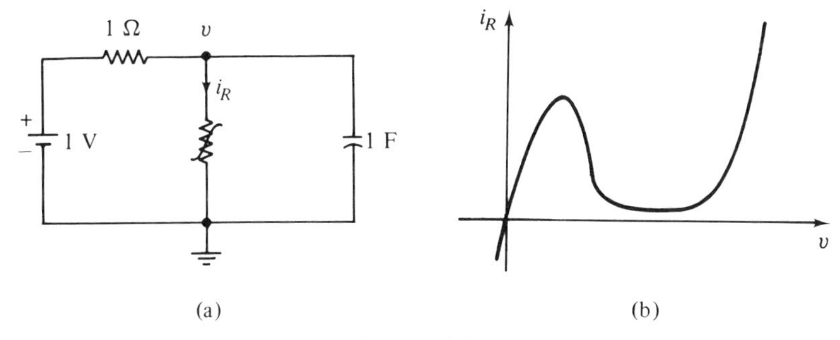

Figure P10.3

10.4 Find the equilibrium points of the network shown in Fig. P10.4 without writing its state equations.

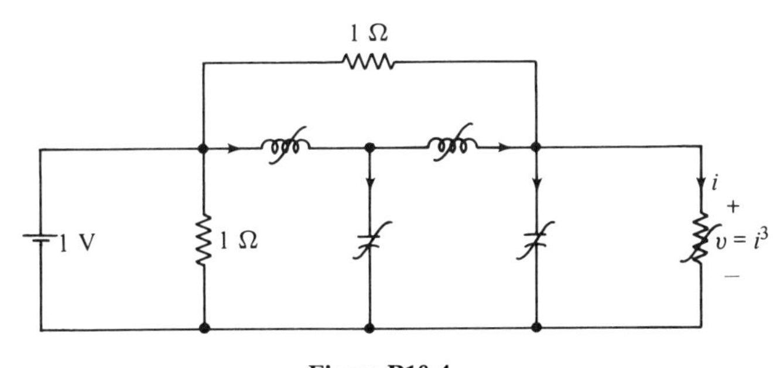

Figure P10.4

10.5 Repeat Problem 10.4 for the network shown in Fig. 10.2-4.

10.6 The state equations of a nonlinear network are

$$\begin{bmatrix} \dot{x}_1 \\ \dot{x}_2 \end{bmatrix} = \begin{bmatrix} 0 & 1 \\ 0 & -1 \end{bmatrix} \begin{bmatrix} x_1 \\ x_2 \end{bmatrix} + \begin{bmatrix} 0 \\ 1 \end{bmatrix} \sin(-x_1)$$

Show that this network has infinitely many equilibrium points.

10.7 For the network of Fig. P10.4 let the q-v relation of the capacitors be

$$q = v^3$$

and the φ-i relation of the inductors be

$$\varphi = i^3$$

Obtain the small-signal equivalent of this network and determine the small-signal capacitors, inductors, and resistors.

10.8 Figure P10.8 shows a common-base p-n-p transistor. The v-i relations of each terminal are given by

$$i_e = a_{11}(e^{kv_{eb}} - 1) + a_{12}(e^{kv_{cb}} - 1)$$
$$i_c = a_{21}(e^{kv_{eb}} - 1) + a_{22}(e^{kv_{cb}} - 1)$$

Determine the small-signal equivalent of this transistor.

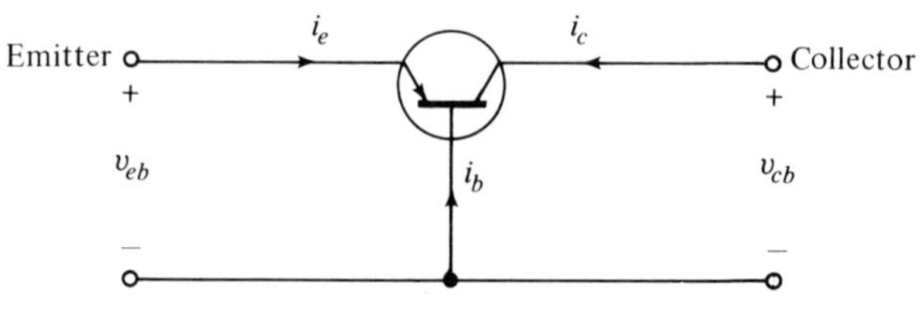

Figure P10.8

10.9 Consider the nonlinear network shown in Fig. P10.9. Let the characteristics of nonlinear resistors, capacitors, and inductors be

$$i = \frac{1}{2 + v}, \qquad q = v^3, \qquad \varphi = \tanh i$$

respectively. Assume that the bias $\bar{u}(t)$ is

$$\bar{u}(t) = 1 + 0.5 \sin \omega t$$

(a) Obtain the frozen operating point of this network.
(b) Find the small-signal equivalent network using the frozen-operating-point method.
(c) Determine the error involved in using the frozen operating point in terms of ω.

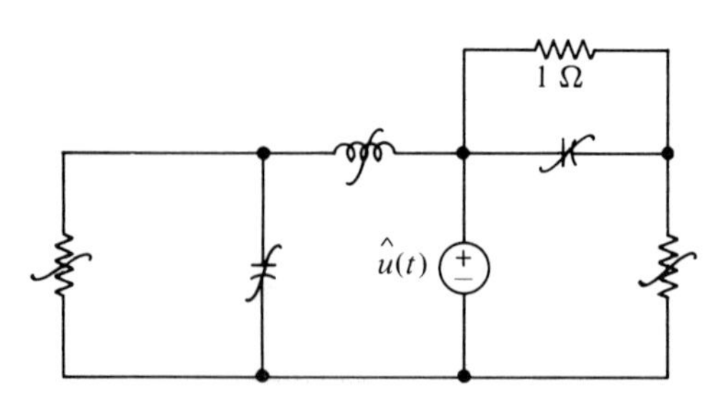

Figure P10.9

10.10 Consider the nonlinear capacitor shown in Fig. P10.10. Find the q-v characteristic of the capacitor such that its small-signal equivalent has a capacitance

$$C(t) = 1 + 0.5 \sin t$$

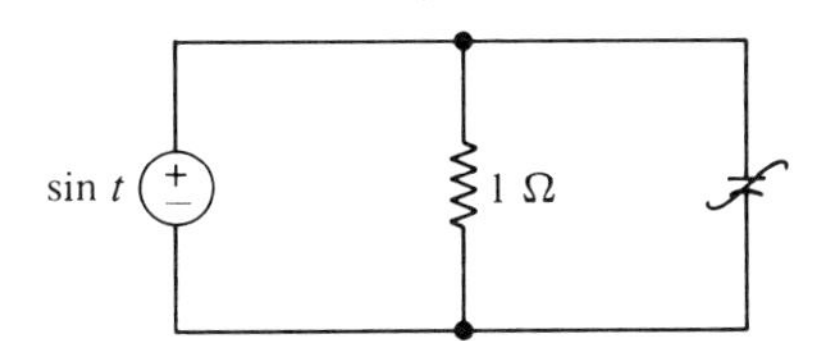

Figure P10.10

10.11 The state equations of a nonlinear network are

$$\begin{bmatrix} \dot{x}_1 \\ \dot{x}_2 \\ \dot{x}_3 \end{bmatrix} = \begin{bmatrix} -1 & 0 & 1 \\ 0 & -1 & -1 \\ -1 & 1 & 0 \end{bmatrix} \begin{bmatrix} f_1(x_1) \\ f_2(x_2) \\ f_3(x_3) \end{bmatrix} + \begin{bmatrix} 0 \\ 1 \\ 0 \end{bmatrix} \hat{u}$$

Let

$$\bar{u}(t) = \sin \omega t$$

and

$$f_1(x_1) = \frac{x_1}{1 - |x_1|}, \qquad f_2(x_2) = \frac{x_2}{1 - |x_2|}, \qquad f_3(x_3) = \frac{x_3}{2 - |x_3|}$$

Use the frozen-operating-point method to determine the state equations of the small-signal equivalent of this network. Check your answer with Example 10.5-1.

10.12 Figure P10.12 shows an *electronic tuning circuit* used in some transistor receivers. The incoming signal is received from the antenna by an ideal transformer. The input signal is assumed to be small (a few millivolts) in comparison to the adjustable bias source E (a few volts). Tuning from one station to another is achieved by changing the value of E. Find the characteristic of the nonlinear capacitor so that the resonant frequency of the circuit is proportional to E.

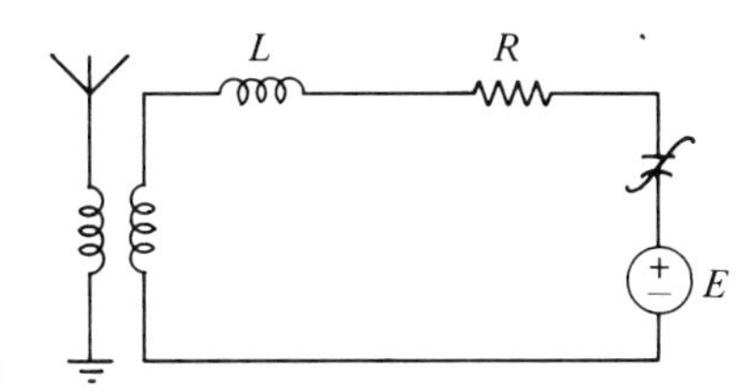

Figure P10.12

Stability Analysis of Linear and Nonlinear Networks

11

11.1 INTRODUCTION

In Chapter 2 we briefly introduced the concept of stability and gave some examples of stable and unstable networks. Since the stability concept plays an increasingly important role in the analysis and design of networks, we devote this entire chapter to various aspects of this topic. In view of the fact that linear time-invariant passive networks are inherently stable, the importance of the stability concept becomes apparent when one considers the design of active networks. Indeed, a major constraint usually imposed on the design of linear active filters as well as other time-varying and nonlinear networks is that they must be stable under all normal operating conditions.

Intuitively, a network is said to be stable if all the voltages and currents remain finite as t goes to infinity. The instability in a network manifests itself as an uncontrollable growth in the amplitude of the voltage and/or current in one or more branches, which eventually results in the destruction (burning up) of the network elements. In this chapter (Section 11.2) we put this intuitive concept in a mathematical framework and give three definitions of stability, two in terms of the equilibrium point (or operating point) of the network and one in terms of its input–output behavior. In Section 11.3 we investigate the necessary and sufficient conditions for the stability of the equilibrium state of a linear time-invariant network and give a simple rule for testing the stability of such networks, the Routh–Hurwitz stability test.

The local and global stability of the equilibrium state of nonlinear networks is considered in Section 11.4, and the Lyaponov theorem concerning the global stability of autonomous networks is discussed. The remainder of the chapter is then devoted to a detailed discussion on the operating-point stability of nonautonomous networks and the bounded input–bounded output stability concept.

Throughout this chapter we confine our discussion to the fundamental concepts and the application of stability results. A more detailed treatment of these topics can be found in various system theory texts and articles [1–6].

11.2 ZERO-INPUT STABILITY

In this section we present two new definitions of stability, Lyaponov stability and asymptotic stability. We then apply these definitions to investigate the stability of linear and nonlinear time-invariant networks whose inputs are identically zero. Furthermore, instead of considering the output stability, we study the stability of the state of the network in the neighborhood of an equilibrium point. This derivation from the usual output stability analysis is justified by the fact that the output and the state of the networks under study are usually related through a bounded nondynamic function (which may be either linear or nonlinear). In other words, state space stability implies output stability.

Thus, let us consider the class of networks whose state equations can be written as

$$\dot{\mathbf{x}}(t) = \mathbf{f}[\mathbf{x}(t)] \qquad \mathbf{x}(0) = \mathbf{x}_0 \tag{11.2-1}$$

where $\mathbf{x}(t)$ is an n-column vector representing the state and $\mathbf{f}(\cdot)$ is a vector-valued nonlinear function. If the network under consideration is linear, the state equation may be represented as

$$\dot{\mathbf{x}}(t) = \mathbf{A}\mathbf{x}(t) \qquad \mathbf{x}(0) = \mathbf{x}_0 \tag{11.2-2}$$

In what follows we show that it is possible to obtain a necessary and sufficient condition for the stability of linear networks. However, only sufficient conditions, as yet, have been obtained regarding the stability of general nonlinear networks.

Note that the stability analysis of networks whose equations are (11.2-1) and (11.2-2) amounts to studying the behavior of the trajectory of the state vector $\mathbf{x}(t)$, which starts from an initial state $\mathbf{x}_0$ in the Euclidean n-dimensional space and goes either to infinity (in the unstable case), to a limit cycle (in the case of Lyaponov stability), or to an equilibrium point (in the case of asymptotic stability). Before taking up stability analysis, let us define the term "trajectory" and give an example.

Definition 11.2-1 (State Trajectory). The geometrical locus of $\mathbf{x}(t)$ in the Euclidean n-dimensional space is called the *trajectory* of the state equations (11.2-1) and (11.2-2).

Typically, to obtain the state trajectory of a network, it is best to eliminate the independent variable t among the components of the state vector $\mathbf{x}(t)$. The resulting equation describes a space *curve*, which is the desired trajectory. This method is illustrated by a simple example.

Example 11.2-1. Obtain the state trajectory of a network whose state equations are

$$\begin{bmatrix} \dot{x}_1 \\ \dot{x}_2 \end{bmatrix} = \begin{bmatrix} 0 & 1 \\ -1 & 0 \end{bmatrix} \begin{bmatrix} x_1 \\ x_2 \end{bmatrix}, \qquad \mathbf{x}(0) = \begin{bmatrix} 0 \\ 1 \end{bmatrix}$$

Solution

Using any of the techniques discussed in Chapter 6, we can solve for $x_1(t)$ and $x_2(t)$; the result is

$$x_1(t) = \sin t \quad \text{and} \quad x_2(t) = \cos t$$

To eliminate the parameter t between x_1 and x_2, simply square each component and add together; it yields

$$x_1^2 + x_2^2 = 1$$

Consequently, the desired trajectory is a circle with radius 1 in the two-dimensional space. In general, a state trajectory can be obtained by computing $\mathbf{x}(t)$ at each point $\mathbf{x}(t_0)$, $\mathbf{x}(t_1)$, . . . , and connecting these points by a smooth curve in the n-dimensional space. This is shown in Fig. 11.2-1 for a two-dimensional vector $\mathbf{x}(t)$. In this figure the trajectory starts at the initial state $\mathbf{x}_0$ and ends up at the equilibrium point $\mathbf{x}_e$ (for the definition of equilibrium point see Definition 10.2-2). The trajectory remains at $\mathbf{x}_e$ unless it is perturbed by an input signal.

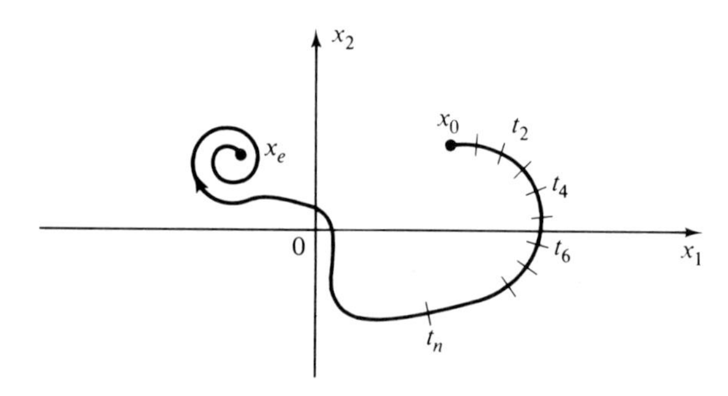

Figure 11.2-1 Trajectory of a state $\mathbf{x}(t)$ starting at the initial state $\mathbf{x}(t) = \mathbf{x}_0$ and ending at the equilibrium state $\mathbf{x}_e$.

Now, loosely speaking, an equilibrium point $\mathbf{x}_e$ is called *stable* if for all initial conditions $\mathbf{x}(t_0)$ in the neighborhood of $\mathbf{x}_e$ the trajectory gets arbitrarily close to $\mathbf{x}_e$. An equilibrium point is called *unstable* if for some initial condition the trajectory will go to infinity. These intuitive ideas are made precise in the definitions that follow.

Consider an autonomous network whose state equation is given by (11.2-1) or (11.2-2). Let $\mathbf{x}_e$ denote an equilibrium point of this network; then

Definition 11.2-2 (Lyaponov Stability). The equilibrium point $\mathbf{x}_e$ of (11.2-1) or (11.2-2) is said to be *stable in the sense of Lyaponov* (i.s.L) if for any given $\epsilon > 0$ there exists a positive number δ such that

$$|\mathbf{x}(t_0) - \mathbf{x}_e| \leq \delta \Longrightarrow |\mathbf{x}(t) - \mathbf{x}_e| \leq \epsilon \quad \text{for all } t \geq t_0 \qquad (11.2\text{-}3)$$

where the norm is usually taken to be the l_2 norm.

This, in effect, means that an equilibrium point is stable (i.s.L.) if there exists a region around $\mathbf{x}_e$ so that trajectories originated from that region remain close to $\mathbf{x}_e$. This idea is illustrated in Fig. 11.2-2 for the trajectory of a first-order network.

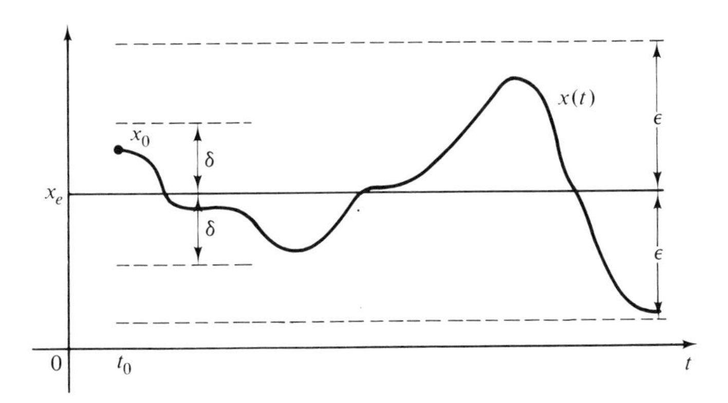

Figure 11.2-2 Trajectories starting in the strip with a radius δ will remain in the strip with radius ϵ.

Another important form of stability is asymptotic stability, which specifies the behavior of the trajectory as t goes to infinity.

Definition 11.2-3. (Asymptotic Stability). The equilibrium point $\mathbf{x}_e$ of (11.2-1) or (11.2-2) is said to be *asymptotically stable* if the following two conditions are satisfied:

1. $\mathbf{x}_e$ is stable in the sense of Lyaponov.
2. $|\mathbf{x}(t) - \mathbf{x}_e| \longrightarrow 0$ as $t \longrightarrow \infty$.

In words, an equilibrium point $\mathbf{x}_e$ is asymptotically stable if any trajectory originating from a point sufficiently close to $\mathbf{x}_e$ remains close for all $t \geq t_0$ and will go to $\mathbf{x}_e$ as $t \longrightarrow \infty$. Figure 11.2-3 shows a stable i.s.L., an asymptotically stable, and an unstable equilibrium point in two-dimensional space.

Let us now work out some simple examples to illustrate these definitions.

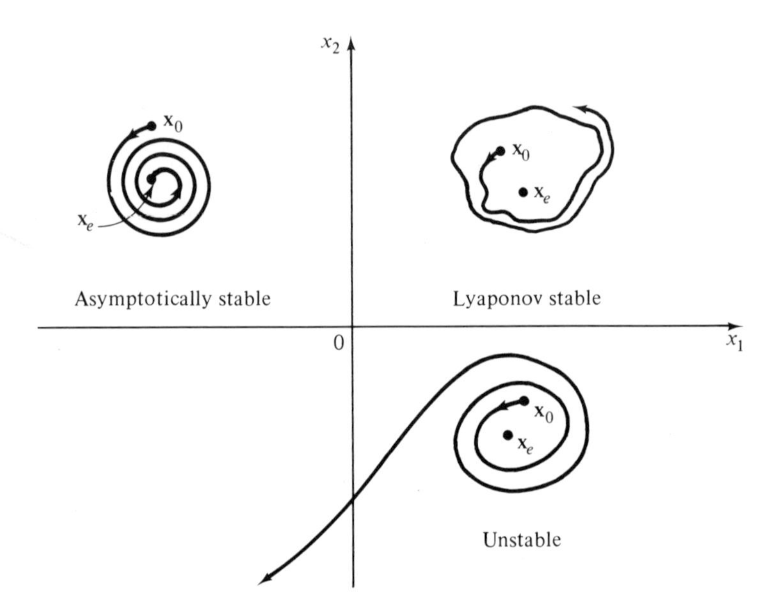

Figure 11.2-3 Behavior of typical trajectories in two-dimensional space.

Example 11.2-2. Determine the Lyaponov stability of the linear time-invariant network shown in Fig. 11.2-4(a). Assume that the initial voltage on the capacitor is v_0.

Solution

Taking the voltage across the capacitor and the current through the inductor as the state variables, the state equations can be written as

$$\begin{bmatrix} \dot{x}_1 \\ \dot{x}_2 \end{bmatrix} = \begin{bmatrix} 0 & 1 \\ -1 & 0 \end{bmatrix} \begin{bmatrix} x_1 \\ x_2 \end{bmatrix}, \qquad \mathbf{x}(0) = \begin{bmatrix} 0 \\ v_0 \end{bmatrix}$$

The state trajectory of this network can be found to be (see Example 11.2-1)

$$x_1^2 + x_2^2 = v_0^2$$

This equation represents the circle with radius v_0 and center at the origin in a two-dimensional space, as shown in Fig. 11.2-4(b). Notice that this network has a unique equilibrium point located at the origin:

$$\mathbf{x}_e = \mathbf{0}$$

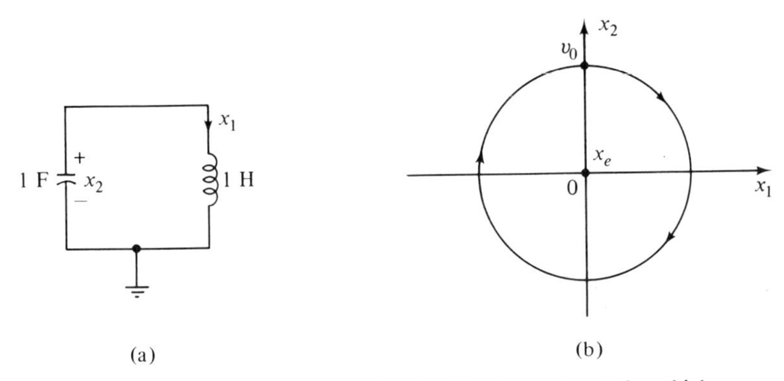

Figure 11.2-4 (a) Linear time-invariant lossless network for which the state variables are x_1 and x_2; (b) state trajectory of the network.

Then, for any given $\epsilon > 0$, we can choose a δ such that

$$|\mathbf{x}(0) - \mathbf{x}_e| \leq \delta \Longrightarrow |\mathbf{x}(t) - \mathbf{x}_e| \leq \epsilon \quad \text{for all } t \geq 0$$

Indeed, for any given ϵ, we can choose $\delta = \epsilon$. Then for all $v_0 \leq \delta$ we have

$$|\mathbf{x}(0)| \leq \delta \Longrightarrow |\mathbf{x}(t)| \leq \epsilon = \delta \quad \text{for all } t \geq 0$$

Hence, the network under consideration is stable i.s.L. Note, however, that it is not asymptotically stable, since for any nonzero initial voltage v_0 the trajectory never reaches the equilibrium point $\mathbf{x}_e = \mathbf{0}$.

Let us now add a resistor to the circuit and observe its behavior.

Example 11.2-3. Consider the linear time-invariant *RLC* network shown in Fig. 11.2-5(a). Let the initial current through the inductor be i_0. Determine the stability of this network.

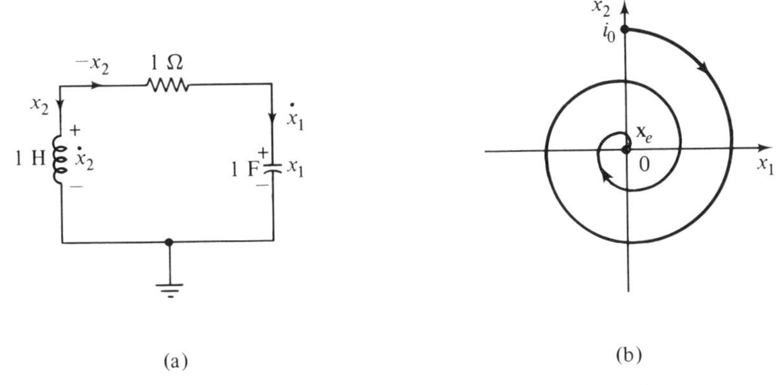

Figure 11.2-5 (a) Linear time-invariant lossy network with state variables x_1 and x_2; (b) state trajectory of the network.

Solution

Taking the state variables as shown in the figure, the state equation can be written as

$$\begin{bmatrix}\dot{x}_1\\ \dot{x}_2\end{bmatrix}=\begin{bmatrix}0 & -1\\ 1 & -1\end{bmatrix}\begin{bmatrix}x_1\\ x_2\end{bmatrix}, \qquad x(0)=\begin{bmatrix}0\\ i_0\end{bmatrix}$$

Since the $\mathbf{A}$ matrix in this state representation is nonsingular, the network has a single equilibrium point $\mathbf{x}_e = \mathbf{0}$ (see Remark 10.2-1). Using any of the methods discussed in Chapter 6, the solution of this equation can be seen to be

$$x_1(t) = -\left(\frac{2}{3}e^{-t/2}\sin\frac{\sqrt{3}}{2}t\right)i_0$$

and

$$x_2(t) = e^{-t/2}\left(\cos\frac{\sqrt{3}}{2}t - \frac{\sqrt{3}}{3}\sin\frac{\sqrt{3}}{2}t\right)i_0$$

Eliminating the parameter t between these equations and plotting the resulting trajectory, we obtain a curve similar to that shown in Fig. 11.2-5(b). Quite clearly, for any given $\epsilon > 0$, if we choose $\delta < \epsilon$ and $|i_0| \leq \delta$, we have

$$|i_0| \leq \delta \Longrightarrow |\mathbf{x}(t)| \leq \epsilon \quad \text{for all } t \geq 0$$

Thus, the network is stable i.s.L. Furthermore, due to the factor $e^{-t/2}$ in both $x_1(t)$ and $x_2(t)$, as $t \longrightarrow \infty$ the trajectory tends to $\mathbf{x}_e = \mathbf{0}$. Hence, the network is asymptotically stable.

Let us now give an example of an unstable network.

Example 11.2-4. Consider a simple first-order nonlinear network whose state equation is

$$\dot{x}(t) = x^2(t), \qquad x(0) = 1$$

Determine the stability of the equilibrium point of this network.

Solution

The equilibrium point of this state equation is at $x_e = 0$. Furthermore, the solution of this equation can be obtained by a simple integration:

$$x(t) = \frac{1}{1-t}, \qquad t \geq 0$$

Hence, as $t \longrightarrow 1$, it is clear that $|x(t)| \longrightarrow \infty$; that is, the network is unstable.

In the last three examples, we first obtained the trajectory and then determined whether the network was stable or not. Since in most practical cases solving the state equations is a time-consuming task and requires much effort, it is desirable to determine the stability or instability of the network without first solving for the trajectory. In the next section we give a necessary and sufficient condition for the equilibrium stability of linear time-invariant autonomous networks.

11.3 EQUILIBRIUM STABILITY OF LINEAR AUTONOMOUS NETWORKS

In this section we first state and prove a theorem concerning the asymptotic stability of linear time-invariant networks, and then give a simple method for checking the stability of a given network by examining its eigenvalues or characteristic roots.

Consider a linear autonomous network whose state equation is

$$\dot{\mathbf{x}}(t) = \mathbf{A}\mathbf{x}(t), \qquad \mathbf{x}(0) = \mathbf{x}_0 \tag{11.3-1}$$

where $\mathbf{x}(t)$ is an n vector and $\mathbf{A}$ is an $n \times n$ constant matrix. In Chapter 10 we showed that if $\mathbf{A}$ is nonsingular, then $\mathbf{x}_e = \mathbf{0}$ is the only equilibrium point. Let us assume that $\mathbf{A}$ is nonsingular and present a theorem concerning the stability of the equilibrium point of (11.3-1).

Theorem 11.3-1. The equilibrium point $\mathbf{x}_e = \mathbf{0}$ of (11.3-1) is asymptotically stable if and only if all the eigenvalues of the matrix $\mathbf{A}$ have negative real parts.

The proof of this theorem will be given in two parts: first, we show that the condition stated above is sufficient for asymptotical stability; second, we show that this condition is also necessary.

Proof (Sufficiency)

The solution of (11.3-1) was found in Chapter 6 to be

$$\mathbf{x}(t) = e^{\mathbf{A}t}\mathbf{x}_0 \tag{11.3-2}$$

Taking the norm of both sides of this equation and using the Schwartz inequality, we get

$$|\mathbf{x}(t)| \leq |e^{\mathbf{A}t}| \cdot |\mathbf{x}_0| \tag{11.3-3}$$

Note that $e^{\mathbf{A}t}$ is an $n \times n$ matrix whose elements are composed of terms like $t^k e^{\lambda_j t}$, where k is a finite integer and λ_j is the jth eigenvalue of $\mathbf{A}$. Since by the hypothesis of the theorem the real part of λ_j is negative, terms like $t^k e^{\lambda_j t}$ remain bounded for all $t \geq 0$ (why?). Consequently, the norm of $e^{\mathbf{A}t}$ will also remain bounded; that is,

$$|e^{\mathbf{A}t}| \leq M < \infty$$

where M is a positive constant. Hence, from (11.3-3) we can write

$$|\mathbf{x}(t)| \leq |e^{\mathbf{A}t}| \cdot |\mathbf{x}_0| \leq M \cdot |\mathbf{x}_0|$$

For any given $\epsilon > 0$, choose

$$\delta = \frac{\epsilon}{M}$$

Then for all $|\mathbf{x}_0| < \delta$ we have

$$|\mathbf{x}_0| < \delta \Longrightarrow |\mathbf{x}(t)| \leq M \cdot \frac{\epsilon}{M} = \epsilon$$

Then by Definition 11.2-2 the network is stable i.s.L.

To show asymptotical stability, observe that since the real part of λ_j is negative, for any finite integer k, we have

$$t^k e^{\lambda_j t} \longrightarrow 0 \quad \text{as} \quad t \longrightarrow \infty$$

Hence, every element of $e^{\mathbf{A}t}$ will go to zero as t goes to ∞; that is,

$$|\mathbf{x}(t)| \longrightarrow 0 \quad \text{as} \quad t \longrightarrow \infty$$

This proves the first part of the theorem.

Proof (*Necessity*)

Suppose that the matrix $\mathbf{A}$ has one eigenvalue, say λ_j, which has a nonnegative real part (i.e., $\text{Re}\{\lambda_j\} \geq 0$). If the real part of λ_j is positive (i.e., $\text{Re}\{\lambda_j\} > 0$), then the term $t^k e^{\lambda_j}$ will grow unbounded as t goes to infinity;

$$t^k e^{\lambda_j t} \longrightarrow \infty \quad \text{as} \quad t \longrightarrow \infty$$

Consequently, by appropriately choosing the initial conditions $\mathbf{x}_0$, we obtain

$$|\mathbf{x}(t)| \longrightarrow \infty \quad \text{as} \quad t \longrightarrow \infty$$

This means that the network is not stable i.s.L. If $\text{Re}\{\lambda_j\} = 0$, terms like $|e^{\lambda_j t}|$ will have a constant magnitude for all $t \geq 0$; more explicitly;

$$\text{Re}\{\lambda_j\} = 0 \Longrightarrow |e^{\lambda_j t}| = 1 \quad \text{for all } t \geq 0$$

As a result, $|\mathbf{x}(t)|$ will tend to a nonzero constant as $t \longrightarrow \infty$. This implies that the network will not be asymptotically stable. Hence, for asymptotical stability it is necessary to have $\text{Re}\{\lambda_j\} < 0$. This proves the theorem.

Let us now introduce a simple method for checking the location of the eigenvalues of the $\mathbf{A}$ matrix in the complex plane. This method will eliminate the need for actually obtaining the eigenvalues of a network.

Routh–Hurwitz Stability Test. Recall that the eigenvalues of $\mathbf{A}$ are the roots of the characteristic polynomial

$$q(\lambda) = \det(\mathbf{A} - \lambda \mathbf{1}) = 0$$

which can be written in a more explicit form:

$$a_0 \lambda^n + a_1 \lambda^{n-1} + \cdots + a_{n-1}\lambda + a_n = 0 \qquad (11.3\text{-}4)$$

where the a_i are real constants and n represents the dimension of the state vector. The eigenvalues are the roots of (11.3-4). These roots are either real or complex numbers. In the latter case they always appear in complex conjugate pairs (why?). The objective of this subsection is to determine the sign of the real part of the eigenvalues without actually solving the corresponding polynomial. This can be easily achieved by using the Routh–Hurwitz stability test. This test basically consists of two steps, which we discuss next.

Step 1. If any of the coefficients a_i of (11.3-4) are zero or negative, at least one of the eigenvalues of the equation is either zero, purely imaginary, or has a positive real part. In such a case, there is no need to pursue the test further; the network under consideration is not asymptotically stable. However, having positive coefficients is only a necessary (but not sufficient) condition for the roots of (11.3-4) to have negative real parts. That is, if all the a_i are nonzero and positive, we cannot conclude that the network is asymptotically stable. This fact can be illustrated by the following polynomial:

$$\lambda^3 + \lambda^2 + \lambda + 21 = 0 \tag{11.3-5}$$

Clearly, all the coefficients in this polynomial are nonzero and positive; but the corresponding eigenvalues are

$$\lambda_1 = -3, \qquad \lambda_2 = 1 + j\sqrt{6}, \qquad \lambda_3 = 1 - j\sqrt{6}$$

Two of the eigenvalues have positive real parts; thus, the corresponding network is unstable. A necessary and sufficient condition for the eigenvalues to have negative real parts is given in the following.

Step 2. If all the coefficients a_i are nonzero and positive, we proceed to form the following table:

λ^n	a_0	a_2	a_4	a_6	$\cdots$
λ^{n-1}	a_1	a_3	a_5	$\cdots$	
λ^{n-2}	b_1	b_2	b_3	$\cdots$	
λ^{n-3}	c_1	c_2	c_3	$\cdots$	
$\cdot$	$\cdot$	$\cdot$	$\cdot$		
$\cdot$	$\cdot$	$\cdot$	$\cdot$		
$\cdot$	$\cdot$	$\cdot$	$\cdot$		
λ^2	d_1	d_2	d_3	$\cdots$	
λ^1	e_1	e_2	e_3	$\cdots$	
λ^0	f_1	f_2	f_3	$\cdots$	

where the first two rows are obtained from the polynomial and the remaining rows are determined from the following equations:

$$b_1 = \frac{a_1 a_2 - a_0 a_3}{a_1}, \qquad b_2 = \frac{a_1 a_4 - a_0 a_5}{a_1}, \qquad b_3 = \frac{a_1 a_6 - a_0 a_7}{a_1}$$

$$c_1 = \frac{b_1 a_3 - a_1 b_2}{b_1}, \qquad c_2 = \frac{b_1 a_5 - a_1 b_3}{b_1}, \qquad c_3 = \frac{b_1 a_7 - a_1 b_5}{b_1}$$

and so on. The stability test then amounts to checking the sign of the elements in the first column of the table. *If all the elements in the first column are positive, all the eigenvalues will have negative real parts.* Furthermore, the number of the *sign changes* in the first column is equal to the number of the eigenvalues with positive real parts.

Example 11.3-1. Let us determine the asymptotic stability of a network whose characteristic equation is given as in (11.3-5). The Routh–Hurwitz table for this polynomial is

λ^3	1	1	0
λ^2	1	21	0
λ^1	−20	0	0
λ^0	21	0	0

Since there are two sign changes in the first column (from +1 to −20 and from −20 to +21), there are two eigenvalues with positive real parts. This result agrees (as it should) with the earlier conclusion about the roots of the polynomial under study.

In the application of this method two specific cases might arise that require special consideration:

1. If an element of the first column becomes identically zero, the remaining rows in the table cannot be found (why?). To remedy this problem the zero in the first column may be replaced by a small positive number ϵ and the process continued. If the signs of the elements in the first column just before and after the ϵ term are the same, the corresponding polynomial has a pair of roots on the $j\omega$ axis. This implies that the network is not asymptotically stable.

Example 11.3-2. Consider a network whose characteristic equation is

$$2\lambda^4 + \lambda^3 + 4\lambda^2 + 2\lambda + 1 = 0$$

The corresponding Routh–Hurwitz table is (the zero term in the first column is replaced by $\epsilon > 0$)

λ^4	2	4	1
λ^3	1	2	0
λ^2	ϵ	1	0
λ^1	$\frac{2\epsilon - 1}{\epsilon}$	0	
λ^0	1		

Since ϵ is assumed to be a small positive number, $(2\epsilon - 1)/\epsilon$ is negative; consequently, there are two sign changes in the first column. Thus, two of the eigenvalues have positive real parts and the corresponding network is unstable.

2. If an entire row in the table becomes zero, we face the same difficulty; the remaining rows cannot be found. In this case we form an auxiliary polynomial $A(\lambda)$ whose coefficients are the elements of the previous row. Once the auxiliary polynomial is obtained, the row of zeros is replaced by a row whose elements are the coefficients of the derivative of $A(\lambda)$. The procedure can then be continued to obtain the remaining elements in the first column. The presence of a row of zeros, however, indicates that the polynomial has either a pair of roots on the $j\omega$ axis or at least one root with positive real part. In either case the network is not asymptotically stable.

Example 11.3-3. Consider the polynomial

$$\lambda^5 + 2\lambda^4 + 3\lambda^3 + 6\lambda^2 + \lambda + 2 = 0 \qquad (11.3\text{-}6)$$

The corresponding Routh–Hurwitz table is

λ^5	1	3	1
λ^4	2	6	2
λ^3	0	0	
λ^2	?	?	

Thus, the third row is identically zero and the test fails. To circumvent this difficulty, the following auxiliary polynomial is formed from the row preceding the zero row.

$$A(\lambda) = 2\lambda^4 + 6\lambda^2 + 2 = 0$$

Hence,

$$\frac{d}{d\lambda}A(\lambda) = 8\lambda^3 + 12\lambda$$

Consequently, the row of zeros is replaced by the coefficients of the derivative of $A(\lambda)$; the resulting Routh–Hurwitz table is

λ^5	1	3	1
λ^4	2	6	2
λ^3	8	12	
λ^2	3	2	
λ^1	$\frac{20}{3}$	0	
λ^0	2		

There are no sign changes in the first column. Consequently, the polynomial has no roots with positive real parts. However, the row of zeros indicates that there are at least a pair of zeros on the $j\omega$ axis. As a result, the corresponding network is stable i.s.L., but not asymptotically stable.

The stability criteria discussed so far require an explicit knowledge of the state equation of the network under study. Next we state two theorems concerning the stability of linear autonomous networks in terms of their element values. The proofs of these theorems, although not difficult, are long and hence are omitted here.

Theorem 11.3-2. The equilibrium state $\mathbf{x}_e = \mathbf{0}$ of any linear autonomous network, comprised of *positive* inductors and capacitors so that each capacitor or inductor is in series or in parallel with some positive resistor, is asymptotically stable.

Note that the conditions of this theorem exclude the zero-valued elements; hence, short circuits or open circuits are not permitted. The usefulness of this theorem is evident: to determine the stability of a linear autonomous network, we first examine the structure of the network; if the conditions of the theorem are satisfied, there is no need for writing the state equations or finding the state trajectory. The network is asymptotically stable.

Remark 11.3-1. Note that the conditions stated in Theorem 11.3-2 are only sufficient conditions—not necessary; a linear autonomous network containing some nonpositive elements may still be asymptotically stable.

Theorem 11.3-3. The equilibrium state $\mathbf{x}_e = \mathbf{0}$ of any linear autonomous network comprised of *nonnegative* inductors, capacitors, and resistors only is stable i.s.L.

The networks considered in Theorems 11.3-2 and 11.3-3 can be shown to be passive. Those considered in Theorem 11.3-2 are sometimes referred to as *strictly passive.* Let us now proceed with a discussion of the stability of nonlinear networks.

11.4 EQUILIBRIUM STABILITY OF NONLINEAR AUTONOMOUS NETWORKS

In discussing the equilibrium stability of nonlinear networks, we need to consider two cases. The first is when the initial condition is near the equilibrium point (local stability), and the second is when the initial condition is relatively far from the equilibrium point (global stability). Before giving a detailed discussion of these cases, let us give a more precise definition of these terms:

Local Stability Versus Global Stability. From the proof of Theorem 11.3-1 it is clear that if a linear autonomous network is stable it will remain stable, regardless of how far or how close its initial state and equilibrium states are from each other. More specifically, in (11.3-3) if $|e^{\mathbf{A}t}| \longrightarrow 0$, then $|\mathbf{x}(t)| \longrightarrow 0$ irrespective of the norm of $\mathbf{x}_0$. In such cases, the network is said to be *globally* asymptotically stable. This property, however, is not shared by nonlinear autonomous networks in general. A particular equilibrium point of a nonlinear network may only be asymptotically stable if the initial state is confined to a certain finite region in the state space. In this case the corresponding equilibrium point is said to be *locally asymptotically stable.* The region in which the initial state must lie is called the *region of asymptotical stability*, or sometimes, for obvious reasons, it is called the *region of attraction* of the corresponding equilibrium point. Let us illustrate this concept by a first-order example:

Example 11.4-1. Consider the simple network shown in Fig. 11.4-1. Let the nonlinear resistor be voltage controlled with *i-v* relations

$$i = v - v^3$$

Determine the region of stability of the equilibrium points of this network.

Solution

Taking the voltage across the capacitor as the state variable, the state equation of the network will be

$$\dot{x}(t) = -x(t) + x^3(t)$$

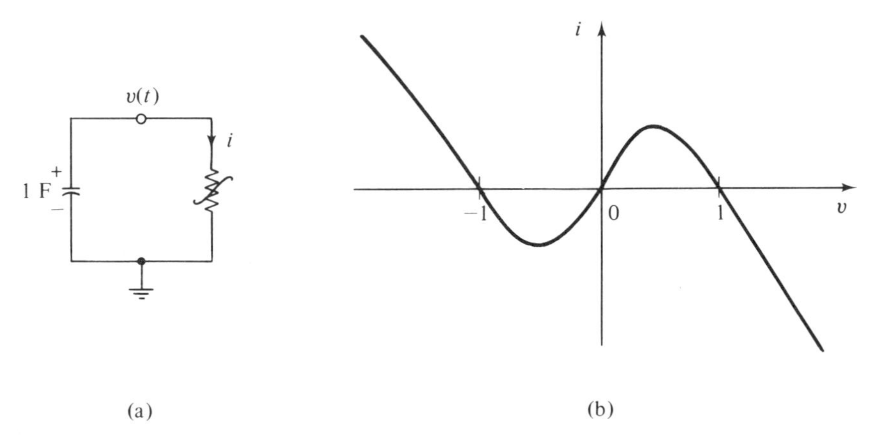

Figure 11.4-1 (a) Parallel *RC* network; the capacitor is linear and the resistor is voltage controlled; (b) characteristics of the nonlinear resistor; equilibrium points with negative slope are unstable and the equilibrium point with positive slope is stable.

The equilibrium points of this network can be easily seen to be

$$x_e = 0 \quad \text{and} \quad x_e = \pm 1$$

If we now find the small-signal equivalent of this network around each equilibrium point, we get

$$\dot{\xi}(t) = -\xi(t) \quad \text{for } x_e = 0$$

and

$$\dot{\xi}(t) = 2\xi(t) \quad \text{for } x_e = \pm 1$$

These equations are linear time invariant, and from Theorem 11.3-1 it is clear that the first equation is asymptotically stable, whereas the second one is unstable. Therefore, it only makes sense to talk about the region of asymptotical stability of the equilibrium point $x_e = 0$. To obtain this region of attraction, let us observe the characteristic of the nonlinear resistor in Fig. 11.4-1(b). From this figure and the linearized equations we can see that if $|x_0| > 1$ then $|x(t)| \longrightarrow \infty$, and if $|x_0| < 1$, then $|x(t)| \longrightarrow 0$. Consequently, the region of asymptotical stability is $|x_0| < 1$. Intuitively, this conclusion is plausible—around the equilibrium point $x_e = 0$ the linearized resistor is positive; the corresponding small-signal equivalent network is passive and hence stable. However, at $x_e = \pm 1$ the linearized resistor is negative; the resulting small-signal equivalent network is active, which causes instability.

The procedure for obtaining the stability of the equilibrium points of this example motivates us to state a general theorem concerning the local stability of nonlinear autonomous networks. This is accomplished in the following.

Local Stability of Nonlinear Autonomous Networks. Consider a nonlinear autonomous network whose state equation is

$$\dot{\mathbf{x}}(t) = \mathbf{f}[\mathbf{x}(t)] \tag{11.4-1}$$

Let us obtain the small-signal equivalent of this network around the equilibrium point $\mathbf{x}_e$; that is, let

$$\mathbf{x}(t) = \mathbf{x}_e + \boldsymbol{\xi}(t) \tag{11.4-2}$$

Then from (11.4-1) and (11.4-2) and the definition of equilibrium point we can write

$$\dot{\boldsymbol{\xi}}(t) = \mathbf{f}(\mathbf{x}_e + \boldsymbol{\xi}) = \mathbf{f}'_x(\mathbf{x}_e)\boldsymbol{\xi}(t) + \mathbf{g}(\boldsymbol{\xi})$$

Neglecting the second- and higher-order terms, $\mathbf{g}(\boldsymbol{\xi})$, the small-signal equivalent state equation becomes

$$\dot{\boldsymbol{\xi}}(t) = \mathbf{f}'_x(\mathbf{x}_e)\boldsymbol{\xi}(t) \tag{11.4-3}$$

where $\mathbf{f}'_x(\mathbf{x}_e)$ is the $n \times n$ constant Jacobian matrix. Note from (11.4-2) that $\mathbf{x}_e + \boldsymbol{\xi}(t)$ describes the trajectory of the network in the neighborhood of $\mathbf{x}_e$.

The equilibrium point $\mathbf{x}_e$ is therefore asymptotically stable if $|\boldsymbol{\xi}(t)| \longrightarrow 0$ as $t \longrightarrow \infty$. Hence, *to study the local stability of* (11.4-1) *around* $\mathbf{x}_e$, *we must study the asymptotical stability of* (11.4-3).

Denote the eigenvalues of $\mathbf{f}'_x(\mathbf{x}_e)$ by $\lambda_1, \lambda_2, \ldots, \lambda_n$; then the following theorem is a natural extension of Theorem 11.3-1 to the present case.

Theorem 11.4-1. Consider the nonlinear autonomous network whose state equation is given by (11.4-1). Let $\mathbf{x}_e$ be an equilibrium point of this network; then

(a) The network is locally asymptotically stable around $\mathbf{x}_e$ if *all* the eigenvalues of $\mathbf{f}'_x(\mathbf{x}_e)$ have negative real parts.
(b) The network is unstable if at least one of the eigenvalues of $\mathbf{f}'_x(\mathbf{x}_e)$ has a positive real part.
(c) If some of the eigenvalues of $\mathbf{f}'_x(\mathbf{x}_e)$ are purely imaginary and the rest have negative real parts, this method is inconclusive; we must employ other methods that take into account the effect of the higher-order terms in $\mathbf{g}(\boldsymbol{\xi})$.

The proof of this theorem is similar to that of Theorem 11.3-1 and hence is omitted here. Instead, let us work out an example to illustrate its application.

Example 11.4-2. The state equations of a nonlinear autonomous network are

$$\dot{x}_1 = -x_1 + x_2^3 + x_1 x_2$$
$$\dot{x}_2 = -2x_2 - x_1^3$$

Determine the stability of the zero state of this network.

Solution

The zero state is indeed an equilibrium point of this network; hence, we can linearize the network around this point and apply Theorem 11.4-1.

$$\mathbf{f}'_x(\mathbf{0}) = \begin{bmatrix} -1 + x_2 & 3x_2^2 + x_1 \\ -3x_1^2 & -2 \end{bmatrix}_{\substack{x_1=0\\x_2=0}} = \begin{bmatrix} -1 & 0 \\ 0 & -2 \end{bmatrix}$$

The eigenvalues of $\mathbf{f}'_x(\mathbf{0})$ are -1 and -2; hence, by Theorem 11.4-1 the network under consideration is locally asymptotically stable around the equilibrium point $\mathbf{x}_e = \mathbf{0}$.

Let us next consider the global stability of nonlinear autonomous networks.

Global Stability of Nonlinear Autonomous Networks. As was shown in Example 11.4-1, an equilibrium point of a nonlinear autonomous network may or may not be globally stable. A sufficient condition for global stability of such networks is given in the Lyaponov theorem stated next. In this theorem, for simplicity, we assume that the network under consideration has an equilibrium point at the origin (i.e., $\mathbf{x}_e = \mathbf{0}$), and we discuss the global stability of this point. The extension of the theorem to other equilibrium points can then be achieved by a simple change of coordinates in the state space.

Theorem 11.4-2 (Lyaponov). Consider a nonlinear network whose state equation is

$$\dot{\mathbf{x}} = \mathbf{f}(\mathbf{x}) \tag{11.4-4}$$

where $\mathbf{x}$ is a column vector and $\mathbf{f}(\mathbf{x})$ is a vector-valued function such that

$$\mathbf{f}(\mathbf{0}) = \mathbf{0} \tag{11.4-5}$$

If there exists a real scalar continuously differentiable function $V(\mathbf{x})$ so that

(a) $V(\mathbf{x}) > 0$ for all $\mathbf{x} \neq \mathbf{0}$ and $V(\mathbf{0}) = 0$,
(b) $(d/dt)V(\mathbf{x}) \leq 0$ and $(d/dt)V(\mathbf{x})$ is not identically zero along any trajectory of (11.4-4),
(c) $V(\mathbf{x}) \longrightarrow \infty$ as $|\mathbf{x}| \longrightarrow \infty$,

then the equilibrium point $\mathbf{x}_e = \mathbf{0}$ of (11.4-5) is globally asymptotically stable.

We prove a more general form of this theorem in the next section when we discuss the stability of nonautonomous networks.

Remark 11.4-1. The function $V(\mathbf{x})$ defined in Theorem 11.4-2 is called a *Lyaponov function.* It should be mentioned at this stage that the existence of a Lyaponov function is a sufficient (but not necessary) condition for the global asymptotical stability of (11.4-5). If we cannot find such a Lyaponov function for a given autonomous network, no conclusions can be drawn regarding its stability by this method. Furthermore, if we find a Lyaponov function for a network, this function may not be unique. However, finding one Lyaponov function that satisfies conditions (a), (b), and (c) is sufficient to guarantee the global asymptotical stability of the network under study. Let us now work out some examples to illustrate this method.

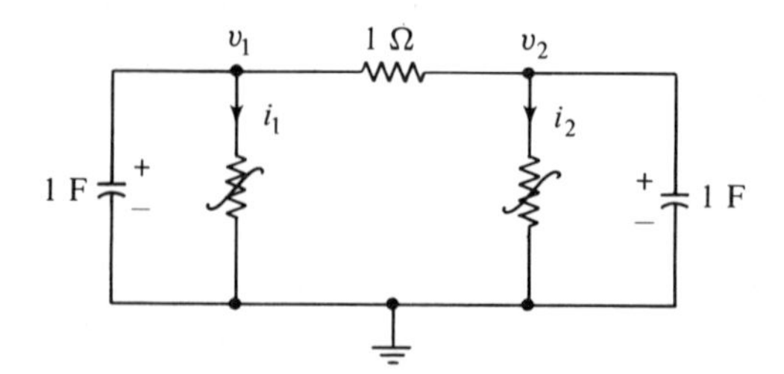

Figure 11.4-2 Nonlinear network with a zero equilibrium state.

Example 11.4-3. Consider the nonlinear network shown in Fig. 11.4-2. The resistors are voltage controlled with i-v relations

$$i_1 = v_1^3 \quad \text{and} \quad i_2 = v_2^5$$

Determine the stability of the equilibrium state $\mathbf{x}_e = \mathbf{0}$ of this network.

Solution

Taking the capacitor charges as the state variables, the state equations of the network can be written as

$$\dot{x}_1 = -x_1 - x_1^3 + x_2$$

$$\dot{x}_2 = x_1 - x_2 - x_2^5$$

Note that $\mathbf{x}_e = \mathbf{0}$ is indeed an equilibrium point of this network. Let us now choose, as a candidate for Lyaponov function, the quadratic function

$$V(\mathbf{x}) = \tfrac{1}{2}(x_1^2 + x_2^2)$$

This scalar function clearly satisfies conditions (a) and (c) of Lyaponov's theorem. To check condition (b), let us take the time derivative of $V(\mathbf{x})$ just given:

$$\dot{V}(\mathbf{x}(t)) = x_1(t)\dot{x}_1(t) + x_2(t)\dot{x}_2(t)$$

Replacing $\dot{x}_1(t)$ and $\dot{x}_2(t)$ from the state equations yields

$$\dot{V}(\mathbf{x}(t)) = -x_1^2 - x_1^4 + x_1x_2 + x_1x_2 - x_2^2 - x_2^6$$

or, by rearranging the right-hand side, it can be written as

$$\dot{V}(\mathbf{x}(t)) = -(x_1 - x_2)^2 - x_1^4 - x_2^6$$

which is negative for all $x_1 \neq 0$ and $x_2 \neq 0$. This implies that condition (b) is also satisfied; hence, $V(\mathbf{x})$ as chosen qualifies as a Lyaponov function for the network under study; consequently, $\mathbf{x}_e = \mathbf{0}$ is globally asymptotically stable.

Example 11.4-4. The state equations of a nonlinear network are

$$\dot{x}_1 = x_2$$

$$\dot{x}_2 = -x_1 - g(x_2)$$

where $g(x_2)$ is a nonlinear function so that $g(0) = 0$. Obtain a sufficient condition on the form of $g(\cdot)$ such that the origin becomes globally asymptotically stable.

Solution

Choose a Lyaponov function as

$$V(\mathbf{x}) = \tfrac{1}{2}(x_1^2 + x_2^2)$$

This function satisfies conditions (a) and (c) of Lyaponov's theorem. To check condition (b), take the time derivative of $V(\mathbf{x})$:

$$\dot{V}(\mathbf{x}(t)) = x_1(t)\dot{x}_1(t) + x_2(t)\dot{x}_2(t)$$

Replacing $\dot{x}_1$ and $\dot{x}_2$ from the state equations, we get

$$\dot{V}(\mathbf{x}(t)) = x_1x_2 - x_1x_2 - x_2g(x_2)$$

or, equivalently;

$$\dot{V}(\mathbf{x}(t)) = -x_2g(x_2)$$

To satisfy condition (b), we must have

$$x_2g(x_2) > 0$$

This implies that the graph of $g(x_2)$ versus x_2 must lie in the first and third quadrant. Hence, *any* function satisfying this condition will make the origin of the network under study globally asymptotically stable.

11.5 LYAPONOV STABILITY OF NONAUTONOMOUS NETWORKS

In the last section we concerned ourselves only with the Lyaponov stability of the equilibrium points of autonomous networks. In this section we show that similar results can be obtained for the Lyaponov stability of operating points of nonautonomous networks. Let us first define the Lyaponov stability of an operating point.

Consider a nonlinear nonautonomous network whose state equation is

$$\dot{\mathbf{x}}(t) = \mathbf{f}[\mathbf{x}(t), \mathbf{u}(t), t] \tag{11.5-1}$$

Let $\bar{\mathbf{x}}(t)$ denote the operating point of (11.5-1) with respect to a bias source $\bar{\mathbf{u}}(t)$; then

Definition 11.5-1. (Lyaponov Stability of Operating Points). The *operating point* $\bar{\mathbf{x}}(t)$ of (11.5-1) is said to be *stable in the sense of Lyaponov* if for any given $\epsilon > 0$ and any t_0 there exists a positive number δ, which may depend on ϵ and t_0, such that

$$|\mathbf{x}(t_0) - \bar{\mathbf{x}}(t_0)| \leq \delta \Longrightarrow |\mathbf{x}(t) - \bar{\mathbf{x}}(t)| \leq \epsilon \quad \text{for all } t \geq t_0 \tag{11.5-2}$$

The operating point $\bar{\mathbf{x}}(t)$ is called *uniformly stable* i.s.L. if the constant δ is independent of the initial time t_0. As in the case of autonomous networks, this definition implies that $\bar{\mathbf{x}}(t)$ is stable i.s.L. if there exists a region of radius δ around $\bar{\mathbf{x}}(t_0)$ so that, if the trajectory originates from that region, it remains close to $\bar{\mathbf{x}}(t)$ for all $t \geq t_0$. This is shown in Fig. 11.5-1 for a scalar case.

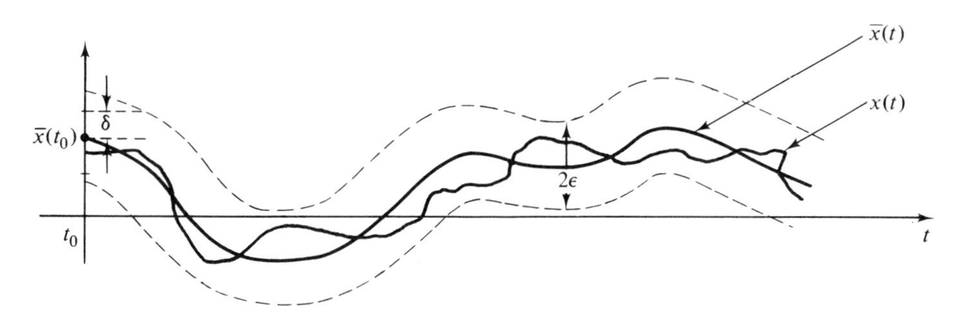

Figure 11.5-1 If the operating point $\bar{x}(t)$ is stable i.s.L., the trajectories starting near $\bar{x}(t_0)$ remain near $\bar{x}(t)$ for all $t \geq t_0$.

Definition 11.5-2. (Asymptotic Stability of Operating Points). The operating point $\bar{\mathbf{x}}(t)$ of (11.5-1) is said to be *asymptotically stable* if the following two conditions are satisfied:

(a) $\bar{\mathbf{x}}(t)$ is stable i.s.L.
(b) $|\mathbf{x}(t) - \bar{\mathbf{x}}(t)| \longrightarrow 0$ as $t \longrightarrow \infty$.

This, in effect, means that if the operating point $\bar{\mathbf{x}}(t)$ is perturbed at t_0 the perturbation will go to zero as t goes to ∞.

Remark 11.5-1. As in the case of autonomous networks, we only consider the Lyaponov stability of the zero operating point, $\bar{\mathbf{x}}(t) = \mathbf{0}$. It is easy to show that with a simple change of variable any operating point $\bar{\mathbf{x}}(t)$ can be transferred to the origin. Indeed, consider the operating point of (11.5-1) with respect to $\bar{\mathbf{u}}(t)$ that is defined by

$$\dot{\bar{\mathbf{x}}} = \mathbf{f}(\bar{\mathbf{x}}, \bar{\mathbf{u}}, t) \tag{11.5-3}$$

Make the following change of variable:

$$\mathbf{x}(t) = \mathbf{y}(t) + \bar{\mathbf{x}}(t) \tag{11.5-4}$$

Replacing $\mathbf{x}(t)$ from (11.5-4) in (11.5-1) yields

$$\dot{\mathbf{y}} = \mathbf{f}(\mathbf{y} + \bar{\mathbf{x}}, \hat{\mathbf{u}}, t) - \mathbf{f}(\bar{\mathbf{x}}, \bar{\mathbf{u}}, t) \triangleq \mathbf{g}(\mathbf{y}, \hat{\mathbf{u}}, t) \tag{11.5-5}$$

The operating point $\bar{\mathbf{y}}(t)$ of $\dot{\mathbf{y}} = \mathbf{g}(\mathbf{y}, \bar{\mathbf{u}}, t)$ is then the solution of

$$\mathbf{y} = \mathbf{f}(\bar{\mathbf{y}} + \bar{\mathbf{x}}, \bar{\mathbf{u}}, t) - \mathbf{f}(\bar{\mathbf{x}}, \bar{\mathbf{u}}, t) \triangleq \mathbf{g}(\bar{\mathbf{y}}, \bar{\mathbf{u}}, t) \tag{11.5-6}$$

Quite clearly, $\bar{\mathbf{y}}(t) = \mathbf{0}$ is a solution of (11.5-6) and, thus, the operating point of (11.5-5). Consequently, to determine the stability of $\bar{\mathbf{x}}(t)$ of (11.5-1), it is sufficient to examine the stability of $\bar{\mathbf{y}}(t) = \mathbf{0}$ of

$$\dot{\mathbf{y}} = \mathbf{g}(\mathbf{y}, \bar{\mathbf{u}}, t)$$

where $\mathbf{y}$ and $\mathbf{g}$ are defined by (11.5-4) and (11.5-5), respectively. With this background let us now present two theorems concerning the Lyaponov stability of nonautonomous networks.

Theorem 11.5-1 (Lyaponov). Consider a nonlinear network whose state equation is

$$\dot{\mathbf{x}} = \mathbf{f}(\mathbf{x}, \mathbf{u}, t) \tag{11.5-7}$$

Assume that $\bar{\mathbf{x}}(t) = \mathbf{0}$ is an operating point of (11.5-7); that is,

$$\mathbf{f}(\mathbf{0}, \bar{\mathbf{u}}, t) = \mathbf{0} \quad \text{for all } \bar{\mathbf{u}} \text{ and } t \tag{11.5-8}$$

Suppose that there exists a scalar function $V(\mathbf{x}, t)$ with continous first partial derivatives with respect to $\mathbf{x}$ and t so that

$$V(\mathbf{0}, t) = 0 \quad \text{for all } t \tag{11.5-9}$$

Furthermore, assume that

(a) There exist continuous nondecreasing functions $\alpha(\cdot)$ and $\beta(\cdot)$ so that

$$0 < \alpha(\|\mathbf{x}\|) \leq V(\mathbf{x}, t) \leq \beta(\|\mathbf{x}\|) \tag{11.5-10}$$

with $\alpha(0) = \beta(0) = 0$.

(b) $\alpha(\|\mathbf{x}\|) \longrightarrow \infty$ as $\|\mathbf{x}\| \rightarrow \infty$.

(c) There exists a continuous nondecreasing function $\gamma(\cdot)$ such that

$$\dot{V}(\mathbf{x}, t) \leq -\gamma(\|\mathbf{x}\|) < 0 \quad \text{for all } \|\mathbf{x}\| > 0 \tag{11.5-11}$$

and $\gamma(0) = 0$.

Then the operating point $\bar{\mathbf{x}}_e(t) = \mathbf{0}$ is globally asymptotically stable.

Proof

Quite clearly, for any function $V[\mathbf{x}(t), t]$ we can write

$$V[\mathbf{x}(t), t] = \int_{t_0}^{t} \dot{V}[\mathbf{x}(\tau), \tau]\, d\tau + V(\mathbf{x}_0, t_0)$$

But since, by (11.5-11), $\dot{V}[\mathbf{x}(\tau), \tau] < 0$, then

$$V[\mathbf{x}(t), t] \leq V(\mathbf{x}_0, t_0) \quad \text{for all } t \geq t_0 \tag{11.5-12}$$

To prove Lyaponov stability, let us refer to Fig. 11.5-2, which shows the relative position of $\alpha(\cdot)$, $\beta(\cdot)$, and $V(\cdot, t)$ versus $\|\mathbf{x}\|$. Note that these three functions are zero only at $\|\mathbf{x}\| = 0$.

To prove stability i.s.L., we must show that for any given $\epsilon > 0$ there exists a δ such that $|\mathbf{x}(t_0)| < \delta \Rightarrow |\mathbf{x}(t)| < \epsilon$. To see how this can be done, choose an arbitrary ϵ on the $\|\mathbf{x}\|$ axis in Fig. 11.5-2 and denote the corresponding point on the range of $\alpha(\cdot)$ as $\alpha(\epsilon)$. Now choose a $\delta > 0$ such that $\beta(\delta) \leq \alpha(\epsilon)$. This can always be done because of (11.5-10). Consequently, for any $\|\mathbf{x}_0\| < \delta$, (11.5-10) and (11.5-12) imply

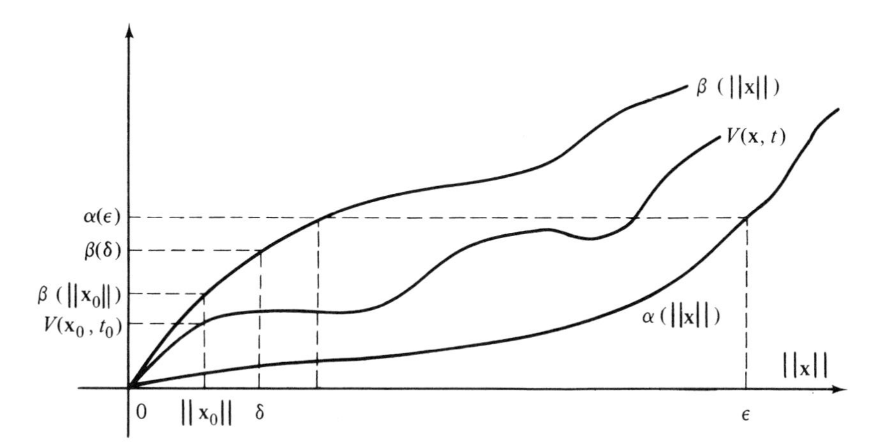

Figure 11.5-2 Graph of Lyaponov function used in the proof of Theorem 11.5-1.

$$\alpha(\|\mathbf{x}\|) \le V(\mathbf{x}, t) \le V(\mathbf{x}_0, t_0) \le \beta(\|\mathbf{x}_0\|) \le \beta(\delta) \le \alpha(\epsilon) \qquad (11.5\text{-}13)$$

Taking the first and last terms in (11.5-13) yields

$$\alpha(\|\mathbf{x}\|) \le \alpha(\epsilon)$$

Since by assumption $\alpha(\|\mathbf{x}\|)$ is a nondecreasing function, this inequality yields

$$\|\mathbf{x}\| < \epsilon$$

Thus,

$$|\mathbf{x}(t)| < \epsilon \quad \text{for all } t \ge t_0$$

Note that δ as chosen depends only on ϵ and not t_0. Hence, $\mathbf{x}(t)$ is *uniformly* stable i.s.L. Asymptotic stability can also be shown using similar arguments. Interested readers are referred to [3].

Let us now state a theorem concerning the asymptotic stability of linear time-varying networks.

Theorem 11.5-2. Consider a linear time-varying network whose state equation is

$$\dot{\mathbf{x}}(t) = \mathbf{A}(t)\mathbf{x}(t) \qquad (11.5\text{-}14)$$

Denote the state transition matrix of this network by $\mathbf{\Phi}(t, t_0)$. Then the zero solution of (11.5-14) is asymptotically stable if and only if there exists a finite constant M such that

$$|\mathbf{\Phi}(t, 0)| < M \quad \text{for all } t \ge 0 \qquad (11.5\text{-}15)$$

and

$$|\mathbf{\Phi}(t, 0)| \longrightarrow 0 \quad \text{as} \quad t \longrightarrow \infty \qquad (11.5\text{-}16)$$

Proof (Sufficiency)

The solution of (11.5-14) can be written as

$$\mathbf{x}(t) = \mathbf{\Phi}(t, t_0)\mathbf{x}_0$$

By the property of the transition matrix (6.3-9), we can write

$$\mathbf{x}(t) = \mathbf{\Phi}(t, 0)\mathbf{\Phi}(0, t_0)\mathbf{x}_0$$

Taking the norm of both sides of this equation, we get

$$|\mathbf{x}(t)| \leq |\mathbf{\Phi}(t, 0)\mathbf{\Phi}(0, t_0)| \cdot |\mathbf{x}_0| \tag{11.5-17}$$

Hence,

$$|\mathbf{x}(t)| \leq |\mathbf{\Phi}(t, 0)| \cdot |\mathbf{\Phi}(0, t_0)| \cdot |\mathbf{x}_0|$$

By (11.5-15) this can be written as

$$|\mathbf{x}(t)| \leq M |\mathbf{\Phi}(0, t_0)| \cdot |\mathbf{x}_0| \leq M^2 \cdot |\mathbf{x}_0|$$

Then for any given $\epsilon > 0$ we can choose $\delta = \epsilon/M^2$; consequently,

$$|\mathbf{x}_0| < \frac{\epsilon}{M^2} \triangleq \delta \Longrightarrow |\mathbf{x}(t)| < \epsilon$$

This implies that the network is stable i.s.L. Asymptotical stability then follows from (11.5-16) and (11.5-17).

Exercise 11.5-1. Prove the necessity part of Theorem 11.5-2. (Hint: Give a proof by contradiction.)

Remark 11.5-2. Note that if the operating point of the linear network under study is asymptotically stable it is then *globally* asymptotically stable.

Remark 11.5-3. In contrast to linear time-invariant networks, even if all the eigenvalues of (11.5-14) have negative real parts, the network may be unstable. This fact is best illustrated by an example.

Example 11.5-1. Consider the linear time-varying network whose state equation is

$$\begin{bmatrix} \dot{x}_1(t) \\ \dot{x}_2(t) \end{bmatrix} = \begin{bmatrix} -1 + 1.5\cos^2 t & 1 - 1.5\sin t\cos t \\ -1 - 1.5\sin t\cos t & -1 + 1.5\sin^2 t \end{bmatrix} \begin{bmatrix} x_1(t) \\ x_2(t) \end{bmatrix}$$

Determine whether the operating point $\bar{\mathbf{x}}(t) = \mathbf{0}$ is stable i.s.L.

Solution

The origin $\bar{\mathbf{x}}(t) = \mathbf{0}$ is obviously an operating point of this network. The eigenvalues can be found to be

$$\lambda_1 = -0.25 + j\sqrt{1.75}, \qquad \lambda_2 = -0.25 - j\sqrt{1.25}$$

both of which have negative real parts. The state transition matrix of this network is (verify this)

$$\mathbf{\Phi}(t, 0) = \begin{bmatrix} e^{0.5t} \cos t & e^{-t} \sin t \\ -e^{0.5t} \sin t & e^{-t} \cos t \end{bmatrix}$$

The solution of the state equation is therefore

$$\begin{bmatrix} x_1(t) \\ x_2(t) \end{bmatrix} = \begin{bmatrix} e^{0.5t} \cos t & e^{-t} \sin t \\ -e^{0.5t} \sin t & e^{-t} \cos t \end{bmatrix} \begin{bmatrix} x_1(0) \\ x_2(0) \end{bmatrix}$$

Note that, due to the term $e^{0.5t}$, $|\mathbf{\Phi}(t, 0)| \longrightarrow \infty$ as $t \longrightarrow \infty$. Hence, a slight perturbation from the origin $\bar{\mathbf{x}}(t) = \mathbf{0}$ will cause $\mathbf{x}(t)$ to go to ∞. This implies that the network under study is unstable even though all of its eigenvalues have negative real parts.

Conversely, a linear time-varying network some of whose eigenvalues have *positive* real parts may still be stable. Consequently, in linear time-varying networks stability or instability cannot be determined solely from the real part of the eigenvalues. It is easy to observe that the stability of (11.5-14) depends on the rate of change of $\mathbf{A}(t)$. If $|\dot{\mathbf{A}}(t)|$ is sufficiently small, and if all the eigenvalues of $\mathbf{A}(t)$ have negative real parts, it can be shown that the origin $\bar{\mathbf{x}}(t) = \mathbf{0}$ of (11.5-11) is asymptotically stable [5].

11.6 BOUNDED INPUT–BOUNDED OUTPUT STABILITY

So far we have discussed the Lyaponov stability of the equilibrium point or operating point of a network. In many practical cases, however, it is essential to know whether a bounded input applied to a given network will generate a bounded output. Throughout this section we assume that the network under consideration is represented by its input–output state equations.

Consider a network with the input vector $\mathbf{u}(t)$ and the output vector $\mathbf{y}(t)$. According to Definition 2.8-2, this network is bounded input–bounded output (b.i.b.o.) stable if there exist finite constants M_1 and M_2 such that

$$||\mathbf{u}|| \leq M_1 < \infty \Longrightarrow ||\mathbf{y}|| \leq M_2 < \infty \qquad (11.6\text{-}1)$$

This definition is in terms of the input and the output vectors only. However, to obtain some conditions for stability of a given network, we must resort to its input–output state representation. For this reason, we consider the b.i.b.o. stability of linear time-invariant, linear time-varying, and nonlinear networks separately. We start by obtaining a set of necessary and sufficient conditions for b.i.b.o. stability of linear time-invariant networks.

Bounded Input–Bounded Output Stability of Linear Time-Invariant Networks. Consider a linear time-invariant network whose input–output state equations are

$$\dot{\mathbf{x}}(t) = \mathbf{A}\mathbf{x}(t) + \mathbf{B}\mathbf{u}(t), \qquad \mathbf{x}(t_0) = \mathbf{x}_0 \tag{11.6-2}$$

$$\mathbf{y}(t) = \mathbf{C}\mathbf{x}(t) + \mathbf{D}\mathbf{u}(t) \tag{11.6-3}$$

where **A**, **B**, **C**, and **D** are constant matrices with appropriate dimensions. If the network under study is a proper network (see Chapter 5), $\mathbf{u}(t)$ represents the input vector and $\mathbf{y}(t)$ represents the output vector. If, on the other hand, the network is not proper, the vector $\mathbf{u}(t)$ will represent the inputs and some of their derivatives. For simplicity, throughout the rest of this section we assume that the network under consideration is proper. We now give a necessary and sufficient condition for the b.i.b.o. stability of such networks. The extension of this result to the general case can then be achieved with little difficulty.

Denote the state transition matrix of (11.6-2) by $\mathbf{\Phi}(t - t_0)$; that is, let

$$\mathbf{\Phi}(t - t_0) = e^{\mathbf{A}(t-t_0)} \tag{11.6-4}$$

Then solving (11.6-2) and (11.6-3) for $\mathbf{y}(t)$ yields

$$\mathbf{y}(t) = \mathbf{C}\mathbf{\Phi}(t - t_0)\mathbf{x}_0 + \mathbf{C}\int_{t_0}^{t} \mathbf{\Phi}(t - \tau)\mathbf{B}\mathbf{u}(\tau)\, d\tau + \mathbf{D}\mathbf{u}(t) \tag{11.6-5}$$

The following theorem, therefore, gives a sufficient condition for stability of such networks.

Theorem 11.6-1. The linear time-invariant network described by (11.6-2) and (11.6-3) is b.i.b.o. stable if there exist numbers α_1 and α_2 such that

(a) $|\mathbf{\Phi}(t - t_0)| < \alpha_1 < \infty$ for all $t \geq t_0$, (11.6-6)

(b) $\int_{t_0}^{t} |\mathbf{\Phi}(t - \tau)|\, d\tau < \alpha_2 < \infty$ for all $t \geq t_0$, (11.6-7)

where $|\cdot|$ represents either the l_1, l_2, or l_∞ norm of a matrix.

Proof

We must show that $\mathbf{y}(t)$ is bounded for all bounded $\mathbf{u}(t)$. To show this, let us take the norm of both sides of (11.6-5) and use the Schwartz inequality to get

$$|\mathbf{y}(t)| \leq |\mathbf{C}|\cdot|\mathbf{\Phi}(t - t_0)|\cdot|\mathbf{x}_0| + |\mathbf{C}|\int_{t_0}^{t} |\mathbf{\Phi}(t - \tau)|\cdot|\mathbf{B}|\cdot|\mathbf{u}(\tau)|\, d\tau + |\mathbf{D}|\cdot|\mathbf{u}(t)|$$

Since matrices **B**, **C**, **D**, and $\mathbf{x}_0$ are constants, their norms are fixed numbers; denote these norms by β_1, β_2, β_3, and β_4, respectively. That is, let

$$|\mathbf{C}| \triangleq \beta_1, \qquad |\mathbf{B}| \triangleq \beta_2, \qquad |\mathbf{D}| \triangleq \beta_3, \qquad |\mathbf{x}_0| \triangleq \beta_4$$

Also, since by assumption the input $\mathbf{u}(t)$ is bounded, there exists a constant γ

such that

$$|\mathbf{u}(t)| < \gamma \quad \text{for all } t \geq t_0$$

Using (11.6-6) and (11.6-7) together with this inequality, we get

$$|\mathbf{y}(t)| \leq \beta_1\alpha_1\beta_4 + \beta_1\alpha_2\beta_2\gamma + \beta_3\gamma \quad \text{for all } t \geq t_0$$

This implies that the output is bounded; hence, the theorem is proved.

Remark 11.6-1. If condition (b) of Theorem 11.6-1 is replaced by

$$\int_{t_0}^{t} |\mathbf{\Phi}(t-\tau)|_1 \, d\tau < \alpha_2 < \infty \quad \text{for all } t \geq t_0 \tag{11.6-8}$$

where $|\mathbf{\Phi}(t-\tau)|_1$ denotes the l_1 norm of $\mathbf{\Phi}(t-\tau)$, then (11.6-6) and (11.6-8) represent *necessary* and *sufficient* conditions for b.i.b.o. stability of the linear time-invariant network under study. A proof of this fact can be found in [2]. Let us now work out a simple example to illustrate the result.

Example 11.6-1. Consider a linear time-invariant network whose input–output state equations are

$$\begin{bmatrix} \dot{x}_1 \\ \dot{x}_2 \end{bmatrix} = \begin{bmatrix} -1 & 0 \\ 1 & -2 \end{bmatrix} \begin{bmatrix} x_1 \\ x_2 \end{bmatrix} + \begin{bmatrix} 1 \\ 0 \end{bmatrix} u(t), \qquad \begin{bmatrix} x_1(0) \\ x_2(0) \end{bmatrix} = \begin{bmatrix} 1 \\ 1 \end{bmatrix}$$

$$\begin{bmatrix} y_1(t) \\ y_2(t) \end{bmatrix} = \begin{bmatrix} 1 & 0 \\ 0 & 1 \end{bmatrix} \begin{bmatrix} x_1 \\ x_2 \end{bmatrix} + \begin{bmatrix} 1 \\ 1 \end{bmatrix} u(t)$$

Determine the b.i.b.o. stability of this network.

Solution

The state transition matrix of this network is easily found to be

$$\mathbf{\Phi}(t) = \begin{bmatrix} e^{-t} & 0 \\ e^{-t} - e^{-2t} & e^{-2t} \end{bmatrix}$$

Let us now find a norm, say the l_1 norm, of this matrix:

$$|\mathbf{\Phi}(t)|_1 \triangleq \max_j \left(\sum_{i=1}^{n} |\Phi_{ij}| \right) = \max\{e^{-t}, 2e^{-2t} - e^{-t}|\} = e^{-t}$$

Consequently,

$$|\mathbf{\Phi}(t)|_1 \leq 1$$

Furthermore, integrating $|\mathbf{\Phi}(t)|_1$ yields

$$\int_0^t |\mathbf{\Phi}(t-\tau)|_1 \, d\tau = \int_0^t e^{-(t-\tau)} \, d\tau = (1 - e^{-t}) \leq 1$$

As a result, conditions (a) and (b) of the theorem are satisfied; hence, the network under study is b.i.b.o. stable.

Next we introduce a corollary of Theorem 11.6-1, which gives a more stringent sufficient condition for b.i.b.o. stability of linear time-invariant networks. This sufficient condition turns out to be quite easy to check.

Corollary 11.6-1. Consider a linear time-invariant network whose input–output state equations are given by (11.6-2) and (11.6-3). Assume that *all* the eigenvalues of **A** have negative real parts; then the network is b.i.b.o. stable.

The usefulness of this corollary is the simplicity of its implementation; if all the eigenvalues of the **A** matrix have negative real parts, no further computation is needed—the network under study is b.i.b.o. stable. The Routh–Hurwiz criterion discussed in Section 11.3 can of course be used to check the location of the eigenvalues of **A**.

Bounded Input–Bounded Output Stability of Networks Represented by Their Transfer Function. In many practical applications, the linear time-invariant networks are represented by their transfer functions. In what follows we present a simple method for checking the b.i.b.o. stability of such networks. We only consider the single input–single output networks; the extension of the results to multiport networks is then a relatively simple exercise.

Consider a single input–single output network whose transfer function is

$$H(s) = \frac{N(s)}{D(s)}$$

where $D(s)$ and $N(s)$ are polynomials of degree n and m. Denote the poles of this transfer function by $p_1, p_2, \ldots, p_n$; that is, let

$$D(s) = (s - p_1)(s - p_2) \ldots (s - p_n)$$

where the p_i are real or complex numbers. The b.i.b.o. stability of the network is then related to the location of the p_i in the complex plane; more precisely,

Theorem 11.6-2. A linear time-invariant network is b.i.b.o. stable if and only if all the poles of its transfer function have negative real parts.

The proof of this theorem is similar to the proof of Theorem 11.3-1 and hence is left as an exercise for the reader.

Exercise 11.6-1. Construct a proof for Theorem 11.6-2. [Hint: Write the input–output relation of the network as

$$y(s) = \frac{N(s)}{D(s)} \cdot u(s)$$

Expand the right-hand side by a partial fraction expansion, and take the inverse Laplace transform.]

Note that location of the poles of $H(s)$ can easily be determined by the Routh–Hurwitz criterion.

Example 11.6-2. The transfer function of a low-pass filter is

$$H(s) = \frac{k}{s^3 + 4s^2 + 3s + 2}$$

Determine the b.i.b.o. stability of this filter.

Solution

To obtain the location of the poles of the transfer function, the following Routh–Hurwitz table is formed:

s^3	1	3
s^2	4	2
s^1	2.5	0
s^0	2	

The elements in the first column are all positive. Consequently, the poles of the transfer function all have negative real parts, and the filter is b.i.b.o. stable.

Bounded Input–Bounded Output Stability of Linear Time-Varying Networks. The conditions for b.i.b.o. stability of linear time-varying networks are similar to those stated in the previous case. Consider a linear, proper, time-varying network whose input–output state equations are

$$\dot{\mathbf{x}}(t) = \mathbf{A}(t)\mathbf{x}(t) + \mathbf{B}(t)\mathbf{u}(t), \qquad \mathbf{x}(t_0) = \mathbf{x}_0 \tag{11.6-9}$$

$$\mathbf{y}(t) = \mathbf{C}(t)\mathbf{x}(t) + \mathbf{D}(t)\mathbf{u}(t) \tag{11.6-10}$$

where $\mathbf{A}(t)$, $\mathbf{B}(t)$ $\mathbf{C}(t)$, and $\mathbf{D}(t)$ are *continuous* and *bounded* matrices with appropriate dimensions. Let $\mathbf{\Phi}(t, t_0)$ denote the state transition matrix of this network; the output $\mathbf{y}(t)$ can then be written as

$$\mathbf{y}(t) = \mathbf{\Phi}(t, t_0)\mathbf{x}_0 + \mathbf{C}(t)\int_{t_0}^{t} \mathbf{\Phi}(t, \tau)\mathbf{B}(\tau)\mathbf{u}(\tau)\,d\tau + \mathbf{D}(t)\mathbf{u}(t)$$

or, equivalently,

$$\mathbf{y}(t) = \mathbf{\Phi}(t, t_0)\mathbf{x}_0 + \int_{t_0}^{t} \mathbf{W}(t, \tau)\,\mathbf{u}(\tau)\,d\tau + \mathbf{D}(t)\mathbf{u}(t) \tag{11.6-11}$$

where $\mathbf{W}(t, \tau)$ denotes the weighting function of the network; that is,

$$\mathbf{W}(t, \tau) = \mathbf{C}(t)\mathbf{\Phi}(t, \tau)\mathbf{B}(\tau) \tag{11.6-12}$$

We can now state the following theorem, which gives a necessary and sufficient condition for the stability of a linear time-varying network.

Theorem 11.6-3. The linear time-varying network represented by (11.6-9) and (11.6-10) is b.i.b.o. stable if and only if there exist constant numbers M_1

and M_2 such that

(a) $|\mathbf{\Phi}(t, t_0)| \leq M_1 < \infty$ for all $t \geq t_0$. (11.6-13)

(b) $\int_{t_0}^{t} |\mathbf{W}(t, \tau)|_1 \, d\tau \leq M_2 < \infty$ for all $t \geq t_0$. (11.6-14)

The proof of this theorem is parallel to the proof of Theorem 11.6-1 and hence is omitted.

Example 11.6-3. Consider the simple network shown in Fig. 11.6-1. The input is the current source and the output is the voltage across the terminating resistor. Determine the b.i.b.o. stability of this network.

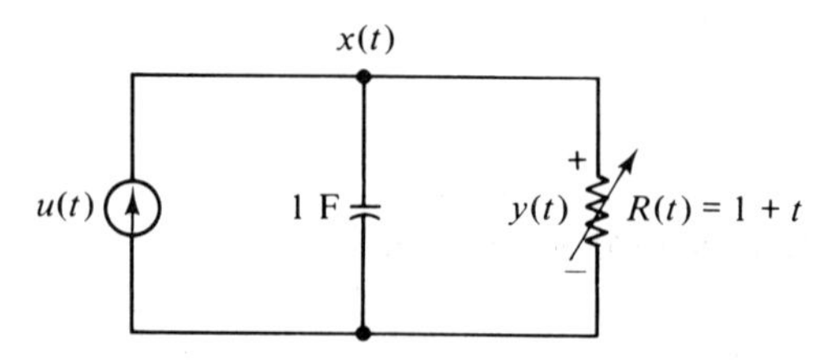

Figure 11.6-1 Linear time-varying *RC* network; the resistance $R(t)$ will go to ∞ as $t \to \infty$.

Solution

The input–output state equations of this network are

$$\dot{x} = -\frac{1}{1+t}x + u$$

$$y = x$$

Hence, $$\mathbf{\Phi}(t, t_0) = \exp\left[\int_{t_0}^{t} \frac{-1}{1+\tau}\, d\tau\right] = \exp\left[-\log\frac{1+t}{1+t_0}\right] = \frac{1+t_0}{1+t}$$

The l_1 norm of $\mathbf{\Phi}(t, t_0)$ is

$$|\mathbf{\Phi}(t, t_0)|_1 = \frac{1+t_0}{1+t} \leq 1 \quad \text{for all } t \geq t_0$$

Condition (a) of Theorem 11.6-2 is therefore satisfied. To check condition (b), observe that

$$\int_{t_0}^{t} \left|\frac{1+\tau}{1+t}\right| d\tau = \frac{1}{1+t}[t + 0.5t^2 - t_0 - 0.5t_0^2]$$

which grows unbounded as $t \longrightarrow \infty$. Consequently, the network under consideration is not b.i.b.o. The same conclusion can be obtained by inspection.

Observe from Fig. 11.6-1 that as $t \longrightarrow \infty$ the resistor behaves as an open circuit. Consequently any finite dc current source, say $u(t) = 1\text{A}$, will increase the voltage across the capacitor indefinitely.

At the conclusion of this chapter it should be mentioned that, unfortunately, there are no simple methods for determining the b.i.b.o. stability of general nonlinear time-varying networks; each individual problem should be considered separately. In many cases it is possible to represent a given network in the form of a feedback system and use the existing methods in system theory to determine its b.i.b.o. stability. This method, however, is more in the realm of abstract system theory and, hence, will not be pursued here.

REFERENCES

[1] STERN, T. E., *Theory of Nonlinear Networks and Systems*, Addison-Wesley Publishing Company, Inc., Reading, Mass., 1965.

[2] ZADEH, L. A., and C. A. DESOER, *Linear System Theory*, McGraw-Hill Book Company, New York, 1963.

[3] KALMAN, R. E., and J. E. BERTRAM, "Control System Analysis and Design via the Second Method of Lyaponov," *Trans. ASME Ser. D, J. Basic Eng.*, **82**, ser. 2 (1960), 371–400.

[4] OGATA, K., *State Space Analysis of Control Systems*, Prentice-Hall, Inc., Englewood Cliffs, N.J., 1967.

[5] DESOER, C. A., "Slowly Varying System $\dot{\mathbf{x}} = \mathbf{A}(t)\mathbf{x}$," *Trans. IEEE Automatic Control*, **AC-14** (Dec. 1969), 780–781.

PROBLEMS

11.1 Find and plot the trajectory of a linear time-invariant network whose state equation is

$$\begin{bmatrix} \dot{x}_1 \\ \dot{x}_2 \end{bmatrix} = \begin{bmatrix} -1 & 0 \\ 1 & -2 \end{bmatrix} \begin{bmatrix} x_1 \\ x_2 \end{bmatrix}, \qquad \begin{bmatrix} x_1(0) \\ x_2(0) \end{bmatrix} = \begin{bmatrix} 1 \\ 1 \end{bmatrix}$$

11.2 For a given second-order network it is possible to sketch the state trajectory by the *isocline* method. Let the network's state equations be

$$\dot{x}_1 = a_{11}x_1 + a_{12}x_2$$

$$\dot{x}_2 = a_{21}x_1 + a_{22}x_2$$

Then

$$\frac{dx_1}{dx_2} = \frac{a_{11}x_1 + a_{12}x_2}{a_{21}x_1 + a_{22}x_2}$$

The left-hand side of this equation represents the slope of the state trajectory in the x_1-x_2 plane. Hence, starting at $\mathbf{x}(t_0) = \mathbf{x}_0$, we can use this relation to obtain the approximate location of $\mathbf{x}(t_1)$ and $\mathbf{x}(t_2) \ldots \mathbf{x}(t_n)$. Use this procedure to sketch the state trajectory of the following state equations:

$$\dot{x}_1 = 2x_1, \qquad x_1(0) = 1$$
$$\dot{x}_2 = x_1 - 2x_2, \qquad x_2(0) = 0$$

11.3 Plot the trajectory of the following nonlinear state equation using the isocline method:

$$\dot{x}_1 = x_1 x_2, \qquad x_1(0) = x_2(0) = 1$$
$$\dot{x}_2 = x_1^2 x_2 + 4x_2$$

11.4 Find all the equilibrium points for the following state equations and plot the trajectories using the isocline method. Discuss the effect of the initial condition on the state trajectory.

$$\dot{x}_1 = 2x_1 - x_1 x_2, \qquad x_1(0) = 2, \quad x_2(0) = 1$$
$$\dot{x}_2 = 2x_1 x_2 - 8x_2$$

11.5 Repeat Problem 11.4 for the following state equations:

$$\dot{x}_1 = 2x_1 - x_1 x_2$$
$$\dot{x}_2 = 8x_2 - 2x_1 x_2, \qquad x_1(0) = 2, \quad x_2(0) = 1$$

11.6 Determine the stability of the equilibrium point of the network discussed in Problem 11.1. If the network is stable, check for asymptotic stability.

11.7 Determine the stability of the equilibrium point of the network discussed in Problem 11.2.

11.8 Determine the asymptotic stability of the equilibrium point $\mathbf{x}_e = \mathbf{0}$ of the linear network given by

$$\begin{bmatrix} \dot{x}_1 \\ \dot{x}_2 \\ \dot{x}_3 \end{bmatrix} = \begin{bmatrix} 0 & 1 & 0 \\ 0 & 0 & 1 \\ -6 & -11 & -6 \end{bmatrix} \begin{bmatrix} x_1 \\ x_2 \\ x_3 \end{bmatrix}$$

11.9 Assume that $a > 0$ and show that

$$\lim_{t \to \infty} t^k e^{-at} = 0$$

for all finite k.

11.10 Use the Routh–Hurwitz test to check the asymptotic stability of the networks whose characteristic polynomials are

(a) $\lambda^3 + 6\lambda^2 + 11\lambda + 6 = 0$
(b) $\lambda^6 + 5\lambda^4 + 2\lambda^3 + 3\lambda^2 + 6\lambda + 1 = 0$
(c) $\lambda^3 + 2\lambda^2 - \lambda + 6 = 0$
(d) $\lambda^4 + 3\lambda^3 + 3\lambda^2 + 3\lambda + 2 = 0$
(e) $\lambda^6 + 4\lambda^5 + 2\lambda^4 + 8\lambda^3 + 3\lambda^2 + 2\lambda + 1 = 0$
(f) $\lambda^7 + \lambda^6 + 2\lambda^5 + 2\lambda^4 + 3\lambda^3 + 3\lambda^2 + \lambda + 1 = 0$

11.11 The state equations of a nonlinear autonomous network are

$$\dot{x}_1 = -x_1 + x_1^2 + 2x_2^2$$
$$\dot{x}_2 = 2x_1 - 2x_2 + x_2^3$$

Determine the local stability of the equilibrium point $\mathbf{x}_e = \mathbf{0}$ of the network.

11.12 Consider a nonlinear first-order network whose state equation is

$$\dot{x} = f(x)$$

where $f(x)$ is a nonlinear function whose plot is given in Fig. P11.12. Find all the equilibrium points of this network and determine the local stability of each equilibrium point.

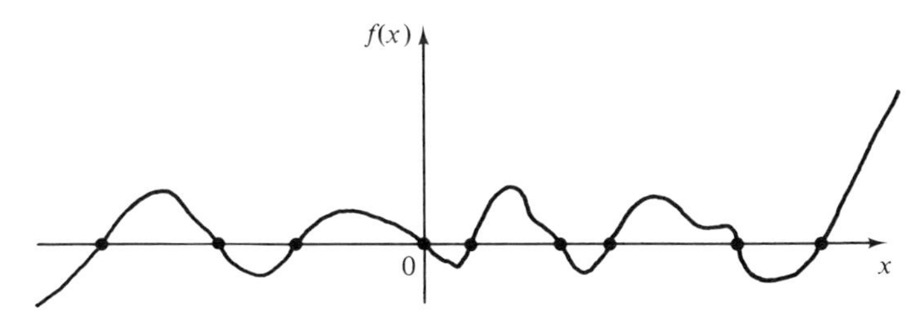

Figure P11.12

11.13 The state equations of a nonlinear autonomous network are

$$\dot{x}_1 = x_2 - 2x_1(x_1^2 + x_2^2)$$
$$\dot{x}_2 = -x_1 - 2x_2(x_1^2 + x_2^2)$$

Show that the equilibrium state $\mathbf{x}_e = \mathbf{0}$ of this network is globally asymptotically stable. [Hint: Choose a Lyaponov function $V(\mathbf{x}) = \frac{1}{4}(x_1^2 + x_2^2)$.]

11.14 Repeat Problem 11.13 for the network whose state equations are

$$\dot{x}_1 = x_2$$
$$\dot{x}_2 = -x_2 - x_1^3$$

[Hint: Choose a Lyaponov function $V(\mathbf{x}) = x_1^4 + 2x_2^2$.]

11.15 Repeat Problem 11.13 for the network whose state equations are

$$\dot{x}_1 = -x_1^3 + 2x_1x_2$$
$$\dot{x}_2 = -x_2$$

[Hint: Choose a Lyaponov function $V(\mathbf{x}) = \frac{1}{2}(x_1^2 + x_2^2)$.]

11.16 Prove that if a linear time-varying network is locally asymptotically stable it is globally asymptotically stable.

11.17 The input–output state equations of a linear time-invariant network are

$$\begin{bmatrix} \dot{x}_1 \\ \dot{x}_2 \\ \dot{x}_3 \end{bmatrix} = \begin{bmatrix} -1 & 0 & 0 \\ 0 & 1 & 0 \\ 2 & 0 & -2 \end{bmatrix} \begin{bmatrix} x_1 \\ x_2 \\ x_3 \end{bmatrix} + \begin{bmatrix} 1 \\ 1 \\ 1 \end{bmatrix} u(t)$$

$$y = [1 \quad 1 \quad 0]\mathbf{x}$$

Determine the b.i.b.o. stability of this network.

11.18 The transfer function of a two-port linear time-invariant network is

$$H(s) = \frac{s^2 + 7s + 12}{s^4 + 3s^3 + 4s^2 + 6s + 4}$$

Determine the b.i.b.o. stability of this network.

11.19 The transfer function of a certain active low-pass filter is

$$H(s) = \frac{2}{s^3 + (K - 1)s^2 + 2Ks + 1}$$

Determine the range of the gain K such that the resulting filter is b.i.b.o. stable.

11.20 Determine the b.i.b.o. stability of a linear time-varying first-order network whose input–output state equations are

$$\dot{x} = (-1 + 0.5 \sin t)x + u(t)$$

$$y = x$$

11.21 Determine the b.i.b.o. stability of a linear time-varying network whose input–output state equations are

$$\begin{bmatrix} \dot{x}_1 \\ \dot{x}_2 \end{bmatrix} = \begin{bmatrix} 0 & -t \\ -1 + t & 0 \end{bmatrix} \begin{bmatrix} x_1 \\ x_2 \end{bmatrix} + \begin{bmatrix} 1 \\ 0 \end{bmatrix} u$$

$$y = [1 \quad 1]\mathbf{x}$$

Appendix
Laplace Transform

A.1 INTRODUCTION

In this appendix we present a summary of the basic definition and properties of the Laplace transform as applied in the analysis of lumped linear time-invariant networks and systems. Since the primary objective of this appendix is to provide a supplement for the material covered in the main body of the text, no attempt has been made to give any rigorous proof of the theorems or a detailed derivation of the results. The reader interested in these topics may consult any of the texts cited in the references.

In Section A.2 we introduce the definition of unilateral Laplace transform and discuss its basic properties. The inverse Laplace transform and partial fraction expansion are discussed in Section A.3.

A.2 BASIC DEFINITION AND CONVERGENCE PROPERTIES OF LAPLACE TRANSFORM

Let s denote a complex variable with real part σ and imaginary part ω; that is,

$$s = \sigma + j\omega$$

Then

Definition A.2-1 (Laplace Transform). The one-sided *Laplace transform* of a time function $f(t)$ is denoted by $F(s)$ and is defined by

$$F(s) = \mathcal{L}\{f(t)\} = \int_{0^-}^{\infty} f(t)e^{-st}\,dt \tag{A.2-1}$$

Note that the lower limit of the integral starts at 0^- in order to take into account the case when $f(t)$ contains an impulse at $t = 0$. The definition is valid only when the integral remains finite. The following theorem provides a sufficient condition for the existence of the Laplace transform of a given function.

Theorem A.2-1. If there exist finite numbers M and σ_0 such that

$$|f(t)| < Me^{\sigma_0 t} \quad \text{for all } t \geq 0 \tag{A.2-2}$$

then $f(t)$ has a Laplace transform.

The proof of this theorem is straightforward; since $|e^{j\omega_0 t}| = 1$, from (A.2-1) we have

$$\left|\int_{0^-}^{\infty} f(t)e^{-st}\,dt\right| \leq \int_{0^-}^{\infty} |f(t)|\,e^{-\sigma t}\,dt \leq M\int_{0^-}^{\infty} e^{-(\sigma-\sigma_0)t}\,dt$$

Thus, for all $\sigma = \mathrm{Re}\{s\} > \sigma_0$ the integral in the right-hand side of this inequality remains finite. Hence, the Laplace transform of $f(t)$ is defined only for $\mathrm{Re}\{s\} > \sigma_0$. For this reason, σ_0 is called the *abscissa of convergence* of $F(s)$. Functions satisfying condition (A.2-2) are said to be of *exponential order*.

An example of a function for which (A.2-2) is not satisfied is $f(t) = e^{t^2}$ (why?). Since most of the functions considered in the analysis and design of electrical networks are of exponential order, in the remainder of this appendix we assume that the time functions under consideration all have Laplace transforms.

Example A.2-1. Find the Laplace transform of $f(t) = e^{-at}1(t)$, where $1(t)$ denotes the unit step function.

Solution

Clearly, $f(t)$ is of exponential order and its abscissa of convergence is $-a$. Thus, we have

$$\mathcal{L}\{e^{-at}1(t)\} = \int_{0^-}^{\infty} e^{-at}e^{-st}\,dt = \frac{1}{s+1}$$

Let us now outline some of the basic properties of the Laplace transform:

1. Linearity. The Laplace transformation is a linear transformation. That is, for any scalar constants α and β the following equation holds:

$$\mathcal{L}\{\alpha f_1(t) + \beta f_2(t)\} = \alpha\mathcal{L}\{f_1(t)\} + \beta\mathcal{L}\{f_2(t)\} \tag{A.2-3}$$

The proof of this property follows directly from (A.2-1).

2. Time Differentiation. The Laplace transform of the derivative of a function can be expressed in terms of the Laplace transform of that function as

$$\mathcal{L}\left\{\frac{d}{dt}f(t)\right\} = sF(s) - f(0^-) \tag{A.2-4}$$

The proof of this result can be obtained by using (A.2-1) and integration by parts:

$$\mathcal{L}\left\{\frac{d}{dt}f(t)\right\} = \int_{0^-}^{\infty}\frac{d}{dt}f(t)e^{-st}\,dt = f(t)e^{-st}\Big|_{0^-}^{\infty} + s\int_{0^-}^{\infty}f(t)e^{-st}\,dt \tag{A.2-5}$$

But, assuming that $f(t)$ is of exponential order, we get

$$\lim_{t\to\infty} f(t)e^{-st} = 0$$

Consequently, (A.2-5) yields the desired result.

In general this property can be expanded to include the nth derivative of $f(t)$:

$$\mathcal{L}\left\{\frac{d^n}{dt^n}f(t)\right\} = s^n f(s) - s^{n-1}f(0^-) - s^{n-2}f^{(1)}(0^-) - \cdots - sf^{(n-2)}(0^-) - f^{(n-1)}(0^-) \tag{A.2-6}$$

where $f^{(k)}(0^-)$ denotes the kth derivative of $f(t)$ evaluated at $t = 0^-$.

Example A.2-2. The Laplace transform of $f(t) = \cos\omega t$ can be seen to be

$$\mathcal{L}\{\cos\omega t\} = \frac{s}{s^2 + \omega^2}$$

Obtain the Laplace transform of $\sin\omega t$.

Solution

Using (A.2-4) and the homogeneity property of the Laplace transform, we get

$$\mathcal{L}\{\sin\omega t\} = -\frac{1}{\omega}\mathcal{L}\left\{\frac{d}{dt}\cos\omega t\right\} = -\frac{1}{\omega}\left[\frac{s^2}{s^2+\omega^2} - 1\right]$$

$$= \frac{\omega}{s^2 + \omega^2}$$

3. Time Integration. The Laplace transform of the integral of a function can be expressed in terms of the Laplace transform of that function as

$$\mathcal{L}\left\{\int_0^t f(\sigma)\,d\sigma\right\} = \frac{1}{s}F(s) \tag{A.2-7}$$

This result can be proved by a direct application of the definition of Laplace transform and integration by parts [1].

Example A.2-3. The Laplace transform of $f(t) = 1(t)$ is given as

$$\mathcal{L}\{1(t)\} = \frac{1}{s}$$

Find the Laplace transform of $f(t) = t^2$.

Solution

Let us first obtain the Laplace transform of t using (A.2-7):

$$\mathcal{L}\{t\} = \mathcal{L}\left\{\int_0^t 1(\sigma)\, d\sigma\right\} = \frac{1}{s} \cdot \frac{1}{s} = \frac{1}{s^2}$$

Then using (A.2-7) once more, we get

$$\mathcal{L}\{t^2\} = 2\mathcal{L}\left\{\int_0^t \sigma\, d\sigma\right\} = 2 \cdot \frac{1}{s} \cdot \frac{1}{s^2} = \frac{2}{s^3}$$

Indeed, this procedure can be repeated n times to obtain the Laplace transform of t^n; the result is

$$\mathcal{L}\{t^n\} = \frac{n!}{s^{n+1}}$$

4. Multiplication by Time. If the Laplace transform of $f(t)$ exists, the Laplace transform of $tf(t)$ can be obtained from

$$\mathcal{L}\{tf(t)\} = -\frac{d}{ds}F(s) \tag{A.2-8}$$

To prove this relationship, notice that

$$te^{-st} = -\frac{d}{ds}e^{-st}$$

Then

$$\mathcal{L}\{tf(t)\} = \int_{0^-}^{\infty} f(t)te^{-st}\, dt = -\int_{0^-}^{\infty} f(t)\frac{d}{ds}e^{-st}\, dt$$

$$= -\frac{d}{ds}\int_{0^-}^{\infty} f(t)e^{-st}\, dt = -\frac{d}{ds}F(s)$$

5. Time Shift. Let $f(t)$ be a function that is zero for $t \leq 0$; the Laplace transform of the shifted function, $f(t - t_0)1(t - t_0)$, can be written as

$$\mathcal{L}\{f(t - t_0)1(t - t_0)\} = e^{-st_0}F(s) \tag{A.2-9}$$

This result may be proved by direct application of (A.2-1).

6. Scaling the Time Axis. For a given positive number α and any function $f(t)$ whose Laplace transform $F(s)$ exists, it can be shown that

$$\mathcal{L}\{f(\alpha t)\} = \frac{1}{\alpha}F\left(\frac{s}{\alpha}\right)$$

7. Division by t. Using the relation

$$\int_s^{\infty} e^{-st}\,ds = \frac{1}{t}e^{-st}$$

it can easily be shown that

$$\mathcal{L}\left\{\frac{1}{t}f(t)\right\} = \int_s^{\infty} F(s)\,ds \tag{A.2-10}$$

8. Multiplication by an Exponential. For any given constant α the Laplace transform of $e^{-\alpha t}f(t)$ can be shown to be

$$\mathcal{L}\{e^{-\alpha t}f(t)\} = F(s+\alpha) \tag{A.2-11}$$

9. Convolution Integral. Let $f_1(t)$ and $f_2(t)$ be two time functions defined on the interval $[0, \infty)$ whose Laplace transforms are $F_1(s)$ and $F_2(s)$, respectively. The Laplace transform of the convolution integral of $f_1(t)$ and $f_2(t)$ is given by

$$F(s) = \mathcal{L}\left\{\int_0^t f_1(t)f_2(t-\tau)\,d\tau\right\} = F_1(s)F_2(s) \tag{A.2-12}$$

A proof for (A.2-12) can be constructed using the basic definition of the Laplace transform; more specifically, we can write

$$F(s) = \mathcal{L}\left\{\int_0^t f_1(t)f_2(t-\tau)\,d\tau\right\} = \int_{0^-}^{\infty}\left[\int_0^t f_1(t)f_2(t-\tau)\,d\tau\right]e^{-st}\,dt$$

or, equivalently,

$$F(s) = \int_0^{\infty}\left[\int_0^{\infty} f_1(t)f_2(t-\tau)1(t-\tau)\,d\tau\right]e^{-st}\,dt$$

where $1(t-\tau) = 0$ for $\tau > t$ and $1(t-\tau) = 1$ for $\tau \leq t$. The change of the order of integration in this equation together with the change of variable $\sigma = t - \tau$ yields

$$F(s) = \int_0^{\infty} f_1(\tau)\left[\int_0^{\infty} f_2(\sigma)e^{-s(\sigma+\tau)}\,d\sigma\right]d\tau$$

or, equivalently,

$$F(s) = \left[\int_0^{\infty} f_1(\tau)e^{-st}\,d\tau\right]\cdot\left[\int_0^{\infty} f_2(\sigma)e^{-s\sigma}\,d\tau\right] = F_1(s)F_2(s)$$

which proves equation (A.2-12).

10. Initial- and Final-Value Theorems. Let $F(s)$ be the Laplace transform of $f(t)$, assume that $\lim_{s\to\infty} sF(s)$ exists, and if $f(0^+)$ is defined, then

$$f(0^+) = \lim_{s\to\infty} sF(s) \tag{A.2-13}$$

This result is known as the *initial-value theorem*.

If $sF(s)$ has no poles on the imaginary axis or in the right half of the s

plane, the final value of $f(t)$ can be obtained from

$$f(\infty) = \lim_{s \to 0} sF(s) \tag{A.2-14}$$

This result is known as the *final-value theorem.*

Example A.2-4. Let

$$F(s) = \frac{s}{s^2 + 1}$$

Find $f(0^+)$ and $f(\infty)$.

Solution

Since $\lim_{s \to \infty} sF(s)$ is defined, we use (A.2-13) to get

$$f(0^+) = \lim_{s \to \infty} sF(s) = 1$$

To obtain $f(\infty)$, note that $sF(s)$ has a pair of poles on the imaginary axis and therefore (A.2-14) does not apply; the final value of $f(t)$ is not well defined. Indeed, it can be seen that $f(t) = \cos t$; hence, $f(\infty)$ is somewhere between -1 and $+1$; otherwise unspecified.

Using these properties and the basic definition, the Laplace transform of a large class of functions can be obtained.

Example A.2-5. The Laplace transform of $\sin \omega t$ is given as

$$\mathcal{L}\{\sin \omega t\} = \frac{\omega}{s^2 + \omega^2}$$

Obtain the Laplace transform of $e^{-2t} \cos [2\omega t - (\pi/8)]1[t - (\pi/8)]$ using properties 1 through 10 discussed earlier.

Solution

Since $\cos \omega t = (1/\omega)(d/dt) \sin \omega t$ and $\sin 0 = 0$, using property 2 we get

$$\mathcal{L}\{\cos \omega t\} = \frac{1}{\omega} \cdot \frac{s\omega}{s^2 + \omega^2} = \frac{s}{s^2 + \omega^2}$$

Choosing $t_0 = \pi/8$ and using property 5, we obtain

$$\mathcal{L}\left\{\cos\left(\omega t - \frac{\pi}{8}\right)1\left(t - \frac{\pi}{8}\right)\right\} = \frac{se^{-s(\pi/8)}}{s^2 + \omega^2}$$

By property 6, we get

$$\mathcal{L}\left\{\cos\left(2\omega t - \frac{\pi}{8}\right)1\left(t - \frac{\pi}{8}\right)\right\} = \frac{1}{2}\,\frac{(1/2)se^{-s(\pi/16)}}{(s^2/4) + \omega^2} = \frac{se^{-s(\pi/16)}}{s^2 + 4\omega^2}$$

and, using property 8, we obtain

$$\mathcal{L}\left\{e^{-2t}\cos\left(2\omega t - \frac{\pi}{8}\right)1\left(t - \frac{\pi}{8}\right)\right\} = \frac{(s + 2)e^{-(s+2)(\pi/16)}}{(s + 2)^2 + 4\omega^2}$$

The Laplace transforms of some of the commonly used time functions are given in Table A.2-1. More complete tables of Laplace transformation may be found in standard mathematical tables.

Table A.2-1
UNILATERAL LAPLACE TRANSFORM PAIRS

	$f(t);\ t \geq 0$	$F(s)$
1	$\delta(t)$	1
2	$\frac{d^n}{dt^n}\delta(t)$	s^n
3	$1(t)$	$\frac{1}{s}$
4	$t^n 1(t)$	$\frac{n!}{s^{n+1}}$
5	$e^{-at} 1(t)$	$\frac{1}{s+a}$
6	$t^n e^{-at} 1(t)$	$\frac{n!}{(s+a)^{n+1}}$
7	$\sin(\omega t) 1(t)$	$\frac{\omega}{s^2+\omega^2}$
8	$\cos(\omega t) 1(t)$	$\frac{s}{s^2+\omega^2}$
9	$\sinh(at) 1(t)$	$\frac{a}{s^2-a^2}$
10	$\cosh(at) 1(t)$	$\frac{s}{s^2-a^2}$
11	$e^{-at}\sin(\omega t) 1(t)$	$\frac{\omega}{(s+a)^2+\omega^2}$
12	$e^{-at}\cos(\omega t) 1(t)$	$\frac{s+a}{(s+a)^2+\omega^2}$
13	$te^{-at}\sin(\omega t) 1(t)$	$\frac{2\omega(s+a)}{[(s+a)^2+\omega^2]^2}$
14	$te^{-at}\cos(\omega t) 1(t)$	$\frac{(s+a)^2+\omega^2}{[(s+a)^2+\omega^2]^2}$

A.3 INVERSE LAPLACE TRANSFORM

One of the main applications of the Laplace transform is in the solution of linear time-invariant differential equations. Given an nth order linear time-invariant differential equation in $y(t)$, we apply the Laplace transform to obtain an algebraic equation that can easily be solved for $Y(s)$. The solution of the differential equation then reduces to finding the inverse Laplace transform of $Y(s)$. In this section we introduce and discuss various methods of finding inverse Laplace transforms. To motivate the discussion, let us first work out an illustrative example.

Example A.3-1. Solve the following differential equation by Laplace transform:

$$\ddot{y}(t) + 4\dot{y}(t) + 3y(t) = \delta(t), \qquad y(0) = 0 \quad \text{and} \quad \dot{y}(0) = 0$$

Solution

Taking the Laplace transform of both sides of the equation and using properties 1 and 2, we get

$$Y(s) = \frac{1}{s^2 + 4s + 3}$$

Thus, $y(t)$ can be obtained by taking the inverse Laplace transform of $y(s)$; that is,

$$y(t) = \mathfrak{L}^{-1}\{Y(s)\} = \mathfrak{L}^{-1}\left\{\frac{1}{s^2 + 4s + 3}\right\} = \mathfrak{L}^{-1}\left\{\frac{1}{(s+1)(s+3)}\right\}$$

or, equivalently,

$$y(t) = \mathfrak{L}^{-1}\left\{\frac{1/2}{s+1} - \frac{1/2}{s+3}\right\} = \frac{1}{2}(e^{-t} - e^{-3t})1(t)$$

Consequently, determining the inverse Laplace transform of the rational function $Y(s)$ yields the solution of the differential equation under consideration.

Generally, the inverse Laplace transform of a function $F(s)$ can be obtained from the equation

$$f(t) = \mathfrak{L}^{-1}\{F(s)\} = \frac{1}{2\pi j}\int_{\sigma_1 - j\infty}^{\sigma_1 + j\infty} F(s)e^{st}\, ds \tag{A.3-1}$$

where σ_1 is a real constant number chosen larger than the real parts of all the poles of $F(s)$. The justification of (A.3-1) is beyond the intended scope of this appendix and, therefore, is omitted. The interested reader can refer to [7–10].

In many practical applications the Laplace function $F(s)$ is a rational function of s and thus can be written as the quotient of two polynomials in s. In this case one can use the *partial fraction expansion* method discussed next to obtain the inverse Laplace transform.

Partial Fraction Expansion. Let $F(s)$ be given as the quotient of two polynomials $N(s)$ and $D(s)$:

$$F(s) = \frac{N(s)}{D(s)} \tag{A.3-2}$$

Denote the roots of $D(s)$ by $p_1, p_2, \ldots, p_n$, where the p_i may be real or complex, simple or multiple. We consider two distinct cases:

Case 1 (Simple Poles). Without any loss of generality, assume that s^n in $D(s)$ has unity coefficient; then $F(s)$ can be written as

$$F(s) = \frac{N(s)}{(s - p_1)(s - p_2)\ldots(s - p_n)} \tag{A.3-3}$$

If the degree of $N(s)$ is less than the degree of $D(s)$, the function $F(s)$ can be written as

$$F(s) = \frac{r_1}{s - p_1} + \frac{r_2}{s - p_2} + \cdots + \frac{r_k}{s - p_k} + \cdots + \frac{r_n}{s - p_n} \tag{A.3-4}$$

where r_k, $k = 1, 2, \ldots, n$, is a real or complex constant known as the *residue* of the pole at $s = p_k$. To obtain r_k, we can multiply both sides of (A.3-4) by $s - p_k$ and let $s \longrightarrow p_k$. Since by assumption $p_j \neq p_k$, this yields

$$r_k = \lim_{s \to p_k} (s - p_k)F(s) \tag{A.3-5}$$

To illustrate this procedure, let us consider the function discussed in Example A.3-1. In this case the roots of the denominator polynomial are $p_1 = -1$ and $p_2 = -3$. Thus, $Y(s)$ can be written as

$$Y(s) = \frac{1}{(s + 1)(s + 3)} = \frac{r_1}{s + 1} + \frac{r_2}{s + 3}$$

Using (A.3-5), we get

$$r_1 = \lim_{s \to -1} (s + 1)Y(s) = \tfrac{1}{2}$$

and

$$r_2 = \lim_{s \to -3} (s + 3)Y(s) = -\tfrac{1}{2}$$

Hence, as before, we get

$$Y(s) = \frac{1/2}{s + 1} - \frac{1/2}{s + 3}$$

Using Table A.2-1, the inverse Laplace transform of $Y(s)$ can be found as

$$y(t) = \tfrac{1}{2}(e^{-t} - e^{-3t})1(t)$$

Remark A.3-1. If p_k is a complex root of $D(s)$, then its complex conjugate $\bar{p}_k$ is also a root (why?). In this case such complex roots can be combined to form a quadratic polynomial, which can be represented as a shifted sinusoidal function. This procedure can be best illustrated by an example.

Example A.3-2. Find the inverse Laplace transform of the function

$$F(s) = \frac{1}{s^3 + 5s^2 + 11s + 7}$$

Solution

The roots of the denominator polynomial are $p_1 = -1$, $p_2 = -2 + j\sqrt{3}$, and $\bar{p}_2 = -2 - j\sqrt{3}$. We can then write

$$F(s) = \frac{1}{s^3 + 5s^2 + 11s + 7} = \frac{1}{(s+1)(s^2 + 4s + 7)}$$
$$= \frac{r_1}{s+1} + \frac{r_2 s + r_3}{s^2 + 4s + 7}$$

The residue of the pole at $s = -1$ can be found by multiplying both sides of this equation by $s + 1$ and letting $s = -1$. The result is $r_1 = \frac{1}{4}$. The coefficients r_2 and r_3 can also be obtained by taking the common denominator for the extreme right-hand side of the equation. This yields

$$\tfrac{1}{4}(s^2 + 4s + 7) + (s + 1)(r_2 s + r_3) = 1$$

This equation must hold for all s. Hence, we must have $r_2 = -1$ and $r_3 = -3$. The function $F(s)$ can then be written as

$$F(s) = \frac{1/4}{s+1} + \frac{s+3}{s^2 + 4s + 7}$$

or, alternatively, as

$$F(s) = \frac{1/4}{s+1} + \frac{s+2}{(s+2)^2 + 3} + \frac{1}{(s+2)^2 + 3}$$

Using Table A.2-1, the inverse function can then be obtained as

$$f(t) = \left[\frac{1}{4}e^{-t} + e^{-2t}\cos\sqrt{3}\,t + \frac{\sqrt{3}}{3}e^{-2t}\sin\sqrt{3}\,t\right]1(t)$$

Case 2 (Multiple Poles). Suppose that one of the roots of $D(s)$, say p_1, is of multiplicity m and the remaining roots are simple. Furthermore, assume that the degree of $N(s)$ is less then the degree of $D(s)$. We can then write $F(s)$ as

$$F(s) = \frac{N(s)}{(s - p_1)^m (s - p_2) \ldots (s - p_n)} \tag{A.3-6}$$

The partial fraction expansion of this function is

$$F(s) = \frac{r_{11}}{(s - p_1)^m} + \frac{r_{12}}{(s - p_1)^{m-1}} + \cdots$$
$$+ \frac{r_{1m}}{s - p_1} + \frac{r_2}{s - p_2} + \cdots + \frac{r_n}{s - p_n} \tag{A.3-7}$$

The residue of the simple poles, $r_2, r_3, \ldots, r_n$, can be determined as in Case 1.

To obtain the constants r_{1k}, $k = 1, 2, \ldots, m$, multiply both sides of (A.3-7) by $(s - p_1)^m$, differentiate $k - 1$ times, and put $s = p_1$. This yields

$$r_{1k} = \frac{1}{(k-1)!}\frac{d^{k-1}}{ds^{k-1}}\{(s - p_1)^m F(s)\}\bigg|_{s=p_1}, \qquad k = 1, 2, \ldots, n \tag{A.3-8}$$

Example A.3-3. Find the inverse Laplace transform of

$$F(s) = \frac{s+2}{(s+1)^3(s+4)}$$

Solution

We can write

$$F(s) = \frac{r_{11}}{(s+1)^3} + \frac{r_{12}}{(s+1)^2} + \frac{r_{13}}{s+1} + \frac{r_2}{s+4}$$

where

$$r_2 = (s+4)F(s)|_{s=-4} = \frac{2}{27}$$

Using (A.3-8), the remaining coefficients can be found as

$$r_{11} = (s+1)^3 F(s)|_{s=-1} = \frac{1}{3}$$

$$r_{12} = \frac{1}{(2-1)!}\frac{d}{ds}\{(s+1)^3 F(s)\}\bigg|_{s=-1} = \frac{2}{9}$$

$$r_{13} = \frac{1}{(3-1)!}\frac{d^2}{ds^2}\{(s+1)^3 F(s)\}\bigg|_{s=-1} = -\frac{2}{27}$$

Thus, using Table A.2-1, we get

$$f(t) = \tfrac{1}{27}[2e^{-4t} - 2e^{-t} + 6te^{-t} + \tfrac{9}{2}t^2 e^{-t}]1(t)$$

Remark A.3-2. In both Cases 1 and 2 we assumed that the degree of the numerator polynomial $N(s)$ is *less* than the degree of $D(s)$. If the degree of $N(s)$ is equal to or greater than that of $D(s)$, the partial expansions given in (A.3-4) or (A.3-7) must be modified.

Let the degree of $N(s)$ be m and the degree of $D(s)$ be n, and assume that $m > n$; then $F(s)$ can be written as

$$F(s) = \frac{N(s)}{D(s)} = k_{m-n}s^{m-1} + k_{m-n-1}s^{m-n-1} + \cdots + k_1 s + k_0 + \frac{\hat{N}(s)}{D(s)}$$

where $\hat{N}(s)$ is a polynomial whose degree is less than the degree of $D(s)$. Thus, the procedures of Cases 1 and 2 can be used to obtain a partial fraction expansion of $\hat{N}(s)/D(s)$.

Note that the inverse Laplace transform of $k_0, k_1 s, \ldots, k_{m-n}s^{m-1}$ are $k_0\delta(t), k_1(d/dt)\delta(t), \ldots, k_{m-n}(d^{m-n}/dt^{m-n})\delta(t)$.

REFERENCES

[1] DESOER, C. A., and G. S. KUH, *Basic Circuit Theory*, McGraw-Hill Book Company, New York, 1969.

[2] KARNI, S., *Intermediate Network Analysis*, Allyn and Bacon, Inc., Boston, Boston, 1971.

[3] CHIRLIAN, P. M., *Basic Network Theory*, McGraw-Hill Book Company, New York, 1969.

[4] HUELSMAN, L. P., *Basic Circuit Theory with Digital Computation*, Prentice-Hall, Inc., Englewood Cliffs, N.J., 1972.

[5] GUPTA, S. C., J. W. BAYLESS, and B. PEIKARI, *Circuit Analysis with Computer Application to Problem Solving*, International Textbook Company–College Division, Scranton, Pa., 1972.

[6] BALABANIAN, N., and T. A. BICKART, *Electrical Network Theory*, John Wiley & Sons, Inc., New York, 1969.

[7] CHURCHILL, R. V., *Operational Mathematics*, McGraw-Hill Book Company, New York, 1958.

[8] LEPAGE, W. R., *Complex Variables and the Laplace Transform for Engineers*, McGraw-Hill Book Company, New York, 1961.

[9] SPIEGEL, M. R., *Theory and Problems of Laplace Transforms*, McGraw-Hill Book Company, New York, 1965.

[10] SCHWARZ, R. J., and B. FRIEDLAND, *Linear Systems*, McGraw-Hill Book Company, New York, 1965.

Index

A

abscissa of convergence, 484
active filter, 365
    synthesis of, 365-87
Adams-Moulton formula, 327
adder, 40, 367
admissible signal pair, 14, 49, 65
algorithm, 360
ampere, 1
amplifier:
    nonlinear, 41
    operational, 36-42
    parametric, 20
angular frequency, 156
autonomous network, 419
    equilibrium point of, 420
    perturbation analysis, 423
    small-signal equivalent of, 423
    small-signal response of, 425

B

back substitution, 288
Balabanian, N., 179, 234, 494
Bartel, R. G., 275
Bashkow, T. R., 233
Beightler, C. S., 415
Bellman-Gronwall Lemma, 273
Berge, C., 122
Berry, R. D., 333
Bertram, J. E., 479
bias, 9
    slowly varying, 431
Bickart, T. A., 179, 234, 494
biquad realization, 371
Bott, R., 382
boundedness, 85
branch, 91
    voltage-current relationship, 129
Branin, F. H., Jr., 333
Brayton, R. K., 234, 333
Brown, G. C., 333
Broyden, C. G., 333
Brune, O., 382
Bryant, P. R., 233, 234

C

Calahan, D. A., 234, 414, 415
capacitance, 16
    incremental, 21
capacitor, 15
    active, 18
    charge-controlled, 15
    energy stored on, 18
    linear, 16
    -only loop, 190, 222
    monotone, 15
    multiterminal, 30
    nonlinear, 15
    passive, 18
    small-signal behavior, 20
    time-varying, 16
    two terminal, 15
    v-i relationship, 16
    voltage-controlled, 15
Cayley-Hamilton Theorem, 248
Chan, S. P., 122
characteristic impedance, 174

charge, 1
Chen, C. T., 234
Chen, W. H., 382
Chirlian, P. M., 180, 494
chords, 94
Chua, L. O., 43, 234, 333, 415, 446
Churchil, R. V., 494
circuits, (*see* Networks)
clipping circuit, 70
Coddington, E. A., 276
cofactor, 54
column vector, 50
computer-aided design, 387, 410
computer program for:
analysis of linear resistive networks, 292, 298
analysis of nonlinear resistive networks, 308, 311
computation of steady state response, 300
design of active filter, 399
plotting a graph, 322
solution of state equations, 321, 324, 328
conjugate gradient method, 406
connected graph, 92
controlled sources, 36
convertors, 40
current-inversion negative-impedance, 42
voltage-inversion negative-impedance, 42
convexity, 388
convolution integral, 487
cotree, 94
coulombs, 1
current, 1
scattering matrix, 178
transfer functions, 169
current source:
current-controlled, 36
independent, 4
voltage-controlled, 36
cutset, 94
fundamental, 94
cutset matrix:
augmented, 101
fundamental, 103

D

damped natural frequency, 366
damping ratio, 366
dc source, 8
dependent sources, 36, 148
Dervishoglu, A., 234
Desoer, C. A., 87, 179, 234, 275, 414, 415, 446, 479
differentiator, 39
diode:
ideal, 12
tunnel, 11
Director and Rohrer method, 406
Director, S. W., 419
distributed element, 2
driving point admittance matrix, 136
driving point impedance, 337
matrix, 162
Duffin, L. J., 382
Dunker, 122

E

eigenvalues, 248
multiple, 249
electric:
energy, 1
power, 1
element:
admittance matrix, 134
impedance matrix, 129
equilibrium state, 420
stability of, 453-62
error function, 389
convex, 415
mean squared, 389
unimodal, 415
weighted function, 389
Euler integration method, 315
existence and uniqueness theorem, 269
explicit integration schemes, 317

F

feedback:
effect on sensitivity, 380
field effect transistor, 31
filter, 366
Butterworth, 372, 375, 397, 444
Chebyshev, 384
Gaussian, 385
low-pass, 169
final value theorem, 488
finite escape time, 269
Fletcher-Powell method, 405
Fletcher, R. M., 415
frequency multiplier, 13, 28
Friedland, B., 494
function of a matrix, 250
fundamental cutset, 94
fundamental loop, 94, 108, 143

G

Gaussian elimination method, 287
Gauss-Seidel method, 295
generating solution, 427
Ghausi, M. S., 382
global stability, 463
gradient, 393
Grame, J., 43, 382
graph, 91
connected, 93
oriented, 91
Gupta, S. C., 179, 494
Gustavson, F. G., 333
gyrator, 30
circuit for simulating of, 33
inductor-simulating, 68
network, 137, 145, 198
stability, 85

H

Hajj, I. N., 333
Hatchel, G. D., 333
Hazony, D., 382
Hohn, F. E., 87

Huelsman, L. P., 43, 180, 382, 494
hybrid parameters, 171

I

ideal diode, 12
ideal transformer, 34, 79
implicit integration formulas, 316
incidence matrix, 99
  augmented, 96
  rank of, 97
inductor, 23
  coupled, 33
  current-controlled, 24
  energy stored in, 26
  flux-controlled, 23
  linear, 25
  multiterminal, 30
  -only cutset, 190, 222
  reciprocal, 34
  simulation by gyrator, 32
  small-signal analysis, 28
  time-invariant, 25
  two-terminal, 23
  v-i relationship, 25
initial value theorem, 487
input-output state equations, 217
integration formulas, 317
integrator, 39, 367
integro-differential equation, 64
  additive, 65
  homogeneous, 65
inverse Laplace transform, 489
Isaacson, E., 333
isocline, 479
iterative algorithm, 390
  methods, 388

J

Jacobian matrix, 309
joules, 1

K

Kalman, R. E., 479
Kaplan, W., 87
Karni, S., 180, 382, 494
Katzenelson, J., 234, 333
Keller, H. B., 333
Key, E. L., 382
Kirchhoff's current law (KCL), 98, 100
Kirchhoff's voltage law (KVL), 105
Kuh, E. S., 87, 179, 233, 234, 333
Kuo, F. F., 415
Kuratowski nonplanar graphs, 110

L

Laplace transform, 483, 484
least upper bound, 62
Lee, H. B., 333
LePage, W. R., 494
Levinson, N., 276
links, 94
Lipshitz condition, 269
local stability, 463
loop, 93
  fundamental, 94, 108, 143
  impedance matrix, 143, 151
  matrix, augmented, 104
  self, 93
lossless network, 78
lumped elements, 3
LU transformation, 295
Lyaponov:
  function, 466
  Theorem, 466, 470

M

magnetic flux, 1
Magnuson, W. G., Jr., 415
matched networks, 175
matrices, multiplication of, 51
matrix:
  algebra, 50
  determinant of, 53
  diagonal, 51
  differentiation, 55
  dominant, 297
  fundamental, 259
  Hessian, 403
  identity, 51
  integration, 55
  inversion, 52
  Jacobian, 211
  minor of, 53
  multiplication by a scalar, 51
  nonsingular, 52
  norm of, 61
  partitioning, 54
  positive definite, 211
  positive real, 341
  positive semi-definite, 211
  postmultiplied, 52
  premultiplied, 52
  rank of, 57
  sparse, 295
  square, 50
  state transition, 244
  transpose, 51
  zero, 51
mesh equations, 109
  matrix representation of, 112
Milne's integration formula, 327
minimal polynomial, 255, 257
minimum function, 356
Mitra, S. K., 382
Moser, J. K., 234
multipliers, 367
mutual inductance, 34

N

negative impedance convertors, 42
network:
  active, 80
  adjoint, 406
  autonomous, 419
  B.I.B.O. stability of, 474
  causal, 81
  characterization, 48
  computer-aided analysis, 284

network *(cont.)*
computer solution of, 303
direct analysis method of, 127
impulse response, 265
linear, 67
loop analysis, 142
lossless, 78
mesh analysis, 142
nodal analyses, 139
nonanticipative, 82
nonautonomous, 419
noncausal, 82
nonplanar, 110
n-port, 49
passive, 77
phase variable representation, 186
planar, 110
portwise linear, 70
portwise nonlinear, 70
portwise time-invariant, 73
portwise time-varying, 73
proper, 199
sensitivity consideration, 378
small-signal equivalent, 429
stability analysis, 450-79
state variable formulation, 221-41
state variable representation, 186
steady state analysis, 156
steady state response, 300
time-invariant, 76
time-invariant *n* ports, 72
time-varying, 72, 76
unstable, 85
network element, 1
active, 5
distributed parameter, 2
impulse response, 265
lumped, 2, 3
multiterminal, 29
passive, 4
two terminal, 4-29
network functions, 162
network synthesis, 337-86
active, 365
Bott-Duffin synthesis, 361
Brune's method, 352, 358
Cauer's methods, 345
Foster methods, 343, 348
frequency domain, 388
Newcomb, R. W., 43, 87, 382
Newton-Raphson Method, 305
nodal analysis, 139-42, 284
node, 91
admittance matrix, 140
degree of, 93
isolated, 93
simple, 93
nonlinear amplifier, 41
nonlinear networks:
computer-aided design, 439
global stability, 465
iterative solution, 305
local stability, 464
Lyaponov stability, 468
perturbation analysis, 423, 427
small-signal analysis, 419, 427
small-signal response, 423
nonlinear resistor, 5
application of, 11
small signal behavior, 8
norm:
induced, 62
of a matrix, 61, 63
of a vector, 59, 61
normal tree, 221
numerical solution of:
nonlinear resistive networks, 303-14
state equation, 314-32
steady state response, 300
numerical stability, 317

O

Ogata, K., 275, 479
ohm, 7
open-circuit impedance matrix, 162
operating point, 427
asymptotical stability, 468
frozen, 431
Lyaponov stability, 468
operational amplifiers, 36
application of, 39-42
feedback, 38
noninverting, 374
realization, 37
operator, integro-differential, 64
optimization:
bang-bang, 401, 404
general method, 402
iterative method, 390, 411
mean-squared error, 392
order of complexity, 216, 218, 219
steepest descent method, 403

P

Palais, R. S., 333
parametric amplifier, 20
partial fraction expansion, 491
passive:
element, 4
network, 77-80
path, 93
Peikari, B., 43, 87, 179, 415, 446, 494
Penfield, P. Jr., 43, 122
phasor, 156, 389
Picard iteration, 269
planar network, 110
port, 48
portwise linear, 66
positioning for size, 290
positive real functions, 338
Powel, J. C., 415
predictor-corrector method, 316, 325, 327
proper tree, 200
pulse generator, 11
Purslow, E. J., 234

R

Rafuse, R., 43
Ralston, A., 33
Ramo, S., 180

reactance function, 343
synthesis of, 342
rectifier, 12
Reed, M. B., 122
Reeves, C. M., 415
reference convention, 3
reflection coefficient, 174
region of attraction, 463
relative error, 284
resistance, 7
resistor, 5
current-controlled, 5
linear, 7
linear time-invariant, 7
monotonic, 6
multiterminal, 30
nonlinear, 5
time-invariant, 7
time-varying, 6, 10
voltage-controlled, 6
Richards, P. I., 382
Rohrer, R. A., 87, 179, 233, 234, 414, 415
rotor, 42
round-off error, 284
row, 50
row vector, 50
Runge-Kutta method, 318

S

Sallen, R. P., 382
Sandberg, J. W., 33
Sato, N., 333
scattering matrix, 178
scattering parameters, 173
Schwarz, R. J., 494
Seitelman, L. H., 333
self inductance, 34
sensitivity function, 379
Seshu, S., 122, 179, 234
short-circuit admittance matrix, 163
signal generator, 13
simultaneous equations, 58
linear independence of, 58
source transformation, 131, 152
Spence, R., 122
Spiegel, M. R., 494
Spivak, M., 446
stable, bounded input–bounded output, 85
stability:
asymptotic, 415
bounded input–bounded output, 473
equilibrium, 415
global, 463
local, 463
Lyaponov, 453
Routh-Hurwitz test, 458
sufficient conditions for, 457
zero input, 451
state:
concept of, 216
of a network, 217
state equations:
computer solution, 314
linear time-invariant networks, 204
linear time-varying networks, 200
nonlinear networks, 209
normal form, 190, 193, 196
state equations *(cont.)*
solution of, 243, 329
solution of linear time-varying, 259
state trajectory, 452
state transition matrix, 244, 260
computation, 245
state variable, 186
state vector, 180
steepest descent method, 396, 403
Stern, T. E., 446, 479
subgraph, 93
superposition principle, 66
supremum, 62

T

tandem connection:
of two ports, 173
Tellegen's theorem, 118
Temes, G. C., 414
time-invariant, portwise, 75
time-varying, portwise, 73
Tinny, W. F., 333
Tobey, G., 43, 382
transfer function, 168
matrix, 169
transistor, 30
field effect, 31
transmission line, 43
transmission parameters, 171
trapezoidal integration rule, 316
tree, 94
branches, 94
truncation error, 283
tunnel diode, 6
two-port parameters, 171

U

unimodality, 388

V

vacuum tube, 30
Vanduzer, T., 180
Van Valkenburg, M. E., 382
varactor diode, 21
vectors:
linear dependence of, 57
norm of, 59
voltage, 1
inversion negative-impedance convertor, 42
scattering matrix, 178
voltage source:
current-controlled, 36
independent, 4
voltage-controlled, 36
voltage transfer function, 169

W

Walker, J. W., 333
watts, 1
Weber, 1
weighting function, 264, 265
Whinery, J. R., 180

White, J. V., 446
Wilde, D. J., 415
Wilson, Jr., A. N., 33
Wing, O., 180
Wong, K. K., 87

Y

y parameters, 164, 171

Z

z parameters, 163, 171
Zadeh, L. A., 275, 479
zero input response, 245
zero state response, 245